DESIGN OF ANALOG-DIGITAL VLSI CIRCUITS FOR TELECOMMUNICATIONS AND SIGNAL PROCESSING

Second Edition

José E. Franca and Yannis Tsividis, editors

Prentice Hall, Englewood Cliffs, New Jersey 07632

Library of Congress Cataloging-in-Publication Data

Design of analog-digital VLSI circuits for telecommunications and
  signal processing / José Franca and Yannis Tsividis, editors. -- 2nd
  ed.
      p.   cm.
    Rev. ed. of: Design of MOS VLSI circuits for telecommunications.
  c1985.
    Includes biblingraphical references and index.
    ISBN 0-13-203639-8
    1. Integrated circuits--Very large scale integration.
  2. Semiconductors. 3. Telecommunication--Equipment and supplies.
  I. Franca, José. II. Tsividis, Yannis. III. Title: Design of MOS
  VLSI circuits for telecommunications.
  TK7874.D475 1994
  621.39'5--dc20   93-1982
                                                             CIP

Publisher: Alan Apt
Production Editor: Mona Pompili
Copy Editor: Shirley Michaels
Prepress Buyer: Linda Behrens
Manufacturing Buyer: Dave Dickey
Editorial Assistant: Shirley McGuire

 © 1994 by Prentice-Hall, Inc.
A Simon & Schuster Company
Englewood Cliffs, New Jersey 07632

The author and publisher of this book have used their best efforts in preparing this book. These efforts
include the development, research, and testing of the theories and programs to determine their effectiveness.
The author and publisher shall not be liable in any event for incidental or consequential damages in
connection with, or arising out of, the furnishing, performance, or use of these programs.

Printed in the United States of America

10 9 8 7 6 5 4 3 2

ISBN 0-13-203639-8

PRENTICE-HALL INTERNATIONAL (UK) LIMITED, London
PRENTICE-HALL OF AUSTRALIA PTY. LIMITED, Sydney
PRENTICE-HALL CANADA, INC., Toronto
PRENTICE-HALL HISPANOAMERICANA, S.A., Mexico
PRENTICE-HALL OF INDIA PRIVATE LIMITED, New Delhi
PRENTICE-HALL OF JAPAN, INC., Tokyo
SIMON & SCHUSTER ASIA PTE. LTD, Singapore
EDITORA PRENTICE-HALL DO BRASIL, LTDA., Rio de Janeiro

To

Leonor, Felicia, and Pedro

Preface

What LSI did for computers in the 1970's, VLSI is doing for telecommunications and signal processing in the 1990's. Entire systems of high complexity, often incorporating both analog and digital functions, are being put on a single chip, thus attaining reduced size, cost, and power dissipation, while at the same time improving performance or even enabling new applications. This volume is devoted to techniques that make this possible. It is divided into two parts, one devoted to circuits, and another devoted to systems. Many representative applications will be found in the systems part.

The precursor of this book, entitled *Design of MOS VLSI Circuits for Telecommunications* and edited by Tsividis and Antognetti in 1985, reflected the contents of a course by the same title, offered for the first time in Italy in 1985. Since then, the course was offered four more times (in Italy, Spain, Finland, and Portugal), with the lectures each time being updated to reflect new advances, and new lectures being introduced. By the time the course was offered in 1990 at the Instituto Superior Técnico in Lisboa, Portugal, it had changed to such an extent that the publication of a new book was in order. This volume reflects the contents of this latest offering, and in fact includes further updating changes made by the authors all the way to early 1992. Reflecting the advanced nature of the course and its attendees, the book is addressed to experienced readers.

We want to thank our authors for making available their expertise to the course attendees and our readers. We also want to thank the course attendees for helping shape the course and the book. Finally, we want to thank Miss Ana Marcelino and Miss Salvina Ribeiro whose tireless work made possible a professional appearance without sacrificing the timeliness of the volume.

J.E. Franca
Y. Tsividis

Contributors

ADAMS, P.S., BT Laboratories, Ipswich, England

FRANCA, J.E., Instituto Superior Técnico, Lisboa, Portugal

GOPINATHAN, V., Texas Instruments, Dallas, Texas, U.S.A.

GRAY, P.R., University of California, Berkeley, California, U.S.A.

GRÜGER, K., Deutsche Thomson Brandt GmbH, Villingen, Germany

HEATLEY, D.J.T., BT Laboratories, Ipswich, England

KLAASSEN, F.M., Philips Research Laboratories, Eidhoven, The Netherlands

MALOBERTI, F., University of Pavia, Pavia, Italy

MARTINS, R.P., Instituto Superior Técnico, Lisboa, Portugal

NEFF, R.R., University of California, Berkeley, California, U.S.A.

PIRSCH, P., University of Hannover, Hannover, Germany

RAPELI, J., Nokia Mobile Phones, Ouw, Finland

ROTHERMEL, A., Deutsche Thomson Brandt GmbH, Villingen, Germany

SCHEMANN, H., Deutsche Thomson Brandt GmbH, Villingen, Germany

SCHWEER, R., Deutsche Thomson Brandt GmbH, Villingen, Germany

SEDRA, A., University of Toronto, Ontario, Canada

SENDEROWICZ, D., Synchrodesign Inc., Berkeley, California, U.S.A.

SNELGROVE, M., University of Toronto, Ontario, Canada

TEMES, G.C., Oregon State University, Corvallis, Oregon, U.S.A.

TSIVIDIS, Y., National Technical University of Athens, Athens, Greece

VITTOZ, E.A., Centre Suisse d'Electronique et de Microtechnique (CSEM),
    Neuchatel, Switzerland

WEHBERG, T., University of Hannover, Hannover, Germany

WEINBERGER, G., Siemens A.G., Munich, Germany

Contents

1. Circuit-Level Models for VLSI Components   **1**
F.M. Klaassen

Introduction   1

1.   MOSFET Modelling,   1
    1.1.   Threshold voltage,   1
    1.2.   Drain current at strong inversion,   4
    1.3.   Subthreshold current,   6
    1.4.   Saturation mode,   7
    1.5.   Charges and capacitances,   9
    1.6.   Parasitics and model extensions,   10

2.   Bipolar Modelling,   11
    2.1.   Carrier transport,   11
    2.2.   Collector current,   12
       2.2.1.   Charge control approach,   12
       2.2.2.   Carrier control approach,   13
    2.3.   Base current,   14
    2.4.   Charges and capacitances,   14
    2.5.   Quasi-saturation,   16
    2.6.   Parasitics and model extensions,   17

3.   Parameters,   20
    3.1.   Parameter acquisition,   20
    3.2.   Geometry and process dependence,   21
    3.3.   Statistics of parameters,   22

References,   23

2.   CMOS Operational Amplifiers   27

Daniel Senderowicz

1.   General Considerations for VLSI of Amp Design,   27

2.   The Differential Circuit Approach,   28

3.   The Number of Internal Stages,   31

4.   Cascode Op Amps,   31
    4.1.   The folded-cascode OTA,   34
    4.2.   The telescopic-cascode OTA,   35
    4.3.   Comparisons between the folded and telescopic cascode OTAs,   36
        4.3.1.   Simulations,   37
    4.4.   Single-ended versions of the cascode OTAs,   38

5.   Multiple Stage Op Amps,   39
    5.1.   The differential class-A two stage op amp,   40
    5.2.   The resistor driving class-AB op amp, version I,   42
    5.3.   The resistor driving class-AB op amp, version II,   44

6.   Bias Circuits,   47

7.   Layout Techniques,   49

References,   51

3.   Micropower Techniques   53

Eric A. Vittoz

1.   Introduction,   53

2.   Technology,   54

3.   Devices,   55
    3.1.   MOS transistors,   55
    3.2.   CMOS compatible bipolar transistors,   67
    3.3.   Passive devices,   69

    3.4. Parasitic effects, 70

4. Basic Circuit Techniques, 70
    4.1. Current mirrors, 70
    4.2. Differential pair, 71
    4.3. Voltage gain cells, 72
    4.4. Low-voltage cascoding, 75
    4.5. Current references, 77
    4.6. Voltage references, 78
    4.7. Analog switches, 79
    4.8. Dynamic CMOS gates, 80

5. Examples of Building Blocks, 81
    5.1. Operational transconductance amplifiers (OTAs), 81
    5.2. Quartz crystal oscillators, 87
    5.3. Sequential logic building blocks, 91

6. Conclusions, 92

7. References, 93

4. Dynamic Analog Techniques   97
Eric A. Vittoz

1. Introduction, 97

2. Sample-and-Hold, 97
    2.1. Principle, 97
    2.2. Transfer function, 98
    2.3. Noise sampling, 99
    2.4. Charge injection by the switch, 101
    2.5. Sample-and-hold used as a time differentiator (autozero circuit), 105

3. Offset Compensation of Operational Amplifiers, 108
    3.1. First order compensation, 108
    3.2. Compensation by a low-sensitivity auxiliary input, 108

4. Dynamic Biasing of CMOS Inverter-Amplifiers, 112

5. Dynamic Comparators, 114
   5.1. Basic circuit, 114
   5.2. Multistage comparators, 117

6. Dynamic Current Mirrors, 118
   6.1. Principle and limitations, 118
   6.2. Examples of practical implementations, 120

7. Conclusion, 122

8. References, 122

5. Optical Receivers   125
David J.T. Heatley

1. Introduction, 125

2. Receiver Design, 126

3. Receiver bandwidth, 127
   3.1. Low impedance receiver, 127
   3.2. High impedance receiver, 128
   3.3. Transimpedance receiver, 130

4. Receiver Noise, 132
   4.1. Photodiode noise, 132
   4.2. Resistor noise, 133
   4.3. Amplifier noise, 134
   4.4. 2nd Stage noise, 136
   4.5. Total noise power spectral density, 136
      4.5.1. Low impedance receiver, 137
      4.5.2. High impedance receiver, 140
      4.5.3. Transimpedance receiver, 143
      4.5.4. Comparison of all receiver configurations, 144

5. Signal-to-Noise Ratio, 147
   5.1. SNR from the low impedance receiver, 148
   5.2. SNR from the high impedance receiver, 151

5.3. SNR from the transimpedance receiver, 151

5.4. Comparison of SNRs, 152

6. Receiver Sensitivity, 155

6.1. Analogue transmission system, 155

6.2. Digital transmission system, 156

6.3. Transmission distances, 161

7. Dynamic Range, 164

8. Developments in Receiver Design, 166

8.1. Coherent detection receivers, 167

8.2. Optically pre-amplified direct detection receivers, 170

8.3. The transmission engineer's dream, 171

References, 171

Appendix A, 175

6. **Continuous-Time Filters** **177**

Yannis Tsividis and Venugopal Gopinathan

1. Introduction, 177

2. MOSFET-C Filters, 179

2.1. The MOSFET as a voltage-controlled resistor, 179

2.2. Cancellation of nonlinearities, 181

2.3. MOSFET-capacitor filters based on balanced structures, 183

2.3.1. Principle, 183

2.3.2. Filter synthesis, 185

2.3.3. Accuracy, 188

2.3.4. Distortion and dynamic range, 189

2.3.5. Parasitic capacitance, 189

2.4. Worst-case design, 191

3. Transconductor-C Filters, 192

3.1. Introduction, 192

3.2. Transconductor design, 193

3.2.1. Designs that are based on MOSFET characteristics in the saturation region, 194

3.2.2. Designs that are based on MOSFET characteristics in the non-saturation region, 196

3.3. Gm-C filter synthesis, 199

4. MOSFET-C vs Transconductor-C Filters, 201

5. On-Chip Tuning Schemes, 202

6. A Warning Concerning Computer Simulations, 206

7. Conclusion, 206

References, 207

7. Switched-Capacitor Filter Synthesis   **213**
Adel S. Sedra and Martin Snelgrove

1. Introduction, 213

2. First-Order Building Blocks, 215
    2.1. Basic operation, 215
    2.2. Exact transfer functions and the LDI variable, 217
    2.3. Damped integrators, 221
    2.4. General form of the first-order building block, 223
    2.5. Switch design, 224
    2.6. Differential circuits, 225

3. Biquadratic Sections and Cascade Design, 225
    3.1. Circuit generation via equivalent resistors, 226
    3.2. Exact transfer function: LDI design, 228
    3.3. Capacitive damping, 230
    3.4. Biquads for exact design: the bilinear variable, 231
    3.5. Biquad circuits for exact design, 232
    3.6. Differential biquads, 234
    3.7. Cascade design, 236

4. Ladder Filters, 237

4.1.   Circuit generation,   238

4.2.   Design,   241

4.3.   A general exact design method,   242

5.   Concluding Remarks,   246

6.   References,   247

8.   Multirate Switched-Capacitor Filters   **251**
José E. Franca and R.P. Martins

1.   Introduction,   251

2.   FIR SC Decimators and Interpolators,   252
2.1.   Direct-form polyphase SC structure for decimation,   252
2.2.   Direct-form polyphase SC structure for interpolation,   258
2.3.   Multi-amplifier FIR SC decimators,   262

3.   IIR SC Decimator Building Blocks,   266
3.1.   First-order SC decimator building block,   268
3.2.   Second-order SC decimator building block,   271
3.3.   Nth.-order SC decimator building block,   277

4.   IIR SC Interpolator Building Blocks,   281

5.   Conclusions,   285

References,   286

**9.   Analog-Digital Conversion Techniques for Telecommunications
Applications**   **289**
Paul Gray and Robert R. Nepp

1.   Introduction,   289

2.   The Role of A/D Converters in Telecommunications Systems,   290

3.   Classification of A/D Conversion Techniques,   292

4. Circuit Building Blocks for A/D Converters, 296
    4.1. High-speed DACs in MOS technology, 296
        4.1.1. Current-switched MOS DACs, 296
        4.1.2. Resistor-string DACs, 297
        4.1.3. Charge-redistribution DACs, 297
        4.1.4. Component mismatch effects in monolithic DACs, 302
    4.2. Monolithic sample/hold amplifiers, 307
    4.3. Monolithic voltage comparators, 309

Summary, 313

References, 314

10. Delta-Sigma Data Converters   **317**
Gabor C. Temes

1. Introduction, 317

2. Noise-Shaping A/D Converters, 320
    2.1. Quantization noise, 320
    2.2. The delta-sigma modulator, 322
    2.3. Higher-order single-stage converters, 325
    2.4. Multi-stage (cascade) converters, 327
    2.5. Modulators with multibit quantizers, 329

3. Noise-Shaping D/A Converters, 335

4. Conclusions, 337

Acknowledgement, 338

5. References, 338

11. Layout of Analog and Mixed Analog-Digital Circuits   **341**
Franco Maloberti

1. Introduction, 341

2. Differences Between Layout and Circuit, 342

3. Absolute and Relative Accuracies, 343

4. Layout of MOS Transistors, 344

5. Layout of Resistor, 348

6. Layout of Capacitors, 351

7. Stacked Layout for Analog Cells, 354

8. Digital Noise Coupling, 356

9. Floor Planning of Mixed Analog-Digital Blocks, 363

10. Concluding Remarks, 366

11. References, 366

12. System Architectures and VLSI Circuits for Telecommunications   **369**
   Guenter Weinberger

1. Introduction, 369

2. System Structure, 370

3. Communication ICs for Digital Exchange Systems, 371
   3.1. PCM switching network components, 371
   3.2. Serial communication controllers, 372
   3.3. Line card functions and architectures, 374
      3.3.1. Line card functions, 375
      3.3.2. Line card architectures, 375
   3.4. Line card controller, 376
   3.5. Analog subscriber line card devices, 378
      3.5.1. Line card functions, 378
      3.5.2 Codec and filter devices, 380
      3.5.3. Subscriber line interface circuit (SLIC), 384
   3.6. Digital subscriber line card devices, 386
      3.6.1. Line card functions, 386
      3.6.2. ISDN transceiver devices, 387

4. Analog Telephone Sets, 387

5. ISDN Subscriber Equipment, 390
   5.1. ISDN terminals and terminal adaptors, 391
         5.1.1. ISDN layer-2 terminal controller, 391
         5.1.2. ISDN codec filter, 392
         5.1.3. ISDN terminal adaptors, 392
   5.2. ISDN network terminations (NT1), 393

6. DSP in Telecommunications, 394
   6.1. Introduction, 394
   6.2. DSP design methodology, 396

7. Conclusion, 397

8. References, 398

13. **Integrated Circuits for the ISDN**   **401**
   Peter Adams

   1. Introduction, 401

   2. ISDN Fundamentals, 401
      2.1. ISDN access functional blocks, 403

   3. S/T-Bus Transceivers, 404

   4. D-Channel Controllers, 405

   5. Digital Subscriber Loop Transceivers, 406
      5.1. Key loop characteristics, 408
      5.2. 2-wire duplex operation, 410
      5.3. Choice of line code, 410
      5.4. Transceiver architecture, 411
      5.5. Transmitter implementation, 413
      5.6. Receiver implementation, 414
      5.7. Echo canceller implementation, 418
      5.8. DSP implementation issues, 421
      5.9. Transceiver examples, 423

6. Conclusion, 425

7. References, 425

14. VLSI Architectures and Circuits for Visual Communications   **429**
       P. Pirsch and T. Wehberg

1. Introduction, 429

2. Source Coding Algorithms, 430
       2.1. Algorithms for redundancy reduction, 430
       2.2. Quantization, 433
       2.3. Variable word length coding, 434
       2.4. Source coding for different applications, 435

3. VLSI Implementation Strategies, 437

4. Programmable Multiprocessor Systems, 441
       4.1. Architecture of SIMD and MIMD multiprocessors, 442
       4.2. SIMD system based on multiprocessor IC, 443
       4.3. MIMD system based on multiprocessor IC, 445

5. Key Components for Functional Blocks, 450
       5.1. Dedicated realizations of discrete cosine transform, 451
       5.2. Dedicated realizations of block matching, 455

    References, 459

15. Digital Television   **463**
       A. Rothermel

1. Introduction, 463

2. TV Tuner, 464

3. Audio Processing, 467

4. Video Signal Processing, 470
       4.1. Coding principles, 470

4.2.   Implementation issues,   473

    4.2.1.   The luminance channel,   475

    4.2.2.   The chrominance channel,   477

5.   Scan Control Processing,   481

    5.1.   Tasks and problems,   481

    5.2.   Deflection processor,   482

    5.3.   Picture geometry correction,   483

6.   Quality Improvements,   484

    6.1.   Improved luma/chroma separation,   484

    6.2.   Flicker reduction,   487

    6.3.   Format control,   490

7.   Conclusion,   492

8.   References,   493

16.   VLSI Architectures and Circuits for Digital Coding of High Definition Television   **495**

P. Pirsch and K. Grüger

1.   Introduction,   495

2.   Source Coding Algorithms for HDTV,   496

    2.1.   HD-MAC coding,   496

    2.2.   DPCM coding,   498

    2.3.   DCT coding,   499

    2.4.   Subband coding,   500

3.   VLSI Implementation Aspects,   502

4.   VLSI Architecture for 2D DPCM,   502

    4.1.   Implementation problems of DPCM,   502

    4.2.   Parallel DPCM processors,   503

    4.3.   Modified DPCM structure,   505

    4.4.   Parallel processing and delayed decision,   506

    4.5.   Comparison of DPCM architectures,   507

    4.6.   CMOS circuit technique for DPCM,   508

4.6.1. Adder realization, 508

4.6.2. Realization of XOR, 509

4.7. Table look-up realization, 509

4.8. Realization of line delay, 511

4.9. Example of a DPCM processing element for HDTV, 514

5. Subband Filterbanks, 515

5.1. Hardware requirements, 515

5.2. Nonrecursive wave digital filter (NRWDF), 516

5.3. Transversal FIR filter, 517

5.4. Polyphase structure, 517

5.5. Expansion in time by subsampling, 520

5.6. Architectures for quadrature mirror filterbanks (QMFs), 521

5.6.1. Filter characteristics, 521

5.6.2. Filter structures for QMFs, 521

5.7. Realization of arithmetic blocks, 522

5.7.1. Realization of multipliers, 522

5.7.2. Multioperand adders for FIR-filter, 524

5.8. Complexity of filterbank VLSI-implementation, 525

6. References, 525

17. IC Solutions for Mobile Telephones   **529**
Juha Rapeli

1. Introduction, 529

2. Mobile Communication Systems, 529

3. Functional Blocks of a Cellular Telephone, 533

3.1. Receiving Function (RX), 537

3.2. Frequency synthesis, 539

3.3. TX-function, 547

3.4. Signalling function, 548

3.5. Speech and audio processing, 548

3.6. Control functions, 548

4. Digital Mobile Phones, 549

4.1. Speech coding, 549

4.2. Modulation, 550

4.3. Symbol decoder (demodulator) and channel codec, 552

5. Future developments in mobile communication, 556

5.1. CDMA technology, 558

5.2. 2-GHz developments, 561

5.3. Integration of new features into mobile phones, 561

6. IC Technology, 565

6.1. FDMA, TDMA and CDMA from integrations point of view, 565

6.2. Integration technologies, 565

6.3. ASICs versus Standard ICs, 566

7. Conclusions, 566

References, 567

Index   **569**

Chapter 1

Circuit-Level Models for VLSI Components

F.M. Klaassen

Philips Research Laboratories, P.O. Box 80000, 5600 JA Eindhoven, The Netherlands

Introduction

Essentially compact or circuit-level models give a description of the currents and charges of electronic devices in terms of the applied terminal voltages. Generally three different types can be distinguished: physics-based models, empirical models and table look-up models. In the first type the above relations are analytical expressions, which have been derived via an in-depth understanding of the physical mechanisms underlying current transport, etc. In the second type the relations are of a curve fitting nature and in the last type the characteristics are reconstructed via tables of measured or simulated data.

Usually in physics-based models the relations are of a type $I_D=I_D(V_i,P_j)$, etc., where V_i is a terminal voltage and P_j is a parameter. Owing to its physical base, the parameters obey geometrical scaling rules, when the device dimensions are altered. Furthermore realistic statistical modelling of circuit properties is feasible. Naturally, in order to avoid excessive CPU-time, a trade-off between complexity and accuracy has to be made.

Despite their short development time, owing to the huge data storage required for the description of small devices, until now table models have not realized a breakthrough. In addition this approach has no predicting, scaling or statistics capability. Obviously, the latter facts also apply to empirical models.

This review is limited to physics-based modelling. The modelling of the MOSFET, the modelling of the bipolar transistor and the acquisition, process dependence and statistics of parameters are discussed in turn. In addition mainly basic approaches are given. For detailed formulae we refer to [1].

1. MOSFET Modelling

1.1. Threshold voltage

Since the threshold voltage has a pivot function in MOSFET characteristics, its description is discussed first. In order to understand the approach in Figure 1, the cross section of a small n-type MOSFET is given in a direction parallel and perpendicular to current flow. Practically the onset of inversion occurs in this structure, if a voltage drop across the depleted substrate layer is induced, which approximately equals the diffusion voltage ϕ_F (N_B). Under this condition the threshold voltage V_T is determined from a balance of the applied gate charge

$$Q_G = C_{ox} ZL(V_T - V_{FB} - \phi_F) \tag{1}$$

and the depletion charge

$$Q_B = C_{ox}(Z + 2\delta Z)(L - 2\delta L) q \int_{y_d}^{0} N_B(y) \, dy. \tag{2}$$

In these equations V_{FB} is the flatband voltage, C_{ox} is the unit area gate capacitance, $N_B(y)$ is the substrate doping, $y_d(\phi_F)$ is the depletion width and L,Z are the actual channel length and width, respectively.

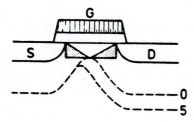

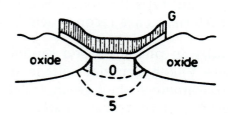

Fig. 1a: Shape of the depletion region in a short-channel MOSFET at zero and high drain bias. The grey areas indicate the charge shared by the gate and the junctions at zero drain bias.

Fig. 1b: Shape of the depletion region in a narrow-width MOSFET at zero and high back-bias. The grey areas indicate excess charge, which is formed under the LOCOS region.

The Q_B expression can be evaluated using several simplifying assumptions. Since V_T is the result of an integration, the implanted Gaussian-type doping profile $N_B(y)$ is replaced by an equivalent box profile [2]. The increase Z of the depletion layer width is attributed to its spreading into the channel stop area under the field insulator. Finally the decrease in length L is associated with charge sharing by the junctions and the channel area. The latter effects have been calculated either by assuming trapezoidal forms of the depletion region [3] or by making use of pseudo-2D solutions of Poisson's equation [4,5].

Naturally according to the above approach V_T is expressed in terms of the substrate doping [2]. Since however the latter quantity cannot be extracted directly from the I-V characteristics, it is more practical [1] to rewrite the result in terms of the substrate coefficients γ_i and γ, which are the asymptotic values of

$\partial V_T / \partial V_S$ (compare Figure 2)

$$V_T = V_{T0} - \gamma_i \, \phi_F^{1/2} + u(U_S = (V_S + \phi_F)^{1/2}). \tag{3}$$

If $\qquad 0 < U_S < U_{SX} = (V_{SX} + \phi_F)^{1/2}$

$\qquad\qquad u = \gamma_i \, U_S$

If $\qquad U_{SX} \le U_S$

$$u = \left\{ 1 - \left(\frac{\gamma}{\gamma_i}\right)^2 \right\} \gamma_i \, U_{SX} + \gamma \left[U_S^2 - \left\{ 1 - \left(\left|\frac{\gamma}{\gamma_i}\right|\right)^2 \right\} U_{SX}^2 \right]^{1/2}.$$

Figure 2 gives a comparison between measured and modelled results. For the γ-parameters the following scaling relations apply

$$\gamma = \gamma_\infty - \gamma_L / L + \gamma_Z / Z. \tag{4}$$

Although a buried channel causes the evaluation of the dope integral (2) to be more complicated for p-channel transistors, it has been shown [6] that eqs. (3) and (4) still apply.

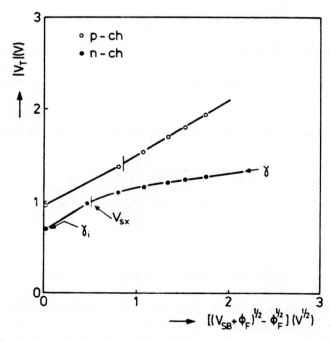

Fig. 2: Comparison of measured values of the threshold voltage with calculated data (fully drawn lines). Relevant parameter values are:

$V_{T0N} = 0.70V$, $\gamma_i = 0.60 \ V^{1/2}$, $\gamma = 0.15 \ V^{1/2}$, $V_{SX} = 1.05V$; $V_{T0P} = -0.95V$,

$\gamma_i = 0.55 \ V^{1/2}$, $\gamma = 0.65 \ V^{1/2}$, $V_{SX} = -2.00 \ V$, $\phi_F = 0.70 \ V$

1.2. Drain current at strong inversion

In the strong inversion regime current transport is mainly by drift. Consequently the channel current (at position x with a potential V) is given by

$$I_D = Z\mu_s \left\{ C_{ox} (V_G - V_{FB} - V) - q_B(V) \right\} \frac{dV}{dx}, \tag{5}$$

where μ_s is the local surface mobility, the first term between brackets represents the applied gate charge and q_B is the charge in the depletion layer. Generally the mobility is affected by the normal and lateral field [7]. In practice this can be satisfactory modelled writing [6, 8, 9]

$$\mu_s = \mu_{so} [1 + \theta_A(V_G - V_T) + \theta_B V_S]^{-1} (1+E_x/E_c)^{-1}, \tag{6}$$

where θ_A, θ_B and E_c have to be considered as parameters and the last term represents velocity saturation. For gate insulators thinner than 15 nm, owing to surface roughness an additional term $\theta_D V_{GS}^2$ becomes necessary [1]. By calculating q_B for a box-type doping profile the drain current is obtained from an integration of eq. (5). Since however for an implanted substrate the result is complicated [2] and in the case of short-channel devices leads to a quadratic equation for the saturation voltage, it is useful to approximate the depletion charge

$$q_B(V) \approx q_B(V_S) + \delta C_{ox}(V - V_S).$$

If $\gamma_i = \gamma$, $\delta \approx 0.3 \gamma (V_S + \phi_F)^{-1/2}$; however for $\gamma_i > \gamma$, $\delta(\gamma_i, \gamma)$ becomes more complicated [1].

Substituting the above approximation and eq. (6) in eq. (5), we finally obtain

$$I_{DS} = \frac{Z}{L} \mu_{so} C_{ox} \frac{(V_G-V_T) V_{DS} - \frac{1}{2} (1+\delta) V_{DS}^2}{[1 + \theta_A(V_G-V_T) + \theta_B V_S] (1+\theta_C V_{DS})}, \tag{7}$$

where $\theta_C = (LE_c)^{-1}$.

Usually, the product $\mu_{so} C_{ox}$ is considered as a parameter β_0.

Since for higher drain bias current saturation is induced by reversal of the normal field at the drain end of the channel, formally the saturated drain current can be written

$$I_{DSAT} = Z C_{ox} [V_G-V_T-(1+\delta) V_{DSAT}] \frac{\mu_{so} E_{xd}}{1+E_{xd}/E_c}, \tag{8}$$

where E_{xd} is the actual lateral field and V_{DSAT} is the saturation voltage. By equating eqs. (5) and (8) for this saturation condition, I_{DSAT} and V_{DSAT} are obtained from a square-law equation.

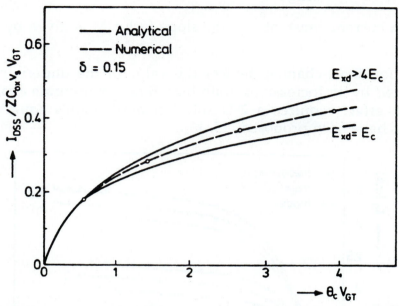

Fig. 3:   Modelled value of saturated drain current as a function of normalized gate driving voltage at several values of short-channel parameter E_{xd}/E_c. The broken line represents values calculated numerically using a 2-D device simulator.

Since E_{xd} is not known exactly, MOSFET models have been based on the assumption $E_{xd}=E_c$ [10, 11] or $E_{xd} \gg E_c$ [12, 13]. In reality as shown by Figure 3 the truth lies somewhere in between. In this figure the normalized saturation current according to the above approaches has been compared with results of 2-D numerical device simulation [14]. Since in practice $\theta_C V_{GT} < 1.5$ and $\theta_c = (LE_c)^{-1}$ is a measurable parameter, the error of both models can be made sufficiently small. Finally Figure 4 gives a comparison between the modelled and measured characteristics of a submicron n-channel MOSFET. In order to obtain a good fit of the drain conductance in the saturation region, the above model has been extended. This is discussed in section 14.

1.3. Subthreshold current

When a gate bias well below threshold is applied, a potential barrier exists between the channel area and the source. In this case current flow is possible by means of carriers which are able to pass the above barrier, and transport takes place by diffusion rather than by drift. Therefore the current can be written in the simple form $I_D=ZqD_n n(o)/L$, where D_n is a diffusion constant and $n(o)$ is the number of carriers that can pass the above barrier. By solving the charge balance equation discussed in 1.1, it can be provided that

$$I_D = \frac{Z}{L} \, I_0 \exp \frac{q(V_G - V_T)}{M k_B T} \, , \tag{9}$$

in which I_0 is a current constant and the slope factor M is given by $M = 1 + m(V_S + \phi_F)^{-1/2}$.

Unfortunately for short-channel devices the value of the underlying potential barrier is affected by an increase of drain bias. Since an accurate analysis of this so-called DIBL effect requires a 2-D solution of Poisson's equation, several approximations have been proposed [15, 16].

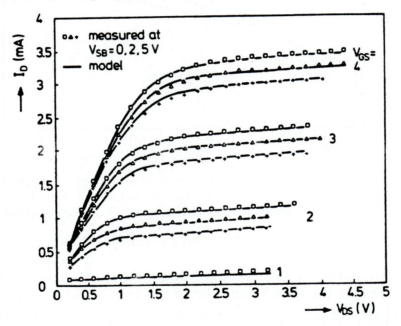

Fig. 4:   Measured and modelled characteristics of 0.5 μm n-type MOSFET. Relevant parameter values are $V_{T0} = 0.70$ V, $\beta = 2.45$ mA/V^2, $\gamma = 0.20$ V$^{1/2}$, $\theta_A = 0.40$ V^{-1}, $\theta_B = 0.040$ V^{-1}, $\theta_C = 0.36$ V^{-1}, $\theta_D = 0.019$ V^{-2}, f = 0.095.

Following the latter one, in which the effect of the drain-induced field is transformed into an apparent decrease of the substrate doping, it can be shown [1] that the DIBL effect can be expressed as a decrease of threshold voltage

$$\Delta V_T \approx -s(V_S) \, V_{DS} \tag{10}$$

In the case of a uniformly doped substrate $s(V_S) \approx s(V_S + \phi_F)^{1/2}$.

For the weak inversion region around V_T transport by drift and diffusion equally contribute to the current [17]. Therefore presently no satisfactory analytical results are available. However the expressions for strong inversion and subthreshold mode can be successfully merged either by adding eq. (9) after multiplication with a limiting function to eq. (5) [12, 13] or by defining a generalized gate driving voltage [9]. Figure 5 gives a comparison of the measured and modelled characteristics below inversion. Note the increase of the DIBL effect at higher values of back bias.

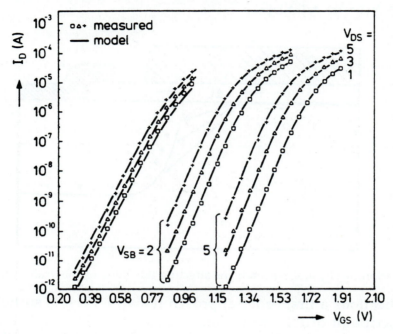

Fig. 5: Subthreshold characteristics for 0.7 μm n-type MOSFET. Relevant parameter values are V_{T0}=0.95 V, γ=0.49 $V^{1/2}$, β=1.70 mA/V^2, m=0.57 $V^{1/2}$, s=0.014.

1.4. Saturation mode

As already shown in Figure 4, the drain current increases slightly in the saturation region. Generally this is caused by three physical mechanisms. First an increase of drain bias causes the point of normal field reversal to shift towards the source, next for shorter channels near the same point excess mobile charge is induced and finally for very short channels weak avalanche multiplication will occur. The first effect, which can be interpreted as modulation of effective channel length, has to be calculated via a 2-D solution of Poisson's equation, including mobile carrier space charge (compare Figure 6). Therefore few published analytical models sustain a comparison with experimental results [1]. After a slight adaptation satisfactory results are obtained [16] for a model [15], which expresses the effective length by

$$L_{eff} \approx L\left[1 + \alpha \ln\left(1 + \frac{V_D - V_{DSAT}}{\alpha V_p}\right)\right]^{-1}, \tag{11}$$

where α and V_p are parameters. The second effect, which is known as static feedback, can successfully be taken into account by adding a term to the gate drive [12]

$$V_{GT} = V_G - V_T + f(V_{DS}) \,. \tag{12}$$

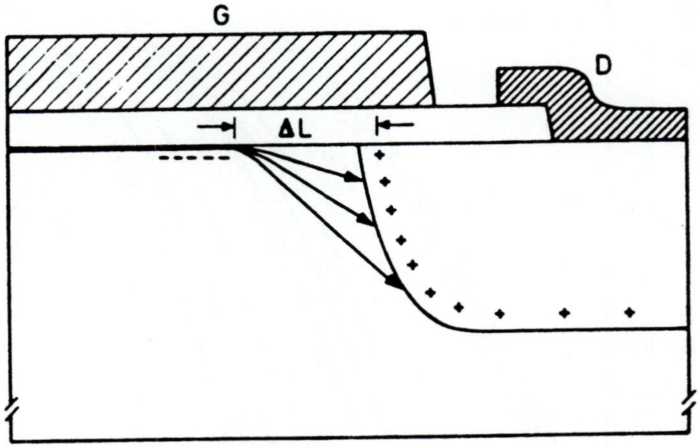

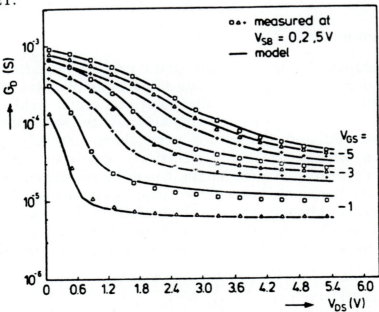

Fig. 6:   Illustration of physical mechanisms affecting the value of the drain conductance.

For digital applications a first-order approach $f(V_{DS}) = fV_{DS}$ is sufficient, but for an accurate description of the drain conductance in analog applications a more refined approach is necessary [16]. Figure 7 gives a comparison of the modelled and measured characteristics of the drain conductance of a submicron p-channel MOSFET.

Fig. 7:   Drain conductance characteristics of 0.8 μm p-type MOSFET. Relevant parameters: β=0.45 mA/V^2, f=0.11, α=0.038, V_p=4.0 V.

1.5. Charges and capacitances

Compared with the notice given to the MOSFET current, the modelling of charges and capacitances lags behind. For a long time models have been used taking only a few small signal capacitances derived for a long channel device. This approach not only leads to errors in some capacitances of short-channel devices, but generally the charge conservation principle

$$Q_G + Q_S + Q_D + Q_B = 0 \qquad (13)$$

is violated [17, 18]. In order to maintain the latter a model has to be based on three independent terminal charges and, since such charges depend on other node voltages too, nine independent capacitances have to be taken into account.

For the calculation of the charges Q_S and Q_D a physical partitioning of the charge in the channel is required. Such a split has been derived along two different lines of approach [19, 20]. For Q_D then we have

$$Q_D = -qZ \int_L^0 n(x,t) \frac{x}{l} dx. \qquad (14)$$

Similarly for Q_S a weighting factor $(1-x/L)$ has to be used. In order to simplify the evaluation of (14) quasi-static operation is assumed, which means that $n(x,t)$ only depends on the instantaneous value of node voltages. In practice this implies that the application is limited to transients in a time step larger than twice the channel transit time [1]. Since n is known only as a function of the potential V, the variable x has to be replaced by V. This can be achieved via the current relation (5). For a short-channel device the expression for Q_S and Q_D, although complicated, can be given in a closed form [21]. In addition, effects of velocity saturation, static feedback, etc., can be taken into account and the continuity of the resulting capacitances across boundaries between possible modes of operation is assured. As an example Figure 8 gives a plot of $C_{DG}=\partial Q_D/\partial V_G$ for all possible modes. Generally, the capacitances are non-reciprocal.

Finally similar to the V_T calculation, the bulk charge is obtained from a solution of Poisson's equation for a box-type doping profile.

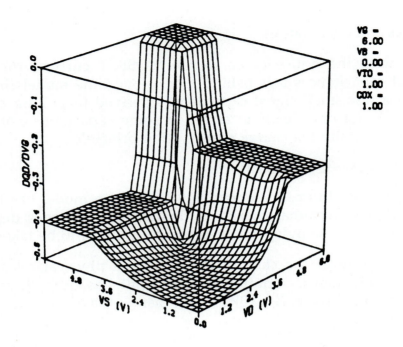

Fig. 8:   The normalized drain-gate capacitance vs drain and source bias at a fixed value of gate bias.

1.6. Parasitics and model extensions

In addition to the intrinsic properties discussed previously, modelling of parasitics becomes increasingly important. Within the limited scope of this review, only a few remarks are made. Owing to the necessity to reduce hot carrier effects via graded junction or offset gate structures, series resistances have a large impact on the characteristics of submicron MOSFETS. Although these resistances can be added to a circuit model, it is more practical to avoid the use of additional nodes and to include the resistance effect implicitly in the model equations. Generally this is achieved by generalizing the θ-parameters [22].

Since the gate-junction and the bulk-junction capacitance do not scale in the same manner as the intrinsic parts and additionally their voltage dependence becomes more complicated, an accurate description is required [1]. A complete MOSFET model is shown in Figure 9 [39].

For wide applications such a model has to be extended with a description of the geometry and temperature dependence of parameters, the weak-avalanche induced substrate current and the noise sources [1].

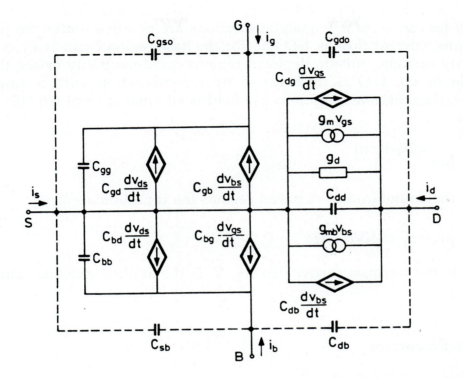

Fig. 9: Equivalent circuit of a MOSFET, showing all intrinsic capacitances (with fully drawn interconnections) and parasitic capacitances (with dashed interconnections).

2. Bipolar Modelling

2.1. Carrier transport

In addition to a diffusion current, owing to the built-in field associated with doping gradients, in bipolar devices transport by drift has to be taken into account as well. Consequently the electron current density is given by

$$J_n = q\mu_n nE + qD_n \nabla_r n \tag{15}$$

with a similar equation for holes. Furthermore under transient conditions the continuity equation applies:

$$\frac{\partial n}{\partial t} - q^{-1} \nabla_r \cdot J_n = G\text{-}R, \tag{16}$$

where G and R are the generation and recombination rate, respectively. Generally the above equations can only be solved numerically. However with almost no loss in accuracy, analytic solutions are possible under the following almost no loss in accuracy, analytic solutions are possible under the following assumptions: current flow is 1-D and for the base region exclusively controlled by minority carriers; outside depleted regions quasi-neutrality exists; the right-hand side of eq. (16) is determined by recombination with a single time constant. Using these assumptions the field is eliminated from eq. (15) yielding the result

$$J_n = \frac{qD_n}{p} \frac{d(pn)}{dx}.$$ (17a)

Finally at the boundaries of depleted regions the condition applies

$$pn = n_i^2 \exp(V_j/U_T),$$ (17b)

where n_i is the intrinsic carrier density, V_j is the applied junction voltage and $U_T = k_B T/q$.

2.2. Collector current

2.2.1. Charge control approach

By integrating eq. (17a) from emitter to collector and using the boundary condition (17b), the electron current density in an npn transistor (compare Figure 10) is given by [23]

$$J_n = qn_{io}^2 G_b^{-1} \{\exp(V_{BE}/U_T) - \exp(V_{BC}/U_T)\},$$ (18)

in which the Gummel number

$$G_b = \int_e^c \frac{p(x)}{D_n(x)} \left(\frac{n_{io}}{n_i(x)}\right)^2 dx.$$

In these equations V_{BE}, V_{BC} are the applied junction voltages and n_i is considered to be subject to bandgap narrowing [24]. Fortunately in the latter integral $D_n(n_i/n_{io})^2 \cong 25$ cm^2 s^{-1} in a wide range of doping densities [1]. Therefore the collector current is inversely proportional to the total hole charge Q_b between emitter and collector. Consequently the above current is written as

$$I_C = I_s (Q_{bo}/Q_b) [\exp(V_{BE}/U_T) - \exp(V_{BC}/U_T)],$$ (19a)

where I_s is a process-related current constant.

In principle Q_b can be divided into parts:

$$Q_b = Q_{bo} + Q_{Te} + Q_{Tc} + Q_{be} + Q_{bc}.$$ (19b)

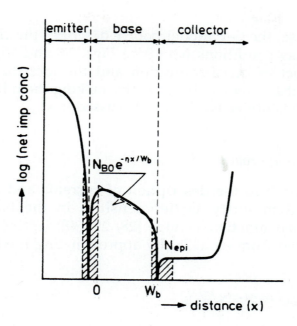

Fig. 10:   Typical doping profile in an npn transistor. Hatched areas denote depletion regions. The dashed line gives the doping profile approximation of eq. (21).

Q_{bo} is the zero bias base charge. Q_{Te} and Q_{Tc} are the depletion charges of the junctions, and Q_{be} and Q_{bc} are the stored charges of the minority carriers, injected from the emitter and collector, respectively. Disregarding all charges except Q_{bo} in the denominator, eq. (19a) forms the base for the well-known Ebers-Moll model [25]. In order to describe better the characteristics at higher currents, Q_b is further evaluated in the current Gummel Poon model [26]. Then the depletion charges are calculated by integration of a bias-dependent depletion capacitance (*see* 2.4) and the stored charges are related to the injected currents by means of the charge control principle, e.g.,

$$I_f = I_s \left[\exp(V_{BE}/U_T) - 1\right] = Q_{be}\, \tau_f^{-1} \tag{20}$$

where τ_f is called the forward base transit time. For Q_{bc} a similar relation can be given. In the above concept stored charge moving across the base is considered as the driving force of current. Since all charges in (19b) are now modelled in terms of junction voltages, Q_b can be eliminated from eq. (19a) and a closed form expression

$$I_C = I_C(V_{BE}, V_{BE}, Q_{bo}, I_s, \tau_f, \tau_r)$$

is obtained, which is claimed to be valid for all operation modes.

In spite of its wide use, the G-P model poorly describes the characteristics under

In spite of its wide use, the G-P model poorly describes the characteristics under high current, low bias conditions (compare Figs. 14 and 15). Generally this is caused by the neglect of quasi-saturation and the fact that τ_f and τ_r are not constants. Although the latter can be cured by making them bias dependent [27], a basic new approach is preferred. This is discussed next.

2.2.2. Carrier control approach

An alternative approach to the description of currents and charges is to relate them to the injected minority carrier densities in the base, which in turn depend on the applied junction voltage [28, 29, 30]. In this approach the hole density p is eliminated from eq. (17a) by approximating the base doping profile (*see* Figure 10)

$$N_B(x) = N_{BO} \exp(-\eta x / W_B), \tag{21}$$

where the parameter η also is a measure for the built-in field. In this way for the neutral base region from eq. (17a) the basic equation

$$I_n = qD_n \left\{ \frac{2n + N_B}{n + N_B} \frac{dn}{dx} + \frac{dN_B}{dx} \right\} \tag{22}$$

is obtained. Generally this equation can be solved analytically for the low injection case ($n \ll N_B$) and the high injection case ($n \gg N_B$). By writing $I_c = I_f(V_{BE}) - I_r(V_{BC})$, for instance in the latter case we have

$$I_f = (I_s I_k)^{1/2} [\exp(V_{BE}/2U_T) - 1], \tag{23}$$

where the parameter I_k (onset current value of high injection) is determined by $I_k = 4 D_n W_B^2 Q_{bo}$. For the general case, where no closed form solution is possible, an accurate fit formula is obtained by smoothly merging both asymptotic solutions [28]. In this way the bias dependence of the transit time parameters is obtained naturally. Figure 11 gives the normalized value of τ_f as a function of injected current, clearly showing the asymptotic values.

2.3. Base current

By integrating the continuity equation (16) between emitter and collector and interpreting the resulting terms, the base current is expressed in three parts

$$I_B(t) = I_{EP} + \frac{Q_b - Q_{bo}}{\tau_p} + \frac{dQ_b}{dt},$$

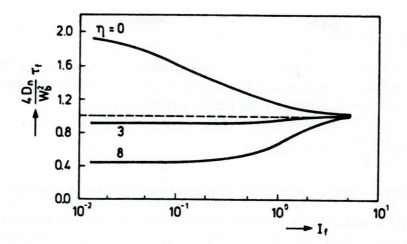

Fig. 11: Current dependence of the forward (τ_f) base transit time.

where τ_p is the hole lifetime and the hole current I_{EP} arriving at the emitter contact has a form similar to eq. (18) with a characteristic Gummel number (G_e) for the emitter. Owing to surface recombination in the depletion regions, usually the second right-hand side term dominates at low forward bias. On the other hand I_{EP} becomes dominant at higher bias conditions. Therefore, introducing a current gain parameter β_f and a nonideality factor m_f, the dc base current is satisfactorily described by

$$I_B(V_{BE}) = \frac{I_s}{\beta_f} \left\{ \exp\left(\frac{V_{BE}}{U_T}\right) - 1 \right\} + I_{bf} \left\{ \exp\left(\frac{V_{BE}}{m_f U_T}\right) - 1 \right\} \tag{24}$$

with a similar equation for the collector bias dependent part.

2.4. Charges and capacitances

Since the total stored base charge is proportional to the integrated minority carrier density $n(x)$ and the latter is obtained by solving eq. (22), the charge Q_{BE} and Q_{BC} can be calculated in principle [28]. Here we give only the high injection result

$$Q_{BE} = \tau_f I_f = (W_B^2/4D_n) I_f.$$

Similar to the current, the general charge stores are obtained by merging the low-injection and high-injection solutions. As τ_f, τ_r are not suitable parameters (compare Figure 11), in the practical model I_C, Q_{BE} and Q_{BC} are expressed in terms of the current constants I_s, I_k, the zero-bias base charge Q_{bo} and the parameter η, e.g.,

$$Q_{BE} = Q_{BE}(V_{BE}, I_s, I_k, Q_{bo}, \eta) \cdot$$

Usually the depletion charges are calculated by integrating the junction depletion capacitance expression $C_T = C_0 [1-(V_j/\phi_F)^p]$, where the parameter p is the grading coefficient. Since the resulting expression

$$Q_T = \frac{C_0 \phi_F}{1-p} [1-(1-V_j/\phi_F)^{1-p}] \tag{25}$$

has a singularity at $V_j = \phi_F$, in practice a slight modification is applied [1].

Under high forward injected current conditions a complication arises at the collector junction. First the moving minority carrier space charge modulates Q_{Tc} and secondly charge may be injected in the collector. This is further discussed in the next section.

2.5. Quasi-saturation

Due to an internal voltage drop in the collector region, the base-collector junction may become forward biased at high currents although the external voltage constitutes a reverse bias. This effect, which occurs mainly under the emitter, is called quasi-saturation. The locally forward biased junction causes an increase of reverse current I_r and the injection of an excess charge Q_{epi} in the collector epilayer (base push-out effect). As a result the current gain and the cut-off frequency will decrease. However, depending on the combination of current injected and applied collector bias, a number of different cases have to be considered [31]. This is illustrated in Figure 12, where the typical field and electron density distribution are given.

The main effect on the model described previously is that the boundary condition (17b) at the collector-side is changed and that the charge Q_{epi} has to be taken into account. Using the same assumptions as in section 2.1, the current in the quasi-neutral collector region is controlled by

$$J_n = qD_n \left(1 + \frac{n}{p}\right)\frac{dn}{dx} = qD_n \left(2 + \frac{N_{epi}}{p}\right)\frac{dp}{dx} \tag{26a}$$

with the approximate boundary condition (*see* Figure 12b)

$$p(x_i) \approx n_i. \tag{26b}$$

Solving this equation, the charge Q_{epi} can be expressed explicitly in terms of the injected carrier density p(o) at the junction [31]. Since however p(o) according to (26b) depends on x_i and the latter, via the voltage drop, on I_C and V_{CB}, finally I_C and the product $Q_{epi}*I_C$ become an implicit function of the external V_{CB}

voltage. In practice Q_{epi} and I_C are solved in a few steps using first an estimated value of I_C.

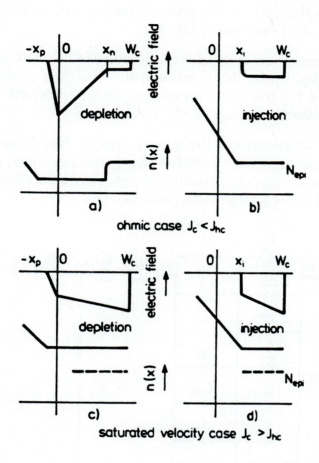

Fig. 12: The field and electron density in the collector epilayer (n-type) for the various depletion and injection conditions.

2.6. Parasitics and model extensions

Since an increase of depletion charge occurs at the cost of the neutral base width, the currents I_f, etc., are affected via a modulation of the zero bias base charge Q_{bo} (Early effect). Usually this effect is taken into account by putting $Q_{bo} + Q_{Tc} = Q_{bo} (1+\zeta)$ instead of Q_{bo}. It is easily shown that in the forward mode

$$\zeta \approx (V_{BE}/V_{ear}) - (V_{CB}/V_{eaf}), \qquad (27)$$

where the parameters V_{ear}, V_{eaf} are called Early voltages.

As with the above correction the most important intrinsic properties have been discussed. Next an equivalent circuit [32] is given in Figure 13. However in this scheme already parasitic elements such as series resistances and the substrate transistor have already been added. Usually the base resistance is split in a part below and adjacent to the emitter. To account for current crowding a diode has to be put in parallel to the first resistor [33, 34].

In contrast to the Gummel-Poon model the above MEXTRAM model describes well the characteristics under high current-low voltage conditions. This is illustrated in Figure 14 for the common emitter dc characteristics and in Figure 15 for the cut-off frequency.

Naturally for wider applications the above model has been extended [1]. Without further discussion we mention here temperature effects, charge storage in the emitter, avalanche multiplication, nonlinear Early effect, noise sources and non-quasi static charge behaviour [35, 36].

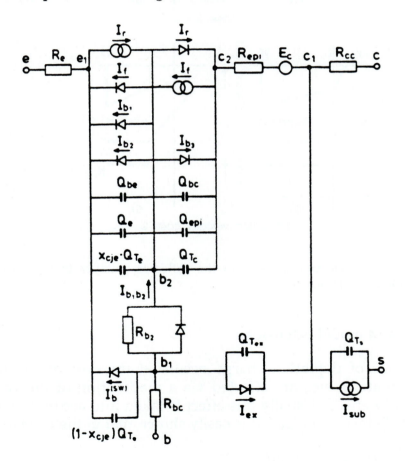

Fig. 13: Equivalent circuit diagram of the MEXTRAM model.

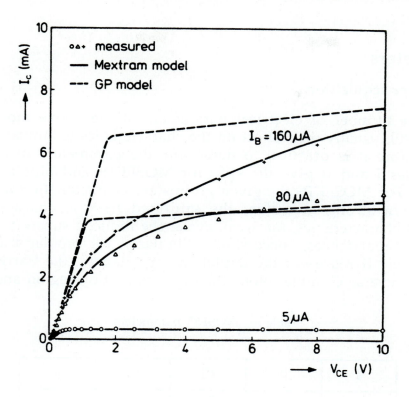

Fig. 14: Measured and modelled dc characteristics of bipolar transistor.

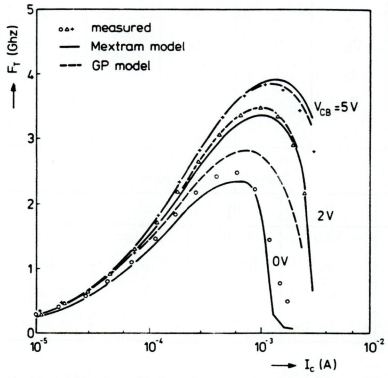

Fig. 15: Measured and modelled cut-off frequency of bipolar transistor.

3. Parameters

3.1. Parameter acquisition

Although the number of electrical parameters of the above models is large (typically > 20), usually their effect on the characteristics is limited to a specific range. Therefore it is practical to determine the parameters in characteristic groups. Tables I and II give the split for MOSFETS and bipolar transistors, respectively. For MOSFETS the given groups are characteristic for the channel conductance in strong inversion, the saturated I_D-V_{DS} characteristics, the subthreshold characteristics, the weak avalanche induced substrate current and the intrinsic capacitances, successively. Similarly, for bipolar transistors the groups of Table II represent the depletion capacitances, the Early effects, the forward and reverse Gummel plots, the quasi-saturation region and the cut-off frequency.

Table I: MOSFET parameters

I	II	III	IV	V
V_{T0}	θ_C	m	a	C_{ox}
γ_i	f	(I_0)	V_a	V_{FB}
γ	α	s		
V_{SX}	V_p	s_i		
β_0		(t)		
θ_A				
θ_B				

Table II: Bipolar parameters

I	II	III	IV	V	VI
C_{0e}, P_e, V_{de}	Q_{b0}	I_s, β_f	I_{ss}, β_{rx}	R_{epi}	η
C_{0c}, P_c, V_{dc}	x_{cjc}	I_{bf}, V_{if}	I_{br}, V_{lr}	I_{hc}	τ_{ne}, m_τ
C_{0s}, P_s, V_{ds}	(x_{cje})	I_k	I_{ks}	(V_{dc})	η
		R_e			

3.2. Geometry and process dependence

Owing to junction sidewall effects or MOSFET short-channel effects (compare section 1.1), most electrical parameters are geometry dependent. Usually simple geometrical rules apply such as [1]

$$P = p_b \, ZL + p_1 \, L + p_z \, Z \tag{28}$$

or
$$P = p_0 + p_1 \, L^{-1} + p_z \, Z^{-1} , \tag{29}$$

in which p_b, p_1, etc. are called unity parameters.

The electrical length (L) and width (Z), which generally differ from the drawn values L_M and Z_M owing to process deficiencies, can be measured via a specific parameter, like for instance the MOSFET gain factor

$$\beta = \beta_0 \frac{Z}{L} = \beta_0 \frac{Z_M - \Delta Z}{L_M - \Delta L} . \tag{30}$$

As an example, Figure 16 illustrates this for a 0.5 μm CMOS process. Once ΔL and ΔZ are known, the unity parameters of eqs. (18) or (29) can be determined from a set of devices with different Z_M, L_M values.

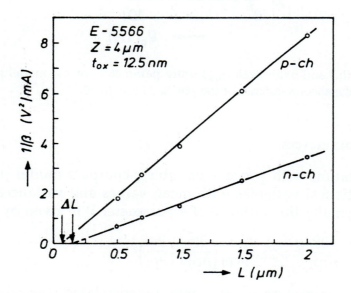

Fig. 16:   Determination of channel length correction by plotting the (inverse) gain factor as a function of drawn gate length.

In addition to the geometry, the parameters also may depend on specific process quantities. For instance, the bipolar current constant I_s will depend on the active base sheet resistance ρ_{b1}. Figure 17 gives the result for the bottom part and the oxide sidewall part. For other reasons the MOSFET threshold voltage may depend on the gate insulator thickness and the threshold implantation dose.

Using the above dependences on geometry and process quantities, the electrical parameters can be calculated from process data [1].

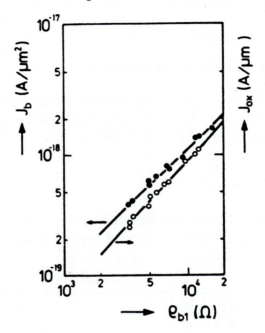

Fig. 17:   The bottom (J_b) and oxide wall (J_{ox}) unity parameters of the model parameter I_s as a function of the sheet resistance of the pinched base. $J_{si}=0$.

3.3. Statistics of parameters

Owing to their relation with process variables, compact model parameters can be regarded as statistical variables with mean values and variances as a measure of their spread. Formally the variance of a parameter P is given by

$$\sigma_P^2 = \Sigma\sigma_{pi}^2\left(\frac{\partial P}{\partial p_i}\right)^2 + 2\Sigma\sigma_{pi\ pj}^2\left(\frac{\partial P}{\partial p_i}\right)\left(\frac{\partial P}{\partial p_j}\right) \tag{31}$$

Since the process parameters p_i, etc., may be correlated, the covariance is not necessarily zero.

For bipolar devices the major process variables to characterize spread are the geometry corrections (ΔL, ΔZ), the active and non active base sheet resistances (ρ_{b1}, ρ_{b2}) and the breakdown voltage V_{cbo} [1]. For MOSFETS in addition to ΔL, ΔZ, the gain factor β_0 and the threshold voltage V_{TO} of a large area device and the unit area interconnect capacitance C_{int} have been used [38]. In the latter case

$\sigma_{\beta 0}V_{T0} \approx -1$. As an example Figure 18 gives the measured probability density of the collector current parameter I_s together with a curve calculated from process spread data. Since the agreement for other parameters is satisfactory too, the above calculation can be further extended to predict the spread of a designed electronic function via circuit simulation [37, 38].

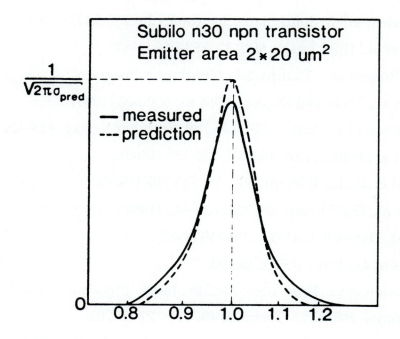

Fig. 18:   Propability density curve for collector current constant I_s.

References

[1]   H.C. de Graaf and F.M. Klaassen, *Compact transistor modelling for circuit design*, Springer Verlag, Vienna (1990).

[2]   E. Demoulin, F. v.d. Wiele, *Process and Device Modelling for IC Design*, Noordhoff, Leyden, 617-675 (1977).

[3]   L.D. Yau, Solid State Electronics 17, 1059-1063 (1974).

[4]   K.N. Ratnakumar et al., IEEE Journ. SSC-17, 937-947 (1982).

[5]   T. Skotnicki et al., Proc. ESSDERC 87, North Holland, Amsterdam, 543-546 (1987).

[6]   F.M. Klaassen et al., Solid State Electronics 28, 359-373 (1985).

[7]   T. Ando et al., Rev. Modern Physics 54, 437-672 (1982).

[8]   G. Merckel et al., IEEE Trans. ED-19, 681-690 (1972).

[9]   G.T. Wright, IEEE Trans. ED-34, 823-833 (1987).

[10] B. Hoeflinger et al., IEEE Trans. ED-26, 513-520 (1979).

[11] T. Poorter et al., Solid State Electronics 23, 765-772 (1980).

[12] F.M. Klaassen et al., Solid State Electronics 23, 237-242 (1980).

[13] B.J. Sheu et al., IEEE Journ. SSC-22, 558-566 (1987).

[14] CURRY, Proprietary Philips 2-D Simulation Program.

[15] P.K. Ko et al., Technical Digest IEDM 81, 600-603 (1981).

[16] F.M. Klaassen et al., Proc. ESSDERC 89, Springer Verlag, 418-422 (1989).

[17] P. Yang et al., IEEE Journ. SSC-18, 128-138 (1983).

[18] D.E. Ward et al., IEEE Journ. SSC-13, 703-707 (1978).

[19] S.Y. Oh et al., IEEE Journ. SSC-15, 636-643 (1980).

[20] M.F. Sevat, Digest ICCAD87, 208-210 (1987).

[21] T. Smedes et al., Proc. ESSDERC 90

[22] F.M. Klaassen et al., Proc. ESSDERC 88, Journ. Physique C4, 257-260 (1988).

[23] H.K. Gummel, Bell Syst. Techn. J. 49, 115-122 (1970).

[24] J.W. Slotboom et al., Solid State Electronics 19, 857-862 (1976).

[25] J.J. Ebers, J.L. Moll, Proc. IRE 42, 1761-1771 (1954).

[26] H.K. Gummel, H.C. Poon, Bell Syst. Techn. J. 49, 827-844 (1970).

[27] H.M. Rein et al., IEEE Trans. ED-34, 1741-1761 (1987).

[28] H.C. de Graaff et al., IEEE Trans. ED-32, 2415-2419 (1985).

[29] J.G. Fossum, Proc. IEEE Bip. Circuits and Technology Meeting, 234-241 (1989).

[30] K.W. Michaels, A.J. Strojwas, Bip. Circuits and Technology Meeting, 242-244 (1989).

[31] H.C. de Graaf, in *Process and Device Modelling for IC Design*, Noordhoff, Leyden, 419-442 (1977).

[32] H.C. de Graaff et al., 18th Conf. Solid St. Devices, Tokyo, 287 (1986).

[33] G. Rey, Solid State Electronics 12, 645-655 (1969).

[34] H. Groendijk, IEEE Trans. ED-20, 329-333 (1973).

[35] J. te Winkel, Adv. Electronics/Electron Physics 39, 253-289 (1976).

[36] J.G. Fossum et al., IEEE EDL-7, 652-654 (1986).

[37] P. Yang et al., Techn. Digest IEDM 82, 286-289 (1982).

[38] M.J.B. Bolt et al., ISMSS 90, 1989.

[39] Y.P. Tsividis, *Operation and Modeling of the MOS Transistor*, McGraw-Hill, 1987.

Chapter 2
CMOS Operational Amplifiers

Daniel Senderowicz

SynchroDesign Inc., 2140 Shattuck Ave, Suite 605, Berkeley, CA 94704-1210

1. General Considerations for VLSI of Amp Design

Until the availability of very-large-scale integrated circuits, analog-system designers had very little interaction with integrated op amp designers. Because of this, IC manufacturers kept the number of types within reasonable bounds by designing op amps capable of performing under a wide range of configurations, e.g., inverting, noninverting, with high and low gain, etc. The current trend in VLSI implementation of assembling the system with the aid of a cell library also dictates guidelines restricting the number of different types. In principle, it is also possible to follow similar strategies in the design of op amps to suit cases with broad applications. However, there is a fundamental cost difference between the former discrete and highly integrated systems. The impact on the incremental cost of an discrete system for a slightly larger op amp is not as significant as in the case of a VLSI in which the die area, and in turn, yield play crucial roles in determining the cost. The small area criterion becomes of high priority, even at the expense of sacrificing the functional scope of the cells.

In the process of cell simplification, some of the burdens assigned to discrete op amp designers are shifted to the system designers. The latter have to understand not only the capabilities of the simplified cells, but also the performance limitations of ICs, such as power-supply noise rejection (PSRR), reduced power-supply voltages, etc. Treatment methods for PSRR improvement differ substantially between discrete and integrated systems. In the former, power lines are decoupled between critical points by means of RC filters, but for integrated circuits the low value of capacitance achievable makes this approach impractical. On the other hand, techniques such as differential circuitry can help to alleviate some of the impairments found in integrated versions. These differential techniques however are not practical in discrete designs, which aim at reducing the component count.

For those undertaking the task of designing an analog subsystem, it is very useful to follow a small set of guidelines in choosing the configurations, restricting them to those optimizing the utilization of the simple op amps, which can be regarded as the fundamental cells. Listed below is a set of such rules. The reasons for some of these rules may not appear obvious at this point, but will later become evident after further insight into the transistor level circuitry.

(1) Avoid the use of internal resistors. These can appear as op amp loads, forcing the need for complex buffer stages.

(2) Constrain the function around an op amp to an inverting configuration, avoiding such functions as voltage followers. The noninverting configuration requires an amplifier with large input common-mode range.

(3) Avoid configurations which use MOS switches around the feedback path of an op amp. This restriction is based on similar arguments as those applying to rule (2).

(4) Design the systems in such a way as to accommodate for a wide spread in the DC voltage gain. This uncertainty is mostly due to inaccuracies in the modelling of the MOS transistors in the simulators.

(5) In the case of analog-to-digital (ADC) or digital-to-analog converters (DAC), choose *sequential* techniques whenever possible. This approach, which is somehow opposed to the idea of parallelism, generally simplifies the amount of analog circuitry at the expense of some added digital complexity and reduced speed of operation.

(6) Design the systems using differential circuitry whenever possible. The op amps will be simpler, and most importantly, this may be the only way to achieve the desired specifications of dynamic range. The cost for this is an increased amount of internal elements and busing.

2.   The Differential Circuit Approach

Digital circuitry generates a large amount of switching noise which appears in the power lines, the common substrate, and as parasitic capacitive coupling between the internal interconnections. This scenario imposes as one of the main goals for the designer that all the necessary precautions be taken in order to make the analog circuit highly immune to this noise coupling. Differential circuitry is a technique which provides a way to achieve this immunity. Furthermore, differential circuitry doubles the effective signal swing which increases the dynamic range, a very important feature in high-density technology where the maximum operating supply is limited to 5 volts or less. The use of differential circuits for the analog paths generally implies a larger number of components and interconnections. However, as will be seen later, op amps with comparable functionality are considerably simpler. Another advantage of differential circuits is that the signal is immune to noise in the internal reference node (e.g., 1/2(V$_{DD}$ - V$_{SS}$), a feature which is very important in single supply systems.

For certain applications such as those requiring the processing of an input current (or charge), the use of differential circuitry cannot be easily implemented. In these cases, single-ended circuitry is the only practical alternative. However, if there is a significant amount of signal post-processing, one can always implement a single-ended to differential conversion as close to the input as possible, aiming at keeping the amount of single-ended circuitry to a minimum. With this consideration in mind, the need for single-ended op amps also arises, which in general will have to be subject to similar minimization arguments as those applying to their differential counterparts.

In a differential amplifier there is an additional *degree of freedom* over a single-ended amplifier. While the differential output should be proportional to a differential input signal, the average voltage of these two outputs should also be controllable with another input signal. This is usually referred to as the common-mode input. Fig. 1 shows a representation of a differential op amp where the inner amplifier represents the common-mode path.

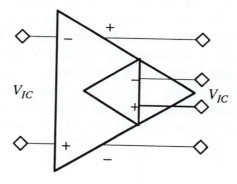

Fig. 1: Differential op amp.

In most applications, the common-mode input is fed with a signal derived from the average of the two outputs However, it could also be possible to superimpose another signal to that average which may or may not be directly related to the differential signal. Fig. 2 shows a differential switched-capacitor integrator with the averaging of the two outputs implemented by means of an additional set of switched-capacitors. C$_U$ is usually the unit capacitor, while C$_X$ has to be large enough to keep the common-mode switching noise within acceptable levels.

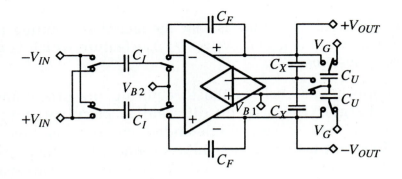

Fig. 2: Differential switched-capacitor integrator.

It should be noted that in Fig. 2 the bias point of the differential switched-capacitors can be chosen arbitrarely. In fact, V_{B2} and V_G do not have to be the same, and they can both be different from V_{B1}, the bias point for the common-mode switched-capacitors. This flexibility in choosing these bias points results in the possibility of superimposing DC voltages to the signal in order to avoid level-shifting operations inside the op amp. A numerical example can illustrate this idea: V_G is chosen to maximize the swing; therefore, for a single voltage supply of 5 volts, $V_G=2.5V$. This implies that the differential voltage V_{OUT} will have an average value of 2.5 volts. The bias voltages V_{B1} and V_{B2} can be chosen in such a way as to fit the internal characteristics of the amplifier or other requirements. At this point it becomes obvious that one of the main goals of level-shifting with the external capacitors is to allow for op amps with limited input common-range. The penalty for having this diversity of bias voltages precludes certain configurations, such as reset switches in parallel with capacitors, which is found in the case of an offset-compensated amplification-module (Fig. 3).

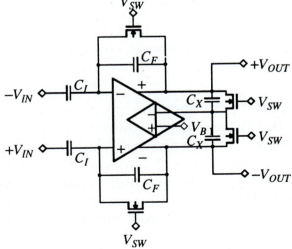

Fig. 3: Amplifier with MOS reset switch.

Fig. 4 describes a simplified schematic for an alternative solution that avoids the use of MOS switches while still performing the discharging of the feedback capacitor C_F. After examining the characteristics of different amplifiers, the reasons for avoiding certain configurations will become more obvious.

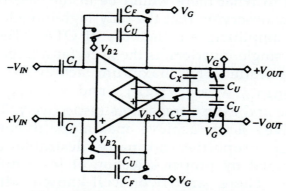

Fig. 4: Feedback reset without switch.

3. The Number of Internal Stages

Most switched-capacitor integrators operate under a condition in which $C_F \gg C_I$ (Fig. 2). This implies that the gain-bandwidth products of the integrator and the op amp are essentially the same. In this case, a single voltage-gain stage is recommended. This type of op amp is sometimes referred to as an operational transconductance amplifier (OTA). OTAs do not require additional compensation capacitors, thus leading to very compact circuits. On the other hand, when the feedback configuration is such that $C_D \ll C_I$ (Fig. 3), such as in high-gain amplification stages, it might become necessary to have a multiple-stage amplifier. Although these amplifiers generally require phase-compensation, this can be reduced to minimum capacitance due to the low amount of feedback.

Another instance in which multiple-stage op amps are required is when heavy resistive and/or capacitive loads need to be driven, as in the case of off-chip drivers. These amplifiers can become very complex because of the limitations of MOS transistors as current drivers. Examples of these will be given later.

4. Cascode Op Amps

Fig. 5 shows a typical implementation of a simple cascode amplifier. It is well known from single amplifier theory that cascode stages increase the voltage gain by means of boosting their effective output impedance. This boost is proportional to the product $g_{M3} \times R_{OUT1}$. The load also has to be cascoded (M_5 and M_7) in order to increase its effective impedance to a level comparable to the driver pair (M_1 and M_3). One of the great advantages of cascode amplifiers is that since they operate in *current* mode, there is very little degradation of the bandwidth compared to what there would be in the case of a simple inverter. This is further seen by observing that the only high-impedance node existing is the output. Cascode amplifiers are the core of OTAs. Their main limitation, when compared with single inverters, is the need for a larger supply voltage for a given signal amplitude. The maximum achievable voltage swing is V_{DD}-$2V_{DSAT_N}$ - $2V_{DSAT_p}$, where V_{DSAT_N} and V_{DSAT_N} are the saturation voltages of the N-type and P-type transistors respectively. Therefore, one of the main design objectives is to achieve maximum dynamic range without sacrificing the other properties normally desirable of op amps. This maximization is obtained by proper selection of the amplifier's internal bias points (V_{B3} - V_{B4}). There are two well-known alternatives for the implementation of OTAs using cascode stages: the *folded*-cascode and the *telescopic*-cascode. These will be described in the following sections.

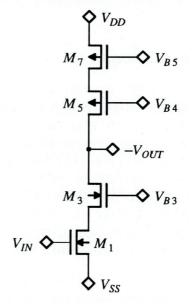

Fig. 5: Simple cascode stage.

The simplicity of cascode op amps implies a high degree of flexibility in choosing the transistor geometries. Therefore other considerations such as the concept of stacked layout (explained later on) can be used to precise these geometries. However, it is interesting to study the dynamics of cascode amplifiers to find the effects of the internal parameters (derived from the geometries and loading conditions) on their global performance, with the aim of introducing other constraints in the design that can lead to optimum performance.

Fig. 6 shows a simplified model for the cascode op amp of Fig. 5. Assuming that this op amp is used as an integrator, there will be a feedback capacitor c_f and an input current source i_{in}, which in reality could take the form of discrete charge-packets injected at a given rate. The voltage v_{in} is the small signal appearing at the gate of M_1 (virtual ground). Capacitor c_1 is a lumped equivalent of the capacitance associated with the node at the source transistor M_3, usually dominated by its gate-to-source capacitance, c_l and r_l represent the capacitive and resistive loading, respectively, of the whole stage.

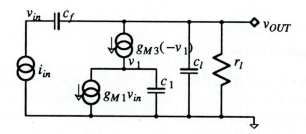

Fig. 6: Model for a cascode OTA.

The natural modes (poles) of the transfer function for the model of Fig. 11 are given by:

$$s_{1,2} = \frac{-\left(r_l c_l + \frac{c_1}{g_{M3}}\right) \pm \sqrt{\left[r_l c_l + \frac{c_1}{g_{M3}}\right]^2 - 4\left(1 + g_{M1} r_l\right)\frac{r_l c_l c_1}{g_{M3}}}}{2\frac{r_l c_l c_1}{g_{M3}}} \tag{1}$$

Assuming that $g_{M1} = g_{M3}$ and $c_1 \ll c_l$, the resulting poles are:

$$s_1 = 0$$

$$s_2 = -\left(\frac{g_{M3}}{c_1} + \frac{1}{r_l c_l}\right) \approx -\frac{g_{M3}}{c_1}$$

This is clearly an over-dumped system. It is interesting to note that if the variables g_{M3} and/or $c1$ are allowed to vary, it is possible to achieve a critically dumped system. This condition occurs whenever the parameter values for (1) results in $s_{1,2}$ being a complex conjugate number. The transitional value of the parameters can be obtained by forcing the term inside the square root to zero:

$$\frac{g_{M3}}{c_1} > 4 \, \frac{g_{M1}}{c_1} \tag{2}$$

where it is assumed that $g_{M3}r_1 \gg 1$ and $g_{M1}r_1 \gg 1$. The significance of (2) is that the settling time can be minimized by dimensioning g_{M3} and/or c_1. Furthermore, since c_1 is essentially the gate capacitance of M_3, the variation of the left member of (2) can be achieved by simply varying the channel length of this transistor. However, if the value of c_1 is too large, it might not be practical to try to achieve critical dumping, mainly because of the resulting large value of c_1. Furthermore, additional parasitics, mainly the source capacitance of M_5 (Fig. 5), increase the number of poles impeding the approach. Therefore, in most practical cases c_1 is minimized and g_{M3} is made equal to g_{M1}.

4.1. The folded-cascode OTA

An implementation of the *folded*-cascode OTA is depicted in Fig. 7. The higher the biasing voltage of the gates of M_3 and M_4, the larger the output voltage swing will be. Therefore, this voltage is set so as to keep the current sources M_9 and M_{10} biased just above their saturation voltage. It turns out that in this operating condition, the common-mode input range is also maximized. The same idea applies to the cascode loads M_5, M_7 and M_6, M_8. That is, the gates of M_5 and M_6 should be biased at a voltage that will maintain M_7 and M_8 biased at their saturation voltage. Given the normally low operating voltages of today's MOS technologies, it is desirable to have this saturation voltage as low as possible. The saturation voltage is:

$$V_{DSAT} = V_{GS} - V_T. \tag{3}$$

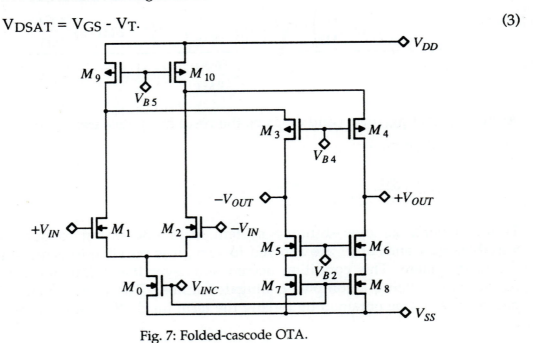

Fig. 7: Folded-cascode OTA.

Noise considerations in the design dictate that the transistors used as loads should have significantly lower transductance than the ones used for the input stage. Complying with this precept implies a trade-off with the output dynamic range, as the following expressions will show. The saturation voltage can also be expressed as:

$$V_{DSAT} = \sqrt{\frac{2I_{DS}}{\beta}} \tag{4}$$

where β is the gain constant of the transistor and I_{DS} is the DC operating current. The transductance g_M expressed as a function of I_{DS} is:

$$g_M = \sqrt{2\beta I_{DS}}. \tag{5}$$

The product of equations (4) and (5) yields the following expression:

$$V_{DSAT} = \frac{2I_{DS}}{g_M}. \tag{6}$$

Equation (6) shows that the maximizing the dynamic range implies degrading the noise performance and vice versa. The best choice of parameters will depend on the design targets. These considerations apply to the dimensioning of transistors pairs M_7, M_8 and M_9, M_{10}.

In the *folded*-cascode op amp, the common-mode input can be implemented by transistor M_0, or alternatively by the pair M_7, M_8, or by the parallel combination of M_0 and the pair M_7, M_8.

4.2. The telescopic-cascode OTA

This amplifier, depicted in Fig. 8, has a configuration that is a straightforward differential implementation of the single-cascode stage shown in Fig. 5. As in the previous case, designing for maximum output swing also implies proper biasing of the cascode stages. The main difference with the *folded*-cascode is that there is a trade-off between the input common-range and the output swing. To achieve the maximum output swing, the pair M_3, M_4 has to be biased in such a way as to leave the input pair M_1, M_2 and the common-mode amplifier M_0 biased just above their saturation voltage. This condition limits the common-mode compliance at the input of the op amp to zero. If the operation of the op amp forces any common-mode input signal, some output swing has to be sacrificed by increasing the value of V_{B3}.

4.3. Comparisons between the folded and telescopic cascode OTAs

Although the two configurations are essentially similar in operating principles, they have some advantages and disadvantages over each other. From the capabilities standpoint, the *folded*-cascode is more flexible than its *telescopic-cascode* counterpart. Its large input common-mode range makes the biasing of the input stage not as restrictive as in the case of the *telescopic*-cascode. Therefore, any existing voltage between V_{B2} and V_{DD} can be used for this purpose. Furthermore, the output swing is slightly larger because there is one fewer transistor in the *lower* part of the amplifier. Nevertheless, if one is willing to sacrifice these two features, simulations show that the performance is essentially similar.

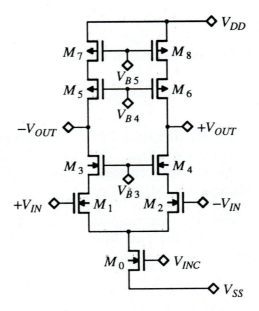

Fig. 8: Telescopic-cascode OTA.

The main basis for the evaluation comparisons is that the input pairs (M_1, M_2) and the common-mode transistor (M_0) have the same dimensions and carry the same bias current in both types of amplifiers; e.g., $I_{0fc} = I_{0tc}$, where fc and tc are suffixes indicating the *folded* and *telescopic* cascodes, respectively.

Power dissipation: In the *folded*-cascode, the pair M_9, M_{10} has to carry at least $2I_0$. Otherwise one of the transistors in the pair M_3, M_4 may turn off under large input signal conditions, thus increasing the recovery time during a transient.

Die area: The mobility, and therefore the β ratio between PMOS and NMOS transistors is approximately one-third. Based on (5), the g_M ratio can be approximated at one-half. Assuming that similar dynamic characteristics are to be achieved between the two types of amplifiers, the transistors in the pair M_3, M_4 of the folded-cascode have to be twice as wide as the corresponding ones M_3, M_4 of the *telescopic*-cascode. The power dissipation condition requiring the pair M_9, M_{10} to carry $2I_0$ has to be satisfied at the expense of doubling the width of this pair. Otherwise the saturation voltage will increase (*see* (4)). Using the argument of the gain difference between the N- and P-channel transistors, the pairs M_5, M_6 and M_7, M_8 of the *folded*-cascode can be one-half the width of the corresponding pairs in the *telescopic*-cascode counterpart. Table 1 summarizes typical size differences of equivalent transistors in the cases of *folded* and *telescopic* cascode amplifiers. The numbers in the table indicate normalized values relative to one transistor unit (1 U).

Table 1

Comparison between OTAs		
Transistor	*Folded*	*Telescopic*
M_0	2 U	2 U
M_1, M_2	1 U	1 U
M_3, M_4	2 U	1 U
M_5, M_6	1/2 U	1 U
M_7, M_8	1/2 U	1 U
M_9, M_{10}	2 U	—
Total (x2 for pairs)	14 U	10 U

4.3.1. Simulations

The Spice simulation setup is shown in Fig. 9, with the transistor sizes as given in Table 1 for a basic transistor unit of $1U = (100\,\mu)/(2\mu)$ and the current $I_0 = 100\,\mu A$.

Figs. 10 and 11 show, respectively, the input and output nodes of the transient response.

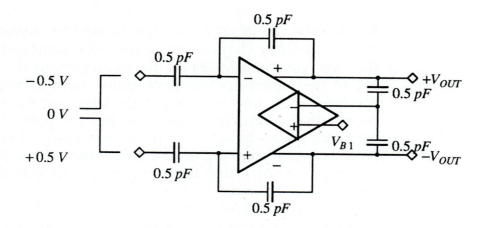

Fig. 9: Simulation setup.

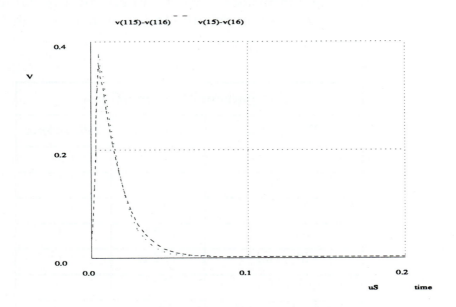

Fig. 10: Input voltage of transient response.

It can be observed that the input and output from the telescopic-cascode [variables v(15)-v(16)] are almost identical to the corresponding variables of the folded-cascode [variables v(115)-v(116)].

4.4. Single-ended versions of the cascode OTAs

The differential *telescopic* and the *folded* cascode OTAs can be modified to operate with a single-ended output by simply converting the loads to current mirrors as shown in Figs. 12 and 13 respectively. It should be noted that the same restrictions for the output swing are in effect for the single-ended versions.

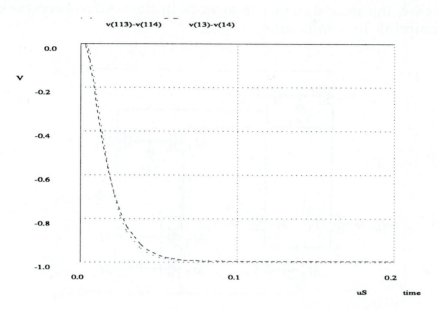

Fig. 11: Output voltage of transient response.

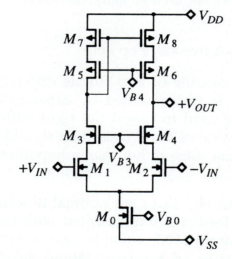

Fig. 12: Single-ended telescopic-cascode OTA.

5. Multiple Stage Op Amps

As was mentioned before, when the application requires that the gain for the stage be much larger than unity (e.g., 10 times), multiple voltage-gain stages may be introduced. More than one stage may also be needed if the application requires driving a resistive load, such as in the case of an analog off-chip driver. In this latter case, the second stage has to work in class-AB to keep the quiescent power consumption to a minimum.

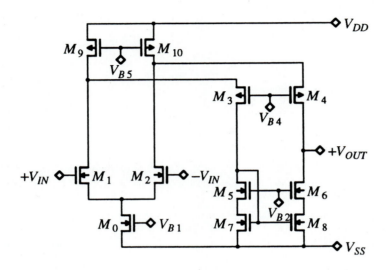

Fig. 13: Single-ended folded-cascode OTA.

5.1. The differential class-A two stage op amp

This op amp is simply a cascade of two simple differential inverters (Fig. 14). The phase compensation is implemented by capacitors C_{F1} and C_{F2} and the resistors R_{C1} and R_{C2} are used to cancel the right half-plane zero. It is worth reiterating here that the value of these capacitors should be calculated based on the actual gain setup for the stage rather than a hypothetical worst case of unity gain.

This amplifier, as shown in Fig. 14, is not optimal in terms of the output swing. The local common-mode feedback implemented with transistor M_0 imposes an operating voltage for the drain of M_7 equal to V_{GS_0}, which is considerably larger than the saturation voltage of M_7 (e.g. 1V vs. 0.3V). Therefore the most negative voltage that any of the outputs can reach is $V_{SS} + V_{GS_0} + V_{DSAT8,9}$. This limitation can be corrected by inserting a switched-capacitor level shifter between the drain of M_7 and the gate of M_0, as shown in Fig. 15.

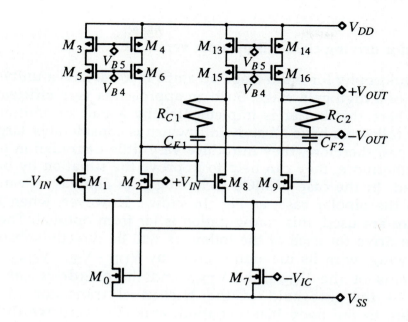

Fig. 14: Class-A, two-stage op amp.

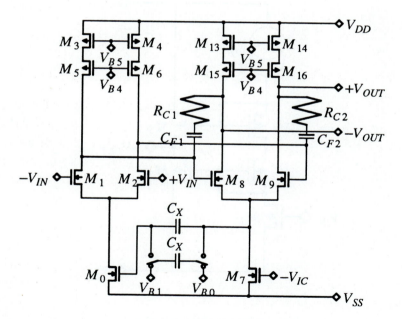

Fig. 15: Extended range Class-A, two-tage op amp.

5.2. The resistor driving class-AB op amp, version I

One design philosophy for multi-stage op amps is to cascade a unity-gain buffer to the output of a high gain stage. Such an approach is very efficient in bipolar technology where this buffer is implemented by a pair of emitter followers. These emitter followers have a bandwidth which is considerably larger than the preceding stages; therefore, they introduce very little degradation of the phase response. Furthermore, they can help to improve the situation by buffering the capacitive load. In the case of MOS technology, an equivalent version can be derived from the bipolar case, as Fig. 16 shows. However, when low-voltage power supplies are used, this configuration is far from optimal. The amount of gate-to-source drive for each of the followers will be directly substracted from the voltage swing, with its maximum given by $V_{DD} - V_{SS} - V_{GS_N} - V_{GS_P}$. The two components of the V_{GS}'s, the V_{TH0} and the overdrive can have large magnitudes — the threshold of the N-channel transistor in an N-well technology due to the body bias variation, and the overdrive due to a high output current. For a reasonable W/L ratio for M_3 and M_5 (<200), the output voltage can be made to reach 1.5V from each supply.

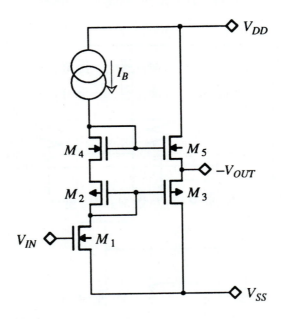

Fig. 16: Source follower stage.

Based on the previous consideration, it becomes more convenient to use the common-source configuration for the output transistors, which limits the output swing to only the saturation voltage of the transistors (equal to the overdrive). This changes the picture of the design approach, since cascading the preceding high-gain portion with these common-source transistors can no longer be considered as adding a wide bandwidth stage. Thus the design of the amplifier has to be carried out as a whole, taking into account the frequency/phase characteristics of each composing stage with the estimated worst-case output loading.

Probably the most critical aspect of designing a class-AB output stage using a common-source stage is the definition of the quiescent output current. Both complementary transistors in the output have to be biased with a signal which establishes the desired operating current. If this current is too low, then the bandwidth will be degraded; if it is too high, the power consumption can become unacceptable. Fig. 17 shows a single-ended amplifier with a class-AB output stage where the output quiescent current is defined by transistor ratios.

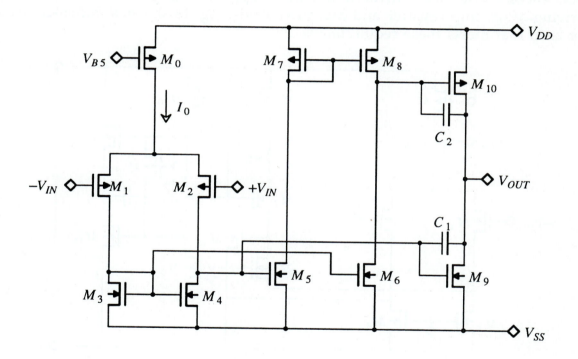

Fig. 17: The resistor driving op amp — Type I.

The operation of the amplifier can be better understood assuming no loading current, so that only the quiescent current flows through M_9 and M_{10}. This also implies that the differential input is zero. Otherwise some driving signal would develop at the gates of the two above-mentioned transistors, thus forcing an output current other than zero. It is also given that M_3, M_4, M_5 and M_6 are all identical, and M_7 is identical to M_8. Under these conditions, $I_{DS6}=I_{DS3}$, to which I_{DS8} must match (otherwise the op amp will not be in the stable quiescent condition) $I_{DS5}=I_{DS6}$ because of the mirror M_7, M_8; and lastly, the absence of signal establishes the following equality: $I_{DS6}=I_{DS3}=I_0/2$. Since the gates of M_9 and M_5 are driven by the same voltage, their currents will have the same proportional ratio as their geometries.

This amplifier uses Miller compensation which is implemented by capacitors C_1 and C_2.

5.3. The resistor driving class-AB op amp, version II

Many applications require a true-ground voltage for the inverting input of the op amp. This imposes design challenges if the chip containing the op amp has to operate off a single 5V supply and yet have an input common-mode range that encompasses 0V. Furthermore, these applications generally require the driving of off-chip resistive and capacitive loads. Fig. 18 shows a folded-cascode op amp that meets these requirements.

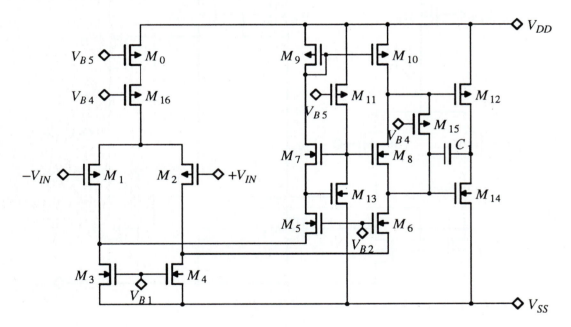

Fig. 18: Resistor driving op amp — Type II.

This op amp achieves the desired input common-mode range by biasing the drains of M_3, M_4 just above their saturation voltage. If the inputs are connected to ground, $V_{TH1,2}$ is the maximum voltage that the drains of M_1, M_2 can sustain with respect to ground while still remaining in the saturation region. This voltage is generally larger than $V_{DSAT3,4}$.

The output stage also operates in class-AB, although using a different configuration than in version I. The constant I_{DS11} bias when applied to M_{13} establishes the operating voltage at its gate by means of the feedback through M_7. Its mirror M_8 will have the same voltage at its source and therefore the bias of the gate of M_{14} will be the same as in the gate of M_{13}. Thus the current I_{DS14} will be the quiescent output current. Under large signal conditions, the two following cases are valid:

(1) For positive going signals, the gate of M_{12} will be strongly driven towards the negative direction while the drive for M_{14} decreases,

(2) For negative going signals, most of I_{DS10} will flow through M_{15} pulling the gate of M_{14} in the positive direction.

This op amp requires a single compensation capacitor C_1 because its feedback current is simultaneously applied to M_{14} or M_{12} through the common-gate M_8. It is interesting to note the nonlinear behavior of the stage; i.e., for small signals the pair M_7, M_8 acts as a differential to single-ended converter, while for large positive going signals M_8 is turned off. This nonlinearity usually causes zero-crossing distortion which can be reduced by adjusting the bias voltage V_{B4} to the gate of M_{15}. This in turn rises the gates M_{12} and M_{14}, leading to a higher quiescent current — a situation quite common in class-AB amplifiers. This adjustment changes the phase response of the amplifier. So the optimum compromise will depend on the particular operating condition.

Fig. 19 shows a simulation setup where the quiescent current is changed by adjusting the bias voltage of M_{15}.

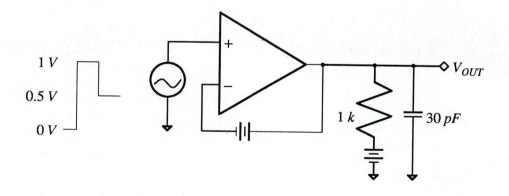

Fig. 19: Driver simulation.

Figs. 20 and 21 show, respectively, the output for the cases where $I_{DS15}=0$ and $I_{DS15}=I_{DS10}/2$.

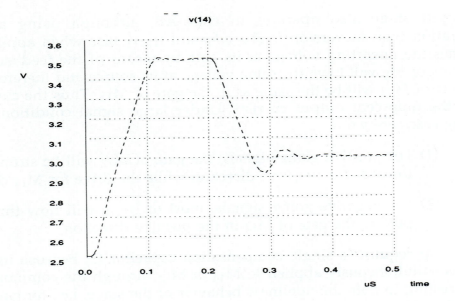

Fig. 20: Transient response with $I_{DS15} = 0$.

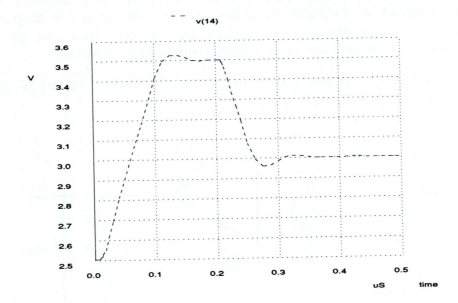

Fig. 21: Transient response with $I_{DS215} = I_{DS10}/2$.

6. Bias Circuits

The op amps previously described do not show the required bias circuitry. This circuitry when properly designed can be far more complex than the op amp itself, adding a significant amount of die area and power consumption. However, the voltages generated by the bias circuit can be shared by several amplifiers in the chip, diminishing the impact of the increase in area and power. In the *telescopic*-cascode amplifier five different bias voltages are required: V_{B1} through V_{B5}. Their values are as follows:

V_{B1}: One V_{GS_N} above V_{SS},

V_{B2}: One V_{DSAT_N} above V_{B1},

V_{B3}: Two V_{DSAT_N} above V_{B1},

V_{B5}: One V_{GS_P} below V_{DD},

V_{B4}: One V_{SDAT_P} below V_{B5},

V_{B1} is generated simply by forcing the desired bias current over a diode-connected transistor with dimensions identical to that of M_0 in the op amp circuit. Some current scaling can be applied in order to reduce the output impedance of this bias generator. This scaling is done by geometry and bias current rationing. V_{B2} is produced by properly changing the current density on another transistor according to the method described next. Fig. 22 shows two transistors, M_1 and M_2, biased with the same current.

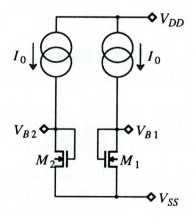

Fig. 22: V_{B2} and V_{B1} generation.

The following two expressions are derived from the simplified second order model:

$$V_{GS_1} = \sqrt{\frac{2I_0}{\beta_1}} + V_{TH}$$

$$V_{GS_2} = \sqrt{\frac{2I_0}{\beta_2}} + V_{TH} \tag{7}$$

V_{GS_2} can be set to be the desired V_{B2}; therefore by definition $V_{GS_2}-V_{GS_1}=V_{DSAT}$. Subtracting the first expression from the second and combining the result with (4), we finally obtain:

$$\beta_2 = \beta_1/4 \tag{8}$$

Equation (8) simply states that the channel length of M_2 has to be four times longer than that of M_1. V_{B5} and V_{B4} are implemented similarly to V_{B1} and V_{B2}, respectively, by changing the polarities of the transistors and bias currents.

The method for generating V_{B2} is not exact even assuming an ideal second order model. In Fig. 8, the sources of M_1 and M_2 have nonzero bias with respect to the substrate (or well, depending on the technology). The derivation of (8) was based on identical theshold voltages for all the transistors regardless of their bias potential. However, since this body bias is equal to only one V_{DSAT}, this error can be neglected. Furthermore, some safety factor can be applied to (8) to avoid having a situation in which V_{B2} results in a value smaller than required. To generate V_{B3} the same design strategy for generating V_{B2} could be used, where $V_{GS3}-V_{GS1}$ would be equal to $2V_{DSAT}$ for deriving the new β ratio. The problem associated with this method for generating V_{B3}, however, are as follows:

(1) The body bias of the pair M_3,M_4 as shown in Fig. 8 is $2V_{DSAT}$ which further increases the error of the method, and

(2) This method does not take into account the actual generated value of V_{B2}. This can lead to a low differential $V_{B3}-V_{B2}$, thus setting the pair M_1,M_2 in the triode region.

Fig. 23 shows a circuit which resembles the *telescopic*-cascode op amp. V_{B3} appears at the drain of M_9, which when combined with M_7 forms the equivalent of M_2 of Fig. 22, here referred to the potential at the drain of M_6. Therefore the sum of the channel lengths of M_7 and M_9 has to be four times longer than M_8, or if M_9 matches M_{10}, M_7 has to be three times longer. It is important to note that this circuit combines both negative and positive feedback, a situation that can lead to two different stable points: a correct one and an incorrect one with zero current through both M_8 and M_7. To avoid the latter, some form of start-up circuit is required, which in this case can be implemented by transistor M_{15} which draws a small start-up current. M_{15} has a longer channel length than M_8. Again, some safety factor can be applied to the channel length of M_7 and M_9 to compensate for body bias errors.

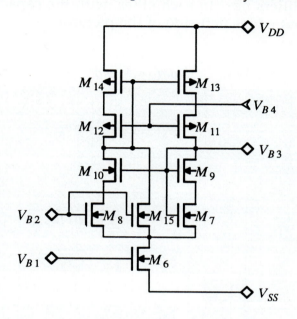

Fig. 23: V_{B3} generation.

7. Layout Techniques

Modern technologies have many interconnecting layers. Currently, it is common to find two levels of metal plus one or two levels of polysilicon. A compact op amp circuit should also be optimized for layout purposes by designing the transistor sizes in such a way as to avoid *breaks* in the diffusion areas and also allow for the metal lines to run over the active area of the op amp. Fig. 24 shows the layout of the *telescopic*-cascode OTA using the transistor sizes indicated in Table 1.

It can be observed that all the N-channel transistors are located at the bottom while the P-channel transistors are at the top. All transistors are stacked in such a fashion as to allow for easy interconnection between their sources and drains using the second level of metal. For the sake of consistency, each level of metal runs in only one direction, i.e., metal-2 in the *north* to *south* direction, while metal-1 in the *east* to *west* direction. The first level is used mainly to strap the sources and drains to reduce their ohmic resistance. The gates are tapped at the edges of the diffusion by the first level of metal which later can be connected to the second level.

This stacked layout allows for an easy interconnection of the different stages composing an analog system; e.g., switched-capacitor filters can be easily laid out by placing these amplifiers next to each other, with the capacitors and switches on the input and output edges (left side of the picture).

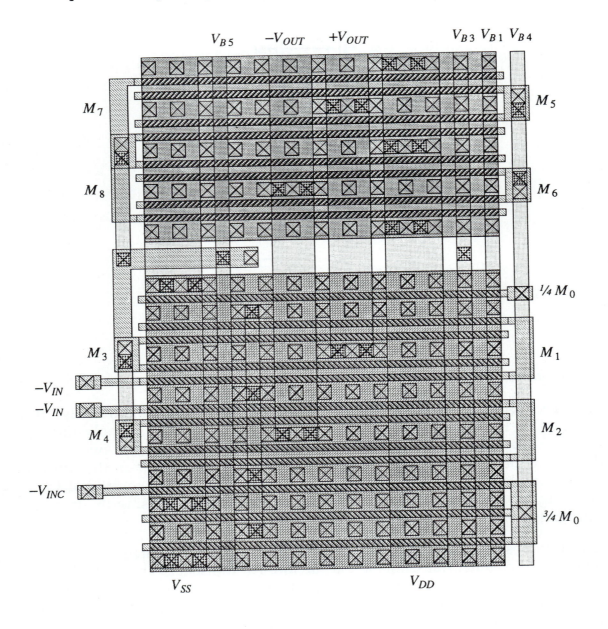

Fig. 24: Telescopic-cascode layout.

References

[1] Y.P. Tsividis and P.R. Gray, "An Integrated NMOS Operational Amplifier with Internal Compensation," *IEEE Journal of Solid-State Circuits*, vol. SC-11, pp. 748-753, December 1976.

[2] D. Senderowicz, D.A. Hodges and P.R. Gray, "A High Performance NMOS Operational Amplifier," *IEEE Journal of Solid-State Circuits*, vol. SC-13, pp. 760-766, December 1978.

[3] P.R. Gray, D. Senderowicz, H. Ohara and B.M. Warren, "A Single-Chip NMOS Dual Channel Filter for PCM Telephony Applications," *IEEE Journal of Solid-State Circuits*, vol. SC-14, pp. 981-991, December 1979.

[4] D. Senderowicz and J.M. Huggins, "A Low-noise NMOS Operational Amplifier," *IEEE Journal of Solid-State Circuits*, vol. SC-17, pp. 999-1008, December 1982.

[5] D. Senderowicz, S.F. Dreyer, J.H. Huggins, C.F. Rahim and C.A. Laber, "A Family of Differential NMOS Analog Circuits for a PCM Codec Filter Chip," *IEEE Journal of Solid-State Circuits*, vol. SC-17, pp. 1014-1023, December 1982.

[6] G. Nicollini, P. Confalonieri and D. Senderowicz, "A Fully Differential Sample-and-Hold Circuit for High-Speed Applications," *IEEE Journal of Solid-State Circuits*, vol. SC-24, pp. 1461-1465, October 1989.

[7] T.C. Choi, R.T. Kaneshiro, R.W. Brodersen, P.R. Gray, W.B. Jett and M. Wilcox, "High-Frequency CMOS Switched-Capacitor Filters for Communications Application," *IEEE Journal of Solid-State Circuits*, vol. SC-18, pp. 652-664, December 1983.

[8] G. Nicollini and D. Senderowicz, "Internal Fully Differential Operational Amplifier for CMOS Integrated Circuits," U.S. Patent N. 918101, June 1987.

Chapter 3

Micropower Techniques

Eric A. Vittoz
Centre Suisse d'Electronique et de Microtechnique (CSEM), Neuchatel, Switzerland

1. Introduction

Micropower integrated circuits techniques and technologies were originally developed for applications in electronic watches [1,2]. Modern watch circuits have a complexity ranging from a few thousand to several tens of thousand of transistors, combine microcontroller architectures with some high-performance analog circuits and are routinely produced by the tens of millions per year with a power consumption below 0.5 µW at 1.5 Volts. The need for very low power consumption extends to other kinds of battery-operated systems such as pocket calculators, implanted biomedical devices, hearing aids, paging receivers and various types of very small size portable instruments. Another interesting niche is that of data acquisition systems that are powered by optical fibers to avoid galvanic coupling.

Most of these applications require low-voltage operation (1 to 3 volts) and a current consumption typically below 50µA, which is a very severe limitation, especially for analog subcircuits. Some tradeoff with dynamic range and speed must and generally is accepted, but specific design techniques had to be developed.

More recently, micropower circuits have received a lot of additional interest for the analog implementation of very highly parallel processing systems based on the neural paradigm [3].

Quite generally, low-voltage techniques become increasingly important as voltages are reduced in scaled-down technologies. A reduction of the power consumption per operation is also necessary to increase the complexity of systems implemented on a single chip. Furthermore, some approaches developed for micropower prove to be very efficient in high-frequency designs.

After mentioning briefly a few particular points related to the fabrication process, this chapter will discuss the devices and circuits aspects of low-current, low-voltage integrated circuits. Some examples of analog and digital subcircuits will then be presented.

2. Technology

CMOS is unquestionably the best technology for very low current operation, and it has been used for that purpose almost exclusively since the early seventies, much before its advent for more general digital and analog circuit applications.

The advantage of CMOS for micropower first appears in digital circuits. A CMOS gate consumes a combination of dynamic and static power given approximately by:

$$P = f C V_B^2 + V_B I_0 \tag{1}$$

where f is the output frequency of the gate, V_B the supply voltage, C the output capacitance and I_0 the average leakage current between transitions. Capacitance C is kept low by having low values of substrate and well doping and by using shallow drain diffusions of minimum length surrounded by local oxidation. All these are standard means to achieve maximum speed in CMOS circuits. In addition, for minimum power, the drain width should always be minimum and the gate oxide should not be too thin, in order to limit the contribution of the gate capacitance.

It can be shown that the minimum possible supply voltage for digital operation is about 200 mV [4]. Meanwhile, if the threshold voltage of the transistors is too low, the residual channel current at zero gate voltage becomes a dominant part of I_0 and the static power consumption is increased. The threshold should not be smaller than about 400 mV. The nominal value thus depends on the spread, which can be kept below ± 100 mV by ion implantation of both well and substrate and by carefully optimized annealing.

Junction leakage currents should be kept below 10 nA / mm^2 at 1.5 V and ambient temperature, which is fulfilled in most modern processes.

Various levels of complexity have been successively introduced for micropower processes. Silicon gate processes have always been preferred for their self-alignment properties, and self-aligned contacts are very interesting for reducing diffusion areas [5]. The polysilicon layer can be used extensively for interconnections since its resistivity is negligible for very low currents. It has also been widely used to form capacitors with the aluminum layer [6]. Some micropower processes allow p and n doping of the same polysilicon layer, which offers interesting devices and circuit techniques, as will be shown later. A second layer of polysilicon has been introduced in low-power processes to implement non-volatile memories and denser capacitors [7]. Double metal processes are exploited to facilitate the hierarchical interconnections in large modern circuits.

3. Devices

3.1. MOS transistors

Many special techniques used in micropower circuits are related to the particular behavior of the MOS transistor at low current, which is very badly modelled in many standard circuit simulators. Therefore, the physics of this device will be briefly revisited in order to provide the necessary understanding of the device operation down to very low currents and to support a simple but effective model usable for analog (and digital) circuit design.

Fig. 1 shows the schematic cross-section and the symbol of an n-channel MOS transistor. In order to express the intrinsic symmetry of the device in the model, the source voltage V_S, the gate voltage V_G and the drain voltage V_D are all referred to the local p-type substrate (p-well or general substrate). This practice is not specifically related to micropower design, but proves to be very convenient in analog circuit design. The symbol and the definitions for the p-channel device are also shown in the figure. In schematics, the connection B to substrate is only represented when it is not connected to the negative supply rail V⁻ (for n-channel) or to the positive rail V⁺ (for p-channel).

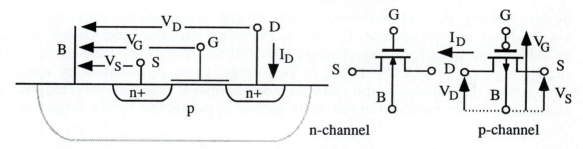

Fig. 1:   Schematic cross-section of an n-channel MOS transistor and symbols for n- and p-channel transistors. All voltages are referred to the local p-type substrate. Positive current and voltages for the p-channel transistor are reversed to maintain the validity of the model derived for the n-channel.

The following definitions will be used in the model:

$U_T = kT/q$	thermal voltage	V
n_i	intrinsic carrier concentration of silicon	m⁻³
ε_S	absolute dielectric constant of silicon	F/m
q	elementary charge	As
C_{ox}	gate capacitance per unit area	F/m²

N_B	net impurity concentration in the channel	m^{-3}
$\phi_F = U_T \ln(N_B/n_i)$	Fermi potential in the substrate	V
V_{T0}	gate threshold voltage for channel at equilibrium	V
μ	mobility of electrons in the channel	m^2/Vs
$\gamma = \dfrac{\sqrt{2qN_B\varepsilon_S}}{C_{ox}}$	substrate factor	$V^{1/2}$
W, L	effective electrical width and length of the channel	m
Q_i	mobile induced charge per unit area in the channel	As/m^2
V	channel "potential"	V

The channel potential V, which depends on the position along the channel, is not the electrostatic potential but a measure of the disequilibrium in the distribution of electrons that is produced by the application of source and drain voltages V_S and V_D. Stated differently, $\phi_F + V$ is the quasi-Fermi potential of electrons in the channel.

By neglecting the second derivatives of the electrostatic potential in the longitudinal and transverse directions, Poisson's equation and Gauss' law can be solved analytically in the channel. The solution yields the density of mobile charge Q_i (V_G,V) [8,9,10]. If V_G is very large, Q_i is dominated by electrons attracted to the surface of the channel. The channel is strongly inverted and the increase of electrostatic potential at its surface is prevented by the large value of Q_i.

In the *strong inversion approximation*, the surface potential is assumed to be independent of V_G and approximately equal to $V + 2\phi_F$. The mobile charge is then controlled by the field in the oxide according to

$$- Q_i = C_{ox} (V_G - V_{TB}) \tag{2}$$

where V_{TB} is the gate threshold voltage for strong inversion, approximately given by

$$V_{TB} = V_{T0} + V + \gamma\left(\sqrt{2\phi_F + V} - \sqrt{2\phi_F} \right) \tag{3}$$

This threshold is also referred to the local substrate potential, whereas the usual definition refers its value to the source potential. It is represented qualitatively in Fig. 2, which also shows how the mobile charge given by (2) can be calculated according to the representation proposed by Memelink [11].

For a given value of V_G, the charge Q_i becomes zero in this approximation when V reaches the particular value V_P called the *pinch-off voltage*. The relation of this pinch-off voltage with the gate voltage can be obtained by combining (2) and (3) for $Q_i = 0$, which yields

$$V_G = V_{TB}(V=V_P) = V_{T0} + V_P + \gamma\left(\sqrt{2\phi_F + V_P} - \sqrt{2\phi_F}\right) \tag{4}$$

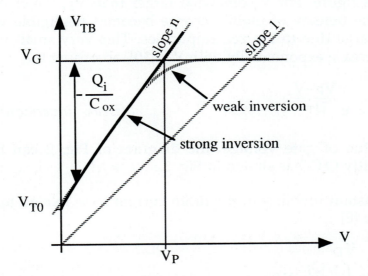

Fig. 2:   Memelink's representation of the strong inversion threshold V_{TB} and the mobile charge density Q_i as functions of the channel potential V.

The slope n of $V_{TB}(V)$ at $V=V_P$ can be obtained by differentiating (3), which results in

$$n = 1 + \frac{\gamma}{2\sqrt{2\phi_F + V_P}} \tag{5}$$

This slope, which is also that of $V_G(V_P)$, is usually smaller than 2 and tends towards 1 for very large values of V_P (and thus of V_G). It can also be expressed as a function of V_G by introducing (4) in (5), which gives

$$n = \left[1 - \frac{\gamma}{2\sqrt{V_G - V_{T0} + \left(\gamma/2 + \sqrt{2\phi_F}\right)^2}}\right]^{-1} \tag{6}$$

As shown by Fig. 2, V_{TB} never departs very much from its tangent at V_P. Thus the mobile charge in strong inversion can be approximated by

$$-\frac{Q_i}{C_{ox}} \cong n\,(V_P - V) \qquad\qquad \textit{(strong inversion)} \tag{7}$$

whereas the value of V_P itself can be approximated by

$$V_P \cong \frac{V_G - V_{T0}}{n} \tag{8}$$

In reality, the charge Q_i does not vanish abruptly when V reaches V_P, but decays smoothly as the channel leaves strong inversion, as represented by the dotted line in the same figure. For V somewhat larger than V_P, the channel is in *weak inversion* and the density of mobile charge becomes negligible with respect to that of fixed charge due to ionized impurities. This very small value of Q_i can be shown to decrease exponentially with V [9,10], according to

$$-\frac{Q_i}{C_{ox}} \sim e^{\frac{V_P - V}{U_T}} \qquad\qquad (weak\ inversion) \tag{9}$$

For a given value of gate voltage, the diagram of Fig. 2 can be redrawn to represent explicitly $Q_i(V)$ as shown in Fig. 3.

Assuming a constant mobility μ, the drain current I_D is related to the density of mobile charge by [8]

$$I_D = B \int_{V_S}^{V_D} -\frac{Q_i}{C_{ox}} dV \tag{10}$$

with $\qquad \beta = \mu C_{ox}\frac{W}{L}$ (11)

and can thus be immediately identified as the shaded area in Fig. 3. Furthermore, it is very convenient to split integral (10) into two components of drain current:

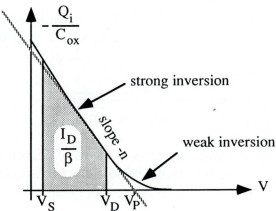

Fig. 3:   Mobile charge Q_i as a function of channel potential V. It decreases approximately linearly in strong inversion (5) and then exponentially in weak inversion (7).

$$I_D = \beta \int_{V_S}^{\infty} -\frac{Q_i}{C_{ox}} dV \; - \; \beta \int_{V_D}^{\infty} -\frac{Q_i}{C_{ox}} dV \tag{12}$$

direct current I_F *reverse* current I_R

(controlled by V_S)   (controlled by V_D)

Then: $I_D = I_F - I_R$ (13)

which is formally identical to the Ebers-Moll model of the bipolar transistors.

Globally, the MOS transistor may operate in several modes, depending on the values of the source voltage V_S and the drain voltage V_D with respect to the pinch-off voltage $V_P(V_G)$, as illustrated schematically in Fig. 4.

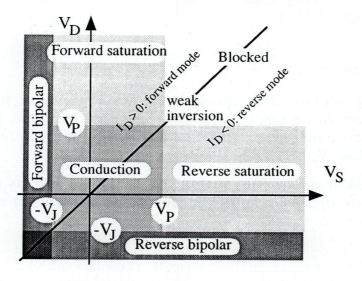

Fig. 4: Modes of operation of a MOS transistor.

Symmetrical forward and reverse modes are possible, depending on the sign of V_D-V_S. When V_S and V_D are both smaller than V_P, the channel is in strong inversion from source to drain and the transistor is in the *conduction* mode. If V_D is increased beyond V_P, the drain end of the channel is pinched-off and the device is in the *forward saturation* mode. If V_S and V_D are both larger than V_P, the whole channel is pinched-off. The device operates in *weak inversion* as long as one of the potentials is still close to V_P, but becomes *blocked* if both of them are sufficiently larger than V_P.

If the drain and/or the source junction is forward biased beyond a junction voltage V_J, a *bipolar mode* is superimposed on the MOS mode.

There is of course no abrupt limit between the various modes, but rather smooth transitions, in particular between weak and strong inversion which are separated by a zone of moderate inversion [8].

An approximation of the drain current in weak inversion is obtained by introducing (9) into (12), which yields

$$I_{F(R)} \sim \beta\, e^{\frac{V_P - V_{S(D)}}{U_T}} \tag{14}$$

or, after introducing (13) and the approximation (8) for V_P:

$$I_D = I_{D0}\, e^{\frac{V_G}{n U_T}} \left(e^{-\frac{V_S}{U_T}} - e^{-\frac{V_D}{U_T}} \right) \tag{15}$$

where

$$I_{D0} \sim \beta e^{-\frac{V_{T0}}{n U_T}} \tag{16}$$

is a very highly unpredictable parameter, which plays a role similar to that of the saturation current of bipolar transistors. As a consequence, MOS transistors in weak inversion should always be biased at a fixed drain current I_D and not at a fixed gate voltage V_G.

In forward weak inversion, the source end of the channel is always close to pinch-off. Therefore, a good approximation of the slope factor n is obtained by replacing V_P by V_S in equation (5):

$$n = 1 + \frac{\gamma}{2\sqrt{2\phi_F + V_S}} \qquad \textit{(weak inversion only)} \tag{17}$$

It can be shown that $1/n$ in weak inversion is the ratio of the capacitive divider between the gate voltage and the surface potential (oxide capacitance-depletion capacitance) and can therefore be found as the relative depth of the dip of the C_G-V_G curve [12].

Equation (15) yields the characteristics represented in Fig. 5.

It can be seen that the drain current saturates to a value I_{Dsat} as soon as V_{DS} exceeds a few U_T. This very low saturation voltage is an advantageous feature of weak inversion, which maximizes the dynamic range for a given supply voltage. The transfer characteristics from the gate are exponential with a slope factor n. The transfer characteristics from the source are identical to those of bipolar transistors and can be shifted by changing the gate voltage. This similarity is more than casual, since both are barrier-controlled devices with diffusion as the dominant charge transport mechanism. The most important difference lies in the fact that carriers diffusing from source to drain at the surface of a MOS transistor are effectively majority carriers with respect to their immediate environment. They can therefore diffuse much further than their diffusion length in the bulk, and weak inversion is possible even with very long channels.

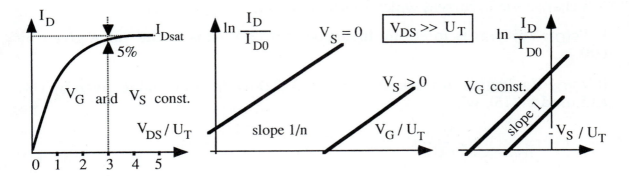

Fig. 5: Characteristics in weak inversion.

An approximation of the drain current in strong inversion is obtained by introducing (7) into (12), which yields

$$I_{F(R)} = \frac{\beta n}{2} (V_P - V_{S(D)})^2 \qquad (18)$$

After introducing the approximation (8) for V_P, the drain current in the direct strong inversion mode ($V_D \geq V_S$) can be expressed as

$$I_D = \beta(V_D - V_S)\left[V_G - V_{T0} - \frac{n}{2}(V_D + V_S) \right] \quad \text{for } V_D \leq \frac{V_G - V_{T0}}{n} \text{ (conduction)} \quad (19)$$

$$I_D = I_{Dsat} = \frac{\beta}{2n}(V_G - V_{T0} - nV_S)^2 \qquad \text{for } V_D \geq \frac{V_G - V_{T0}}{n} \text{ (saturation)} \quad (20)$$

$$I_D = 0 \qquad \text{for } V_S \geq \frac{V_G - V_{T0}}{n} \text{ (blocked)} \quad (21)$$

It is worth pointing out that the saturation value of V_D is equal to V_P and is completely independent of the value of the source voltage V_S.

No closed form expression of the current in moderate inversion can be obtained from physical considerations. Therefore, continuous models from weak to strong inversion must be based on a mathematical interpolation of the current between the two modes. One possibility is to express the direct and reverse components of I_D by [13, 9, 10]:

$$I_{F(R)} = I_S \ln^2\left[1 + e^{\dfrac{V_P - V_{S(D)}}{2U_T}}\right] \tag{22}$$

where $I_S = 2n\beta U_T^2$ $\tag{23}$

is a characteristic current which replaces the transfer parameter β.

If $V_{S(D)} \gg V_P$, $I_{F(R)}$ is in strong inversion and (22) tends to the value given by (18).

If $V_{S(D)} \ll V_P$, $I_{F(R)}$ is in weak inversion and (22) tends to the value given by (14) and used in (15), with

$$I_{D0} = I_S e^{-\dfrac{V_{T0}}{nU_T}} \tag{24}$$

Relation (22) can be introduced into (13) together with the approximation (8) for V_P, which yields an expression of I_D valid for all possible modes of operation of the MOS transistor (except the bipolar mode). This expression uses only three parameters (V_{T0}, n and β) and provides a quite acceptable interpolation in moderate inversion. The slope factor n can be considered constant in a first approximation. More precision is obtained by adapting its value to the gate voltage according to (6). The approximation (8) then provides a slightly too large value for V_P, which in turn yields optimistic values of drain current I_D, especially for V_D and V_S close to V_P. Extracting $V_P(V_G)$ from the exact relation (4) yields slightly pessimistic values of I_D, especially for V_D and $V_S \ll V_P$ (conduction). Furthermore, additional features like a field dependent mobility can be introduced by traditional methods.

The level of inversion of the transistor may be characterized by an inversion coefficient

$$IC = I_{Dsat} / I_S \tag{25}$$

which is much smaller than 1 in weak inversion and much larger than 1 in strong inversion. The value of I_S at room temperature is of the order of 10 to 100 nA for a minimum-size transistor. Therefore, a very wide channel is needed to operate a transistor in weak inversion at currents above a few microamperes.

The minimum operating current is limited by carrier generation in the drain and channel depletion layers. It is therefore roughly proportional to the overall size of the transistor. Minimum-size transistors may be used at a drain current as low as 100 pA at 50 °C.

The gate to drain transconductances in saturation are derived from (15) and (20):

$$g_m = \frac{\partial I_D}{\partial V_G} = \frac{I_D}{nU_T} \qquad \qquad \textit{(weak inversion)} \qquad (26)$$

$$g_m = \frac{2I_D}{V_G-V_{T0}-nV_S} = \sqrt{2\beta I_D/n} = \frac{\beta}{n}(V_G-V_{T0}-nV_S) \quad \textit{(strong inversion)} \qquad (27)$$

For a given transistor, the transconductance is thus proportional to I_D in weak inversion and to $\sqrt{I_D}$ in strong inversion. Relations (15) and (20) show that the effect of V_S on I_D is n-times that of V_G. Therefore, the source to drain transconductance

$$g_{ms} = -\frac{\partial I_D}{\partial V_S} = n\, g_m \qquad (28)$$

is equal in weak inversion to that of a bipolar transistor operated at the same current.

The maximum voltage gain achievable with a single transistor is limited by its non-zero source to drain conductance g_{ds} due to channel shortening. This conductance is approximately proportional to the drain current and can be characterized by an extrapolated voltage V_E roughly proportional to the channel length L:

$$g_{ds} = I_D / V_E \qquad (29)$$

As shown in Fig. 6, the voltage amplification factor $A_0 = g_m/g_{ds}$ reaches a maximum value

$$A_{0max} = V_E / nU_T \qquad (30)$$

in weak inversion, but decreases with $1/\sqrt{I_D}$ in strong inversion. This maximum gain in weak inversion reflects the maximum of the transconductance-to-current ratio g_m/I_D. It may reach a few hundreds even for small values of channel length L and can be further increased by increasing L. Experimental results show that the value of V_E tends to increase somewhat in strong inversion. This effect is thought to be due to the increasing vertical field in the drain area.

It must be pointed out that additional conductances g_d and g_s exist between drain and source and the local substrate, due to the reverse characteristics of the corresponding junctions.

A continuous model for the transconductance in saturation can be obtained by differentiating the saturation current $I_{Dsat} = I_F$ given by relation (22), which yields:

$$g_m = \frac{I_D}{nU_T} \cdot \frac{1 - e^{-\sqrt{IC}}}{\sqrt{IC}} \qquad (31)$$

where IC is the inversion coefficient I_D/I_S. This relation is plotted in Fig. 7. Compared to experimental results, this approximation is found to be a reasonably good model, in particular for the reduction of g_m by about 40% at IC=1 with respect to its asymptotic values in weak and strong inversion. It tends to be slightly pessimistic in weak inversion and optimistic in strong inversion.

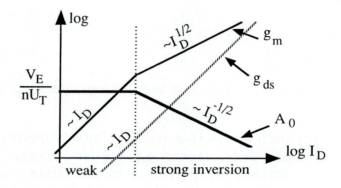

Fig. 6:   Schematic variation of transconductance g_m, output conductance g_{ds} and voltage gain factor A_0 with drain current I_D.

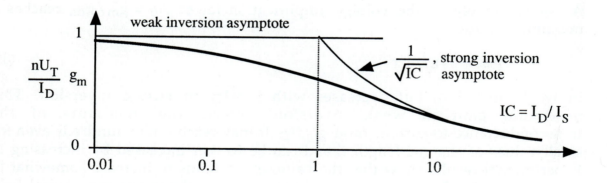

Fig. 7: Continuous transconductance from weak to strong inversion.

The noise behavior of MOS transistors must be modelled by at least two independent noise sources, as shown in Fig. 8 [9,10].

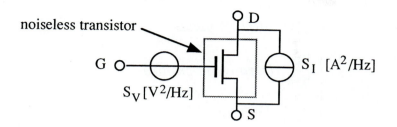

Fig. 8: Noise model of the MOS transistor.

In strong inversion, the noise current source represents the thermal noise of the non-homogeneous resistive channel, with a spectral density in saturation

$$S_I = \frac{8}{3} nkT \, g_m = 2qI_D \frac{8nU_T}{3(V_G-V_{T0}-nV_S)} = \frac{8qI_D}{3\sqrt{IC}} \qquad (strong \; inversion) \qquad (3$$

Since the flow of carriers in weak inversion is controlled by a potential barrier, the noise current source is due to shot noise and has the spectral density

$$S_I = 2qI_D \qquad\qquad (weak \; inversion) \qquad (33)$$

in saturation, that is, when the reverse component I_R of drain current is negligible ($I_D = I_F$) [14,15]. If it is not negligible, in particular for $V_D=V_S$, then $S_I = 2q(I_F+I_R)$ [9].

It must be noticed that the drain noise current for a given value of bias I_D is maximum in weak inversion and decreases progressively in strong inversion, according to (32).

The noise voltage source at the gate is a flicker noise related to the Si-SiO$_2$ interface. Although it can be shown to increase slightly with gate voltage [16,17], one may assume for the sake of simplicity a spectral density given by

$$S_V = 4kT \frac{r}{WL.f} \qquad\qquad (34)$$

where r is a process dependent parameter. This parameter r is often larger for n-channel than for p-channel transistors, and may range between 0.02 and 2 Vs/Am^2.

It is often convenient to refer the total noise to the gate of the transistor and to characterize it with an (frequency-dependent) equivalent noise resistor R_N defined by

$$4kT \, R_N = S_V + S_I \, / \, g_m^2 \qquad\qquad (35)$$

Introducing (32), (33) and (34) into (35) yields

$$R_N = \frac{\rho}{WL.f} + \begin{cases} \dfrac{2n}{3g_m} & \text{(strong inversion)} \\[2mm] \dfrac{n}{2g_m} & \text{(weak inversion)} \end{cases} \tag{36}$$

These relations are illustrated in Fig. 9 for fixed current and for fixed frequency.

At low frequencies, the $1/f$ flicker noise dominates and R_N is independent of I_D. It can only be decreased by increasing the gate area WL.

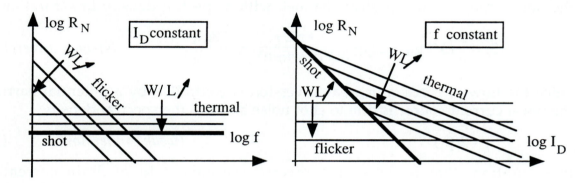

Fig. 9:   Effects of WL and W/L on the equivalent input noise resistance R_N at fixed current and fixed frequency.

At high frequencies, R_N is approximately equal to $1/g_m$. It therefore decreases when I_D increases until it eventually reaches the flicker noise level. However, if I_D is fixed (for limited power consumption), R_N can be reduced by increasing W/L until it reaches its minimum in weak inversion. This minimum equivalent input noise in weak inversion is another advantage of the maximum value of g_m/I_D.

Like noise, the mismatch between two identical transistors must be characterized by two statistical parameters, the threshold mismatch δV_{T0}, with RMS value σ_{VT}, and $\delta\beta/\beta$ with RMS value σ_β. These components of mismatch are usually very weakly correlated and their mean value is zero in a good layout. The RMS mismatch of drain current $\delta I_D/I_D$ of two transistors with the same gate and source voltages is then

$$\sigma_{ID} = \sqrt{\sigma_\beta^2 + \left(\frac{g_m}{I_D}\sigma_{VT}\right)^2} \tag{37}$$

whereas the mismatch δV_G of the gate voltage of two transistors biased at the same drain current and source voltage is given by

$$\sigma_{VG} = \sqrt{\sigma_{VT}^2 + \left(\frac{I_D}{g_m}\sigma_\beta\right)^2} \qquad (38)$$

These results are plotted in Fig. 10 for $\sigma_T = 5mV$ and $\sigma_\beta = 2\%$, by using the continuous transconductance of Fig. 7 with $nU_T = 40mV$.

It can be seen that weak inversion results in a very bad mismatch s_{ID} of the drain currents but a minimum mismatch s_{VG} of the gate voltages, equal to the threshold mismatch σ_{VT}. A very large inversion coefficient is needed to reduce s_{ID} to the minimum possible value σ_β.

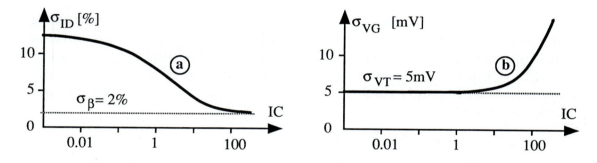

Fig. 10: Matching of drain current (a) and gate voltage (b) as functions of the inversion coefficient of the transistors.

3.2. CMOS compatible bipolar transistors

As was shown in Fig. 4, a MOS transistor operates as a bipolar transistor if the source or the drain is forward-biased in order to inject minority carriers into the local substrate. If the gate voltage is negative enough (for an n-channel device) to have the pinch-off voltage V_P smaller than both V_S and V_D, then no current can flow at the surface and the operation is purely bipolar [18].

Fig. 11 shows the major flows of current carriers in this mode of operation.

Since there is no p+ buried layer to prevent injection to the substrate, this lateral npn bipolar is combined with a vertical npn. The emitter current I_E is thus split into a base current I_B, a lateral collector current I_C and a substrate collector current I_{Sub}. Therefore, the common base current gain $\alpha = -I_C / I_E$ cannot be close to 1. However, due to the very small rate of recombination inside the well and to the high emitter efficiency, the common base current gain $\beta = I_C / I_B$ can be large [18].

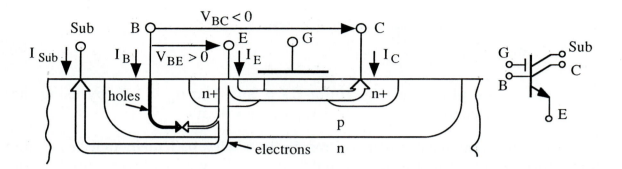

Fig. 11: Bipolar operation of the MOS transistor: carrier flows and symbol.

To favor the lateral device, I_C / I_{Sub} can be maximized by minimizing the emitter area and the lateral base width L, and by having the emitter surrounded by the collector. Values of b of many hundreds can be obtained, with a ranging from .05 to .5, depending on the profile and thickness of the well. The limit of strong injection, beyond which the characteristics of the device start degrading seriously is given by [18]

$$ I_C = \frac{A_l \, q \, \mu \, U_T \, N_B}{L} \tag{39} $$

where A_l is the lateral cross-section of the transistor. This device is therefore limited to low current (10 to 100 μA for a minimum-size transistor) by the low value of impurity concentration N_B in the base (well) required for MOS operation. This low current restriction in turn limits the transition frequency f_T to about 100 MHz [19, 20].

These true bipolar characteristics can be exploited to implement a variety of new DC and low-frequency functions on a CMOS chip [21, 22, 23, 24]. In addition, bipolar operation provides very interesting matching and low-frequency noise properties, both linked to the removal of surface effects dependence. Measurements have shown an improvement in input offset voltage by a factor 10 and a reduction of more than 40dB of the 1/f noise from MOS operation to bipolar operation for the same devices [18, 22, 17].

The negative gate voltage required to prevent surface conduction (for p-well) in low V_{T0} processes can be easily produced on chip. For processes with larger V_{T0}, the gate can simply be connected to the emitter. The correct criterion is to insure a negligible value of gate transconductance $\partial I_C / \partial V_G$. The very small residual gate transconductance can be exploited to compensate mismatch in high-precision circuits [22]. The lateral collector can be removed in common collector configurations, but keeping the whole structure and connecting the lateral collector to the chip substrate improves the overall β-gain.

For analog and low-current applications where the extrinsic base and collector resistances can be accepted, a vertical complementary bipolar transistor can be obtained by a single additional diffusion (n-type for p-well) which is used as the base layer.

3.3. Passive devices

No special requirements are imposed on capacitors for the realization of micropower circuits. The lowering of the values of functional capacitors to help reduce power consumption is bounded by the acceptable kT/C noise limit.

On the contrary, the realization of good resistors is made difficult by the need for high values. The highest sheet resistivity is usually provided by the well diffusion, but its use as a resistor is limited by the large distributed capacitor and the high sensitivity to substrate modulation.

The doped polysilicon gate layer provides excellent resistors, but their values are not compatible with submicroampere current levels. The lightly doped polysilicon resistors available in some technologies do not meet the precision requirements for most analog circuits.

As mentioned in section 2, p and n doping of the same polysilicon layer has been maintained in some micropower processes. In these processes, lateral pn diodes in the polysilicon layer can be realized as shown in Fig. 12 [25].

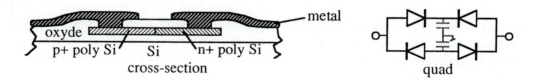

Fig. 12: Lateral diode in the polysilicon layer.

Although its characteristics are far from being ideal, this device provides the great advantage of having no leakage to the substrate. It finds applications in the realization of high impedance on-chip voltage multipliers required for non-volatile memories [7]. It may also be used around the origin of its I-V characteristics as a very high impedance equivalent resistor. Values of many gigaohms can thus be obtained with a negligible area. When symmetrical characteristics are required, a quad configuration has proved to be the best solution to balance the rectification of AC signals by the very small parasitic capacitances at the common point of back-to-back connected diodes.

3.4. Parasitic effects

Parasitic capacitors and noise impose a tradeoff between bandwidth (or speed) and current drain. However, an absolute limit on operating current is given by the leakage currents. Each branch of an analog circuit must operate at a current level sufficiently higher than the local leakage current. This is true for dynamic digital circuits as well, whereas static digital circuits can withstand leakage currents much larger than their nominal average current drain.

It must be remembered that leakage is mostly due to the generation of electron-hole pairs within space charge regions (with a tendency to increase at the Si-SiO$_2$ interface). It is therefore approximately proportional to the volume of these regions. As a consequence, it is impossible to operate a transistor in strong inversion with a very small drain current, as this would require a very long channel and thus a very large space charge region below it, the leakage of which would be comparable to the drain current. For a minimum size transistor, leakage currents can be kept negligible up to 80°C for a drain current as low as 1nA. If the threshold voltage V_{T0} is very low (0.3 to 0.6 V), the residual channel current at $V_{GS}=0$ may be an important contribution to leakage.

Impact ionization causes a component of output conductance which cannot be eliminated by a cascode configuration, since it appears between the drain and the local substrate. Micropower processes tend to have a high value of drain breakdown voltage, because of their low doping of substrate and well. Impact ionization is therefore negligible for 3-volt operation.

The otherwise difficult problem of latch-up has long been solved in a very simple and safe way in micropower. Normal layout procedures insure a value of hold current above 1 mA. Latch-up can thus be prevented by limiting the current available from the power source below this value, by means of a series resistor of a few Kohms. This resistor causes a negligible voltage drop at the nominal current of a few microamperes.

4.   Basic Circuit Techniques

4.1. Current mirrors

The inaccuracy of $\delta I/I$ of a current mirror (Fig. 13) is given by relation (37) and is worst in weak inversion where g_m / I reaches the maximum value $1/nU_T$, as shown by Fig. 10a. Thus, for $nU_T = 40$ mV, a threshold offset $\sigma_{VT} = 5$ mV results in a contribution of 12% to the inaccuracy, which is much larger than usual values of σ_β. The only way to improve the precision (besides using optimum layout and non-minimum size) is to reduce g_m / I by going deeply into strong inversion, where according to (27) and for $V_S=0$

$$\frac{g_m}{I} = \frac{2}{V_G - V_{T0}} \tag{40}$$

The offset is thus improved by a factor $(V_G - V_{T0})/2nU_T$ which is approximately 10 for $V_G - V_{T0} = 1V$. The same is true for the contribution of $1/f$ flicker noise to the output current noise (in $A/\sqrt{Hz}$). According to (32) and (33), the output white noise current is also reduced in strong inversion, but only by a factor $\sqrt{3(V_G-V_{T0})/8nU_T}$.

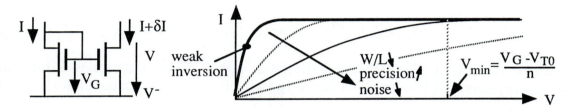

Fig. 13: Current mirror and its output characteristics.

The amount by which V_G can be increased above V_{T0} is obviously limited by the total power supply voltage V_B. However, a more severe limitation is given by the decrease of saturation voltage (minimum output voltage V_{min} as a current source), which may reduce the dynamic range of analog circuits. As shown in Fig. 13, this minimum voltage, which is a few U_T in weak inversion, increases to $(V_G-V_{T0})/n$ in strong inversion. Therefore, for a low value of supply voltage, accuracy and noise must often be traded for maximum voltage amplitude.

Furthermore, it has been shown in section 3 that the presence of leakage allows no other choice but weak inversion when the drain current is very low.

Compatible bipolars offer the possibility of improving the accuracy by one order of magnitude [18] while maintaining a reasonably low value of minimum output voltage.

4.2. Differential pair

The differential pair shown in Fig. 14 is a widely used configuration. The contribution of the mismatch of transistors T_1 and T_2 to the input offset voltage is the difference δV_G of their gate voltages for $I_1 = I_2$. According to (38) and as shown in Fig. 10b, this difference is minimum in weak inversion. Thus, except when some linearity of the transconductance is required, this circuit is normally biased in weak inversion in low-power circuits.

The large signal transfer characteristics in weak inversion are obtained from (15):

$$I_{1(2)} = \frac{I_0}{1 + e^{\frac{-(+)V_{id}}{nU_T}}} \qquad (41)$$

The differential output current (available in the load after mirroring I_1 and/or I_2)

$$I_L = I_1 - I_2 = I_0 \tanh \frac{V_{id}}{2nU_T} \qquad (weak\ inversion) \qquad (42)$$

saturates as soon as $|V_{id}|$ is larger than 3 to 4 nU_T. The small signal differential transconductance at the origin is equal to that of one of the transistors

$$g_m = \frac{\partial I_L}{\partial V_{id}} = g_{m1(2)} = \frac{I_0}{2nU_T} \qquad (43)$$

and is half that of a single transistor biased at the same current I_0. The equivalent input noise resistance is double that obtained with a single transistor at the same current.

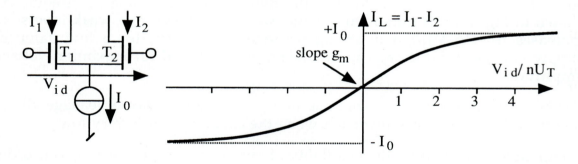

Fig. 14: Differential pair and its transfer characteristics in weak inversion.

4.3. Voltage gain cells

Common source elementary amplifiers may be differentiated by the way the load transistor is connected and by the mode of operation of both the active and the load device.

As shown in Fig. 15, the load transistor T_2 can be connected as a passive dipole which is driven by the active transistor T_1. The voltage gain A_v is the slope of the transfer characteristics $V_o(V_i)$ shown in the figure. It can be expressed from

(26) and (27):

$$-A_v = \frac{g_{m1}}{g_{m2}} = \begin{cases} \sqrt{\dfrac{n_2\beta_1}{n_1\beta_2}} & 1 & T_1 \text{ and } T_2 \text{ in strong inversion.} \\[4mm] \dfrac{1}{n_1 U_T} \cdot \sqrt{\dfrac{n_2 I_0}{2\beta_2}} & 2 & T_1 \text{ in weak and } T_2 \text{ in strong inv.} \\[4mm] \dfrac{n_2}{n_1} & 3 & T_1 \text{ and } T_2 \text{ in weak inversion} \end{cases} \qquad (44)$$

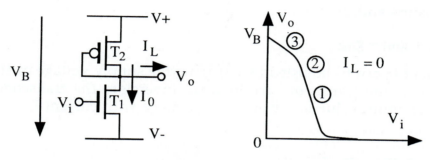

Fig. 15: Gain cell with passive load and its transfer characteristics.

The gain is independent of the current I_0 when both transistors are in strong inversion. It decreases with $\sqrt{I_0}$ when only the active device is in weak inversion and reaches a value close to 1 when the load also enters weak inversion. Thus, no gain can be obtained at very low currents. The equivalent input noise resistance is given by

$$R_N = R_{N1} + R_{N2}\left(\frac{g_{m2}}{g_{m1}}\right)^2 = R_{N1} + \frac{R_{N2}}{A_v^2} \qquad (45)$$

It is reduced to the noise resistance R_{N1} of the active device if the gain A_v is large.

If the load transistor is connected as a current source (output of a current mirror), the DC voltage gain is only limited by the parallel output conductances of the two transistors in saturation. Thus, if the conductances g_d of the reverse biased drain junctions are negligible:

$$-A_v = \frac{g_{m1}}{g_{ds1} + g_{ds2}} = \begin{cases} V_E \sqrt{\dfrac{2\beta_1}{n_1 I_0}} & \text{for } T_1 \text{ in strong inversion} \\[4mm] \dfrac{V_E}{n_1 U_T} & \text{for } T_1 \text{ in weak inversion} \end{cases} \qquad (46)$$

where $\quad V_E = \dfrac{V_{E1} \, V_{E2}}{V_{E1} + V_{E2}}$ (47)

The total input noise resistance is increased by the noise of the current mirror. It is reduced to R_{N1} if T_1 operates close to or in weak inversion and T_2 deep in strong inversion ($IC_2 \gg 1$), which limits the maximum output amplitude.

The most efficient gain cell is the complementary inverter illustrated in Fig. 16.

Around the equilibrium value V_e of input voltage V_i (virtual ground level for which the current to the load $I_L = 0$), the same bias current $I_0 = I_1 = I_2$ flows through the two transistors and the total transconductance is

$$g_m = g_{m1} + g_{m2}$$ (48)

This bias current I_0 can be adjusted by changing the supply voltage or the size of the transistors. If the transistors are in weak inversion, the transconductance reaches the maximum value achievable in one stage with current I_0.

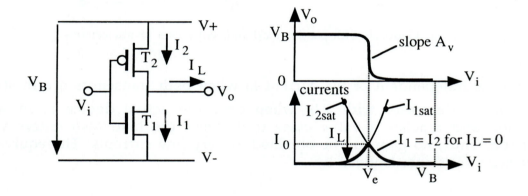

Fig. 16: Complementary CMOS inverter as a voltage gain cell.

$$g_m = \frac{I_0}{U_T} \left(\frac{1}{n_1} + \frac{1}{n_2} \right) \quad = \quad \frac{2I_0}{nU_T} \qquad (weak \;\; inversion)$$ (49)

$$\uparrow$$
$$\text{for } T_1 \equiv T_2$$

The voltage gain $A_v = - \dfrac{g_{m1} + g_{m2}}{g_{ds1} + g_{ds2}}$ (50)

has a value in between the amplification factors A_0 of the two transistors, whereas the equivalent input noise resistance is given by

$$RN = \frac{R_{N1}\, g_{m1}^2 + R_{N2}\, g_{m2}^2}{(g_{m1} + g_{m2})^2} = \frac{R_{N1(2)}}{2} = \frac{n^2 U_T}{4 I_0} \tag{51}$$

$$\uparrow \qquad\qquad \uparrow$$
$$\text{for } T_1 \equiv T_2 \quad \text{weak inv. white noise}$$

Thus, a CMOS inverter operated in weak inversion provides the minimum value of noise resistance achievable with a given current I_0, which is one-half of that of a single transistor and eight times smaller than that of a differential pair biased at the same current. The merits of the various configurations with respect to their transconductances and noise resistances are compared in Table 1.

If some input current can be accepted, the active MOS transistors can be replaced by compatible bipolars. These bipolars provide performances approximately equivalent to those obtained in weak inversion up to higher currents (and thus higher frequencies), and the $1/f$ noise is drastically reduced [18].

Constant current	g_m	R_N	input noise in dB
Weak inversion			
CMOS inverter	1	1	0
Differential pair	1/4	≥ 8	≥ 9
Passive load	1/2	≥ 2	≥ 3
Current source as load	1/2	≥ 2	≥ 3
Strong inversion	divide by	multiply by	add
(for $V_S = 0$)	$IC^{1/2}$	$1.33\, IC^{1/2}$	$1.2 + 5\log IC$

Table 1: Comparison of transconductance and noise of gain cells (normalized to those of the CMOS inverter).

4.4. Low-voltage cascoding

The cascode scheme is a combination of common source and common gate configurations which decreases the output conductance by a factor $g_{ms}/g_{ds} = nA_0$. The transconductance remains that of the common source transistor and no significant noise is added. Single stage cascode amplifiers may thus achieve a DC voltage gain A_v larger than 2-stage non-cascoded amplifiers. The only problem is an increased saturation voltage due to the series configuration of two active transistors which must both remain saturated. Fig. 17 shows a cascode pair biased for minimum saturation voltage, in order to insure a sufficient output amplitude, even with a low supply voltage.

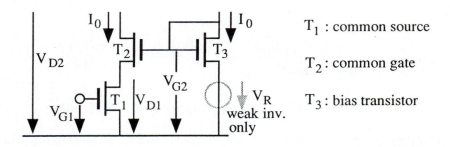

Fig. 17: Low-voltage cascode.

If all transistors are saturated ($I_R = 0$) and in *strong inversion*, relation (18) applied to the three transistors yields:

$$V_{D1} = \sqrt{n_1/n_2} \left(\sqrt{\beta_1/\beta_3} - \sqrt{\beta_1/\beta_2} \right) V_{P1} \tag{52}$$

which can be made just equal to the minimum value $V_{D1} = V_{P1}$ necessary for saturation, independently of the bias current I_0, by choosing

$$\sqrt{\beta_1/\beta_3} = \sqrt{n_2/n_1} + \sqrt{\beta_1/\beta_2} \tag{53}$$

This optimum bias results in

$$V_{D2min} = V_{P2} = V_{D1} \left(1 + \sqrt{n_1/n_2} \sqrt{\beta_1/\beta_2} \right) \tag{54}$$

In particular, since n_1 is approximately equal to n_2, (53) and (54) yield:

For $\beta_2 = \beta_1$: $\beta_3 = \beta_1/4$,   and $V_{D2min} = V_{P2} = 2V_{D1}$ [26].

For $\beta_2 \gg \beta_1$: $\beta_3 \rightarrow \beta_1$,   and $V_{D2min} = V_{P2} \rightarrow V_{D1}$.

This last result is still approximately valid if T_2 alone enters weak inversion.

On the contrary, using the same scheme with both T_1 and T_2 in *weak inversion* would require a prohibitive ratio β_1/β_3 to obtain $V_{D1} = 4$ to $6U_T$ in order to have T_1 saturated. This voltage must be provided by an external reference V_R proportional to U_T, as shown by the dotted line in the figure. A possible implementation of V_R is shown in Fig. 18 [27].

The weak inversion equation (15) applied to transistors T_4 and T_5 gives:

$$V_R = U_T \ln \left(1 + 2 \frac{\beta_4}{\beta_5} \right) \tag{55}$$

which would be replicated at the drain of T_1 if T_2 and T_3 were identical. A larger value of V_{D1} is obtained by choosing $\beta_2 \gg \beta_3$ which, by using (15) again, results in

$$V_{D1} = U_T \ln\left[\frac{\beta_2}{\beta_3}(1 + 2\frac{\beta_4}{\beta_5}) \right] \qquad (56)$$

For example, for $\beta_2/\beta_3 = \beta_4/\beta_5 = 8$, $V_{D1} \approx 5U_T$, which is sufficient to maintain T_1 saturated in weak inversion. The minimum value of output voltage V_{D2min} is then approximately $10\ U_T$.

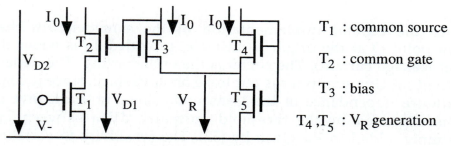

T_1 : common source

T_2 : common gate

T_3 : bias

T_4, T_5 : V_R generation

Fig. 18: Low-voltage cascode in weak inversion.

4.5. Current references

The low power generation of the reference current required to bias analog circuits is made difficult by the lack of good resistors of high value and by the small difference between the threshold V_{T0} and the supply voltage V_B. The solution generally adopted is shown in Fig. 19 [12].

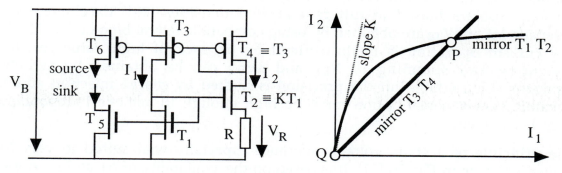

Fig. 19: Low power current reference.

Transistors T_3 and T_4 are identical and preferably operate in strong inversion to improve the mirror accuracy. At low current, the mirror T_1-T_2 has a gain $K = \beta_2/b_1$ larger than 1, but this gain decreases when the current increases, due to the degenerating resistor R. Equilibrium is reached at point P where K>1 is compensated by the voltage V_R across R. If T_1 and T_2 are in weak inversion and located in the same substrate, relation (15) yields

$$RI_2 = V_R = U_T \ln K \tag{57}$$

This result is multiplied by n if T_2 is in a separate well connected to its source. Low currents can be obtained with reasonably low values of R, due to the low value of V_R. For example, for K = 8 ($V_R \approx 2U_T \approx 50mV$), $I_1 = I_2 = 1\mu A$ require R = 50KΩ.

Contrary to the bipolar implementation of the same circuit, the second equilibrium point Q at the origin is unstable, provided T_2 is wider than T_1 (it has a larger leakage current). The circuit is then self-starting. The dependence of output current on U_T usually turns out to be an advantage, since it compensates the temperature dependence of the transconductance in weak inversion. The precision is degraded by the threshold mismatch dV_{T0} between T_1 and T_2, which is simply added to V_R given by (57). The sensitivity to variations of V_B can be reduced by cascoding.

4.6. Voltage references

Micropower voltage references have been proposed in which the variation of the base-emitter voltage V_{BE} of a vertical bipolar to substrate was compensated by a voltage proportional to absolute temperature (PTAT) that was obtained from MOS transistors operated in weak inversion [27, 28]. The precision was limited by the threshold mismatch δV_{T0}, and this limit still exists if the PTAT voltage is obtained from the difference of base-emitter voltages ΔV_{BE} of two vertical bipolars biased at different current densities [29]. Excellent results independent of δV_{T0} are obtained by using compatible lateral bipolars[18, 23], but low current operation is made difficult by the lack of high value resistors. Weighting and summing of V_{BE} and ΔV_{BE} can be obtained by switched capacitor circuits [30], but the precision is limited by charge injection and the dynamic power consumption must be high enough to achieve independence from leakage currents.

The principle of a static voltage reference especially well suited to very low power is shown in Fig. 20 [31]. It is based on the availability of p+ and n+ doping in the same polysilicon layer.

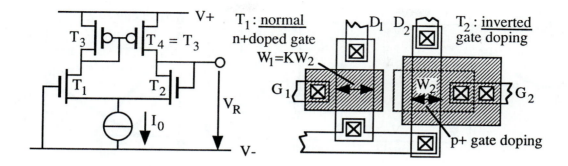

Fig. 20: Principle of a very low power voltage reference using inverted gate doping.

T_1 is a normal n-channel transistor with an n+ doped gate and a threshold V_{T0}. T_2 is also n-channel, but has a p+ doped gate, which shifts its threshold by an amount close to the bandgap voltage V_{gap}. Both transistors are operated in weak inversion at the same drain current. Thus, from equation (15):

$$\frac{W_1}{L_1} \, e^{\frac{V_{G1}-V_{T0}}{nU_T}} = \frac{W_2}{L_2} \, e^{\frac{V_{G2}-V_{T0}-V_{gap}}{nU_T}} \tag{58}$$

and

$$V_R = V_{G2} - V_{G1} = V_{gap} + nU_T \ln \frac{W_1/L_1}{W_2/L_2} \tag{59}$$

After compensation of the variation of V_{gap} with temperature by the second term of this equation, a temperature stability of ± 30 ppm/°C can be obtained. The total current can be as low as 10nA. The precision obtained is about ± 50mV without any adjustment.

4.7. Analog switches

The lowest voltage at which a technology can be used for analog circuits is often limited by the implementation of analog switches. Switches are realized by means of n and/or p-channel transistors. They are switched on and off by connecting their gates to the most positive or most negative potential available, namely V+ or V-, as shown in Fig. 21.

The on-conductance of the switch $g = \partial I_D/\partial V_{DS}$ around $V_D = V_S$ is a strong function of the common mode voltage V_F at which it is "floating." The conductance of the two transistors of Fig. 21 in strong inversion can be calculated from (19)

$$g_n = \beta_n (V_B - V_{T0n} - n_n V_F) \tag{60}$$

$$g_p = \beta_p [V_B - V_{T0p} - n_p(V_B - V_F)] \tag{61}$$

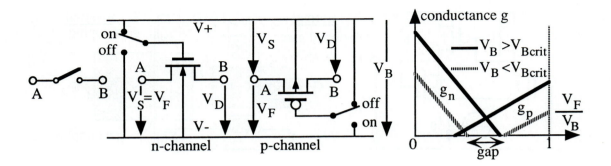

Fig. 21: Transistors used as analog switches and their on-conductances.

and are represented in the same figure as functions of V_F/V_B. If V_B is large enough, a parallel connection of p and n switches insures conduction for any value of V_F. If V_B is smaller than a critical value V_{Bcrit}, a gap appears at mid-range, in which neither switch is conducting. This critical value can be obtained from (60) and (61):

$$V_{Bcrit} = \frac{V_{T0p}\, n_n + V_{T0n}\, n_p}{n_p + n_n - n_p n_n} \tag{62}$$

It can be considerably larger than $V_{T0n} + V_{T0p}$ since n_n and n_p are larger than 1. This important limitation can be circumvented by means of on-chip clock voltage multiplication[32].

4.8. Dynamic CMOS gates

The minimization of the dynamic power consumption of digital CMOS circuits can be obtained by reducing the number of transistors used to implement a given function. This reduction is often possible by using dynamic gates which are an extension of the general CMOS gate shown in Fig. 22 [33].

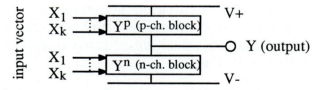

Fig. 22: General CMOS gate, combination of p and n-channel conduction blocks.

A CMOS gate delivering an output variable Y is a series combination of a block of n-channel transistors and a block of p-channel transistors, each driven by the input vector $(X_1...X_k)$. The blocks can be characterized by their conduction functions Y^n and Y^p, which cover respectively the 0 and 1 squares of the

Karnaugh map of Y. In normal CMOS gates:

$$YP . Y^n = 0 \text{ to avoid a direct path from V+ to V-}$$

$$YP + Y^n = 1 \text{ to insure a low output conductance.}$$

In a dynamic gate, $YP + Y^n = 0$ is allowed. It will be shown in section 5 that the application of dynamic gates can be advantageously generalized beyond the clocked CMOS and the pull-up transistor schemes.

5.   Examples of Building Blocks

5.1. Operational transconductance amplifiers (OTAs)

The major problems in the realization of micropower operational amplifiers are obtaining a reasonable speed and an acceptable dynamic range. Battery operation allows relaxed requirements on the power supply rejection, since the power noise only comes from the chip and can be more easily filtered out at very low current.

A key factor for power reduction is the avoidance of any compensation capacitor other than the load itself, which is only possible if the major part of the voltage gain is achieved at the output node, that is by using single-stage operational transconductance amplifiers. A simple OTA with a differential input is shown in Fig. 23.

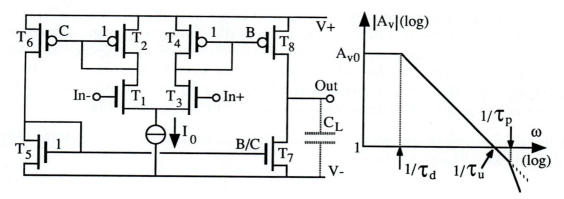

Fig. 23: Simple OTA and its frequency behavior.

Assuming a ratio B of the mirror T_4-T_8, the overall transconductance of the amplifier is

$$g_m = B \ g_{m1(3)} \tag{63}$$

The circuit, loaded by C_L, behaves essentially as an integrator with a time constant

$$\tau_u = C_L / g_m \tag{64}$$

which is the inverse of the unity gain angular frequency ω_u. The very low frequency $1/\tau_d$ of the dominant pole depends directly on the DC gain A_{v0}, as shown in the figure. Let us assume a single non-dominant parasitic pole with time constant τ_p. This time constant may be that of the drain nodes of transistors T_1 and T_3. It may also represent the effect of all the high frequency parasitic poles, which can be shown to be equivalent to a single pole for $\omega < 1/2\tau_p$.

For frequencies much larger than $1/\tau_d$, the voltage gain of the amplifier can be expressed by

$$A_v = \frac{1}{s\tau_u (1 + s\tau_p)} \tag{65}$$

Now, an operational amplifier is usually used with an amount β of negative voltage feedback, with $1 \geq \beta \gg 1/A_{v0}$. The settling time, or the time necessary to reach equilibrium with a relative residual error $\pm \varepsilon$ after application of a small step, can be calculated with the gain given by (65). The result may be approximated by [34]

$$T_s \approx (2\tau_p + \tau_u/\beta) \ln(1/\varepsilon) \tag{66}$$

Even for submicroampere bias currents and large input transistors T_1-T_3, the $1/f$ flicker noise usually still dominates in the audio frequency range. However, in most practical applications, the noise bandwidth exceeds 100KHz and white noise predominates in the total noise power. This is true in particular for switched capacitor circuits clocked at a frequency f_c, for which the whole noise power of the amplifier is transposed below $f_c/2$ by undersampling.

If the equivalent input noise resistance of the amplifier is R_N, the total output noise in closed loop is given by:

$$V_{N0}^2 = 4kTR_N \int_0^\infty \frac{df}{|\,1/A_{v0} + \beta\,|^2} = \frac{kTR_N}{\beta\tau_u} \tag{67}$$

and does not depend on the parasitic time constant t_p [34]. This result can be explained qualitatively by the fact that any reduction of noise at high frequencies due to an increase of τ_p is compensated by some peaking at lower frequencies.

Now, as for a single transistor in (36), the noise resistance of the amplifier may be expressed as a function of its transconductance given by (63):

$$R_N = \gamma \,/\, g_m \tag{68}$$

For the circuit of Fig. 23, the factor of merit g is minimum if the input pair is operated in weak inversion and all the other transistors deeply in strong inversion (IC » 1), which results in

$$\gamma_{min} = nB \tag{69}$$

If the current is too small, all the other transistors eventually reach weak inversion which results in a maximum value of γ:

$$\gamma_{max} = n(2B+1+B/C) = 2 \text{ to } 4 \; \gamma_{min} \tag{70}$$

with B and C defined in Fig. 23. Introducing the values of R_N (68) and t_u (64) in (67) yields

$$V_{N0}^2 = \frac{\gamma}{\beta} \frac{kT}{C_L} \tag{71}$$

The noise power introduced by a single-stage transconductance amplifier is thus inversely proportional to the load capacitance C_L and *independent of the bias current* (except for the small possible variation of γ mentioned above). This result can indeed be verified experimentally. For a given amount of voltage feedback β, the noise can thus only be reduced by increasing C_L, which results in an increase of the settling time T_s given by (66). Introducing the latter in (67) results in

$$V_{N0}^2 \geq \frac{kTR_N}{\beta^2 T_s} \ln(1/\varepsilon) \tag{72}$$

The minimum possible noise for a given settling time T_s is proportional to R_N. It is thus increased at low current but is minimum for a fixed current if the input pair is in weak inversion (g_m maximum). Since 1/b is the closed loop DC gain, the minimum total input referred noise is independent of b.

The output transistors T_7 and T_8 can be replaced by cascode configurations to obtain a high value of DC voltage gain A_{v0}. Practical circuits achieve a DC gain close to 100dB, with a dynamic range larger than 80dB for a total current of 2.5mA at $V_B = 3$ V [35].

If the low frequency noise is not acceptable, all the noisy n-channel transistors can be replaced by compatible bipolars. An equivalent noise density below $100nV/\sqrt{Hz}$ for frequencies as low as 1Hz has been obtained with minimum-size devices and a tail current $I_0 = 2\ \mu A$ [18].

Operating the input pair in weak inversion provides a large transconductance and thus a short settling time even at low bias currents. This fact is only true for small input steps. As soon as the differential input voltage exceeds a few nU_T (see Fig. 14), the output current saturates and the slew rate BI_0/C_L is limited by the very small bias I_0. The calculation of settling time for a large input step ΔV_i and $\tau_p \ll \tau_u/\beta$ yields

$$T_s = \frac{\tau_u}{\beta}\ \frac{\sinh(\Delta V_i/2nU_T)}{\sinh(\varepsilon\Delta V_i/2nU_T)} \qquad \text{\textit{(input pair in weak inversion)}} \qquad (73)$$

This relation is plotted in Fig. 24. The settling time is increased by a factor of 2 to 3 for $\Delta V_i = 1V$.

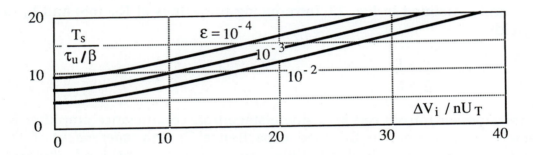

Fig. 24: Large signal settling time for the input pair in weak inversion.

One way to circumvent this fundamental limitation in switched capacitor circuits is to increase the tail current I_0 at the beginning of each clock cycle, that is, when an input step is expected [36]. The drawback of this solution is that some power is wasted by increasing the current systematically, even in standby or with small signals. A better power efficiency is obtained by using class-AB amplifiers, which provide more current to the load only when it is needed.

One possibility is to use an adaptive biasing scheme such as the one shown in Fig. 25 [37].

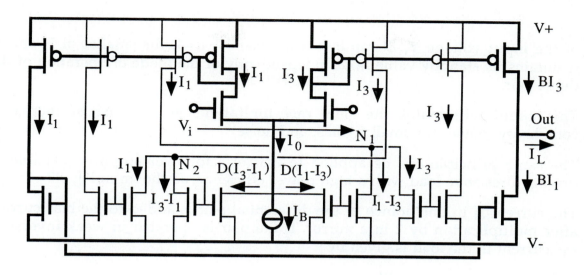

Fig. 25: Amplifier with adaptive bias.

The core of this circuit is represented in thick lines and is identical to the amplifier of Fig. 23. At equilibrium ($V_i = 0$), the tail current I_0 is equal to I_B.

If a differential signal is applied and $I_1 > I_3$, a surplus of current $I_1 - I_3$ is available at node N_1 and is added to the bias current I_B after multiplication by a factor D in a mirror.

If $I_3 > I_1$, no current is available from node N_1, but the surplus current $I_3 - I_1$ at node N_2 is again added to I_B after multiplication by D.

If the input pair is in weak inversion, the application of (15) yields:

$$-I_L = B (I_1 - I_3) = BI_B \frac{\tanh(V_i / 2nU_T)}{1 - D |\tanh(V_i / 2nU_T)|} \tag{74}$$

This relation is plotted in Fig. 26 for various values of the positive feedback D.

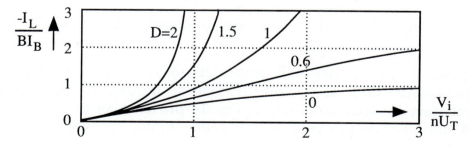

Fig. 26: Output current of the adaptive amplifier in weak inversion.

It can be seen that even though the current in the load never exceeds BI_B for D=0 (normal amplifier), it can become very large for D>1, even for small values of the input voltage V_i. It can be shown that for D=1, the smaller of I_1 and I_3 maintains its standby value $I_B/2$, independently of the mode of operation of the input pair.

This circuit eliminates the slew rate limitations at the cost of increased complexity, power for small signals, and noise.

The same principle can be applied to realize a follower amplifier capable of driving a resistive load with a high power efficiency, as shown in Fig. 27.

The current I_1 in one branch of the differential pair is added to the bias current after multiplication by 2 in a mirror. If the pair is operated in weak inversion, the current in the load is given by

$$I_L = BI_1 = \frac{BI_B}{e^{-\frac{V_{in}-V_{out}}{nU_T}} - 1} \tag{75}$$

This relation is plotted in Fig. 27. It can be seen that if the current I_L in the load is much larger than the bias current BI_B, the output voltage V_{out} follows the input voltage V_{in} with a very small offset. A high power efficiency can be obtained by selecting a very low value of I_B (which may even be limited to the leakage current) and a large value of the mirror ratio B.

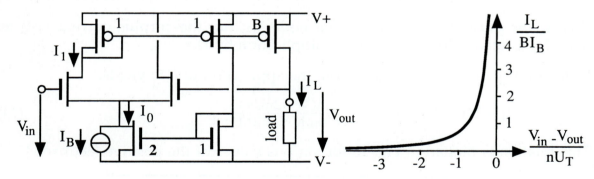

Fig. 27: Adaptive follower amplifier for unipolar load current.

Many other kinds of class-AB amplifiers have been developed to drive capacitive loads [38]. If a non-differential input can be accepted, the complementary inverter amplifier of Fig. 16 provides an excellent and simple class-AB amplifier. As can be seen in the same figure, for large variations of the input voltage V_i around the equilibrium value V_e (virtual ground level), the current I_L supplied to the load can be much larger than the bias current I_0. Some circuitry must be added to control I_0 and to attenuate the effect of power supply variations. Two possible solutions are depicted in Fig. 28 [39].

In Fig. 28a, the supply voltage of all the amplifiers is maintained at the required value V_0 by means of an additional matched inverter biased with I_0. The virtual ground potential V_e is available from this reference inverter. The whole circuit must be fed at a supply voltage higher than V_0 to permit correct operation of the follower-amplifier A. This amplifier can be realized according to Fig. 27. If the transistors are in weak inversion, the value of V_0 is close to the sum of the p- and n-channel thresholds, which limits the maximum possible amplitude at the output of the amplifiers.

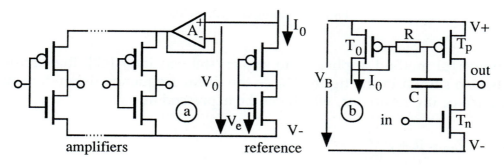

Fig. 28: Bias of CMOS inverter-amplifiers by regulation of V_0 (a) and AC coupling(b).

This limitation is eliminated in the solution shown in Fig. 28b. The bias current I_0 is imposed independently of V_B by the current mirror T_0-T_p. The behavior of this circuit is identical to that of an inverter for frequencies higher than $1/RC$. The high valued resistance R can be realized by a quad of polysilicon lateral diodes (Fig. 12) [40]. The only drawback of this solution is a poor PSRR (approximately 6dB, referred to the output) which is usually not acceptable, even for battery operation. This drawback can be eliminated by using the auto-zeroing property of dynamic bias to replace the RC coupling [41].

5.2. Quartz crystal oscillators

Micropower crystal oscillators have been given much consideration for applications in watches, where long term accuracy is of prime importance. Virtually all such circuits are based on the 3-point oscillator shown in Fig. 29a. Fig 29b shows the equivalent circuit, in which Z_1, Z_2 and Z_3 include all the functional and parasitic components of the real circuit, except the transconductance g_m of the active device, each of which can in principle be nonlinear.

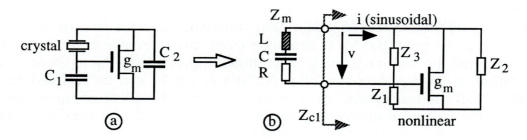

Fig. 29: 3-point crystal oscillator.

As a powerful method applicable to any crystal oscillator [42], the circuit can be split as also shown in the figure. Z_m is the linear frequency dependent motional impedance of the resonator, which can be expressed as

$$Z_m = R + j\frac{2p}{\omega_0 C} \tag{76}$$

where $p = (\omega - \omega_0) / \omega_0$ (77)

is the relative amount of frequency pulling from the natural angular frequency ω_0 of the LCR motional resonator. Because of the high quality factor $Q = 1/\omega_0 RC$ of the resonator, the current $i(t)$ through Z_m may be considered purely sinusoidal, even if the voltage $v(t)$ across it is strongly distorted. The rest of the circuit is nonlinear, but can be considered independent of the frequency in the vicinity of ω_0. It can be characterized by its amplitude-dependent impedance Z_{c1} for the fundamental frequency, since $i(t)$ is purely sinusoidal. Z_{c1} is reduced to the normal linear impedance Z_c for small amplitudes. Sustained oscillation then corresponds to

$$Z_m(p) + Z_{c1}(I) = 0 \tag{78}$$

from which the amplitude I of $i(t)$ and the frequency p can be computed.

The critical condition for start-up of oscillation is obtained by replacing $Z_{c1}(I)$ by $Z_c(g_m)$ in (78). It is then found that Z_c is a bilinear function of g_m. The locus of $Z_c(g_m)$ is therefore a circle in the complex plane. This locus is represented in Fig. 30 for a lossless circuit ($Z_i = 1/\omega C_i$), together with the locus of $-Z_m(p)$ given by (76). The critical condition for oscillation $Z_m + Z_c = 0$ corresponds to the intersections A and B of these loci. It can be shown that only A is stable.

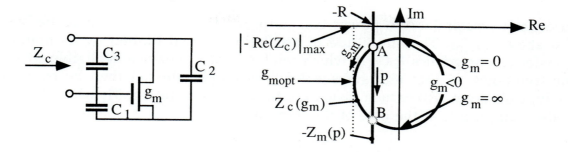

Fig. 30: Linear analysis of the lossless 3-point oscillator.

These intersections exist only if R is not too large, which can be expressed as the condition

$$\frac{QC}{C_3} > 2\left(1 + C_3 \frac{C_1 + C_2}{C_1 C_2}\right) \tag{79}$$

As shown by relation (76), the frequency pulling p is given by the imaginary part of Z_m. Therefore, in order to achieve a frequency of oscillation independent of R (and thus of Q), the slope of Z_c at solution A must be as small as possible. This state is achieved by fulfilling the condition (79) with a large margin (radius of the circle much larger than R). Then, at point A

$$p = \frac{C}{2\left(C_3 + \frac{C_1 C_2}{C_1 + C_2}\right)} \tag{80}$$

$$g_{mcrit} = \omega^2 R \frac{(C_1 C_2 + C_2 C_3 + C_3 C_1)^2}{C_1 C_2} = \frac{\omega C}{Q p^2} \cdot \frac{C_1 + C_2}{4 C_1 C_2} \tag{81}$$

The last expression shows that the critical transconductance is minimum for $C_1 = C_2$ and is inversely proportional to p^2. Thus, decreasing the value of p (to reduce the influence of circuit components) requires a large increase in the transconductance. There is therefore a basic tradeoff between frequency stability and power consumption.

If $g_m > g_{mcrit}$, the circuit provides a surplus of negative resistance and the oscillation grows until it is limited by nonlinear effects according to (78). As can be seen in Fig. 30, this negative resistance reaches a maximum for $g_m = g_{mopt}$. Any further increase of g_m reduces the available negative resistance, until it becomes too small for start-up, beyond solution B.

For a given value of g_{mcrit}, the critical drain bias current for oscillation I_{0crit} is minimized by operating the active transistor of the oscillator in weak inversion, according to Fig. 7.

In order to minimize the power consumption, the transistor should be biased just above g_{mcrit}. The amplitude of oscillation is then limited by its nonlinear transfer characteristics $I_D(V_G)$, which can be shown to have a negligible effect on the frequency so that (80) stays valid. The amplitude can then be computed analytically from (78), which yields the results shown in Fig. 31, where V_1 is the voltage amplitude at the gate of the transistor.

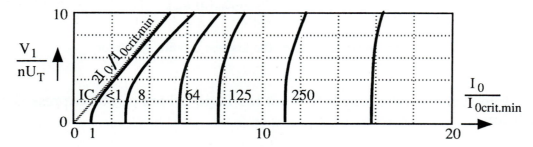

Fig. 31: Amplitude of oscillation for a nonlinear limitation by $I_D(V_G)$.

If $I_0 \gg I_{0crit}$, the amplitude is approximately given by

$$V_1 \approx 2I_0/g_{mcrit} \tag{82}$$

Fig. 32 shows the complete circuit diagram of a very low-power oscillator which has been extensively used for low and medium frequency applications [43].

The core of the circuit is the basic 3-point oscillator represented in thick lines. The active transistor T_1 is biased by the current source I_0 and the diode D_1. Capacitor C_6 insures DC decoupling of the gate from pin Q_1. The regulator is based on weak inversion operation of transistors T_3 and T_5 and provides a sharp drop in I_0 when the amplitude of oscillation reaches a few nU_T [12]. It can be applied in other kinds of oscillators as well [44]. In the absence of oscillation, the regulator is equivalent to the current reference of Fig. 19 and provides the start-up value of I_0. The whole regulator consumes just a few nanoamperes.

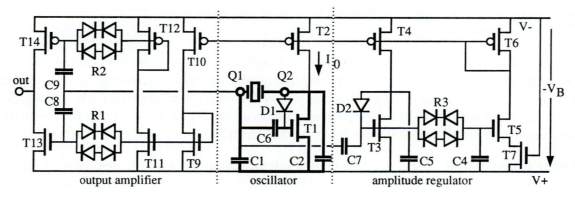

Fig. 32: Very low-power crystal oscillator with amplitude regulation.

The small amplitude of oscillation is efficiently shaped into a logic signal by the AC-coupled CMOS inverter-amplifier T_{13}-T_{14}. All the diodes are lateral polycrystalline diodes as described in Fig. 12. All floating capacitors have values below 2 pF. This circuit may consume as few as 20 nA at 32 kHz, for $V_B = 1$ to 3V, with a start-up current of several microamperes.

If no polycrystalline diodes are available in the process, they can be replaced by combinations of transistors, at the expense of increased power consumption [42].

5.3. Sequential logic building blocks

In order to minimize dynamic power consumption, the average number of nodes which transit during any period of time must be kept as small as possible. High frequency synchronous circuits must therefore be avoided and replaced whenever possible by asynchronous cells. Each of these cells should have a minimum number of nodes changing state to achieve a given function. It should furthermore contain a minimum of transistors, each of minimum size, in order to obtain a small total parasitic capacitance. Logic structures with critical races must therefore be discarded in favor of race-free circuits which work independently of the local delays [45, 47]. Dynamic CMOS gates must be used whenever possible.

As an example, let us consider the asynchronous divide-by-2 cell defined by the following set of equations [46]:

$$A = \overline{IC} \quad B = \overline{(I+D)A} \quad C = \overline{IE+AB} \quad D = \overline{B} \quad E = \overline{C} \tag{83}$$

I is the input variable and each internal variable A to E is produced by a CMOS gate (D and E by simple inverters). The whole circuit comprising 22 transistors is shown in Fig. 33 together with the transition graph of the structure.

This graph shows that no more than one variable tends to transit in any given state; hence there can be no race between variables. Thus, the logic function does not depend on the relative speed of the various gates, which can be optimized independently to minimize power.

Each internal variable transits at half the frequency of the input I and may thus be used as the output of the divider. A method for synthesizing a race-free implementation of any simple sequential function has been developed in [47].

The p- and n-block conduction functions defined in section 4.8 can be obtained from (83):

$$A^n = IC \qquad B^n = IA+DA \quad C^n = IE+AB \qquad\qquad D^n = B \quad E^n = C$$

$$AP = \bar{I}+\bar{C} \quad BP = \bar{A}+\bar{D}\,\bar{I} \quad CP = \bar{I}\,\bar{A}+\bar{I}\,\bar{B}+\bar{E}\,\bar{A}+\bar{E}\,\bar{B} \quad DP = \bar{B} \quad EP = \bar{C} \quad (84)$$

Each term in the above equations corresponds to a series connection of transistors from V+ or V- to the output node of the corresponding gate. Now, in the graph of Fig. 33, the variable which had to transit to reach each state has been identified and is shown in bold type. The term responsible for each transition has been in turn identified in equations (84) and highlighted in bold type. All the other terms are only used to maintain established states against leakage currents and may thus be dropped in high frequency circuits, together with the inverters producing variables D and E. The remaining part of the circuit is shown in thick lines in Fig. 33, and comprises only ten transistors. As a consequence, the power consumption of this dynamic divider is reduced by a factor of 2 to 3 with respect to the static version.

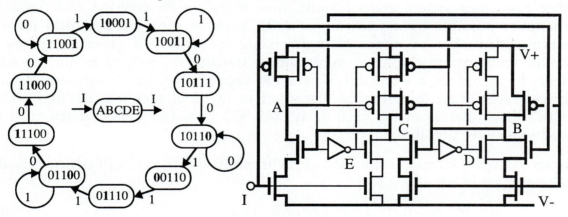

Fig. 33: Transition graph and circuit diagram of a race-free divide-by-2 cell.

This technique amounts to keeping only the transistors necessary to produce the transitions, plus possibly some transistors needed to maintain long-lasting states. It has been efficiently applied to various kinds of logic cells [33, 46, 48], including some low frequency circuits driven by short logic pulses [2]. It has been recently rediscovered for the realization of very high frequency dividers [49].

6.   Conclusions

The realization of micropower analog and digital CMOS circuits requires some tradeoffs with their performance. The low supply voltage affects the maximum signal amplitude and degrades the symmetry of current mirrors. The minimum possible voltage may be limited by the realization of analog switches.

Low current requirements limit the use of resistors, reduce speed and bandwidth and increase the noise voltage spectral densities. However, the possibility of operating transistors in weak inversion helps to counter these limitations. The very low value of saturation voltage permits peak-to-peak amplitudes close to the supply voltage. The maximum value of transconductance-to-current ratio provides a large DC gain per stage, as well as minimum noise for a given current. The well-controlled exponential characteristics permit interesting analog subcircuits.

Other favorable factors are the absence of thermal gradients on the chip, an easy elimination of latch-up, and reduced PSRR requirements. Special design techniques include single stage transconductance amplifiers, class-AB amplifiers to eliminate slew rate limitations, complementary dynamic logic circuits, and the use of special devices, such as the lateral diode in the polysilicon gate layer and compatible bipolar transistors.

7. References

[1] F. Leuenberger, "Solid-State devices in watches," in *Solid-State Devices 1973*, Conference Series 19, The Institute of Physics, London and Bristol, pp. 31-50 (invited paper ESSDERC '73).

[2] E. Vittoz, "LSI in watches," in *Solid-State Circuits 1976*, Ed. Journal de Physique, 1977, Paris. Also published in Pulse, Jan. 1978, pp. 14-20.

[3] C. Mead, *Analog VLSI and Neural Systems*, Addison-Wesley, 1989.

[4] R.M Swanson and J.D.Meindl,"Ion-implanted complementary MOS transistors in low-voltage circuits," IEEE J. Solid-State Circuits, vol. SC-7, pp. 146-153, April 1972.

[5] R. Luescher and J. Solo de Saldivar, "A high density CMOS process," ISSCC Dig. Tech. Papers, pp. 260-261, New-York, 1985.

[6] E.Vittoz and F. Krummenacher, "Micropower SC filters in Si-gate technology," Proc. ECCTD '80, Warsaw, pp. 61-72, 1980.

[7] B. Gerber *et al*, "A 1.5V single-supply one-transistor CMOS EEPROM," IEEE J. Solid-State Circuits, vol. SC-16, pp. 195-199, June 1981.

[8] Y.P. Tsividis, *Operation and Modelling of the MOS Transistor*, McGraw-Hill, New York, 1987.

[9] E. Vittoz, "MOS transistor," Intensive summer course on CMOS VLSI Design, EPFL, Lausanne, 1990.

[10] C. Enz, *High Precision CMOS Micropower Amplifiers*, Ph.D. Thesis No. 802, EPFL, Lausanne, 1989.

[11] J.A. van Nielen and O.W. Memelink, "Influence of substrate upon the DC characteristics of MOS transistors," Philips Research Reports, vol. 22, pp. 55-71, 1967.

[12] E. Vittoz and J. Fellrath, "CMOS analog integrated circuits based on weak inversion operation," IEEE J. Solid-State Circuits, vol. SC-12, pp. 224-231, June 1977.

[13] H. Oguey and S. Cserveny, "MOS modelling at low current density," summer course on Process and Device Modelling, KU-Leuven, 1983.

[14] J. Fellrath and E.Vittoz, "Small signal model of MOS transistors in weak inversion," Proc. Journées d'Electronique 1977, EPF-Lausanne, pp. 315-324.

[15] J. Fellrath, "Shot noise behavior of subthreshold MOS transistors," Revue de Physique Appliquée, Vol. 13, pp. 719-723, Dec. 1978.

[16] M. Aoki *et al*, "Low 1/f noise design in Hi-CMOS devices," IEEE Trans. on Electron Devices, Vol. ED-29, pp. 296-299, Feb. 1982.

[17] J.-P. Bardyn, *Amplificateurs CMOS faible bruit pour applications sonar*, Ph.D. Thesis No 608, Univ. Sc. Tech. of Lille Flandres-Artois, 1990.

[18] E. Vittoz, "MOS transistors operated in the lateral bipolar mode and their application in CMOS technology," IEEE J. Solid-State Circuits, vol. SC-18, pp. 273-279, June 1983.

[19] T.W. Pan and A. Abidi, "A 50dB variable gain amplifier using parasitic bipolars transistors in CMOS," IEEE J. Solid-State Circuits, vol. SC-24, pp. 951-961, Aug. 1989.

[20] H. Ma, *DC and AC characterization of CMOS compatible lateral bipolar transistors*, Ph.D. Thesis, Tsinghua University, Beijing, 1990.

[21] X. Arreguit *et al*, "Precision compressor gain controller in CMOS technology," IEEE J. Solid-State Circuits, vol. SC-22, pp. 442-445, June 1987.

[22] X. Arreguit, *Compatible Lateral Bipolar Transistors in CMOS Technology: Model and Applications*, Ph.D. Thesis No. 817, EPF-Lausanne, 1989.

[23] M. Degrauwe *et al*, "CMOS voltage references using lateral bipolars," IEEE J. Solid-State Circuits, vol. SC-20, pp. 1151-1157, Dec. 1985.

[24] Z. Hong and H. Melchior, "Four-quadrant multiplier core with lateral bipolar transistors in CMOS technology," Electronics Letters, vol. 21, pp. 72-74, Jan. 17, 1985.

[25] M. Dutoit and F. Sollberger, "Lateral polysilicon p-n diodes," J. Electrochem. Soc., Vol. 125, pp. 1648-1651, Oct. 1978.

[26] T.C. Choi *et al*, "High-frequency CMOS switched-capacitor filters for communications applications," IEEE J. Solid-State Circuits, vol. SC-18, pp. 652-664, Dec. 1983.

[27] E. Vittoz and O. Neyroud, "A low-voltage CMOS bandgap reference," IEEE J. Solid-State Circuits, vol. SC-14, pp. 573-577, June 1979.

[28] G. Tzanateas *et al*, "A CMOS bandgap reference," IEEE J. Solid-State Circuits, vol. SC-14, pp. 655-657, June 1979.

[29] Y.P. Tsividis and R.W. Ulmer, "A CMOS voltage reference," IEEE J. Solid-State Circuits, vol. SC-13, pp.774-778, Dec. 1978.

[30] B.S. Song and P.R. Gray, "A precision curvature-compensated CMOS bandgap reference," IEEE J. Solid-State Circuits, vol. SC-18, pp. 634-643, Dec. 1983.

[31] H. Oguey and B. Gerber, "MOS voltage reference based on polysilicon gate work function difference," IEEE J. Solid-State Circuits, vol. SC-15, pp. 264-269, June 1980.

[32] F. Krummenacher *et al*, "Higher sampling rate in SC circuits by on-chip clock-voltage multiplication," Dig. ESSCIRC '83, pp. 240-241, Lausanne, 1983.

[33] E. Vittoz and H. Oguey, "Complementary dynamic MOS logic circuits," Electronics Letters, Vol. 9, Feb. 22, 1973.

[34] E. Vittoz, "Microwatt switched capacitor circuit design," Electrocomponents Science and Technology, Vol. 9, pp. 263-273, 1982.

[35] F. Krummenacher, "High voltage gain CMOS OTA for micropower SC filters," Electronics Letters, Vol. 17, pp. 160-162, 1981.

[36] B.J. Hosticka, "Dynamic CMOS amplifiers," IEEE J. Solid-State Circuits, vol. SC-15, pp. 887-894, Oct. 1980.

[37] M. Degrauwe *et al.*, "Adaptive biasing CMOS amplifier," IEEE J. Solid-State Circuits, vol. SC-15, pp. 887-894, Oct. 1980.

[38] R. Castello, "CMOS buffer amplifiers," CMOS VLSI Design '90, EPF-Lausanne, 1990.

[39] F. Krummenacher, E. Vittoz and M. Degrauwe, "Class AB CMOS amplifiers for micropower SC filters," Electronics Letters, Vol. 17, pp. 433-435, June 25, 1981.

[40] F. Krummenacher, "Micropower switched capacitor biquadratic cell," IEEE J. Solid-State Circuits, vol. SC-17, pp. 507-512, June 1982.

[41] E. Vittoz, "Dynamic analog techniques," in *Design of VLSI Circuits for Telecommunications and Signal Processing*, J.E. Franca and Y.P. Tsividis Editors, Prentice Hall, 1991.

[42] E. Vittoz *et al.*, "High-performance crystal oscillator circuits: theory and applications," IEEE J. Solid-State Circuits, vol. SC-23, pp. 774-783, June 1988."

[43] E. Vittoz, "Quartz oscillators for watches," Proc. Int. Congr. of Chronometry, pp. 131-140, Geneva, 1979.

[44] E. Vittoz, "Micropower switched-capacitor oscillator," IEEE J. Solid-State Circuits, vol. SC-14, pp. 622-624, June 1979.

[45] E. Vittoz *et al.*, "Model of the logic gate," Proc. Journées d'Electronique 1977, pp. 455-467, EPF-Lausanne.

[46] E. Vittoz *et al.*, "Silicon-gate frequency divider for the electronic wrist watch," IEEE J. Solid-State Circuits, vol. SC-7, pp. 100-104, April 1972.

[47] C. Piguet, "Logic synthesis of asynchronous circuits," IEEE Custom Integrated Circuits Conference, CICC '90, paper 29.6, Boston, May 13-16, 1990.

[48] H. Oguey and E. Vittoz, "CODYMOS frequency dividers achieve low power consumption and high frequency," Electronic Letters, Vol. 8, Aug. 23, 1973.

[49] J.R. Yuan and C. Svensson, "High speed CMOS circuit technique," IEEE J. Solid-State Circuits, vol. SC-24, pp. 62-70, 1989.

Chapter 4

Dynamic Analog Techniques

Eric A. Vittoz

Centre Suisse d'Electronique et de Microtechnique (CSEM), Neuchatel, Switzerland

1. Introduction

The absence of any control current is one of the most significant features offered by MOS transistors. It allows the gate control voltage to be stored on a capacitor for a period of time limited by the leakage current of the associated pn junctions. This property is used to simplify digital circuits by temporarily maintaining established logic states as voltages across capacitors. Circuits of this kind must have stable states that are not required to last for too long, hence the appellation of "dynamic logic circuits." Dynamic analog circuits are defined by analogy. They exploit the possibility of storing an analog voltage on a capacitor. They must be distinguished from charge exchange circuits (which include switched capacitor circuits) with which they are often combined. They differ from them by their insensitivity to the exact value and to the nonlinearities of the capacitors.

This chapter will first discuss the principle and the features and limitations of the elementary sample-and-hold which is common to all dynamic circuits. Four different examples using the dynamic concept will then be discussed, namely the compensation of offset in operational amplifiers, a biasing scheme for CMOS inverters used as amplifiers, the dynamic comparator and the dynamic current mirror or current copier.

2. Sample-and-Hold

2.1. Principle

Analog dynamic circuits are based on the elementary sample-and-hold circuit represented in Fig. 1.

This circuit is formed by a signal source $V_i(t)$, a sampling switch S and a hold capacitor C. R represents the total resistance in the circuit. The sampling frequency is f_c and corresponds to a sampling period $T_c=1/f_c$. The sampling duration T_s is assumed to be long enough to permit full sampling, that is,

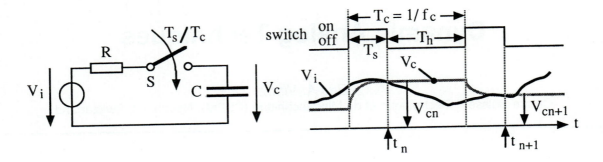

Fig. 1:   Elementary sample-and-hold, principle and definitions.

$$T_s \gg \tau = RC \tag{1}$$

After switch S opens at time t_n, the voltage V_c across capacitor C has a constant value

$$V_{cn} = V_{ib}(t_n) \tag{2}$$

during the subsequent hold period of duration T_h, where V_{ib} is the input signal V_i bandlimited by the low-pass transfer function

$$G(f) = \frac{1}{1 + j2\pi f\tau} \tag{3}$$

due to the RC time constant. The switch S is implemented by a single transistor or by a complementary pair [1].

2.2. Transfer function

The transfer function of the sample-and-hold can be analyzed by means of the block diagram shown in Fig. 2 in which the component of V_c during sampling is not considered.

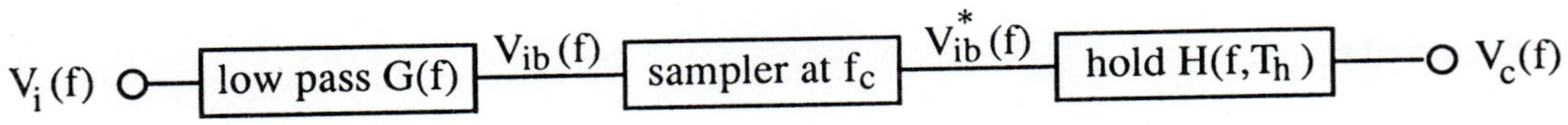

$V_i(f)$ O—[low pass G(f)]—$V_{ib}(f)$—[sampler at f_c]—$V_{ib}^*(f)$—[hold H(f,T_h)]——O $V_c(f)$

Fig. 2:   Block diagram of the sample-and-hold circuit.

The Fourier transform of the band-limited sampled input is periodic and given by [2]

$$V_{ib}^*(f) = \frac{1}{T_c} \sum_{n=-\infty}^{+\infty} V_{ib}(f+nf_c) \tag{4}$$

whereas the transfer function of the hold block is [2]

$$H(f;T_h) = T_h \frac{\sin(\pi f T_h)}{\pi f T_h} e^{-j\pi f T_h} = T_h \mathrm{sinc}(\pi f T_h) e^{-j\pi f T_h} \tag{5}$$

One may define an overall transfer function for each value of n in (4):

$$W_{cn}(f) = \frac{V_c(f)}{V_i(f+nf_c)} = \frac{T_h}{T_c} \frac{\mathrm{sinc}(\pi f T_h) e^{-j\pi f T_h}}{1+j2\pi(f+nf_c)\tau} \tag{6}$$

in which the output signal V_c during the sampling time T_s is not considered. The transfer function $W_{c0}(f)$ for n=0 corresponds to the baseband (same input and output frequencies). The transfer functions $W_{cn}(f)$ for n≠0 are those of the foldover bands due to undersampling.

2.3. Noise sampling

When the switch is closed, the noise of the circuit may be represented by an equivalent noise component V_{iN} of the input voltage V_i, of spectral density

$$S_{Vi} = dV_{iN}^2 / df = 4kTR_N \tag{7}$$

where $R_N = \gamma R \tag{8}$

is the equivalent noise resistance of the circuit [1]. When the noise is only due to the thermal noise of R, $R_N = R$ and $\gamma = 1$. The total noise voltage V_{cN} across C has the mean squared value

$$V_{cN}^2 = \int_0^\infty S_{Vi} |G(f)|^2 \, df \tag{9}$$

Introducing (3), (7) and (8) into (9) for γ independent of frequency (white noise) yields

$$V_{cN}^2 = \gamma \frac{kT}{C} \tag{10}$$

When the switch opens at instant t_n, the instantaneous value of noise voltage V_{cN} is stored on capacitor C. A noise sample δV_{cn} is thus added to the signal sample V_{cn}, as illustrated in Fig. 3. The successive noise samples are not correlated and their variance is equal to the mean squared value V_{cN}^2 of the continuous noise given by (10).

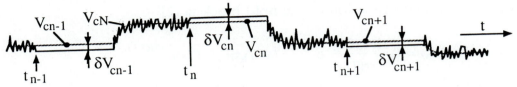

Fig. 3:   Continuous and sampled noise.

Figure 4 shows the power spectral density of the continuous noise ($T_s = T_c$, switch always closed), and that of the sampled noise ($T_h = T_c$, very short sampling duration T_s) calculated from (4) and (5) for a bandwidth $1/\tau$ much larger than the sampling frequency f_c.

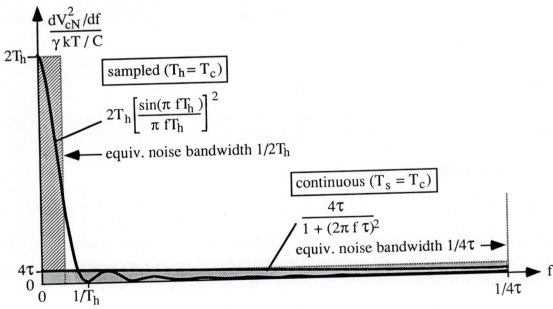

Fig.4:   Power spectral densities of continuous and sampled noise.

The total noise remains constant, but the equivalent noise bandwidth is reduced from $1/4\tau$ in the continuous case to $1/2T_h$ in the sampled case. Most of the noise is thus transposed to low frequencies by the sample-and-hold process.

When neither T_h nor T_s is negligible, the continuous and sampled spectra are approximately weighted by T_s / T_c and T_h / T_c, respectively.

2.4. Charge injection by the switch

Another limitation on the precision of a sample-and-hold circuit is the disturbance of the sampled voltage V_c when the transistor implementing the switch is turned off. Fig. 5 shows a more detailed equivalent circuit of the basic scheme of Fig. 1. The switch is implemented by an n-channel transistor T driven by its gate voltage V_G. R_i is the resistance of the signal source and C_i is the capacitance to ground at the source side of the switch, which includes the junction capacitance of the transistor.

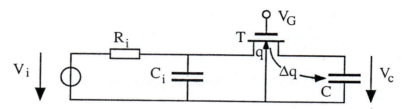

Fig. 5:   Detailed schematic of the sample-and-hold circuit.

The switch is turned on by applying a gate voltage $V_G = V_{Gon}$ and turned off by reducing V_G below the threshold of the transistor. During this process, the charge q stored in the channel is released and may go to the source and drain diffusions and to the substrate. The fraction Δq of q which flows onto the storage capacitor C causes an error of V_c

$$\Delta V_c = \Delta q/C \qquad (11)$$

It may be shown that the fraction of q that goes to the substrate can be neglected in most practical situations [3]. The analysis can be simplified by assuming a time constant R_iC_i much larger than the switch-off time. The charge q is thus shared between C_i and the storage capacitor C. If the transistor is perfectly symmetrical and $C_i = C$, then half the charge q flows onto each capacitor. But if $C_i \neq C$, a voltage appears across the transistor during the descent of V_G, which may result in an uneven sharing of the released charge.

This problem can be analyzed by means of the model of Fig. 6.

V_{Te} is the effective gate threshold voltage below which the channel is blocked:

$$V_{Te} = V_{T0} + nV_{cn} \quad [1] \qquad (12)$$

where V_{T0} is the threshold for zero source voltage, V_{cn} is the particular value of V_c during the sampling period being considered, and n represents the effect of bulk modulation.

For $V_G > V_{Te}$, the channel conductance g is given by [1]

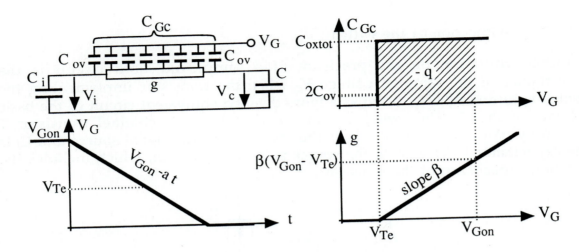

Fig. 6:   Model for analysis of charge injection.

$$g = \beta \, (V_G - V_{Te}) \tag{13}$$

with $\beta = \mu C_{ox} W/L \tag{14}$

and the gate coupling capacitance C_{Gc} has the value C_{oxtot}, which is the sum of the gate oxide distributed capacitance to the channel WLC_{ox} and of the two overlap capacitances C_{ov}. For $V_G < V_{Te}$, $g = 0$ (if weak inversion is neglected) and C_{Gc} is reduced to $2C_{ov}$.

The gate voltage V_G is assumed to decrease linearly with time from its on-value V_{Gon}:

$$V_G = V_{Gon} - a \, t \tag{15}$$

Now $\Delta V_c \ll V_{Gon} - V_{Te}$ and therefore the potential at each end of the channel conductance g may be considered constant. It will be assumed that this condition is true as well for any point in the channel, hence that the current flowing through the distributed capacitor during the descent of V_G causes a negligible voltage drop even at the middle of the conductance. The validity of this assumption is supported by a more detailed analysis of the switching process [3]. The linear decrease of V_G across $C_{Gc} = C_{oxtot}$ is thus equivalent to a distributed current source of total value aC_{oxtot}, which flows symmetrically towards both ends of the channel. The model of Fig. 6 can therefore be reduced to that of Fig. 7 [4].

Introducing relations (13) and (15) into this model yields the following equation:

$$\frac{dU}{dT} = (T - B)\left[\left(1 + \frac{C}{C_i}\right) U + 2 \frac{C}{C_i} T\right] - 1 \tag{16}$$

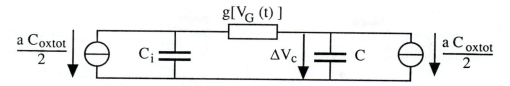

Fig. 7:   Final sample-and-hold model for the analysis of charge injection.

with the normalized variables

$$U = \Delta V_c \Big/ \left[\frac{C_{ox}}{2} \sqrt{\frac{a}{BC}} \right]$$
(normalized voltage error)   (17)

$$T = t \Big/ \sqrt{\frac{C}{a\beta}}$$
(normalized time)   (18)

and $$B = (V_{Gon} - V_{Te}) \sqrt{\frac{\beta}{aC}}$$
(switching parameter)   (19)

For $U=0$ at $T=0$, the numerical integration of (16) from $T=0$ to $T=B$ yields the value of U when the transistor starts to be blocked ($V_G = V_{Te}$). The corresponding value of $\Delta q/q$ is then easily calculated. This value is plotted as a function of the switching parameter B in Fig. 8, for various values of the capacitance ratio C/Ci [3, 4, 5].

The chart of Fig. 8 is very general as it covers a wide range of practical situations. Its validity is confirmed by a more detailed analysis of the charge flows in the transistor and by measurements with perfectly symmetrical transistors [3]. It is distorted by any asymmetry in the transistor, including that of the overlap capacitances which should be kept symmetrical by adequate layout means. Results of numerical and breadboard simulations of the effect of R_i have been reported [6], as well as an analytical solution of equation(16) [7]. This normalized chart can be used to devise various strategies for minimizing charge injection.

A first possibility is to choose $C_i \gg C$ and B very large, in order to let all the released charge flow back onto C_i (which behaves as a voltage source). This high value of B can be obtained by reducing the slope a of the gate voltage decay. A qualitatively equivalent solution is possible by first stepping down the gate voltage to a value just above V_{Te}. Most of the charge q is then released but is still able to flow back onto C_i before the transistor is fully blocked in a second step. If the main switch requires a wide transistor to achieve a sufficient value of on-conductance, a minimum width transistor may be placed in parallel with it, which is switched off in a second step only after the charge released by the main transistor has flown back onto C_i. All these solutions reduce the maximum possible frequency of operation. Furthermore, the residual effect due

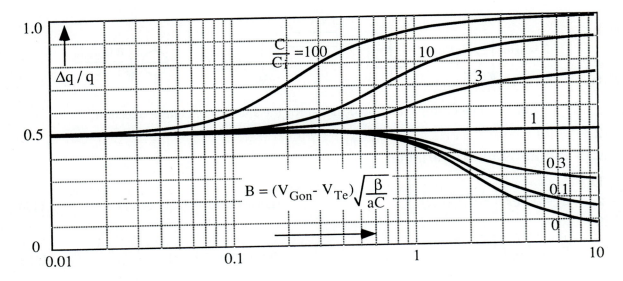

Fig. 8:   General chart of charge injection Δq in the storage capacitor C.

A second strategy is to choose $C_i = C$ [8]. If the transistor is itself perfectly symmetrical, half the charge q flows onto C and can be compensated by half-sized dummy switches which absorb the released charge, as shown in Fig. 9 [9]. In this scheme, the amount of compensation not only depends on the matching of the transistors, but also on that of C_i and C which is usually difficult to achieve.

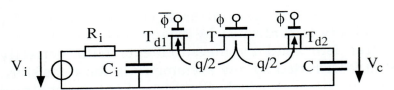

Fig. 9:   Compensation of charge injection by half-sized dummy switches.

A third strategy, which is usually the best, is to choose a small value of B (large slope *a* of the gate voltage decay) to permit the compensation by dummy switches even for $C_i \neq C$ (and/or $R_i \Rightarrow 0$). The validity of the above chart requires a sufficient value of C_i, which can be shown to be obtained in all realistic situations where a junction capacitance is present [3]. The dummy switch T_{d1} is then usually not needed anymore if C_i / C is not too small.

If the switch is made up of complementary transistors, they partially compensate each other's charge q. However, no real matching exists between p- and n-channel devices, and their respective channel charges depend on the analog level V_{cn} through relation (12). Furthermore, the relative phase jitter of the complementary gate clock voltages may produce a randomly variable compensation which amounts to additional noise [10]. It is therefore preferable to avoid any compensation of this kind by turning off one of the transistors before the other.

Each of the above strategies can be combined with a differential approach which turns the charge injected by the switches into common mode perturbations [11].

2.5. Sample-and-hold used as a time differentiator (autozero circuit)

The sample-and-hold circuit of Fig.1 may be redrawn as shown in Fig. 10. During each sampling period T_s, the voltage V_d across the switch is forced to zero. The evolution of voltages V_i, V_c and V_d are shown in the figure for $\tau \ll T_s \ll T_c$.

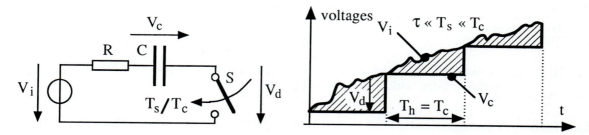

Fig. 10: Differentiation by sample-and-hold (autozero circuit).

During the hold period T_h, the voltage V_d across the switch is the difference between the input voltage V_i and its previously sampled value V_c:

$$V_d = V_i - V_c \tag{20}$$

Therefore, the baseband transfer function (same input and output frequencies) is

$$W_{d0}(f) = \frac{V_d(f)}{V_i(f)} = 1 - W_{c0}(f) \tag{21}$$

where $W_{c0}(f)$ is given by relation (6) with $T_c = T_h$ and $n=0$. The magnitude of this transfer function for $f \ll 1/2\pi\tau$

$$|W_{d0}(f)| = \sqrt{\left(1 - \frac{\sin(2\pi f T_h)}{2\pi f T_h}\right)^2 + \left(\frac{1 - \cos(2\pi f T_h)}{2\pi f T_h}\right)^2} \tag{22}$$

is plotted in Fig. 11.

At low frequencies and for the baseband signal, the circuit behaves like a time differentiator with time constant $T_h/2$. This is the result of taking the difference between the direct input signal V_i and its sampled version V_c. Low frequencies are attenuated and the DC component is cancelled by the zero at the origin.

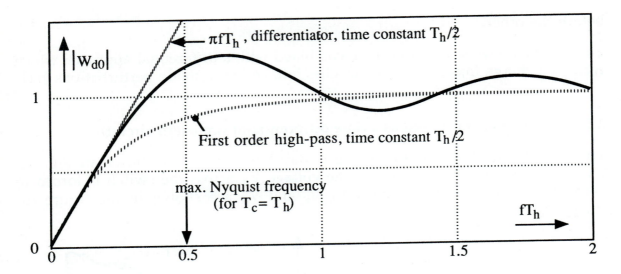

Fig. 11:   Autozero transfer function for baseband signals. At low frequencies, the circuit behaves as a differentiator with time constant $T_h/2$.

The foldover components due to undersampling are not present in V_i. Thus:

$$W_{dn}(f) = \frac{V_d(f)}{V_i(f+nf_c)} = -W_{cn}(f) \qquad\qquad \text{for } n \neq 0 \qquad\qquad (23)$$

and these components are not differentiated. The noise power transfer functions $|W_{d0}|^2$ (for baseband) and $|W_{dn}|^2$ (for foldover bands) are represented in Fig. 12 for $|f+nf_c| \ll 1/2\pi\tau$.

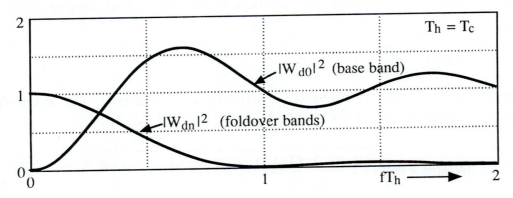

Fig. 12:   Noise power transfer function of the autozero circuit for baseband and foldover bands.

The effect on white noise is almost the same as that of the sample-and-hold itself, since the result is dominated by the accumulation of the foldover components.

For 1/f flicker noise, the double zero of $|W_{d0}|^2$ at f=0 cancels out the dominant low-frequency noise component. The foldover components remain and their accumulation produces a residual noise spectral density that is approximately constant (white noise) for $f \ll f_c$.

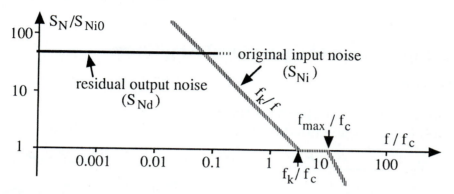

Fig. 13: Residual low-frequency noise of the autozero circuit. In this example, $f_k=3f_c$ and $f_{max}=10f_c$. It results in a total low-frequency white noise density ratio $S_{Nd}/S_{Ni0} = 48.8$ due to undersampling (31.4 due to the original white noise and 17.4 due to the 1/f noise).

Assuming a combination of white and 1/f input noise bandlimited by a single pole at $f=f_{max}$, with a spectral density

$$S_{Ni} = \frac{S_{Ni0}(1 + f_k/f)}{1 + (f/f_{max})^2} \tag{24}$$

illustrated asymptotically in Fig. 13 for particular values of f_k (corner frequency) and f_{max}, the output noise spectral density S_{Nd} of the autozero circuit can be approximated for $f \ll f_c$ by [12,13]

$$\frac{S_{Nd}}{S_{Ni0}} \cong \pi \frac{f_{max}}{f_c} + 2 \frac{f_k}{f_c}(0.6 + \ln\frac{f_{max}}{f_c}) \tag{25}$$

The first term is the effect of the original white noise whereas the second term is the additional white noise due to the accumulation of the foldover components of the 1/f noise .

If the input noise is only the unavoidable white noise $4\gamma kTR$ of the resistance R bandlimited at $f_{max} = 1/2\pi RC$, relation (25) yields $S_{Nd} = 2\gamma kTT_c/C$, which is equal, for $T_c=T_h$, to the low-frequency spectral density of the sample-and-hold given in Fig. 4.

When 1/f noise is present, the autozero circuit reduces the total noise spectral density for frequencies lower than a value obtained by comparing (24) and (25):

$$f < \frac{f_c}{\pi f_{max}/f_k + 1.2 + \ln(f_{max}/f_c) - f_c/f_k} \tag{26}$$

that is equal to $0.073f_c$ in the numerical example of the figure.

3.   Offset Compensation of Operational Amplifiers

3.1. First order compensation

The usual scheme used to compensate the offset of an operational amplifier is shown in Fig. 14, for a single-ended and for a fully differential circuit.

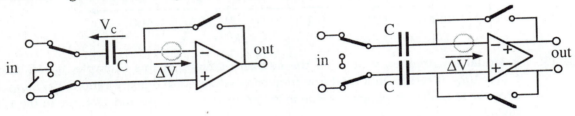

Fig.14:   First order compensation of the offset voltage of an operational amplifier.

During the compensation phase (opposite to that shown in the figure), no signal is applied to the input and the amplifier is connected in a unity gain configuration. If the amplifier has a very large open-loop gain, its differential input voltage is very small, except for the error ΔV which is thus sampled as a voltage V_c on capacitor C. During the amplifying phase (shown in the figure), V_c is held on C and subtracted from the input signal, whereas ΔV is added. The error ΔV is thus autozeroed according to the principle of Fig. 10. Its DC component (offset voltage) is cancelled and its low frequency components (1/f noise or spurious signals coming through the power supply) are attenuated as discussed in section 2.5. The fact that the amplifier is only available during the amplifying phase is often acceptable, or can be avoided by using two time-shared circuits.

The performance of this compensation scheme is bounded by the the very limitations of the sample-and-hold circuit. The noise in the bandwidth of the amplifier is folded down below the clock frequency, and the offset compensation is limited by charge injection from the feedback switch onto C.

3.2. Compensation by a low-sensitivity auxiliary input

The precision of the offset compensation can be drastically improved by using the scheme shown in Fig. 15.

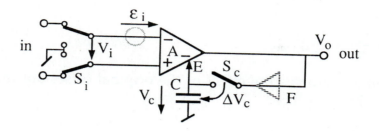

Fig. 15: Offset compensation by low-sensitivity auxiliary input.

The block A is an ordinary operational amplifier. It has a DC gain A_i and an offset ε_i, which could also be the residual offset after first order compensation according to Fig. 14. A compensation input E of reduced sensitivity is added to the main amplifier A. It is controlled by the compensation voltage V_c stored on capacitor C, so that the output voltage V_o is given by

$$V_o = -A_i (V_i - \varepsilon_i) - A_c (V_c - \varepsilon_c) \tag{27}$$

where A_c is the gain from the auxiliary input E and ε_c is the corresponding offset voltage.

During the compensation phase, the main input is short-circuited by switches S_i (thus $V_i=0$) and the compensation loop is closed by switch S_c. The output voltage V_o and the compensation voltage V_c therefore settle to a common equilibrium value V_{c0}. If

$$A_i \gg A_c \gg 1 \tag{28}$$

and if ε_c is not much larger than ε_i, then

$$V_{c0} = \varepsilon_i \frac{A_i}{A_c} \tag{29}$$

When the switch S_c is opened, a perturbation ΔV_c of the compensation voltage is produced by charge injection and sampled noise and results in a modified value of the output voltage

$$V_o = V_{c0} - A_c \Delta V_c \tag{30}$$

This value corresponds to an equivalent residual offset (input voltage V_i required to obtain $V_o=0$)

$$\varepsilon_{iR} = \frac{V_{c0} - A_c \Delta V}{A_i} = \frac{\varepsilon_i}{A_c} - \frac{A_c}{A_i} \Delta V_c \tag{31}$$

The worst case occurs when ε_i and ΔV_c have maximum absolute values ε_{im} and ΔV_{cm} but opposite signs, which results in

$$\varepsilon_{iR} = \frac{\varepsilon_{im}}{A_c} + \frac{A_c}{A_i} \Delta V_{cm} \qquad \textit{(worst case)} \qquad (32)$$

A large value of A_c reduces the effect of the original offset but increases that of the perturbation by switch S_c.

The optimum value is

$$A_{copt} = \sqrt{\frac{\varepsilon_{im} A_i}{\Delta V_{cm}}} \qquad (33)$$

and provides the minimum residual offset in the worst case

$$\varepsilon_{iR} = 2\sqrt{\frac{\varepsilon_{im} \Delta V_{cm}}{A_i}} \qquad (34)$$

The effectiveness of the scheme can be appreciated by means of a numerical example:

For $\qquad \varepsilon_{im} \qquad = 10 \text{ mV}$

$\qquad\qquad \Delta V_{cm} \quad = 10 \text{ mV}$

$\qquad\qquad A_i \qquad = 10^4 \qquad A_c = A_{copt} = 10^2 \text{ provides } |\varepsilon_{iR}| \leq 0.2 \text{ mV}$

Fig. 16 shows a practical implementation of this scheme.

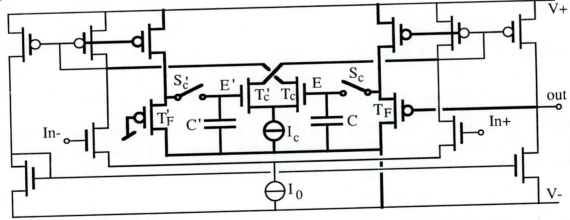

Fig. 16:   Example of implementation of the low-sensitivity auxiliary input E.

It is a standard single-stage operational transconductance amplifier to which compensation circuitry, shown in thicker lines, has been added. An additional differential pair T_c–T_c' connected in parallel with the main pair has a much smaller transconductance in order to achieve $A_c/A_i \ll 1$. This small transconductance ratio is obtained by choosing the tail current ratio I_c/I_0 as small as possible while still ensuring compensation of the maximum possible current asymmetry, and by operating T_c and T_c' at a large value of gate-to-source voltage (ultimately limited by the supply voltage) in order to minimize their transconductance-to-current ratio. In order to speed up the compensation process, a source follower T_F (represented in dotted lines in Fig. 15) is added to drive the storage capacitor C.

The symmetrical implementation of C', S_c' and T_F' allows charge injection ΔV_c from S_c onto C to be partially compensated by that of S_c' onto C'. It must be realized that this additional compensation is very imperfect, because the large voltage stored across C ($\varepsilon_i A_i / A_c$, which can be as large as 1 V for the numerical example above) results in different amounts of charge released by S_c and S_c'.

This implementation has been used to realize a low-offset controlled-gain amplifier as illustrated in Fig. 17 [14].

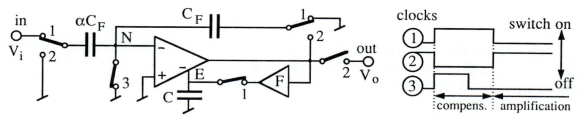

Fig. 17:   Low-offset controlled-gain amplifier.

During the compensation phase, the input node N of the amplifier is grounded by the switch driven by clock 3. The compensation loop is closed by clock 1, which also discharges the feedback capacitor C_F and stores the input signal V_i onto capacitor αC_F. When clock 3 returns to zero, charge injection onto node N produces an error signal ε_i which is corrected by the compensation loop. When clock 1 returns to zero, the compensation loop is opened and the residual equivalent error at the input ε_{iR} is bounded by (34). The charge stored onto αC_F is poured onto C_F which results in an output voltage $V_o = \alpha V_i$.

The application of the same scheme to an accurate voltage integrator is illustrated in Fig. 18.

Fig. 18a is a resistor-capacitor integrator based on an operational transconductance amplifier. Its output can be reset to zero before integration by means of switches S_z. In order to cancel the total equivalent offset at the input

$$\varepsilon_i = \Delta V + R\Delta I \tag{35}$$

due to the voltage offset ΔV and the input current ΔI, a compensation phase is carried out by means of the loop closed at the auxiliary input E, with the input of the integrator grounded through switch S_0. The compensation switch S_c is then opened while the input is still grounded. Integration of the input voltage

V_{in} is only started after equilibrium is reached (compensation of charge injection) by moving the input switch S_0.

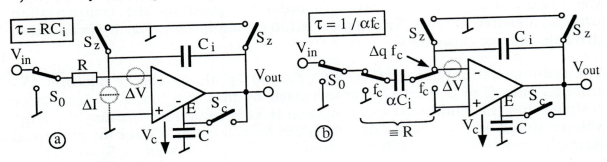

Fig. 18:   Low-offset integrator
           a: resistor-capacitor with compensation of offset voltage ΔV and current ΔI
           b: switched capacitor with compensation of ΔV and charge injection Δq.

Fig. 18b shows the same integrator in which the resistor has been replaced by capacitor αC_i switched at frequency f_c [15]. The circuit operates as that of Fig. 18a and compensates the total equivalent input offset

$$\varepsilon_i = \Delta V - \frac{\Delta q}{\alpha C_i} \tag{36}$$

where Δq is the charge injected at frequency f_c onto the input node of the operational amplifier.

Other means can be used to implement the low-sensitivity input E, such as variable current mirrors inside the amplifier [16]. Furthermore, it can be shown that replacing the compensation gain A_c by a quadratic law can in principle further improve the precision of the compensation [16]. This advantage has not yet been proven experimentally.

4.   Dynamic Biasing of CMOS Inverter-Amplifiers

A standard CMOS inverter is an excellent amplifier [1]. It offers the maximum possible transconductance and the minimum possible noise for a given bias current. Its inherent capability of operating in class AB eliminates all slew rate limitations. The only problem consists in finding a biasing scheme that is insensitive to the values of threshold and power supply voltages, and that provides an acceptable power supply rejection. A good solution is to use a dynamic bias scheme as shown in Fig. 19 [17].

During the biasing phase, all the switches are in the position opposite to that represented in the figure. The input is grounded by S_2 and the output is disconnected by S_3. The transistor T_p is a current source which mirrors the bias current I_0 into the diode-connected transistor T_n.

During the amplification phase represented in the figure, T_n and T_p behave as a CMOS inverter, with an equilibrium value of input voltage V_i (value for $I_L=0$) equal to zero, except for a small offset due to charge injection and to sampled noise.

This circuit behaves as does the first order compensation of the amplifier shown in Fig. 14, and therefore reduces the low frequency components of errors produced in the inverter, including the $1/f$ noise and low-frequency perturbations coming from the power supply. It permits the choice of the virtual ground levels, with an offset limited to the effect of charge injection onto C_2.

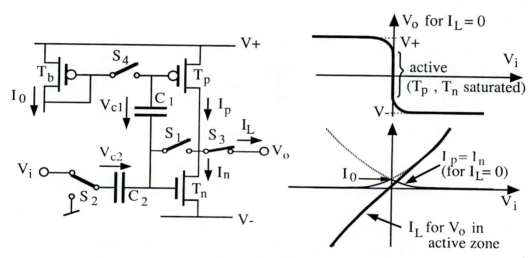

Fig. 19:   CMOS inverter-amplifier with dynamic bias.

Since it is a sampled scheme, the high frequency components of noise are folded down below the clock frequency. This approach is therefore only interesting when sampling is required for other reasons, as in switched capacitor circuits. Sampled noise and charge injection can be minimized by first switching off S_4 at the end of the biasing phase. The disturbance of V_{c1} is compensated by a small variation of V_{c2} when equilibrium is reached. The sampled noise is then only that produced by the opening of S_1. According to (10), the total noise voltage referred to input is then

$$V_{iN}^2 = \gamma \frac{kT}{C_2} \tag{37}$$

where γ can be slightly less than 1 in the best case [1]. The DC gain of this amplifier can be boosted by cascoding transistors T_p and T_n.

This amplifier can be easily transformed into a switched-capacitor integrator by connecting an integrating capacitor between the output node and the gate of T_n. Capacitor C_2 then becomes the switched capacitor of the integrator [18].

If the compensation of offset and 1/f noise is not required, the scheme shown in Fig. 20 can be adopted [19].

The voltages across capacitors C_1 and C_2 that are required to bias T_{p1} and T_{n1} in their active modes for $V_i \cong V_R$ (virtual ground level) are obtained during the biasing phase by means of transistors T_{p2} and T_{n2} through which flows the bias current I_0. The mismatch of pairs T_{p1}-T_{p2} and T_{n1}-T_{n2} results in an input offset, and differentiation by the sample-and-hold process only reduces the low-frequency components of noise coming from the power supply, but not the 1/f noise generated in the transistors.

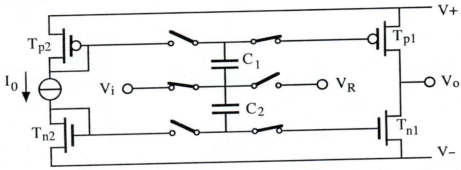

Fig. 20:   Dynamic bias of the CMOS inverter-amplifier for high frequencies.

This solution has the advantage of not disturbing the output voltage during the biasing phase. It is therefore better suited for high frequencies. Furthermore, quasi-continuous operation is possible by using a second pair of capacitors combined with an additional set of switches activated in counterphase. Inverter T_{p1}-T_{p2} then amplifies during both phases.

5.   Dynamic Comparators

5.1. Basic circuit

Another interesting dynamic analog circuit is the comparator based on the principle illustrated in Fig. 21 [20].

The transconductance amplifier A (transconductance g_m) is assumed to have a single significant pole due to the output node loaded by capacitor C_L. In the initialization phase shown in the figure, the circuit reaches equilibrium exponentially with time constant

$$\tau_u = (C + C_L)/g_m \tag{38}$$

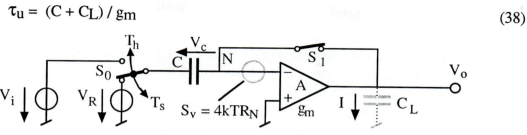

Fig. 21:   Basic dynamic comparator.

At equilibrium, the voltage V_c across capacitor C is exactly equal to the difference between the equilibrium voltage at node N (virtual ground level of the amplifier) and the reference voltage V_R.

When S_1 is then opened, no change occurs except for noise and charge injection. When S_0 is then switched to the input voltage V_i, the difference V_i-V_R is transmitted to node N, and the output voltage starts slewing at rate

$$\frac{dV_o}{dt} = - \frac{g_m}{C_L}(V_i - V_R) \tag{39}$$

The effective gain A_e of the comparator therefore depends on the time T_h allotted for comparison:

$$A_e = \frac{g_m}{C_L} T_h \tag{40}$$

It is ultimately bounded by the DC gain of the amplifier. Since the circuit can accommodate to any virtual ground level at node N, the amplifier does not need to be differential.

The precision of the comparator is limited by various independent phenomena.

Each time the switch S_1 is opened, a noise sample is stored on capacitor C. It will correspond during the comparison phase to a random input error of variance δV_{i1}^2. The application of (10) results in

$$\delta V_{i1}^2 = \frac{\gamma kT}{C + C_L} \qquad \textit{(sampled noise)} \tag{41}$$

where $\gamma = R_N g_m \tag{42}$

characterizes the noise of the amplifier and is assumed to be constant (white noise).

During each comparison phase, the output noise current of spectral density

$$S_I = 4\,kT\,R_N\,g_m^2 = 4\,\gamma\,kTg_m \qquad (43)$$

is integrated on the load capacitance C_L. It can be shown that the variance of the output noise voltage obtained after time T_h is

$$V_{oN}^2 = \frac{T_h}{2C_L^2}\,S_I \qquad (44)$$

The combination of (44), (43) and (40) gives the variance δV_{i2}^2 of the resulting equivalent input error

$$\delta V_{i2}^2 = \frac{2\gamma kT}{g_m T_h} \qquad \textit{(continuous noise)} \qquad (45)$$

The comparison of the two noise contributions

$$\delta V_{i2}^2 / \delta V_{i1}^2 = \frac{2(C + C_L)}{g_m T_h} = \frac{2\tau_u}{T_h} \qquad (46)$$

shows that the sampled noise usually dominates.

The charge Δq released onto capacitor C when the switch S_1 is blocked causes an additional input voltage error

$$\Delta V_{i3} = -\frac{\Delta q}{C} \qquad \textit{(charge injection)} \qquad (47)$$

It can be shown that the analysis of the charge injection phenomenon given in section 2.4 is not invalidated by the presence of the amplifier A, provided the transconductance g_m is smaller than the on-conductance of switch S_1. Since the load capacitance C_L must be kept as small as possible, the best strategy for reducing ΔV_{i3} is to use the compensation by half-size dummy switches while ensuring the equipartition of the total charge q by an adequate choice of the switching parameter B (Fig. 8), and/or to use fully differential circuits [11].

Any charge leaking from node N results in a contribution to the input error. In particular, the leakage current I_{leak} of the junctions associated with the switch S_1 causes an error

$$\Delta V_{i4} = \frac{I_{leak}T_h}{C} \qquad \textit{(leakage current)} \qquad (48)$$

which limits the maximum value of comparison time T_h.

A particularly deleterious effect may occur if many comparisons are carried out after the same initialization phase, as is the case in some A/D converters [11]. A large value of V_i-V_R may bootstrap the potential of node N beyond that of the substrate (or well) and forward bias a junction of the switch. An important loss of charge will result and alter all subsequent comparisons.

5.2. Multistage comparators

Multistage comparators are usually required to improve sensitivity and speed beyond that achievable with the basic single stage circuit of Fig. 21. Fig. 22 shows a possible implementation represented in its initialization phase.

During the comparison phase, the n stages realize an nth order time integration of the input step V_i-V_R, that corresponds to an equivalent gain

$$A_{en} = \frac{1}{n!}\left(\frac{g_m T_h}{C_L}\right)^n \qquad (49)$$

ultimately limited by the product of the DC gains of the n stages.

In addition to increasing the sensitivity of the comparator, the multistage structure can drastically reduce the effects of charge injection and sampled noise if the following adequate sequence of operation of the switches is used [21,22].

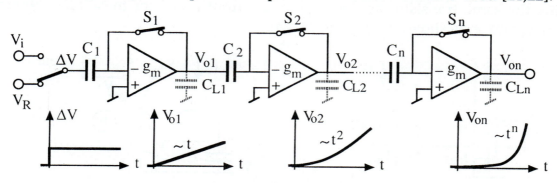

Fig. 22:   Multistage comparator.

At the end of the initialization phase shown in the figure, switch S_1 only is opened first. Charge injection and sampled noise produce an error voltage across C_1. Since S_2 is still closed, the resulting error on voltage V_{o1} is stored on C_2. The perturbations caused by S_1 are thus compensated, provided the first stage does not saturate. The same is true when switches S_2 to S_n are then sequentially opened. The only residual error seen at the input is that due to the last switch S_n, divided by the total gain of the first n-1 stages. As the last operation, the input switch S_0 is moved from V_R to V_i, which leaves no error after equilibrium is reached if the corresponding sources have a finite DC impedance.

This very efficient scheme can virtually eliminate offset. Values as low as 5µV have been reported for a differential implementation [23]. The errors accumulated during the comparison phase (continuous noise δV_{i2} and charge leakage ΔV_{i4}) are not compensated.

The DC gain of each stage must be limited to avoid saturation by the error produced by charge injection. In practical realizations, this gain limitation is obtained by loading the output by a diode-connected transistor or by a resistor. Fully differential implementations (differential input and output) are preferred to further reduce charge injection, to partially compensate charge leakage and to improve the PSRR.

6. Dynamic Current Mirrors

6.1. Principle and limitations

The main limitation on the use of current mirrors in their application to high performance circuits is the error between input and output currents due to the mismatch of the two transistors. This error can be eliminated by sequentially using the same transistor as the input and output transistor to build a dynamic current mirror as shown in Fig. 23 [24, 25].

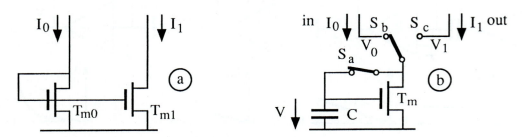

Fig. 23:   Static current mirror (a) and principle of dynamic current mirror (b).

In the first phase (shown in the figure) the single transistor T_m operates as the input device of the mirror, with its gate and drain connected to the input current source I_0. When equilibrium is reached, the value of I_0 is stored as a voltage V across capacitor C. During the second phase, T_m operates as the output device of the mirror, with its drain disconnected from its gate and connected to the output node. It sinks a current I_1 that is controlled by the unchanged gate voltage V and is thus equal to I_0.

This very simple and elegant circuit is also called a current copier [26]. The drain current of the transistor is reset to the value I_0 at each clock cycle, and therefore the low-frequency components of its noise are attenuated by differentiation as discussed in section 2.5, whereas the high frequency components of internal noise and input current are folded down by undersampling. The simple version shown in the figure suffers from various limitations [13, 25].

One limitation is due to the charge injected by switch S_a onto capacitor C. Charge injection can be limited by restricting the total charge q stored in the conducting transistor-switch S_a. This charge is linked to the on-conductance g of the switch by

$$q = gL^2/\mu \tag{50}$$

where L is the channel length of the transistor and μ the carrier mobility in the channel. Since the settling time in the first phase is increased by a reduced value of g, the reduction of charge injection must be traded off with the speed of the mirror. The required value of on-conductance g can be obtained by some additional circuitry [13].

For a given value of q, the amount of charge injection can be reduced by using dummy switches as discussed in section 2.4. It can also be compensated by a method similar to that described in section 3.2 for the compensation of offsets in amplifiers. One possible implementation is shown in Fig. 24 [26].

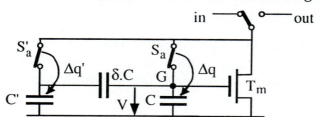

Fig. 24: Reduction of charge injection by low-sensitivity auxiliary input.

When the switch S_a is opened, it dumps a charge Δq onto C. If the circuit is allowed to settle to equilibrium, this charge is eliminated by the compensation loop through capacitor δC. When this loop is then opened, a charge $\Delta q'$ is released by switch S_a' onto C', but the resulting voltage variation at the critical node G is attenuated by the capacitive divider ratio δ.

The analysis of this circuit is the same as that of section 3.2 with $\varepsilon_i = \Delta q/C$, $\Delta V_c = \Delta q'/C'$ and $A_c/A_i = \delta$. Assuming the worst case with uncorrelated values of Δq and $\Delta q'$, the application of (34) and (33) then yields:

$$\Delta V \leq 2\sqrt{\frac{|\Delta q/C|_{max}\ |\Delta q'/C'|_{max}}{A_i}} \tag{51}$$

for $$\delta = \delta_{opt} = \sqrt{\frac{|\Delta q/C|_{max}|}{A_i\ |\Delta q'/C'|_{max}}} \tag{52}$$

where A_i is the DC gain of transistor T_m.

The effect of the variation of V on the output current I_1 can be minimized by operating T_m deeply in strong inversion in order to minimize its transconductance g_m. Further reduction of g_m/I_0 can be obtained by forcing a fraction α of I_0 to flow outside of transistor T_m by connecting a current source of value αI_0 in parallel [27].

A second important limitation to the precision of the elementary dynamic mirror of Fig. 23 is due to the variation of drain voltage V_0-V_1 from the first to the second phase of operation. This variation results in three independent unwanted effects:

A current difference I_1-I_0 due to the nonzero output conductance of the transistor.

A variation of V due to capacitive coupling from drain to gate, which results in an additional current difference I_1-I_0.

A transient output current (glitch) due to the parasitic capacitance at the drain node.

These effects can be drastically reduced by using a cascode configuration that reduces the variation of the drain voltage of T_m.

Other limitations are the leakage currents which limit the minimum clock frequency, and the settling time which limits the maximum clock frequency and therefore the maximum signal frequency which can be handled by this sampled circuit.

6.2. Examples of practical implementations

With the elementary cell shown in Fig. 23, the output current I_1 is only available in the copying phase. This restriction may be accepted in some applications but a continuous output current is often required. Continuous operation can be obtained by using two identical cells working in counterphase as shown in Fig. 25.

The self-bias cascode transistors T_{c0} and T_{c1} avoid large voltage variations at the drains of the mirror transistors T_{m0} and T_{m1} due to the difference V_0-V_1 between the input and output voltages. Drain voltage variations are limited to the small difference of the gate-source voltages of the cascode transistors due to their mismatch, and are responsible for small residual glitches on the otherwise continuous output current I_1. A reproducible precision on the order of 0.1% has been obtained experimentally with this structure [13].

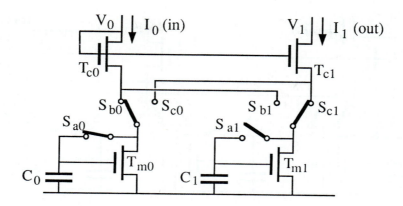

Fig. 25: Dynamic mirror for continuous operation.

This circuit can be extended to provide several copies of the input current I_0 by simply repeating the output cell [13, 25]. As shown in Fig. 26, n+1 phases are needed to provide n continuous copies. The voltage stored across the capacitor of each output cell is refreshed or updated during one phase per cycle. The cell being refreshed is replaced by cell 0; therefore n switches S_{c0} are required.

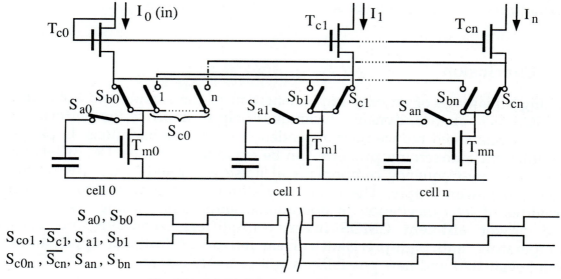

Fig. 26: Multiple output dynamic mirror.

These n continuous output currents may obviously be added to realize a 1 to n multiplying mirror. As shown in Fig. 27, the cascode transistors T_{c1} to T_{cn} can be shared by all the output cells in order to avoid the need to split S_{c0} and to reduce the number of interconnections.

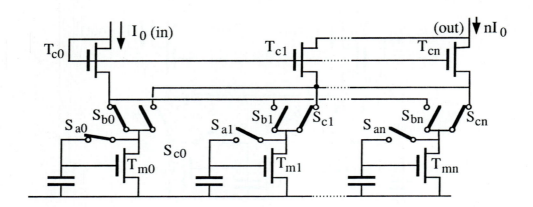

Fig. 27:   One to n multiplying dynamic current mirror.

A dividing dynamic mirror cannot be implemented in a direct way, but requires an algorithmic process to divide evenly the input current among n cells [28].

The idea of using the same transistor to replace sequentially two transistors in order to avoid the effect of mismatch in a pair can in principle be applied to implement more complicated circuits such as adjustable mirrors, normalization circuits, voltage sources or current conveyors [25].

7.   Conclusion

Taking advantage of an easy implementation of the sample-and-hold function in MOS technologies, dynamic techniques help to improve the performance of analog circuits and reduce their dependence on process variations. In particular, the inherent autozeroing property can be exploited to cancel DC offsets and to attenuate low-frequency components of noise generated in devices or coming from the power supply. The temporary storage of voltages across capacitors provides offset cancellation in operational amplifiers, biasing means for fast and power efficient CMOS inverter-amplifiers, sensitive and accurate comparators, and high-precision current mirrors. Additional applications are being evaluated and new possibilities will certainly emerge.

8.   References

[1]   E. Vittoz, "Micropower Techniques," in *Design of MOS VLSI Circuits for Telecommunications and Signal Processing*, Editors J. Franca and Y.P. Tsividis, Prentice Hall, 1991.

[2]   J.T. Tou, *Digital and Sampled-data Control Systems*, McGraw-Hill, New York, 1959.

[3]   G. Wegmann *et al*, "Charge injection in analog MOS switches," IEEE J. Solid-State Circuits, vol. SC-22, pp. 1091-1097, Dec. 1987.

[4]   E. Vittoz, "Microwatt switched capacitor circuit design," Electrocomponents Science and Technology, vol. 9, pp. 263-273, 1982.

[5]   E. Vittoz, "The design of high-performance analog circuits on digital CMOS chips," IEEE J. Solid-State Circuits, vol. SC-20, pp. 657-665, June 1985.

[6]   R.C.T. Yen, "High precision analog circuits using MOS VLSI technology," UC-Berkeley, 1983.

[7]   B.J. Sheu and C.M. Hu, "Modeling of the switch-induced error voltage on a switched capacitor," IEEE Trans. on Circuits and Syst., vol. CAS-30, pp. 911-913, Dec. 1983.

[8]   L. Bienstman and H.J. De Man, "An eight-channel 8-bit microprocessor compatible NMOS converter with programmable scaling," IEEE J. Solid-State Circuits, vol. SC-15, pp. 1051-1059, Dec. 1980.

[9]   E. Suarez *et al.*, "All-MOS charge redistribution analog-to-digital conversion techniques - part II," IEEE J. Solid-State Circuits, vol. SC-10, pp. 379-385, Dec. 1975.

[10] P. Van Peteghem, *Accuracy and Resolution of Switched Capacitor Circuits in MOS Technology*, Ph.D. Thesis, Katholieke Universiteit Leuven, June 1986.

[11] D.J. Alstott, "A precision variable-supply CMOS comparator," IEEE J. Solid-State Circuits, vol. SC-17, pp. 1080-1087, Dec. 1982.

[12] C. Enz, *High Precision CMOS Micropower Amplifiers*, Ph.D. Thesis No. 802, Swiss Federal Institute of Technology, Lausanne (EPFL), 1989.

[13] G. Wegmann, *Design and Analysis Techniques for Dynamic Current Mirrors*, Ph.D. Thesis No. 890, Swiss Federal Institute of Technology, Lausanne (EPFL), 1990.

[14] M. Degrauwe, E. Vittoz and I. Verbauwhede, "A micropower CMOS instrumentation amplifier," IEEE J. Solid-State Circuits, vol.SC-20, pp. 805-807, June 1985.

[15] J. Robert *et al.*, "Offset and clock-feedthrough compensated switched-capacitor integrators," Electronics Letters, vol. 21, pp. 941-942, Sept. 26. 1985.

[16] E. Vittoz, "Dynamic analog techniques," in *Design of MOS VLSI Circuits for Telecommunications*, editors Y. Tsividis and P. Antognetti, Prentice-Hall, 1985.

[17] F. Krummenacher, E. Vittoz, and M. Degrauwe, "Class AB CMOS amplifier for micropower SC filters," Electronics Letters, vol. 17, pp. 433-435, June 25, 1981.

[18] M. Degrauwe and F. Salchli, "A multipurpose micropower SC-filter," IEEE J. Solid-State Circuits, vol. SC-19, pp. 343-348, June 1984.

[19] S. Masuda *et al.*, "CMOS sampled differential push-pull cascode operational amplifier," Proc. ISCAS '84, p. 1211, Montreal, May 1984.

[20] Y.S. Lee *et al.*, "A 1mV MOS comparator," IEEE J. Solid-State Circ., SC-13, pp. 294-297, June 1978.

[21] R. Poujois *et al.*, "Low-level MOS transistor amplifier using storage techniques," ISSCC Dig. Tech. Papers, 1973, pp. 152-153.

[22] A.R. Hamade, "A single-chip all-MOS 8-bit A/D converter," IEEE J. Solid-State Circuits, vol. SC-14, pp. 785-791, Dec. 1978.

[23] R. Poujois and J. Borel, "A low-drift fully integrated MOSFET operational amplifier," IEEE J. Solid-State Circuits, vol. SC-13, pp. 499-503, August 1978.

[24] H. Oguey, private communication, 1978.

[25] E. Vittoz and G. Wegmann, "Dynamic current mirrors," in *Analog IC Design: the current-mode approach*, edited by C. Toumazou, F. Lidgey and D. Haigh, Peter Peregrinus, London, 1990.

[26] S.J. Daubert *et al.*, "Current copier cell," Electronics Letters, vol. 24, pp. 1560-1562, Dec. 8, 1988.

[27] W. Groeneveld *et al.*, "A self-calibration technique for monolithic high-resolution D/A converters," IEEE J. Solid-State Circuits, vol. 24, pp. 1517-1522 , Dec. 1989.

[28] J. Robert *et al.*, "Very accurate current divider," Electronics Letters, vol. 25, pp. 912-913, July 6 1989.

Chapter 5

Optical Receivers

David J.T. Heatley
British Telecom Laboratories, Martlesham Heath, IPSWICH, England, IP5 7RE

1. Introduction

Optical fibre transmission systems offer a number of important advantages over metallic cable systems, advantages which are vital to present day telecommunication networks. For example, the ~40,000GHz bandwidth of a typical optical fibre is at least 4 orders of magnitude broader than that of a good quality coaxial cable. This is clearly a major benefit with regard to the growing world wide demand for telecommunication capacity, particularly with the introduction of broadband customer services, and it accommodates future capacity upgrades using time and wavelength division multiplexing (TDM and WDM). The low insertion loss of optical fibres: typically 0.35dB/km and 0.2dB/km in the 1300 ηm and 1500 ηm windows, respectively, is a further advantage since it permits significantly longer repeater spacings on transmission routes, and in some cases obviates repeaters entirely, all of which reduces hardware costs and increases reliability. The small physical size of fibres and multi-fibre cables is a further benefit, simplifying installation and easing congestion in underground duct-ways. Indeed, the cost of manufacturing and installing optical fibre cables is now significantly less than for metal cables. Because of these advantages some 55% of the world's telecommunication traffic is now carried of over fibres [1].

The advantages of particular interest in this chapter are the greater section length and traffic capacity afforded by optical fibre systems, realised in part through the use of low noise, broadband optical receivers. This chapter examines in detail the design and performance of a broad selection of receivers, paying particular attention to such key parameters as bandwidth, signal-to-noise ratio and sensitivity. A mathematical methodology is developed for computing these performance parameters for the most commonly used receivers, following which comparisons are drawn between these receivers and other more recent designs not covered in the initial analysis.

The objectives of this chapter are principally to outline the fundamentals of receiver design and performance analysis, and to highlight the key part played by the receiver in the overall performance of the system.

2. Receiver Design

An optical receiver, in its simplest form, comprises a semiconductor photodiode followed by a low noise electronic amplifier (Figure 1). In providing gain the amplifier ensures that noise produced in the receiver dominates that produced by later circuitry, and therefore ensures that receiver noise alone determines the overall noise performance of the terminal equipment. The amplifier also acts as an impedance buffer, providing an interface between the high impedance of the photodiode and the low impedance, usually 50Ω, of the signal processing circuitry that follows.

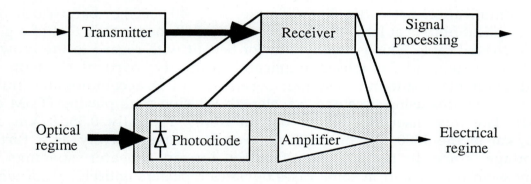

Fig. 1: Principal constituents of an optical receiver.

Receiver designs most commonly embody one of the following three configurations (Figure 2): low impedance, so called because of the low value of the photodiode bias resistor; high impedance, so called because of the high value of the bias resistor; and transimpedance, which is characterised by a feedback resistor whose value is midway between that of the previous two.

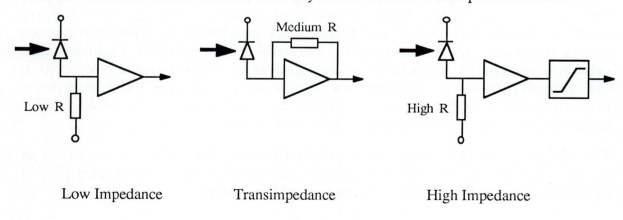

Fig. 2: Principal receiver configurations.

These three configurations each display different characteristics which are beneficial to particular applications. For example, the low impedance configuration most readily ensures that the receiver has a very broad bandwidth and dynamic range. On the other hand, the low bias resistor produces a high noise level. In contrast, the high impedance configuration produces significantly less noise, but the dynamic range of the receiver is relatively narrow and equalisation is necessary after the amplifier in order to achieve a useful bandwidth. The properties of the transimpedance configuration provide a useful compromise, giving a relatively broad bandwidth and dynamic range, as well as a relatively good noise performance.

The performance of all three configurations is analysed in detail in the sections that follow. All practical device parameters assumed in the computations are listed in Appendix A.

3. Receiver Bandwidth

The bandwidth of the opto-electronic conversion process within a photodiode is determined by the transition time and lifetime of carriers in the active region of the device structure [2]. For modern devices this bandwidth can extend to several tens of GHz [3,4]. However in practice the realisable bandwidth is generally somewhat less than this, limited by the time constant of the photodiode's own electronic impedance and the circuitry to which it is terminated. The frequency response of an optical receiver as a whole is therefore adequately described by that of its electronic circuit, which in turn can be reduced for analytical convenience to a signal equivalent circuit.

3.1. Low impedance receiver

Figure 3 shows the signal equivalent circuit of the low impedance receiver. Although greatly simplified, it is nevertheless valid for this study. R is the photodiode bias resistor; C is the total input capacitance produced by the photodiode, bias resistor and amplifier input capacitances; and R_a is the effective input resistance of the amplifier. The signal gain of the amplifier is not important at this stage and so is taken for convenience to be unity. It is often the case that the bandwidth of the amplifier exceeds that of the input C-R network, in which case the frequency response of the receiver, expressed as a power transfer function, is merely,

$$|H(f)|^2 = \frac{R^2}{1 + (\omega C R)^2} \tag{1}$$

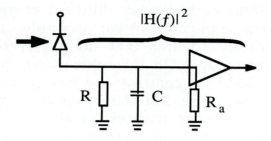

Fig. 3: Signal equivalent circuit of the low impedance receiver.

for $R_a \gg R$. For the typical values of C and R given in Appendix A, namely 0.6pF and 200Ω, respectively, (1) predicts a receiver bandwidth of 1.3GHz. This bandwidth can be increased by reducing C or R. However, C is constrained by the devices used whereas reducing R increases receiver noise (as is demonstrated later in this chapter), so in practice C is predetermined and R is chosen to give the best compromise between bandwidth and noise.

3.2. High impedance receiver

The signal equivalent circuit of the high impedance receiver (Figure 4) is essentially the same as that for the low impedance receiver (Figure 3) but with the addition of an equaliser network after the amplifier. In order to appreciate the action of the equaliser, consider first the frequency response of the receiver prior to the equaliser, as described by (1). For a C of 0.6 pF as before, but now with 1 MΩ taken to be a typical value of R in a high impedance receiver, the bandwidth is only 265 kHz, which is clearly of little practical use. The equaliser is introduced to broaden receiver bandwidth without recourse to reducing R.

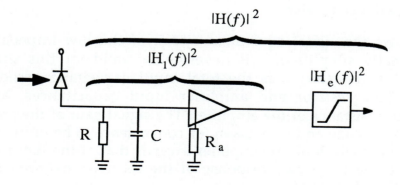

Fig. 4: Signal equivalent circuit of the high impedance receiver.

Figure 5 shows that the frequency response of the equaliser is, in its simplest form, an inverted image of the frequency response of the unequalised receiver. This gives the overall frequency response an unlimited bandwidth. In practice the response of the equaliser cannot increase with frequency unbounded as shown because this would require a circuit with unlimited gain. Practical constraints cause the frequency response of the equaliser to level-off at high frequencies (shown dotted), which in turn limits the overall bandwidth.

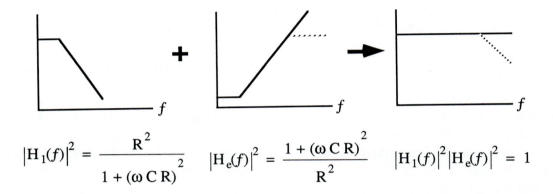

$$|H_1(f)|^2 = \frac{R^2}{1 + (\omega C R)^2} \qquad |H_e(f)|^2 = \frac{1 + (\omega C R)^2}{R^2} \qquad |H_1(f)|^2 |H_e(f)|^2 = 1$$

Fig. 5: Principle of equalisation to increase bandwidth.

The equaliser function in Figure 5 is realised by the simple R-C network in Figure 6, the transfer function of which is,

$$|H_e(f)|^2 = \left[\frac{R_i}{R_e}\right]^2 \left[\frac{1 + (\omega C_e R_e)^2}{1 + (\omega C_e R_i)^2}\right] \qquad (2)$$

for $R_e > R_i > R_0$. It then follows from (1) and (2) that the power transfer function of the equalised receiver is,

$$|H(f)|^2 = \left[\frac{R}{R_e}\right]^2 \left[\frac{R_i^2}{1 + (\omega C_e R_i)^2}\right] \qquad (3)$$

for an amplifier gain normalised again to unity for convenience. Note in (3) that the bandwidth of the high impedance receiver is independent of C and R at the input but is now determined by C_e, the equaliser capacitor, and R_i, the input resistance of the circuitry after the equaliser. C and R cannot however be ignored because they strongly influence receiver noise and sensitivity, as is shown later in this chapter.

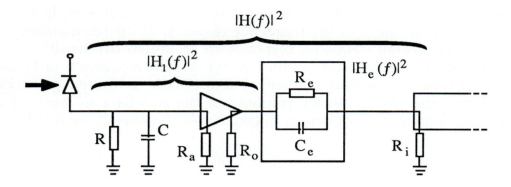

Fig. 6: Signal equivalent circuit of the high impedance receiver.

For the typical values of C_e and R_i given in Appendix A, (3) predicts the same 1.3GHz bandwidth achieved with the low impedance receiver. This bandwidth can be increased by reducing C_e or R_i. However, it may be difficult in practice to reduce R_i below the 50Ω assumed in this computation since this is the input impedance of commercial high frequency components. C_e on the other hand can be reduced but then attention must be paid to the increased insertion loss of the equaliser and hence the stronger influence that noise originating after the receiver has on overall system performance. C_e is therefore chosen in practice to give the best compromise between bandwidth and noise.

3.3. Transimpedance receiver

The signal equivalent circuit of the transimpedance receiver (Figure 7) differs from those previously in the use of a feedback resistor instead of an input bias resistor. The power transfer function of this circuit is,

$$|H(f)|^2 = \frac{R^2}{1 + \left[\omega R \left(C_f + C/A \right) \right]^2} \tag{4}$$

where amplifier gain $A \gg 1$ and C_f is the stray capacitance of the feedback circuit. Note firstly that the bandwidth of the receiver is now determined in part by the amplifier gain. This contrasts with the previous cases where gain could be ignored. Note also the importance of minimising C_f in order to approach the ideal bandwidth (i.e, with $C_f = 0$). For example, with C = 0.6pF as before and now with A = 20 (as listed in Appendix A), (4) shows that a C_f even as low as 0.03pF reduces the ideal bandwidth by a factor of 2, and C_f in practice is rarely better than 0.1pF. If C_f is indeed taken to be 0.1pF, (4) predicts that a feedback resistor of 1kΩ must be used to achieve the same 1.3GHz bandwidth as in the other designs. This compares with ~4kΩ if $C_f = 0$, which produces less noise. Nevertheless, 1kΩ is sufficiently larger than the 200Ω required by the low impedance receiver to produce a useful improvement in noise, as will be shown later.

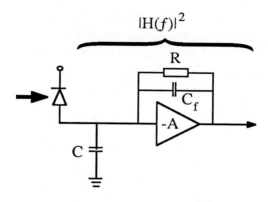

Fig. 7: Signal equivalent circuit of the transimpedance receiver.

It is clear from (4) that the bandwidth of the transimpedance receiver may be increased by in turn increasing A. However, this is only worthwhile if C_f is reduced at the same time, which in practice may be difficult or even impossible. Reducing R will also increase bandwidth but then noise becomes a consideration, so in practice R is once again chosen as the best compromise between bandwidth and noise.

At this stage it is worth noting certain key points in the above analyses of receiver bandwidth. For example, the frequency responses of the three receiver configurations is a simple 1st order filter function with a 1.3GHz bandwidth, despite the different circuit topologies and values of R. Embodied within this observation is the assumption throughout that the gain of the amplifier is constant over the bandwidth of the receiver as a whole. This is a reasonable assumption given the bandwidth of modern day devices, but it is by no means general. Indeed there are certain advantages to overall system performance in designing amplifier bandwidth close to that required from the receiver as a whole.

For the purposes of this study, and in the interests of analytical simplicity, it is adequate to continue with the assumption that amplifier gain is constant over the bandwidth of the receiver.

4. Receiver Noise

Receiver noise arises in three principal locations: the photodiode; the input bias resistor (or feedback resistor); and the amplifier. These noise sources and their spectral densities may be represented by statistically independent current and voltage generators in a noise equivalent circuit (Figure 8). Conventional circuit analysis techniques may then be employed to compute the total noise spectral density at the receiver output. For mathematical convenience, the analyses in this chapter assume all noise spectral densities to be in single sided form (i.e, positive frequencies only).

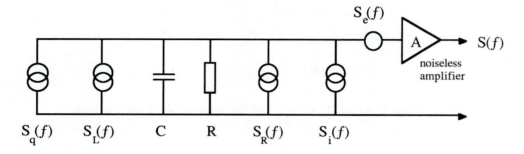

Fig. 8: Equivalent noise circuit for a generalised optical receiver.

4.1. Photodiode noise

$S_q(f)$ in Figure 8 is the spectral density of signal dependent shot (or quantum) noise produced in the photodiode by the packet nature of photonic detection, and is given by [5],

$$S_q(f) = 2 \, e \, M^2 \, F_e \, R_p \, P \tag{5}$$

where,   e is the electron charge,

M is the avalanche gain of the photodiode (M=1 for a PIN photodiode),

P is the mean optical power at the surface of the photodiode,

R_p in (5) is the responsivity of the photodiode, given by,

$$R_p = \frac{\eta \, e \, \lambda}{h \, c} \tag{6}$$

where, η is the quantum efficiency of the photodiode,

λ is the wavelength of the detected light,

h is Planck's constant,

c is the speed of light in free space.

F_e in (5) is a noise figure which, for avalanche photodiodes (APD), takes account of the excess noise produced by the gain process thus [6,7],

$$F_e = k_w M + (1 - k_w)(2 - 1/M) \qquad (7)$$

where k_w is an empirical quantity related to the ratio of the ionisation rates of holes and electrons in the avalanche region of the device. F_e tends to be cumbersome in computations, particularly when it is desirable to optimise avalanche gain M, and so is often replaced by the close approximation M^x where x is the excess noise index. (5) is then simplified to [6],

$$S_q(f) = 2 e M^{2+x} R_p P \qquad (8)$$

$S_L(f)$ in Figure 8 is the spectral density of leakage shot noise produced in the photodiode by bulk effects in the semiconductor material and surface effects around the active region, and is given in simplified form by [8],

$$S_L(f) = 2 e M^{2+x} I_{bk} + 2 e I_d \qquad (9)$$

where, I_{bk} is the mean bulk leakage current ($I_{bk}=0$ for a PIN photodiode),

I_d is the mean dark current due to surface leakage.

Other sources of noise in a photodiode (e.g, series resistor noise) generally have a negligible influence; hence the noise behaviour of a photodiode may be taken without loss of generality to be fully described by (8) and (9).

4.2. Resistor noise

$S_R(f)$ in Figure 8 is the spectral density of thermal noise produced by the input bias resistor R, and is given by,

$$S_R(f) = \frac{4 k T}{R} \qquad (10)$$

where k is Boltzmann's constant and T is temperature. Although R is shown in Figure 8 in a low/high impedance configuration, it is readily shown [9] that the noise produced by the feedback resistor in a transimpedance configuration is modelled in the same manner. (10) is therefore valid for all the receiver configurations covered in this chapter.

4.3. Amplifier noise

Amplifier noise is modelled in Figure 8 by referring all internal noise sources to two uncorrelated sources at the input with spectral densities $S_i(f)$ and $S_e(f)$. Signal amplification may then be regarded as noise free. In order to quantify these two spectral densities, consider first the 2-source noise model in Figure 9 for a single transistor.

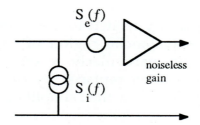

Fig. 9: Noise equivalent circuit for a single transistor.

For a field effect transistor (FET) the noise spectral densities are defined by [10],

$$S_i(f) = 2 e I_g \tag{11}$$

$$S_e(f) = \frac{4 k T \Gamma}{g_m} \left[1 + \frac{f_b}{f} \right] \tag{12}$$

where, I_g is the mean gate leakage current,

g_m is the transconductance of the device,

Γ is a factor that takes account of partial correlation between internal noise sources,

f_b is the $1/f$ noise break frequency,

For a bipolar transistor (BPT) the noise spectral densities are defined by [11],

$$S_i(f) = 2 e I_b \tag{13}$$

$$S_e(f) = 4 k T R_{bb} + 2 e \left[\frac{I_c}{g_m} \right] \tag{14}$$

where, I_b is the mean base current,

R_{bb} is the base spreading resistance,

I_c is the mean collector current.

Now consider the effect of incorporating these transistors in the typical amplifier circuits in Figure 10.

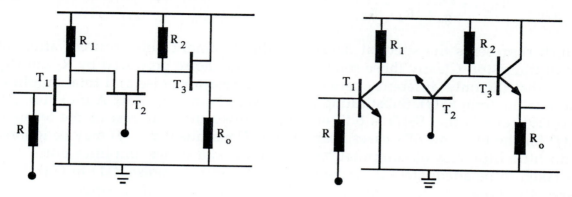

Fig. 10. Cascode amplifier circuits used in low impedance, high impedance and transimpedance receivers.

Amplifiers may utilise FET's or BPT's throughout, as shown in Figure 10, or some combination of both. In either case the circuit is often a cascode topology followed by a buffer output stage. The second transistor in each circuit forms the cascode stage which presents a low impedance to the load R_1 of the input transistor, thereby ensuring that the bandwidth of the circuit as a whole is determined largely by the CR time constant at the load R_2 of the second transistor. By careful design this time constant can be arranged to be small, and so the bandwidth is broad. The third transistor in each circuit provides buffering and impedance matching between the amplifier and the following circuitry.

Superposition can be used to reduce the noise equivalent circuits of these two designs to the 2-source model used in Figure 8 to represent the amplifier [12-14]. For the FET amplifier it can be shown that the noise spectral densities for the 2-source model are,

$$S_i(f) = 2\,e\,I_g \tag{15}$$

$$S_e(f) = \frac{4\,k\,T\,\Gamma}{g_m}\left[1 + \frac{f_b}{f}\right] + S_2(f) \tag{16}$$

while for the BPT amplifier they are,

$$S_i(f) = 2\,e\,I_b \tag{17}$$

$$S_e(f) = 4\,k\,T\,R_{bb} + 2\,e\left[\frac{I_c}{g_m}\right] + S_2(f) \tag{18}$$

where in each case,

$$S_2(f) = \frac{4\,k\,T}{g_m{}^2}\left[\frac{1}{R_1} + \frac{1}{R_2}\right] \tag{19}$$

which represents the spectral density of the dominant noise sources after the input transistor. Clearly those sources are thermal noise produced by R_1 and R_2. Note that the only difference between input transistor noise and total amplifier noise is the term $S_2(f)$. Substituting the parameters in Appendix A into (16), (18) and (19) reveals that $S_2(f)$ is 8dB and 12dB lower than the rest of the terms in $S_e(f)$ for the FET and BPT cases, respectively. Consequently $S_2(f)$ may be ignored with little little loss of computational accuracy, and hence amplifier noise may be taken to be adequately described by input transistor noise, (11) to (14). Note carefully however that this is not a general conclusion since it relates specifically to the parameter values assumed. Although these values have been chosen to be typical, they are by no means general.

4.4. 2nd stage noise

The term $S_2(f)$ can also be used to take account of 2nd stage noise arising in the circuitry that follows the receiver. To a reasonable approximation this is given by [13],

$$S_2(f) = \frac{50\,k\,T_n}{A^2}\,(F - 1) \tag{20}$$

where T_n is the reference noise temperature of 290°K, A is amplifier gain, and F is the noise figure of the circuitry after the receiver. A typical noise figure may be around 5dB, depending on the quality of the devices used. However, even if a rather poor noise figure of 10dB is assumed, (20) predicts 2nd stage noise to be about 20dB lower than input transistor noise, so once again it can be ignored.

In the analyses of receiver performance presented later in this chapter, the reader might find it a useful exercise to include the effect of $S_2(f)$ given by (19) and (20) and observe its influence on the final result. Indeed it will be found that $S_2(f)$ given by (19) in particular cannot be ignored for certain amplifier designs not considered here. A more detailed treatment of $S_2(f)$ appears in a variety of references, for example [8,13,14].

4.5. Total noise power spectral density

With the various individual noise sources in a receiver fully defined by the above formulae, a detailed study of the total noise spectral density from a receiver can now be undertaken. For convenience within the text, PIN and APD fronted receivers which incorporate an FET amplifier will henceforth be labelled PIN-FET and APD-FET respectively. Similarly, a BPT amplifier will be implicit in PIN-BPT and APD-BPT. All computations use the practical device parameters listed in Appendix A.

4.5.1. *Low impedance receiver*

From Figure 11 it is readily shown that the total noise power spectral density from a low impedance receiver is given by,

$$S(f) = [S_q(f) + S_L(f) + S_R(f) + S_i(f)] |Z(f)|^2 + S_e(f) \tag{21}$$

where,

$$|Z(f)|^2 = \frac{R^2}{1 + (\omega C R)^2} \tag{22}$$

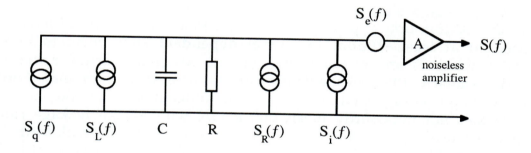

Fig. 11: Equivalent noise circuit for a low impedance optical receiver.

Figure 12 shows the spectral densities predicted by (21) for a PIN-FET configuration. Resistor and photodiode noise come under the influence of $Z(f)$ and hence display the 1st order filter function and 1.3GHz signal bandwidth of the receiver. Take careful note however that amplifier noise extends beyond this bandwidth because the computations assume amplifier gain is constant. Clearly this is unrealistic but does not affect the conclusions drawn from the results since the inclusion of a finite amplifier bandwidth would affect *all* noise sources in the same way. Comparisons between different receiver configurations are similarly unaffected for the same reasons. This applies to *all* noise spectral densities shown in this section for *all* receiver configurations.

It is clear from Figure 12 that noise produced by the bias resistor dominates all other sources by several dB over the 1.3GHz signal bandwidth of the receiver.

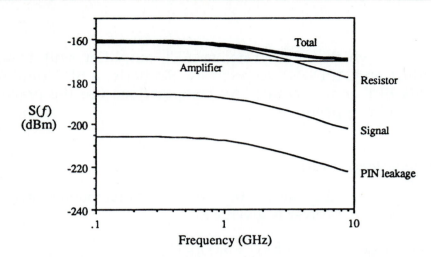

Fig. 12: Noise spectral density of the low impedance PIN-FET receiver.

Recall from (10) that resistor noise is inversely proportional to the resistor value itself, in which case Figure 12 substantiates statements in earlier sections that the total noise of a low impedance receiver increases as the bias resistor value is reduced to increase signal bandwidth. Signal dependent shot noise, and particularly leakage shot noise, are sufficiently low in level to be ignored. Indeed, the -30dBm optical power that is assumed at the PIN would have to be increased to at least -10dBm for signal dependent shot noise to become influential, which is an unusually high optical power by normal operational standards. Similarly, PIN leakage would have to increase by more than 4 orders of magnitude for leakage shot noise to become influential. Take note, however, that this does not necessarily mean that a PIN with a high leakage can be tolerated because this is sometimes an indication of device breakdown. The message from Figure 12 is merely that the noise performance of a low impedance PIN-FET receiver is unaffected by relatively high signal and PIN leakage currents.

The corresponding noise spectral densities for a low impedance PIN-BPT receiver are identical to those shown in Figure 12 with one exception: amplifier noise is about 2.5dB higher, which in turn increases total noise by about 0.5dB across the 1.3GHz signal bandwidth. Nevertheless the comments above for the PIN-FET receiver apply equally in this instance to the PIN-BPT receiver.

Consider now a low impedance APD-FET receiver. Figure 13 shows that signal dependent shot noise now dominates, and by a margin of nearly 10dB. This is purely a consequence of avalanche gain, taken for the moment to be 15. It is therefore clear that, despite the 2.5dB difference in noise levels from the FET and BPT amplifiers, the total noise spectral density in Figure 13 applies equally to APD-FET and APD-BPT receivers. Furthermore, reducing the bias resistor to increase bandwidth will, up to a point, have little affect on noise. Similarly, high APD leakage currents can once again be tolerated, although the above statement regarding the relationship between leakage and device reliability still applies.

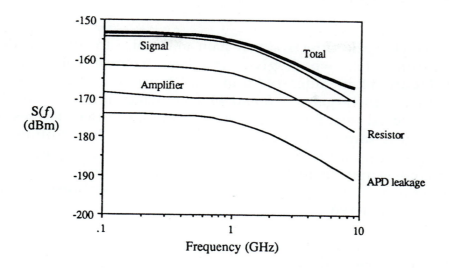

Fig. 13: Noise spectral density of the low impedance APD-FET receiver for M=15.

Figure 14 compares the total noise spectral densities for all the low impedance receivers. In order to make this comparison meaningful, the differing signal gains of the receivers have been normalised out, such that all spectral densities are relative to unity gain. For example, the 200Ω bias resistor produces a signal gain of 46dB, in which case the noise spectral densities for the PIN receivers have been reduced by 46dB. The avalanche gain of 15 for the APD receivers must also be taken into account, so their noise spectral densities have been reduced by a total of 69dB. This normalisation ensures that the noise level of the APD receivers is, in relative terms, substantially lower than that of their PIN equivalents, lower in this case by about 15dB. It is also evident from Figure 14 that, for low impedance receivers at least, the choice of amplifier transistor technology is unimportant. However, this conclusion, like many others drawn throughout this chapter, must be treated with caution since it relates to the circuit topologies and device parameters used in the computations. Although these details have been chosen to be typical, they are by no means general.

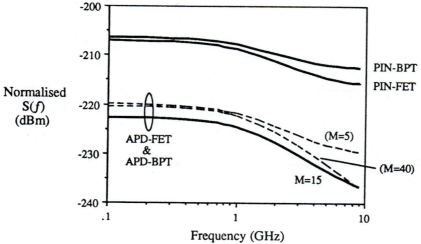

Fig 14. Comparison of noise spectral densities for the low impedance receivers.

Because avalanche gain affects signal gain and noise to different degrees, an optimum value exists which gives the best balance between these two quantities. Figure 14 includes plots of total noise spectral density for APD receivers with avalanche gains of 5 and 40, from which it may be concluded that the value of 15 used thus far is reasonably optimum in that it gives the lowest noise and largest improvement over PIN receivers. In addition, the small changes in noise that ensue from large changes in avalanche gain show that the optimum value is not critical. A wholly analytical approach to optimising avalanche gain will be developed later when signal-to-noise ratio and sensitivity are considered.

4.5.2. *High impedance receiver*

From Figure 15 it is readily shown that the total noise power spectral density from a high impedance receiver is given by,

$$S(f) = \left[S_q(f) + S_L(f) + S_R(f) + S_i(f) \right] |Z(f)|^2 |H_e(f)|^2 + S_e(f) |H_e(f)|^2 \tag{23}$$

where,

$$|Z(f)|^2 |H_e(f)|^2 = \left[\frac{C_e}{C} \right]^2 \left[\frac{R_i^2}{1 + \left(\omega C_e R_i \right)^2} \right] \tag{24}$$

and,

$$|H_e(f)|^2 = \frac{\left(\omega C_e R_i \right)^2}{1 + \left(\omega C_e R_i \right)^2} \tag{25}$$

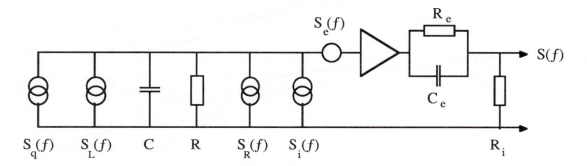

Fig. 15: Equivalent noise circuit for a high impedance optical receiver.

Figure 16 shows the spectral densities predicted by (23) for a PIN-FET configuration. The high value of the bias resistor means that its noise contribution is now negligible, in stark contrast to the low impedance case (Figure 12. The dominant source of noise over the 1.3GHz signal bandwidth of the receiver is now amplifier noise, although signal dependent shot becomes dominant towards low frequencies. Note the strong dependence of total noise on frequency, whereas total noise thus far has been largely flat. It will become evident later that this has important consequences when the relationship between bit rate and receiver sensitivity in digital transmission systems is considered.

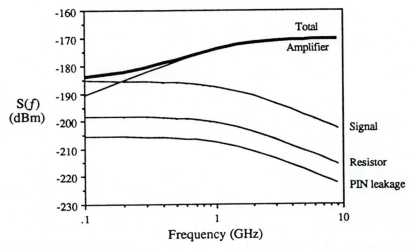

Fig. 16: Noise spectral density of the low impedance PIN-FET receiver.

BPT amplifiers are generally unsuited to high impedance configurations because of the need for an input impedance higher than that of the bias resistor: 1MΩ in this instance. Accordingly it is inappropriate to consider the noise performance of high impedance PIN-BPT and APD-BPT configurations.

Consider next a high impedance APD-FET receiver. With an avalanche gain of 15 still assumed, Figure 17 shows a similar outcome to that for the low impedance configuration (Figure 13): the total noise spectral density is determined by signal dependent shot noise, while all other sources can be ignored. Note that the total noise spectral density is no longer the strong function of frequency as occurs with the PIN configuration (Figure 16), but instead is flat over the signal bandwidth. This is a fundamental characteristic of shot noise limited operation.

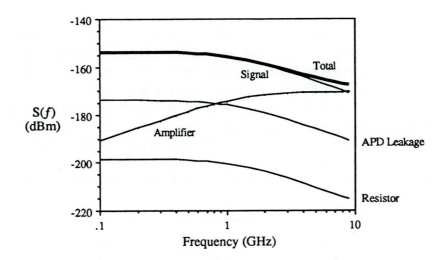

Fig. 17:   Noise spectral density of the low impedance APD-FET receiver for M=15.

Upon comparing the total noise spectral densities of the high impedance PIN-FET and APD-FET receivers, it is clear from Figure 18 that the outcome is entirely different from that observed with the low impedance configuration (Figure 14). There now exists a crossover point below which the PIN receiver gives a better noise performance, and above which the APD receiver is better. On the other hand, if an avalanche gain in the range 2-5 is assumed instead of 15, it is then reasonable to conclude from Figure 18 that the APD receiver gives the lowest noise over the 1.3GHz signal bandwidth of the receiver. Later analysis will confirm that the optimum avalanche gain in this instance is 3. The reason behind the lower optimum gain for the high impedance configuration compared with the low impedance configuration is the lower intrinsic noise level: less noise needs less avalanche gain to overcome it.

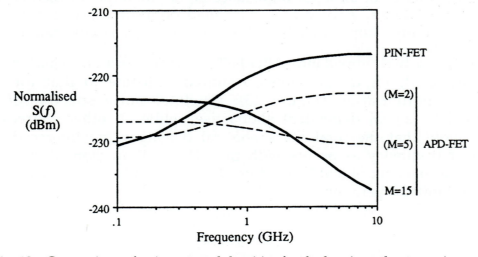

Fig. 18:   Comparison of noise spectral densities for the low impedance receivers.

4.5.3. Transimpedance receiver

From Figure 19 it is readily shown that the total noise power spectral density from a transimpedance receiver is given by,

$$S(f) = [S_q(f) + S_L(f) + S_R(f) + S_i(f)] |H(f)|^2 + S_e(f) \frac{|H(f)|^2}{|H_f(f)|^2} \qquad (26)$$

where,

$$|H(f)|^2 = \frac{R^2}{1 + [\omega R(C_f + C/A]^2} \qquad (27)$$

and,

$$|H_f(f)|^2 = \frac{R^2}{1 + [\omega R(C_f + C]^2} \qquad (28)$$

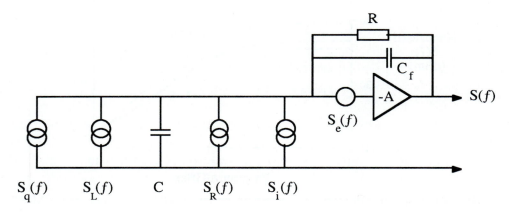

Fig. 19: Equivalent noise circuit for a transimpedance optical receiver.

Figures 20 and 21 show the spectral densities predicted by (26) for PIN-FET and APD-FET configurations, respectively. The relative disposition of the individual plots is broadly similar to that for the low impedance configuration (Figures 12 and 13), except that amplifier noise displays the same frequency dependence as observed with the high impedance configuration (Figures 16 and 17). This is due to the frequency dependence of the feedback path. For the PIN receiver this feature causes resistor *and* amplifier noise to play a part in determining the total noise spectral density, but only towards the upper end of the 1.3GHz signal bandwidth of the receiver. The total noise from the APD receiver is of course unaffected.

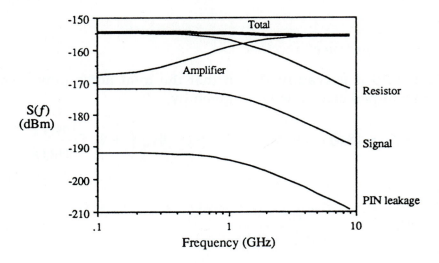

Fig. 20: Noise spectral density of the transimpedance PIN-FET receiver.

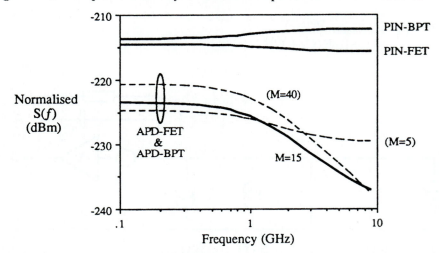

Fig. 21:   Noise spectral density of the transimpedance APD-FET receiver for M=15.

Upon recalling that the noise level of the BPT amplifier is only 2.5dB higher than that of the FET amplifier, it is clear that the total noise spectral density for a transimpedance PIN-BPT configuration will be very similar to that in Figure 20 for the PIN-FET configuration. Indeed, the two levels differ by only 2dB. Once again the APD receiver is unaffected: transimpedance APD-FET and APD-BPT configurations give identical total noise spectral densities.

A comparison of the total noise spectral densities of the transimpedance receivers reveals that the APD receivers give a substantially better noise performance than the PIN receivers (Figure 22), as was observed for the low impedance receivers (Figure 14). However, it is evident that the optimum avalanche gain is now close to 5, because of the lower intrinsic noise level of the transimpedance configuration, although this value as before is not critical. The choice of transistor technology in the APD receivers is again unimportant. However, a small but nevertheless useful noise advantage is obtained with FET technology in the PIN receiver.

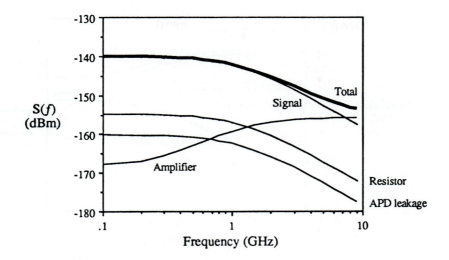

Fig. 22: Comparison of noise spectral densities for the transimpedance receivers.

4.5.4. *Comparison of all receiver configurations*

In the analysis of receiver noise thus far, comparisons have been drawn for a particular receiver configuration, the variables within each comparison being the type of photodiode and the transistor technology in the amplifier. This section draws these comparisons together to examine the relative noise levels of all the receiver configurations. In order to conduct this meaningfully, all noise spectral densities are normalised as before.

Consider first PIN receivers. Figure 23 substantiates earlier statements that the low impedance configuration gives the poorest noise performance, the high impedance configuration gives the best noise performance, and the transimpedance configuration falls approximately mid-way. Note the very broad spread in noise levels: about 10dB at 1GHz, increasing to nearly 25dB at 100MHz. Note also the different shapes of the spectral densities. It will be shown later that this gives each receiver a different relationship between bit rate and sensitivity in a digital transmission system. The noise spectral densities converge at high frequencies because amplifier noise is then dominant: one point of convergence for each transistor technology. This serves to illustrate the importance of normalising out signal gain, without which this important convergent behaviour would not be observed.

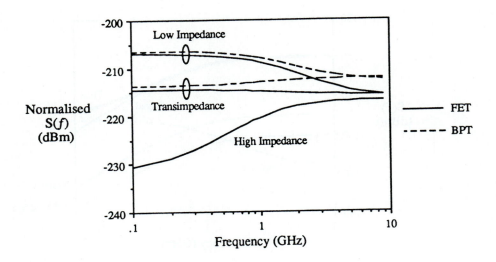

Fig. 23:   Comparison of noise spectral densities for the PIN receivers.

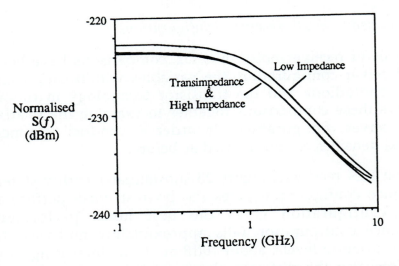

Fig. 24:   Comparison of noise spectral densities for the APD receivers (M=15).

Figure 24 draws the corresponding comparisons for all the APD receivers. Because signal dependent shot noise dominates in each case, it is to be expected that all configurations and both transistor technologies give virtually the same noise spectral densities. The small noise penalty associated with the low impedance configuration is due to the marginally stronger influence of receiver noise. There is therefore little noise advantage in using a high impedance or transimpedance configuration over a low impedance configuration for an APD. This contrasts strongly with the PIN case in Figure 23.

5. Signal-to-Noise Ratio

With the noise spectral density at the output of the various receivers now fully defined, it is a simple task to compute the signal-to-noise ratio (SNR). However, a further mathematical step must first be taken. Receiver noise has been analysed thus far by determining its power spectral density at the output. It is often convenient in analyses of SNR (and receiver sensitivity) to refer the total noise spectral density to the input of the receiver, whereupon the signal portion of the SNR is given directly by the signal photocurrent. For the relatively simple receiver designs considered thus far in this study, it is arguable that such a simplification is unnecessary; nevertheless the following text adopts this convention in order to develop a generalised analysis.

The SNR is defined for all receivers by,

$$SNR = \frac{(m_o\, M\, R_p\, P)^2}{2\displaystyle\int_0^B S(f)\, df} \tag{29}$$

where, m_o is the optical modulation index,

M is the avalanche gain of the photodiode (M=1 for a PIN),

R_p is the responsivity of the photodiode,

P is the mean detected optical power at the surface of the photodiode,

B is the bandwidth of the signal channel from the receiver,

$S(f)$ is the total noise spectral density of the receiver referred to its input.

5.1. SNR from the low impedance receiver

It follows from (21) that the total noise spectral density referred to the input of the receiver is,

$$S(f) = [S_q(f) + S_L(f) + S_R(f) + S_i(f)] + \frac{S_e(f)}{|Z(f)|^2} \tag{30}$$

where as before,

$$|Z(f)|^2 = \frac{R^2}{1 + (\omega\, C\, R)^2} \tag{31}$$

Upon completing the integral in (29) with (30) invoked, the SNR from the low impedance receiver is in full,

$$SNR = \frac{(m_o \, M \, R_p \, P)^2}{2\left[\left\{2eM^{2+x} \, (R_p P + I_{bk}) + 2eI_d + \frac{4kT}{R} + I\right\} B + E\right]} \tag{32}$$

where for an FET amplifier,

$$I = 2eI_g \tag{33}$$

$$E = \frac{4kT\Gamma}{g_m}\left[\frac{B}{R^2} + \frac{(2\pi C)^2}{3} B^3 + \frac{f_b}{R^2} \ln\left[\frac{B}{b}\right] + \frac{(2\pi C)^2}{2} f_b \, B^2\right] \tag{34}$$

and for a BPT amplifier,

$$I = 2eI_b \tag{35}$$

$$E = \left[4kTR_{bb} + \frac{2eI_c}{g_{m^2}}\right]\left[\frac{B}{R^2} + \frac{(2\pi C)^2}{3} B^3\right] \tag{36}$$

The term b in (34) is a non-zero lower limit to the integration bandwidth to avoid $1/f$ noise increasing to infinity (see Appendix A). Figure 25 shows the outcome of substituting the practical parameters in Appendix A into (32). Note that the channel bandwidth B is taken to be 1GHz throughout this study. It is clear that the APD receivers give a substantially better SNR than the PIN receivers, as would be expected from the earlier comparisons of noise levels, but only for received optical powers below -13dBm. Above this level the PIN receivers give a better SNR. The reason for this behaviour rests wholly with the assumption of a fixed avalanche gain of 15, which Figure 15 showed to be reasonably optimum for a received power of -30dBm. However, recalling from (8) that APD noise increases with avalanche gain and optical power, it is reasonable to expect a lower optimum gain at higher optical powers. This can be derived analytically by differentiating (32) with respect to avalanche gain M and equating the result to zero when M equals the optimum value. The resulting formula is,

$$M_{opt} = \left[\frac{N}{e \, x \, B \, (R_p \, P + I_{bk})}\right]^{\frac{1}{2+x}} \tag{37}$$

where,

$$N = \left[2eI_d + \frac{4kT}{R} + I\right] B + E \tag{38}$$

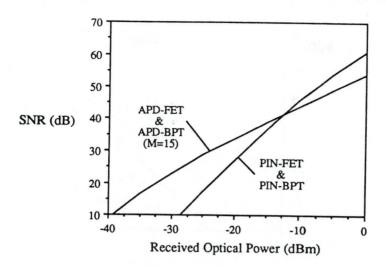

Fig. 25: Comparison of SNR's for the low impedance receivers (M=15).

It will soon become evident that (32) and (37) are valid for low impedance, high impedance and transimpedance receivers: it is merely the terms within E that differ for each.

The optimum avalanche gain predicted by (37) is plotted in Figure 26 which confirms that 15 is reasonably optimum at -30dBm (precisely optimum at -32dBm), but the optimum gain reduces as optical power increases, converging towards unity. With (37) substituted into (32) Figure 27 shows that the APD receivers always give a better SNR that the PIN receivers provided the avalanche gain is optimised to the received power level. This advantage diminishes to zero at powers above -6dBm as the optimum gain approaches the unity gain of the PIN, at which point both types of receiver become equally dominated by shot noise. In practice APD's generally cease to function correctly as their gain approaches unity. In such circumstances PIN receivers are preferred since they also come with the benefits of lower cost and simplicity in use.

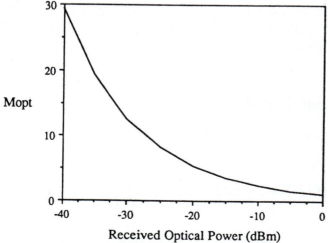

Fig. 26: Optimum APD gain for the low impedance receivers.

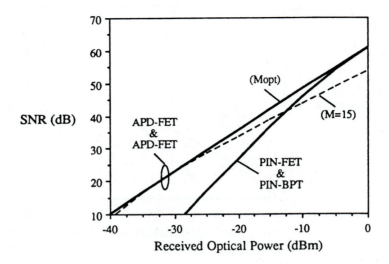

Fig. 27: Comparison of SNR's for the low impedance receivers (optimised M).

The optimised and non-optimised plots in Figure 27 for the APD receivers coincide at a received power of -32dBm because, as is clear from Figure 26, the fixed avalanche gain of 15 is also the optimum value. Take note however of the small size of the SNR penalty to either side of this point. Figure 26 shows that a received power of -40dBm produces an optimum gain of nearly 30, but despite this being double the fixed value assumed, the SNR penalty is only 1dB. Similarly, at -20dBm the optimum gain is 5, but despite this being one-third the fixed value assumed, only produces a 2dB penalty. This serves to emphasise a point made earlier that the optimum avalanche gain is not a critical parameter.

Note that Figure 27 gives some insight to the sensitivity of the receiver. For example, if the target SNR for a particular system application is 30dB, the sensitivity is the corresponding received optical power on the horizontal axis: -25dBm for the APD receivers and -18dBm for the PIN receivers. In this instance it can therefore be concluded that low impedance APD receivers give a sensitivity advantage of 7dB over low impedance PIN receivers, which in a system can either be translated to a longer section length or a better fidelity of signal, or both. Sensitivity is covered in detail in Section 6.

5.2. SNR from the high impedance receiver

It follows from (23) that the total noise spectral density referred to the input of the receiver is,

$$S(f) = [S_q(f) + S_L(f) + S_R(f) + S_i(f)] + \frac{S_e(f)}{|Z(f)|^2} \qquad (39)$$

where as before,

$$|Z(f)|^2 = \frac{R^2}{1 + (\omega C R)^2} \qquad (40)$$

These formulae are identical to those for the low impedance receiver, (30) and (31), (except that only FET amplifiers are relevant in this instance since the lower input impedance of BPT amplifiers is inappropriate to the high impedance configuration). It follows therefore that the SNR and optimum avalanche gain formulae, (32) and (37), are valid for the low impedance and high impedance receivers. There are however certain differences which only become apparent when the practical parameters in Appendix A are introduced. For example, the 1st and 3rd bracketted terms in (34) dominate for the low impedance receiver whereas the 2nd and 4th terms dominate for the high impedance receiver. The effect of this is a substantially lower noise level for the high impedance receiver (Figure 23) which, from (37), dictates a lower optimum avalanche gain (Figure 28). This in turn means that the APD receiver only begins to show an SNR advantage over the PIN receiver at low received power levels, below -20dBm (Figure 29). This, coupled with the inability of most APD's to function correctly at gains close to unity, means that practical high impedance receivers most commonly use PIN's.

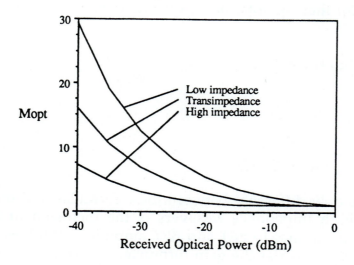

Fig. 28: Comparison of optimum APD gain.

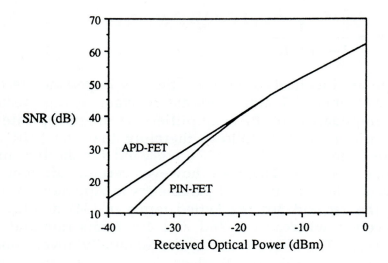

Fig. 29:   Comparison of SNR's for the high impedance receivers (optimised M).

5.3. SNR from the transimpedance receiver

It follows from (26) that the total noise spectral density referred to the input of the receiver is,

$$S(f) = [S_q(f) + S_L(f) + S_R(f) + S_i(f)] + \frac{S_e(f)}{|H_f(f)|^2} \tag{41}$$

where as before,

$$|H_f(f)|^2 = \frac{R^2}{1 + [\omega R(C_f + C]^2} \tag{42}$$

It is clear that the only difference between these formulae and those for the low impedance and high impedance receivers is the inclusion of feedback stray capacitance C_f. This applies to FET and BPT amplifiers. It follows therefore that the SNR and optimum avalanche gain formulae, (32) and (37), are valid for the transimpedance receiver but with C replaced by $(C + C_f)$ throughout. Because the noise level of a transimpedance receiver falls midway between the low impedance and high impedance receivers, its optimum avalanche gain behaves similarly (Figure 28). This in turn means that the APD receivers give a useful SNR advantage over the PIN receivers for received power levels of practical interest (Figure 30).

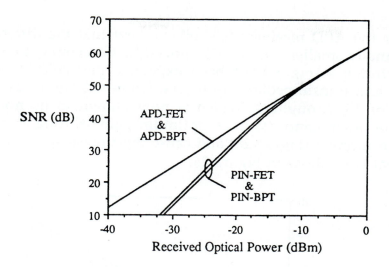

Fig. 30: Comparison of SNR's for the transimpedance receivers (optimised M).

5.4. Comparison of SNR's

Consider now a comparison of the SNR's given by the three receiver configurations. For the PIN receivers Figure 31 shows that the high impedance receiver gives a good SNR advantage over the transimpedance receiver, and similarly so for it over the low impedance receiver. Expressed another way, the high impedance receiver displays the highest sensitivity, followed in descending order by the transimpedance and low impedance receivers. Furthermore, there is little, if anything, to choose between FET and BPT amplifiers. This is all of course entirely consistent with earlier observations of noise levels (Figure 23).

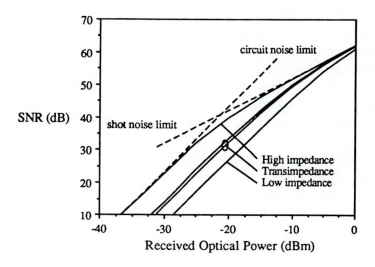

Fig. 31: Comparison of SNR's for PIN receivers.

The outcome for the APD receivers is similar except that the difference in the SNR's is substantially smaller (Figure 32). Indeed, with recourse to Figure 24 an even smaller difference might have been expected, particularly between the high impedance and transimpedance receivers. However, Figure 24 assumes an avalanche gain of 15 throughout, which is now known to be non-optimum. Figure 32 on the other hand assumes optimum gains throughout which are different for each receiver (Figure 28) and cause their noise spectral densities to deviate somewhat from those in Figure 24.

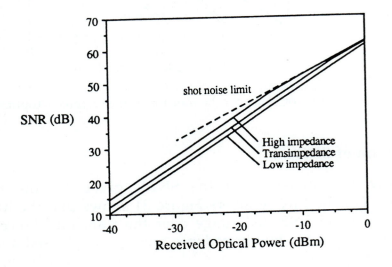

Fig. 32: Comparison of SNR's for ADP receivers.

Note that the plots in Figures 31 and 32 comprise two distinct regions. This is particularly apparent in the case of the PIN receivers where the SNR at the received power levels usually experienced in practice is limited by circuit noise (i.e, bias resistor, amplifier, etc). Under these conditions the plots have a gradient of 2, reducing to unity at high power levels through the dominance of signal dependent shot noise. In contrast, the avalanche gain in the APD receivers ensures that circuit noise and shot noise are experienced in approximately equal proportions at operational received power levels. This gives the plots a gradient midway between the circuit and shot noise limits, only reaching a unity gradient towards high power levels where once again signal dependent shot noise becomes dominant. Note carefully that at high received power levels *all* plots converge to the same point for PIN and APD receivers alike, a fundamental feature of shot noise limited detection.

6. Receiver Sensitivity

Clearly the SNR from a receiver is a critical parameter in designing a transmission system and ensuring its correct operation. Indeed, the design issue that has been addressed in the analysis thus far is, "For a known optical power at the receiver, what is the resulting SNR ?" in practice, however, the system designer is more concerned with the converse question, "For a known target SNR, what must the optical power at the receiver be ?" for digital systems the SNR translates to a bit error rate (BER) in which case the question becomes, "For a known target BER, what must the optical power at the receiver be ?" the answer in both cases is given by the receiver sensitivity which, by definition, is the minimum detected optical power at the receiver which enables the transmitted message to be recovered with a fidelity equal to the target figure, be that an SNR or a BER. A detected power lower than the sensitivity level produces an inadequate fidelity (i.e, a poor SNR or excessive bit errors), whereas a power greater than the sensitivity level guarantees system operation at the target performance with a power margin.

6.1. Analogue transmission system

For the basic analogue system in Figure 33, an expression to compute receiver sensitivity P is readily derived from (32) by rearranging to give,

$$P = \left[\frac{2\,e\,M^x\,B\,SNR}{R_p\,m_o^2}\right]\left[1 + \sqrt{1 + \frac{m_o^2\,(N + 2\,e\,M^{2+x}\,B\,I_{bk})}{2\left(e\,M^{1+x}\,B\right)^2\,SNR}}\right] \qquad (43)$$

where N and the terms within it are given by (38) and (33) to (36). This is valid for all the receivers considered thus far, taking care to replace C with $(C + C_f)$ for the transimpedance configuration. Differentiating (43) with respect to M and equating to zero gives the optimum avalanche gain as,

$$M_{opt} = \left[\frac{m_o^2\,N}{2x(2+x)(eB)^2\,SNR}\right]^{\frac{1}{2(1+x)}}\left[1 - \frac{xeBM_{opt}^{2+x}I_{bk}}{N}\right]^{\frac{1}{1+x}} \qquad (44)$$

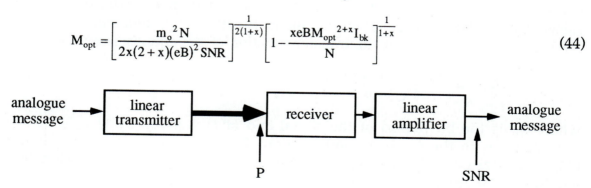

Fig. 33: Analogue transmission system.

The presence of M_{opt} within the argument of (44) renders an analytical solution intractable. However, recalling earlier observations that M_{opt} is not a critical value, (44) can be simplified to,

$$M_{opt} = \left[\frac{m_o{}^2\, N}{2x(2+x)\, (e\, B)^2\, SNR}\right]^{\frac{1}{2(1+x)}} \tag{45}$$

while incurring a negligible error in the sensitivity computed from (43). As confirmation of this, the predictions of (43) with (45) invoked agree precisely with the results presented in Figures 31 and 32, the key difference being that the horizontal axis now charts receiver sensitivity while the vertical axis charts target SNR.

6.2. Digital transmission system

Consider now the sensitivity of the same receivers in the digital transmission system in Figure 34.

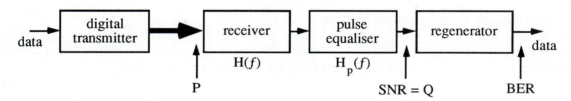

Fig. 34: Digital transmission system.

The equaliser after the receiver gives the pulses at the regenerator input a shape that maximises their tolerance to noise, inter-symbol interference (ISI) and jitter. This in turn minimises the BER delivered by the regenerator. A pulse shape that satisfies these criteria is one whose signal spectral density is a 100% raised cosine. For a receiver with transfer function $H(f)$ detecting rectangular NRZ pulses, it follows that the transfer function of the equaliser must be,

$$H_p(f) = \frac{\pi\, f\, T_b\, [1 + \cos\, (\pi\, f\, T_b)]}{2\, H(f)\, \sin\, (\pi\, f\, T_b)} \tag{46}$$

where T_b is the bit period of the signal and $H(f)$ is given by (1), (3) or (4) in accordance with the receiver configuration. The frequency response defined by (46) is illustrated in Figure 35, from which it is clear that the equivalent noise bandwidth for rectangular filtering is close to $1/2T_b$.

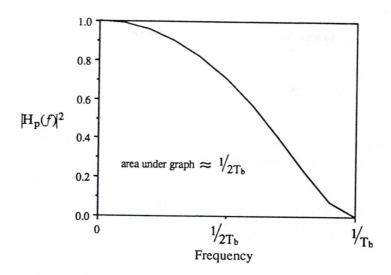

Fig. 35: Frequency response of pulse equaliser.

It is worth noting that although 100% raised cosine equalisation maximises tolerance to ISI and jitter, it may not maximise receiver sensitivity. Theory predicts that receiver sensitivity is maximised from a noise point of view with an equaliser designed as a matched filter [12,15], but then tolerance to ISI and jitter is reduced. However, given that the difference in sensitivity between the two approaches is minor (typically <1dB) and that the 100% raised cosine equaliser is the easier of the two to approximate with practical networks, it is reasonable to assume 100% raised cosine equalisation throughout this study.

The "digital" SNR at the output of the equaliser is defined slightly differently from the analogue case in (29). Commonly labelled Q, it is the ratio of peak signal amplitude to total RMS noise; hence,

$$Q = \frac{m_o \, M \, R_P \, P}{\sqrt{\displaystyle\int_0^{1/T_b} S(f) \, |H_p(f)|^2 \, df}} \tag{47}$$

where $S(f)$ as before is the total noise spectral density of the receiver referred to its input, given by (30), (39) or (41) in accordance with the receiver configuration. Q translates to a BER through the well-known Gaussian error probability function [12,16]. Telecommunication applications generally require a target BER of 1 in 10^9 or better, which corresponds to a Q of 6 or higher. The minimum target of Q = 6 is assumed throughout this study.

Clearly the integral in (47) is more complex than that encountered in the analogue case. However, by assuming $H_p(f)$ to be rectangular with a bandwidth equal to its equivalent noise bandwidth (Figure 35), the integral can be simplified with little loss of generality to give for Q,

$$Q = \frac{m_o \, M \, R_P \, P}{\sqrt{\displaystyle\int_0^{1/2T_b} S(f) \, df}} \tag{48}$$

Following the same steps behind the derivation of (43), it is then readily shown from (48) that the receiver sensitivity P is,

$$P = \left[\frac{e \, M^x \, Q^2}{2 \, T_b \, R_p \, m_o^2}\right] \left[1 + \sqrt{1 + \frac{(2 \, T_b \, m_o)^2 \left(N + \dfrac{e \, M^{2+x} \, I_{bk}}{T_b}\right)}{(e \, M^{1+x} \, Q)^2}}\right] \tag{49}$$

where,

$$N = \left[2 e I_d + \frac{4kT}{R} + I\right]\frac{1}{2T_b} + E \tag{50}$$

and for an FET amplifier,

$$I = 2 e I_g \tag{51}$$

$$E = \frac{4kT\Gamma}{g_m}\left[\frac{1}{2 \, T_b \, R^2} + \frac{(2\pi C)^2}{24 \, T_b^3} + \frac{f_b}{R^2}\ln\left[\frac{1}{2 \, T_b \, b}\right] + \frac{(2\pi C)^2}{8 \, T_b^2} f_b\right] \tag{52}$$

and for a BPT amplifier,

$$I = 2 e I_b \tag{53}$$

$$E = \left[4kTR_{bb} + \frac{2eI_c}{g_m^2}\right]\left[\frac{1}{2 \, T_b \, R^2} + \frac{(2\pi C)^2}{24 \, T_b^3}\right] \tag{54}$$

(49) is valid for all the receivers considered thus far, taking care to replace C with $(C+C_f)$ for the transimpedance configuration. By invoking the same simplifications implicit in (45), it follows from (49) that the optimum avalanche gain is now,

$$M_{opt} = \left[\frac{4 \, (T_b \, m_o)^2 \, N}{x(2+x) \, (e \, Q)^2}\right]^{\frac{1}{2(1+x)}} \tag{55}$$

Figure 36 charts the receiver sensitivity predicted by (49) with (55) invoked. The numbering on the vertical axis is such that sensitivity improves towards the origin. The plots do not extend beyond 2Gbit/s because this is the maximum practical bit rate that can be accommodated within the 1.3GHz bandwidth of the receivers.

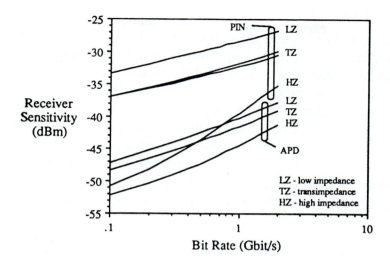

Fig. 36:   Receiver sensitivity for digital transmission system.

Consider first the PIN receivers. It is clear from Figure 36 that the high impedance receiver gives the greatest sensitivity, followed by the transimpedance receiver and then the low impedance receiver. This is of course entirely consistent with observations made earlier regarding SNR (Figure 31). In each case sensitivity degrades towards high bit rates because of the increasing noise bandwidth. Note, however, that the sensitivity of the high impedance receiver degrades more rapidly than the others. This is a direct consequence of the differences observed earlier between their noise spectral densities. The nearly flat spectral densities from the low impedance and transimpedance receivers means that their sensitivities reduce by about 1.5dB for every doubling of bit rate. In contrast, the predominantly f^2 spectral density from the high impedance receiver causes its sensitivity to reduce more rapidly, at between 3 and 4dB for every doubling of bit rate.

The dominance of shot noise in the APD receivers ensures that their sensitivities all reduce with bit rate to the same degree: approximately 1.5dB for every doubling of bit rate. Furthermore, their avalanche gain ensures a sensitivity advantage over their PIN counterparts, the magnitude of the advantage increasing with the intrinsic noise level of the receiver because of the correspondingly higher optimum avalanche gain. However, it is clear from Figure 36 that APD receivers do not necessarily guarantee a sensitivity advantage over all PIN receivers. The f^2 noise spectral density of the high impedance PIN receiver produces a better sensitivity than the low impedance and transimpedance APD receivers at bit rates below about 400Mbit/s. Indeed, even at 1Gbit/s, it is arguable that the few dB sensitivity penalty of the high impedance PIN receiver is a small price to pay for its greater simplicity and potentially lower cost. On the other hand, its dynamic range will soon be shown to be the poorest, and this may limit its scope of applications.

Lastly, Figure 36 confirms yet again that FET and BPT amplifiers give virtually identical sensitivities in the low impedance and transimpedance PIN receivers through the dominance of resistor noise. Similarly, the dominance of shot noise in APD receivers ensures that FET and BPT amplifiers give identical sensitivities. It can therefore be concluded that the choice of transistor technology in low impedance and transimpedance receivers is unimportant, regardless of whether the photodiode is a PIN or an APD. This is not necessarily a general conclusion and the reader is cautioned to take note of this. Neither has it any relevance to the high impedance receivers since, as was explained earlier, they can only be realised with FET technology.

Personick Analysis

It is important to be aware of the accuracy of the above analytical technique in view of the simplifications it incorporates. A more precise analysis of receiver sensitivity for digital systems has been developed by Personick [17] in which the true frequency response of the pulse equaliser (assumed again to be 100% raised cosine) is taken into account. The receiver sensitivity P is now given by,

$$P = \frac{1}{2R_p}\left[\frac{Q}{M}\right]^2\left[\left(\frac{eM^{2+x}I_2}{T_b}\right) + \sqrt{\left(\frac{eM^{2+x}I_2}{T_b}\right)^2 + 2Z\left(\frac{M}{Q}\right)^2}\right] \tag{56}$$

where for an FET amplifier,

$$Z = \frac{I_2}{T_b}\left[2eM^{2+x}I_{bk} + 2eI_d + \frac{4kT}{R} + 2eI_g\right]$$

$$+ \frac{I_3}{T_b{}^3}(2\pi C)^2\frac{4kT\Gamma}{g_m} + \frac{I_4}{T_b{}^2}(2\pi C)^2\frac{4kT\Gamma f_b}{g_m} \tag{57}$$

and for a BPT amplifier,

$$Z = \frac{I_2}{T_b}\left[2eM^{2+x}I_{bk} + 2eI_d + \frac{4kT}{R} + 2eI_b\right]$$

$$+ \frac{I_3}{T_b{}^3}(2\pi C)^2\left[4kTR_{bb} + 2e\frac{I_c}{g_m{}^2}\right] \tag{58}$$

I_2, I_3 and I_4 are normalised integrals, often referred to as Personick integrals, whose solutions are frequency independent and hence allow receiver sensitivity to be computed for any bit rate. By numerical integration these solutions are $I_2 = 1.127$, $I_3 = 0.174$ and $I_4 = 0.367$. The optimum avalanche gain can be determined from (55).

The receiver sensitivity predicted by (56) is charted in Figure 37 from which it is clear that the discrepancy between the simplified analysis conducted in this chapter (Figure 36) and the Personick analysis is small, typically less than 1dB. The high impedance PIN-FET receiver displays a slightly larger discrepancy than the others (~2dB) purely because the simplification in (48) assumes a flat noise spectral density from the receiver and hence does not take full account of the f^2 spectral density from the high impedance PIN-FET receiver. Nevertheless, the discrepancy displayed by all receivers is sufficiently small to vindicate the use of either method of analysis. The two analyses are covered here because the simplified approach follows naturally from the noise and SNR analysis conducted earlier in this chapter, whereas the Personick analysis is widely used by researchers in the field.

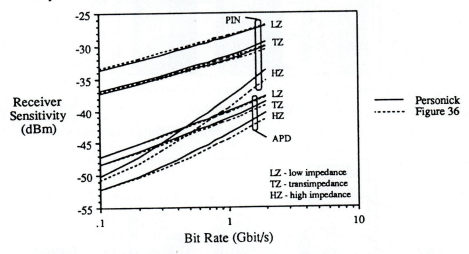

Fig. 37: Receiver sensitivity - comparison of Personick and Fig. 36 analyses.

It should be noted that the sensitivity predictions in Figures 36 and 37 make no allowance for practical system limitations; for example, transmitter bandwidth, fibre dispersion, non-ideal equalisation, etc. As a consequence measured sensitivities will generally be a few dB poorer than those predicted. However, the relative dispositions of the various plots will be largely preserved.

6.3. Transmission distances

The importance of receiver sensitivity can be highlighted by examining its influence on transmission distance. Consider first a single section system (as in Figure 34) with a launch power of 0dBm (1mW) from the transmitter and a 5dB power margin at the receiver. The receiver sensitivities shown in Figures 36 and 37, minus 5dB, then equate to the transmission loss budget of the system, as summarised in Figures 38 and 39 for 140Mbit/s and 1.5Gbit/s respectively. With an optical carrier of 1.55µm wavelength, fibre+splice loss is known in practice to be typically 0.25dB/km, whereupon transmission distance can be computed from the loss budget as shown. At either bit rate it is clear that substantial transmission distances are feasible for all receiver types, typically in excess of 100km for a single section, which compares dramatically with the ~2-4km section length generally required by coaxial cable systems.

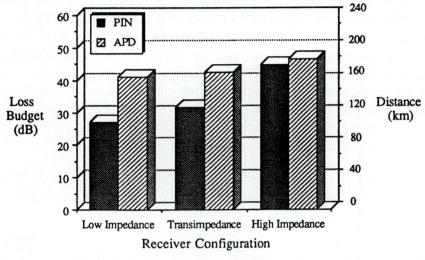

Fig. 38:   Transmission loss budget and distance at 140Mbit/s (1.55 μm wavelength).

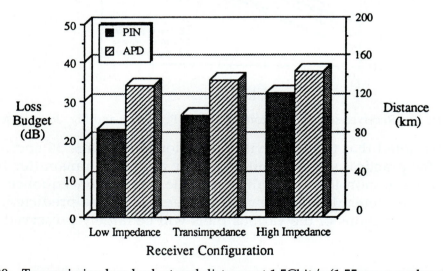

Fig. 39:   Transmission loss budget and distance at 1.5Gbit/s (1.55 μm wavelength).

Note however that Figures 38 and 39 do not take account of fibre dispersion which, for conventional single-mode fibre, is typically 17ps/ηm.km at the 1.55μm wavelength assumed. If the linewidth of the optical carrier is taken to be 50GHz (typical of semiconductor lasers which undergo wavelength chirp when directly modulated [18]) and the total dispersion experienced must not exceed 20% of the bit period, it is readily computed that the distance limits dictated solely by dispersion are ~200km for 140Mbit/s, reducing to only ~20km for

1.5Gbit/s. On the other hand, Gbit capacity systems tend to use externally modulated coherent lasers with linewidths potentially less then a few MHz [19,20], in which case dispersion is determined by the spectral width of the data signal, which for 1.5Gbit/s improves the distance limit to ~200km. Dispersion shifted fibre may be used to realise a similar improvement but then fibre loss is marginally greater [21]. Either way, the distances in Figures 38 and 39, being less than those set by dispersion, show that transmission is predominantly loss limited, and hence they may be taken to be realisable distances in practice.

Of significance in this context is the fact that many of the world's telecommunication networks (particularly those in Europe) utilise power-fed surface buildings every 30km or less; for example, switching centres, repeater stations, etc. It follows therefore from Figures 38 and 39 that single section (i.e, unrepeatered) transmission is achievable throughout the majority of these networks with optical technology, the cable routes being entirely passive. Clearly this impacts significantly on installation and running costs, as well as reliability.

Transmission over multiple sections in tandem can of course realise significantly greater spans than those in Figures 38 and 39. Indeed, installed transoceanic undersea systems are presently operating over spans of 6500km [22,23]. The regenerator linking each section (Figure 40) eradicates any distortion on pulse shape encountered during transmission up to that point and restores the data signal to its original amplitude prior to onward transmission. A received power margin at each regenerator ensures that the accumulated BER at the destination terminal still meets the target figure with a suitable margin.

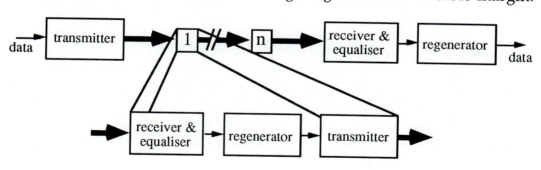

Fig. 40: Multi-section digital transmission system.

Note that at a wavelength of 1.31μm fibre dispersion is virtually zero which ensures that transmission is loss limited for laser linewidths of interest. However fibre+splice loss is typically 0.4dB/km at this wavelength, in which case the single section distances shown in Figures 38 and 39 for 1.55μm are nearly halved. Nevertheless, the distances that can be spanned with multiple sections are in principle the same for either wavelength.

Of course, many system applications do not require great distances to be spanned; for example city centre and local area networks where distances seldom exceed ~10km. In such applications the high sensitivity of optical receivers may be exploited to overcome the insertion loss of optical networking elements such as splitters, combiners and wavelength multiplexers, which are used to realise all-optical transmission between terminal stations and the many benefits that ensue [24-26].

7.   Dynamic Range

The dynamic range of an optical receiver quantifies the range of detected power levels within which system operation at the target fidelity, or better, is ensured. This parameter is important in applications where the spread of possible detected powers is broad. For example, city centre and local area networks comprise point-to-point links of significantly differing lengths and insertion losses, and hence the power levels at receivers may be expected to differ considerably. Similarly, the repair of a long haul multi-section route may involve the insertion of an additional regenerator. The receiver at the adjacent regenerator, instead of detecting power from a transmitter many tens of kilometres distant, may then have to cope with the significantly greater power from a transmitter just a few kilometres distant. For a single receiver design to be deployable in either scenario, which is clearly desirable because of cost reductions through mass usage, its dynamic range must be appropriately broad.

By convention the dynamic range of any receiver is defined as the numerical difference between its sensitivity and overload levels, the latter being the point at which the signal fidelity (SNR or BER) falls below the target level because the detected power is too high for the data signal to be recovered without distortion from the receiver electronics. Most commonly the overload level is associated with the value of the photodiode bias resistor, or feedback resistor. From the signal equivalent circuits given earlier (Figures 3 to 7) it is clear that the signal photocurrent flowing through the bias/feedback resistor produces a voltage which, since it is usually DC coupled to the amplifier, upsets the DC bias conditions of the amplifier. If it is assumed that the amplifier, without active bias compensation, can tolerate a 0.5V potential appearing across the resistor, then for a photodiode responsivity of 1A/W (Appendix A) the overload levels are as listed in Table 7.1. Using the sensitivities plotted in Figure 36, the corresponding dynamic ranges at 1.5Gbit/s are also listed.

Table 7.1: Overload levels and dynamic ranges at 1.5Gbit/s.

Receiver	R	Overload level	Dynamic range	
			PIN	APD
Low impedance	200Ω	2.5mW (+4dBm)	32dB	43dB
Transimpedance	1kΩ	0.5mW (-3dBm)	29dB	38dB
High impedance	1MΩ	0.5μW (-33dBm)	4dB	10dB
(with active bias :-		*15μW (-18dBm)*	*19dB*	*25dB)*

It is clear that, despite the superior sensitivity of the high impedance configuration, its intrinsically poor overload level produces a narrow dynamic range with little, if any, practical use. It is for this reason that high impedance receivers often incorporate the form of active bias compensation shown in Figure 41.

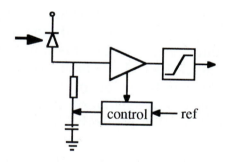

Fig. 41: Active bias control for high impedance receiver.

As the DC voltage at the input to the amplifier increases with detected optical power, the control loop applies an equal but opposite shift in the voltage at the other end of the bias resistor. The voltage at the input to the amplifier is therefore independent of the detected power level, but only within the linear range of the control voltage, which in turn is constrained by the supply rail voltages. It is reasonable to assume that the control voltage is limited in practice to -15V, in which case the overload level is improved by 15dB over the figure for no bias compensation, as shown in Table 7.1. Despite this improvement, the high impedance configuration still delivers the narrowest dynamic range, although it is now of some practical use.

It is clear that two extremes exist regarding receiver performance. At one extreme the high impedance configuration gives the best sensitivity but narrowest dynamic range, while at the other the low impedance configuration gives the broadest dynamic range but poorest sensitivity. From the performance figures for the transimpedance configuration it is therefore reasonable to conclude that for most applications it offers an attractive compromise, and indeed is widely used in practice for this reason.

8.   Developments in Receiver Design

The receivers considered thus far in this chapter have been used in a direct detection configuration, so called because the photodiode recovers the conveyed signal *directly* from the modulated intensity of the optical carrier. Direct detection is the simplest form of optical reception and may be likened to envelope detection in AM radio. It is therefore relatively crude in signal detection terms. However, it offers the key attractions of opto-electronic simplicity and low cost, while the high performance of modern photodiodes and amplifiers ensures that its "crudity" does not detract from the realisation of a good performance.

Figure 42 summarises some of the best reported sensitivities for PIN and APD receivers in digital direct detection systems. Analogue achievements, although not as plentiful, show a similar trend and so need not be shown. It is clear that APD receivers afford a useful 10-15dB sensitivity advantage over their PIN counterparts, a feature observed earlier in Figure 36. However, the APD receiver is the more complex and costly of the two, requiring a high bias potential and gain control at the photodiode to optimise sensitivity and dynamic range. Furthermore, the ~50-70GHz gain bandwidth product of modern APD's [27] means that they run out of useful bandwidth beyond ~8Gbit/s. In contrast, the 25GHz bandwidth of similar quality PIN's [3] is available in full, suggesting a transmission capability approaching ~40Gbit/s. Experimental PIN receivers incorporating such devices have demonstrated a potential to support 20Gbit/s [28], with higher bit rates likely in the near future.

Modern day optical telecommunications frequently require greater sensitivities than those achievable with direct detection receivers, a trend that will increase with the take-up of fibre capacity (using TDM and WDM) and the deployment of all-optical cable routes. A significant increase in the sensitivity of direct detection can be achieved by using coherent detection or optical pre-amplification.

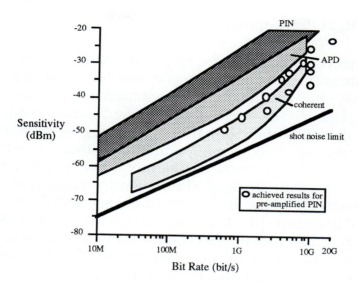

Fig. 42: Best reported receiver sensitivities.

8.1. Coherent detection receivers

Coherent detection is so named because the coherence, or spectral purity, of the transmit laser and a local oscillator (LO) laser is utilised to increase the sensitivity of the receiver. This increase arises from the mixing of the two optical fields that takes place at the surface of the photodiode (Figure 43).

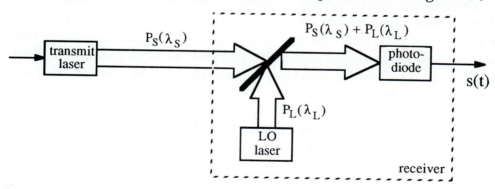

Fig. 43: Principle of optical mixing in a coherent receiver.

If the states of polarisation of the two fields are matched at the surface of the photodiode, it is readily shown for coherent (i.e, single frequency) lasers that the signal from the receiver is given by [29],

$$s(t) = 2 R_p M \sqrt{P_S P_L} [1 + m_1(t)] \cos \left\{ 2\pi (f_S - f_L)t - m_2(t) + \phi \right\}$$

(59)

which compares with,

$$s(t) = R_p \, M \, P_S \, [1 + m_3(t)] \qquad\qquad (60)$$

for a direct detection receiver. In each case information conveyed using amplitude or intensity modulation can be recovered via the $m_1(t)$ and $m_3(t)$ terms, respectively, but note that coherent detection can also utilise frequency modulation through f_S (or wavelength λ_S) and phase modulation through $m_2(t)$. Coherent detection therefore affords a degree of freedom that is denied direct detection, that of chosing the modulation domain (amplitude, phase or frequency) to suit the application. Furthermore, the coherent receiver may be configured as a homodyne or heterodyne detector depending on whether the LO field is respectively phase locked or frequency locked to the signal field. However, the significant practical difficulties associated with homodyne detection mean that heterodyne detection, despite its intrinsic 3dB sensitivity penalty and its additional circuitry to demodulate the intermediate frequency (IF) signal from the receiver (Figure 44), is generally the preferred scheme.

Of most importance in (59) from the point of view of sensitivity is that the amplitude of the signal from a coherent receiver (homodyne or heterodyne alike) is determined by the product of the mean intensities of the signal (P_S) and LO (P_L) fields at the photodiode, whereas the equivalent direct detection signal is determined purely by the mean intensity of the signal field. Increasing LO power will therefore increase the amplitude of the signal from a coherent receiver which, if the noise level is unchanged, will increase SNR and improve sensitivity. Noise, however, is not constant because, as LO power increases, so does its shot noise contribution to overall receiver noise. Indeed, a point is reached (the shot noise limit) when detection becomes LO shot noise limited, beyond which further increases in LO power produce no improvement in SNR and sensitivity because signal and noise increase by equal amounts. This limit is shown in Figure 42. Despite practical sensitivities being several dB poorer than this limit through device and other limitations, it is evident from Figure 42 that coherent receivers can nevertheless realise substantial improvements in sensitivity over direct detection receivers, some 5-15dB over APD receivers, increasing to 20dB or more over PIN receivers.

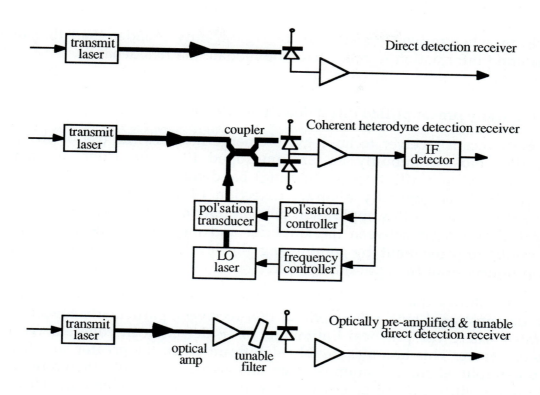

Fig. 44: Principal detection arrangements.

An additional beneficial feature of coherent receivers, particularly where heterodyne detection is employed, is their tunability in multi-channel applications. Multiplexes of channels can be assembled in the optical regime and transported over a single fibre either by using wavelength division multiplexing (WDM) in which individual data channels are conveyed on discrete optical carriers of differing wavelength, or frequency division multiplexing (FDM) in which an electrical multiplex of channels is conveyed on a signal optical carrier. In either case, the wavelength of the LO laser can be tuned to place an individual channel (or narrow group of channels) from the received multiplex within the detection bandwidth of the receiver. The IF detector in a heterodyne receiver can function as a second (fine grain) stage of channel selection in closely packed (dense) WDM applications, with the LO laser affording course grain selection. With the trend in telecommunications towards more efficient utilisation of fibre capacity through channel multiplexing, the tunability of coherent receivers is clearly a key feature.

The sensitivity advantage and tunability of coherent receivers must, however, be balanced with their significantly greater complexity and cost (Figure 44). This ensues from the need for coherent lasers, polarisation control (usually), phase or frequency locking at the LO laser, IF detection, and a receiver bandwidth which, in a heterodyne system, can be significantly broader than that of the transmitted information in order to accommodate the IF sidebands.

8.2. Optically pre-amplified direct detection receivers

The most recent receiver technology to emerge involves the use of an optical amplifier, most commonly an Erbium-doped fibre amplifier, to increase sensitivity without affecting (or limiting) bandwidth (Figure 44) [30,31]. The receiver itself is generally a direct detection PIN receiver since APD and coherent receivers in this application give an inferior performance. A bandpass optical filter placed between the amplifier and the photodiode optimises the sensitivity improvement by minimising the level of the spontaneous noise beat-products from the optical amplifier.

Figure 42 shows that sensitivities rivalling those of coherent receivers have been demonstrated with pre-amplified PIN receivers, particularly towards high bit rates where the practical limitations of IF detection in coherent receivers become severe. This becomes doubly significant when the greater simplicity of the pre-amplified configuration is taken into account. It would therefore appear that pre-amplified PIN receivers offer the more attractive technology. However, coherent receivers still have a unique feature in their favour, that of fine tunability and the resultant ability to select individual channels from a closely packed multiplex. But even here direct detection receivers are rapidly catching up. Optical filters are under development which are already realising a narrow bandwidth (<10GHz) that is tunable over ~5nm (~600GHz) [32]. With these used as the interstitial filter in a pre-amplified PIN receiver, the sensitivity *and* tunability of a coherent receiver can then be closely rivalled.

A further but less frequently proclaimed feature of coherent receivers is that they can be used to overcome the signal distortion caused by fibre dispersion. Because the group delay on the IF signal from a coherent receiver is a perfect replica of the fibre dispersion encountered during transmission, electronic equalisation can be used at that point to flatten the group delay, thereby negating fibre dispersion [33]. However, this feature is once again not necessarily the sole domain of coherent receivers. Work has been reported in which optical interferometers located prior to direct detection receivers can compensate for fibre dispersion over a narrow band [34]. More recently, stable broadband compensation techniques involving pre-distortion and special fibres have been developed for direct detection receivers [35,36]. The technology is therefore becoming available with which to realise a pre-amplified, dispersion compensated, direct detection receiver which rivals *all* the performance features of a coherent receiver, but at a significantly lower cost through its greater simplicity.

8.3. The transmission engineer's dream

Broadband, low noise receivers of the kind discussed in this chapter, together with optical amplifiers and low loss fibre, are the principal enabling technologies for the realisation of all-optical transmission and the "Optical Ether" [37]. This is the transmission engineer's dream: global telecommunications where end-to-end transmission is entirely optical and passive, and where bandwidth is not a limiting criterion to the user or the service conveyed.

References

[1] Heatley, D.J.T. and Cochrane, P., "Future Directions in long haul optical fibre systems," Proceedings of the 3rd IEE Conference on Telecommunications, Edinburgh, March 17-20, 1991, pp. 157-164.

[2] Bowers, J.E. and Burrus, C.A., "Ultra-wideband, long wavelength PIN photodetectors," IEEE Journal of Lightwave Technology, Vol. LT-5, Nº 10, October 1987, pp. 1339-1350.

[3] Wake, D., Walling, R.H., Henning, I.D. and Parker, D.G., "Planar junction, top illuminated GaInAs/InP PIN photodiode with bandwidth of 25GHz," IEE Electronics Letters, Vol. 25, Nº 15, July 1989, pp. 967-968.

[4] Kato, K., Hata, S., Kozen, A., Yoshida, J.I. and Kawano, K., "High efficiency waveguide InGaAs PIN photodiode with bandwidth of over 40GHz," IEEE Photonics Technology Letters, Vol. 3, Nº 5, May 1991, pp. 473-474.

[5] McIntyre, R.J., "Multiplication noise in uniform avalanche diodes," IEEE Transactions on Electron Devices, Vol. ED-13, Nº 1, Jan. 1966, pp. 164-168.

[6] Brain, M.C., "Absolute noise characterisation of avalanche photodiodes," IEE Electronics Letters, Vol. 14, Nº 15, July 1978, pp. 485-487.

[7] Shirai, T., Mikawa, T. and Kaneda, T., "Planar GaInAs avalanche photodiodes for the 1μm region of optical fibre communication systems," Fujitsu Scientific Technical Journal, Vol. 30, Nº 3, Sep. 1984, pp. 303-328.

[8] Muoi, T.V., "Receiver design for high speed optical fibre systems," IEEE Journal of Lightwave Technology, Vol. LT-2, Nº 3, June 1984, pp. 243-267.

[9] Moustakas, S. and Hullett, J.L., "Noise modelling for broadband amplifier design," IEE Proceedings-G, Vol. 128, Nº 2, April 1981, pp. 67-76.

[10] Howes, M.J. and Morgan, D.V. (editors), *Gallium Arsenide Materials, Devices and Circuits*, John Wiley & Sons, 1985.

[11] Watson, J., *Semiconductor Circuit Design*, Adam Hilger, 1977.

[12] Schwartz, M., *Information Transmission, Modulation and Noise*, 3rd Ed, McGraw-Hill, 1981.

[13] O'Mahony, M.J., "The analysis and optimisation of a PIN photodiode, GaAs FET, cascode optical receiver," British Telecom Technology Journal, 1983, Part I in Vol. 1, Nº 1, pp. 38-42, Part II in Vol. 1, Nº 2, pp. 51-56.

[14] Mitchell, A.F. and O'Mahony, M.J., "The performance analysis of PIN-Bipolar receivers," British Telecom Technology Journal, Vol. 2, Nº 2, April 1984, pp. 74-85.

[15] O'Reilly, J.J., "Matched filter receivers," Colloquium digest on Mathematical Topics in Telecommunications - Calculus of Variations, University of Essex, November 25, 1980, pp. 39-46.

[16] Carlson, A.B., *Communication Systems*, 2nd Ed, McGraw-Hill/Kogakusha, 1975.

[17] Personick, S.D, "Receiver design for digital fibre optic communication systems - Parts I and II," Bell System Technical Journal, Vol. 52, Nº 6, July-August 1973, pp. 843-886.

[18] Tucker, R.S., "High speed modulation of semiconductor lasers," IEEE Journal of Lightwave Technology, Vol. LT-3, Nº 6, Dec. 1985, pp. 1180-1192.

[19] Mellis, J., et al., "Miniature packaged external cavity semiconductor laser with 50GHz continuous electrical tuning range," IEE Electronics Letters, Vol. 24, Nº 16, August 1988, pp. 988-989.

[20] Bowers, J.E. and Pollack, M.A., "Semiconductor lasers for telecommunications," in *Optical Fibre Telecommunications II*, Miller, S.E. and Kaminow, I.P. (Eds.), Academic Press, 1988, pp. 509-568.

[21] Ainslie, B.J. and Day, C.R., "A review of single mode fibres with modified dispersion characteristics," IEEE Journal of Lightwave Technology, Vol. LT-4, Nº 8, 1986, pp. 967-979.

[22] Brockbank, R.G., Calton, R.L., Mottram, S.W., Hunter, C.A. and Heatley, D.J.T., "The design and performance of the TAT-8 optical receiver," British Telecom Technology Journal, Vol. 5, Nº 4, October 1987, pp. 26-33.

[23] Carlton, R.L., Heatley, D.J.T., Mottram, S.W., Hunter, C.A. and Brockbank, R.G., "The design and performance of the PTAT-1 optical receiver," British Telecom Technology Journal, Vol. 7, Nº 1, January 1989, pp. 78-82.

[24] Cochrane, P. and Brain, M. C., "Future optical fibre transmission technology and networks," IEEE Communications Magazine, Vol. 26, Nº 11, November 1988, pp. 45-60.

[25] Hill, G.R., "A wavelength routeing approach to optical communication networks," British Telecom Technology Journal, Vol. 6, Nº 3, 1988, pp. 24-31.

[26] Kang, Y., et al, "Fibre-based local access network architectures," IEEE Communications Magazine, Vol. 27, Nº 10, 1989, pp. 64-73.

[27] Imai, H. and Kaneda, T., "High speed distributed feedback lasers and InGaAs avalanche photodiodes," IEEE Journal of Lightwave Technology, Vol. 6, Nº 11, 1988, 1634-1642.

[28] Violas, M.A.R. and Heatley, D.J.T., "Microstrip and coplanar design considerations for microwave bandwidth PIN-HEMT optical receivers," IEE Colloquium Nº 1990/139 on Microwave Opto-electronics, London, October 26, 1990, paper 8.

[29] Hodgkinson, T.G., Smith, D.W., Wyatt, R. and Malyon, D.J., "Coherent optical fibre transmission systems," British Telecom Technology Journal, Vol. 3, Nº 3, July 1985, pp. 5-18.

[30] Smyth, P.P., et al., "152 photons per bit detection at 2.5Gbit/s using an Erbium fibre preamplifier," 16th European Conference on Optical Communication (ECOC '90), Amsterdam, 1990, pp. 91-94.

[31] Gnauck, A.H. and Giles, C.R., "2.5 and 10 Gbit/s transmission experiments using a 137 photon/bit Erbium fibre preamplifier receiver," IEEE Photonics Technology Letters, Vol. 4, Nº 1, January 1992, pp. 80-82.

[32] Miller, C.M., "Field worthy tunable FFP filter," 16th European Conference on Optical Communication (ECOC '90), Amsterdam, 1990, pp. 605-608.

[33] Takachio, N., and Iwashita, K., "Compensation of fibre chromatic dispersion in optical heterodyne detection," IEE Electronics Letters, 1988, Vol. 24, Nº 2, pp. 108-109, and Nº 12, pp. 759-760.

[34] Cimini, L.J., et al., "Optical equalisation for high bit rate optical systems," IEEE Photonics Technology Letters, Vol. 2, Nº 3, 1990, pp. 200-202.

[35] Dugan, J.M., et al., "All optical fibre based 1550ηm dispersion compensation in a 10Gbit/s 150km transmission experiment over 1310ηm optimised fibre," Optical Fibre Communication Conference (OFC '92), San Jose, USA, February 2-6, 1992, Post Deadline Session, pp. 367-370, paper PD14.

[36] Ellis, A.D., et al., "Dispersion compensation in 450km transmission system employing standard fibre," IEE Electronics Letters, Vol. 28, Nº 10, May 1992, pp. 954-955.

[37] Cochrane, P. and Heatley, D.J.T., "Optical fibre systems and networks in the 21st century," Interlink-2000 Journal, February 1992, pp. 150-154.

APPENDIX TO CHAPTER 5

PRACTICAL RECEIVER PARAMETERS ASSUMED IN THIS CHAPTER

Constants :-

$c = 3.0 \times 10^8$ m/s

$e = 1.6 \times 10^{-19}$ C

$h = 6.63 \times 10^{-34}$ Js

$k = 1.38 \times 10^{-23}$ J/K

$B = 1$GHz in computations of SNR

$m_o = 1$

$P = -30$dBm in computations of receiver noise

$Q = 6$ in computations of receiver sensitivity for digital systems

APD :-

$R_p = 1$ A/W

$I_d = 50 \, \eta$A

$I_{bk} = 10 \, \eta$A

$M = 15$ (unless optimised)

$x = 0.7$

PIN :-

$R_p = 1$ A/W

$I_d = 10 \, \eta$A

FET amplifier :-

BPT amplifier :-

$I_g = 5\ \eta A$

$g_m = 40\ mS$

$f_b = 50\ MHz$

$\Gamma = 1.1$

$R_1 = 300\ \Omega$

$R_2 = 300\ \Omega$

$b = 10\ kHz$

$I_b = 10\ \mu A$

$g_m = 40\ mS$

$R_{bb} = 50\ \Omega$

$I_c = 1\ mA$

$R_1 = 300\ \Omega$

$R_2 = 300\ \Omega$

Low impedance receiver :-

$R = 200\ \Omega$

$C = 0.6\ pF$

Transimpedance receiver :-

$R = 1\ k\Omega$

$C = 0.6\ pF$

$C_f = 0.1\ pF$

$A = 20$

High impedance receiver :-

$R = 1\ M\Omega$

$C = 0.6\ pF$

$C_e = 2.5\ pF$

$R_i = 50\ \Omega$

2nd stage noise :-

$T_n = 290\ ^{o}K$

$A = 20$

$F = 10\ dB$

Chapter 6
Continuous-Time Filters

Yannis P. Tsividis[1] and Venugopal Gopinathan[2]

[1]Division of Computer Science, National Technical University of Athens, Zographou 15773
Athens, Greece

[2]Semiconductor Process and Design Center, Texas Instruments, M/S 369, Dallas, TX 7265, USA

1. Introduction

VLSI telecommunication circuits must often incorporate both analog and digital functions, as is apparent from several chapters in this book. In this context, the need for continuous-time filtering arises often; three special cases are illustrated in Figure 1. In (a), the most straightforward use of a continuous-time filter is shown; here a continuous-time signal is processed directly by this filter, with no sampled-data processing involved. In (b) and (c) a switched-capacitor and a digital filter are respectively used in the processing. In these two cases a continuous-time filter is used at the input to attenuate the high-frequency components of the incoming signal, so as to prevent the "aliasing" of these components by the sampling process; at the output, a "smoothing" filter is used to smoothen out the staircase waveform provided by the sampled-data filters, and also to reject high frequency noise. In some cases one or the other of the two continuous-time filters shown in (b) or (c) is not used. For example, in a PCM encoder the sampled-data filter is followed directly by an A/D converter, with no smoothing filter used; similarly, the output of a PCM decoder is fed directly into a switched-capacitor filter with no antialiasing filter in-between. The cases in (b) and (c) have continuous input and output, and thus it is possible to meaningfully compare them to the case in (a). This has been done in detail elsewhere [1], so here we will only summarize the main conclusions.

In applications where programmability and/or large dynamic range are of prime importance, digital filters can be the solution of choice provided the design of the associated peripheral circuitry (Figure 1c) presents no problem. In medium or low dynamic range applications, if programmability is not an issue, analog filters can be advantageous in terms of power dissipation, chip area, and high frequency capability [2]. One is then led to a choice between the solutions in Figures 1a and 1b. At low frequencies, these solutions can offer comparable performance [1]; so far, though, switched-capacitor filters are much more widely used, since they started earliest among integrated analog filters, are "high on the learning curve" and many designers have experience with them. Yet, the design of practical, working switched-capacitor integrated filters is not easy, due to the presence of several nonidealities; these are discussed in the corresponding chapter by Gabor Temes. Some of these nonidealities, notably the clock feedthrough, can become very limiting at high frequencies and have made even experienced switched-capacitor filter designers look to fully continuous-time techniques (Figure 1a). This chapter discusses precision continuous-time filters with on-chip automatic tuning, for both low- and high-frequency applications.[1]

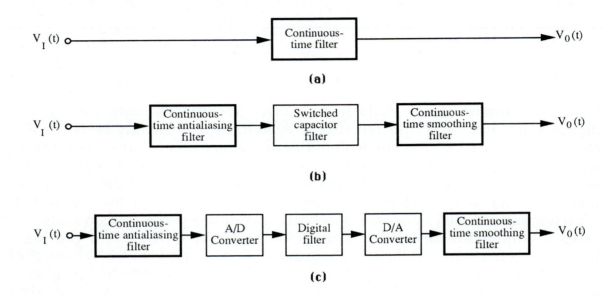

Fig. 1: Three ways of processing a continuous-time signal.

Even when switched-capacitor or digital filters are used, one must face the design of continuous-time filters, as is obvious from Figures 1b and 1c. In many cases, these can be simple passive or active RC circuits. In high-frequency applications, though, where the ratio of input frequency to sampling frequency is not small, the design of such filters becomes critical, and then one can use the techniques described in this chapter to advantage. In such cases, though, one

---

[1] Continuous-time filters can of course be used without automatic tuning in extremely non-critical applications in which wide tolerances can be accepted. In more critical applications trimming is sometimes used to center the response at room temperature and/or temperature effect cancellation is employed. If, for example, the critical frequencies are proportional to the transconductance of a bipolar transistor, they will be propotional to Ic/T, where Ic is the collector current and T is the absolute temperature. If Ic is made proportional to absolute temperature, the most significant temperature dependence of the frequency response is eliminated. For even more critical applications, closed-loop automatic tuning is needed. Due to lack of space, only automatically tuned filters will be discussed in this chapter.

may want to rethink the problem: since precision continuous-time filters must then be used anyway, one should consider again whether the solution of Figure 1a is not the most attractive alternative.

Integrated continuous-time filters with on-chip automatic tuning are finding, at the time of this writing, several commercial applications in video signal processing [3, 4, 5], disk drives [6, 7, 8], and computer communication networks. New applications are envisaged, including personal radio communications. Due to space limitations, of the many techniques already proposed for continuous-time filters (see, for example, Refs. [1-54]), we will concentrate on only a few, chosen for compatibility with CMOS technology. The schemes we will discuss fall into the categories known as MOSFET-C and transconductor-C filters. Several of the principles discussed are valid for other techniques as well. This is notably the case with the schemes presented for on-chip automatic tuning.

2. MOSFET-C Filters

2.1 The MOSFET as a voltage-controlled resistor

Precision RC filters cannot be implemented using capacitors and non-tunable resistors as passive elements, since the resulting RC products vary widely with fabrication process and temperature variations (e.g., by ± 50%). In the techniques to be described in the bulk of this section [1], MOS transistors are used as voltage-controlled resistors which are automatically adjusted to provide precision RC products by an on-chip control system. The operation of transistors in this mode is considered in this section. Unless noted otherwise, all such transistors will be assumed to have their gate connected to a common "control voltage" bus and their substrate to a common substrate voltage bus, as shown in Figure 2(a). N-channel devices will be considered as an example. Only operation in the nonsaturation region will be considered. For a careful investigation of distortion, the commonly used "square law" transistor model is inadequate; a more precise model will instead be used. According to this model, if V_1 and V_2 are the source and drain voltages with respect to ground we have, assuming an n-channel device [9]:

$$I_D = \frac{W}{L} \mu C'_{ox} \left\{ (V_C - V_B - V_{FB} - \phi_B)(V_1 - V_2) \right.$$

$$-\frac{1}{2} \left[(V_1 - V_B)^2 - (V_2 - V_B)^2 \right]$$

$$\left. -\frac{2}{3} \gamma \left[(V_1 - V_B + \phi_B)^{3/2} - (V_2 - V_B + \phi_B)^{3/2} \right] \right\} \tag{1}$$

where W and L are the channel width and length, respectively, and $\mu\, C'_{ox}$, V_{FB}, V_B, ϕ_B and γ are process-dependent quantities [9] (in particular, γ is the body effect coefficient). It is assumed that the source and the drain voltages V_1 and V_2 never become too low to forward-bias the drain/source junctions, and never become too high to drive the device into saturation. The ground potential is defined such that V_1 and V_2 vary around zero. In that case the substrate voltage should be negative in order to keep the drain and the source junctions reversed biased. (Such a definition for ground potential is convenient when two power supplies of opposite values are present.)

If the 3/2 power terms in (1) are expanded in a Taylor series with respect to V_1 and V_2, one obtains [10]:

$$I_D = \frac{W}{L}\mu C'_{ox}\left[b_1(V_1 - V_2) - b_2(V_1^2 - V_2^2) + b_3(V_1^3 - V_2^3) - ...\right] \qquad (2)$$

where $b_1 = V_C - V_T$ $\qquad\qquad\qquad\qquad\qquad\qquad\qquad (3)$

with V_T the threshold voltage (equal to $V_{FB} + \phi_B + \gamma\sqrt{\phi_B - V_B}$, and the b_i's are positive process-dependent terms that are decreasing functions of $|V_B|$ [10]; in particular, b_2 is somewhat larger than 1/2.

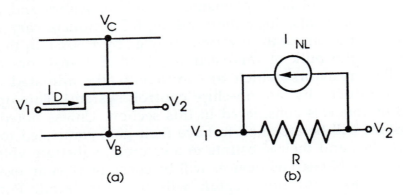

Fig. 2:   (a) MOS transistor with all terminal voltages referred to ground. V_C and V_B are DC biases; V_1 and V_2 are signals. (b) A model for the device showing the linear and non-linear parts.

As seen from (2), the inverse of $(W/L)\,\mu\, C'_{ox}b_1$ is the small-signal resistance of the transistor, R; thus using Eq. 3 we have:

$$R = \frac{L}{W}\,\frac{1}{\mu C'_{ox}(V_C - V_T)} \qquad\qquad\qquad (4)$$

The value of R may be varied with V_C (hence the name "control voltage"); therefore for small signals the MOSFET behaves as a voltage-controlled linear resistor. For large signals the nonlinear terms in (2) must be considered. It can be shown that the main nonlinearity comes from the second-order term [10]. The nonlinearity contributed by the third-order term, for example, can typically be about 0.1% of the linear term, for signal amplitudes of the order of 1 V.

Based on (2) and (4), the MOS transistor in nonsaturation can be represented by the model of Figure 2(b). The resistance value R is given by (4); the current source represents the nonlinearities and is given by:

$$I_{NL} = \mu C'_{ox} \frac{W}{L} \left[b_2 \left(V_1^2 - V_2^2 \right) - b_3 \left(V_1^3 - V_2^3 \right) \right] + \cdots)$$

(5)

If terms of order higher than two are neglected, this can be written:

$$I_{NL} \approx \frac{W}{L} \mu C'_{ox} b_2 \left(V_1^2 - V_2^2 \right)$$

(6)

It will be seen later that the terms neglected in writing (6) have indeed a negligible contribution to the distortion of the proposed filters for many applications, and that this is so even for realistic devices with an effective mobility dependent on the gate field.

2.2. Cancellation of nonlinearities

MOS resistors can be used in lieu of linear resistors, while still maintaining linear circuit operation, if the nonlinearity terms I_{NL} are somehow cancelled. A few possibilities among many [1, 11] are shown in Figure 3, where the connection to V_B is not shown for simplicity. The cancellation of the nonlinearities can be accomplished in a single device [Figure 3(a)] or among more than one devices [Figures 3(b) - (d)]. The results shown directly on the figures can be easily verified by using the model of Figure 2(b) and Eq. 6. In Figure 3(a), I_{NL} is forced to a zero value [assuming (6) is valid] by operating the device with anti-symmetrical terminal voltages. In Figure 3(b) $I_{NL} \neq 0$, but is the same for both devices and cancels out in the difference. The case shown in Figure 3(c) does not provide complete cancellation of the square terms in I_{NL}. This can be improved by modifying the value of the signals applied to the gates, or by the use of substrate signals [1]; in fact, substrate signals have recently been used to cancel the odd-order terms as well [12]. Another way to cancel both even- and odd-order terms is shown in Figure 3(d) [11]. Here we have:

$$R' = \frac{L}{W} \frac{1}{2\mu C'_{ox}(V_{C1} - V_{C2})}$$

(7)

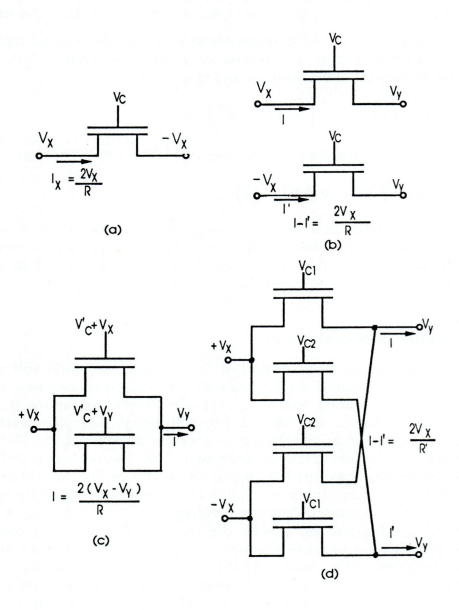

Fig. 3:   Four ways to attain linear resistor effects using MOS transistor (see text for details).

The advantages and disadvantages of this last technique will be discussed in connection to filters later on.

2.3. MOSFET-capacitor filters based on balanced structures

2.3.1. Principle

Throughout this section we will employ topologies resulting from classical filter structures using resistors, capacitors, and op amps. The ideas to be discussed, however, are also applicable to filters with other active elements. A key component in the structures to be presented is a balanced-output op amp, whose symbol is shown in Figure 4. This op amp will be assumed to have the magnitudes of its two outputs matched (which is not necessarily the case for merely `"differential output'" op amps; the latter are often inadequate for our purposes). At low frequencies, a balanced-output op amp could be constructed from two single-ended op amps, one of them being used as an inverter along with two resistors; at high frequencies, though, the balance in such a scheme is destroyed. True balanced-output op amps have been designed and are currently in use in commercial products [13]; in these, balancing is maintained even at high frequencies. A very simple balanced amplifier with a gain-bandwidth product of several GHz has been implemented in BiCMOS technology [14], [15].

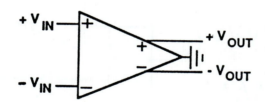

Fig. 4:   A balanced-output op amp.

The balanced-output op amp is shown with other devices in Figure 5(a). For the top output one has:

$$V_{OUT}(t) = V_Z - \frac{1}{C} \int_{-\infty}^{t} I_Z(\tau) \, d\tau \tag{8}$$

where V_Z is the voltage at the input terminals with respect to ground (assuming zero input offset). For the bottom output one has:

$$V_{OUT}(t) = V_Z - \frac{1}{C} \int_{-\infty}^{t} I'_Z(\tau) \, d\tau \tag{9}$$

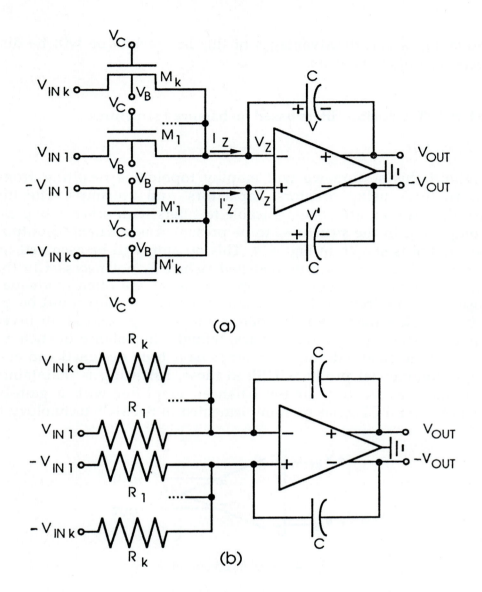

Fig. 5:   (a) A balanced summing integrator using MOS transistor in lieu of resistors. (b) A circuit
          that is input-output equivalent to that in (a).

Subtracting (9) from (8) and dividing by 2 one obtains:

$$V_{OUT}(t) = - \frac{1}{2C} \int_{-\infty}^{t} \left[I_Z(\tau) - I'_Z(\tau) \right] d\tau \qquad (10)$$

The case where M1 is matched to M1', M2 is matched to M2', and so on, is now
considered. In the difference $I_Z - I'_Z$ the nonlinearities of the devices within a

matched pair will cancel out, as suggested by Figure 3(b). Applying for each pair of matched devices the result shown directly on that figure, one obtains:

$$V_{OUT}(t) = -\frac{1}{R_1 C} \int_{-\infty}^{t} V_{IN1}(\tau) d\tau - \frac{1}{R_2 C} \int_{-\infty}^{t} V_{IN2}(\tau) d\tau - \cdots \qquad (11)$$

i.e., a linear summing inverting integrator has been obtained. This integrator is input-output equivalent to the circuit of Figure 5(b). It is emphasized here that the behavior of the circuit in Figure 5(a) is linear from input to output, but not within the circuit; for example, the device currents are not linear functions of the corresponding voltages. Also, V_z at the op amp inputs is in general nonzero. This value is continuously adjusted by feedback action so that $V_z - V$ and $V_z - V'$ are equal in magnitude and opposite in sign, as required by the balanced output op amp (Figure 4). The resulting values for V_z are small (typically no more than 20% of the input peak value); hence the input common mode range requirements on the op amp are modest.

2.3.2. Filter synthesis

Since the circuit of Figure 5(b) is input-output replaceable with that in Figure 5(a), a number of well-known filter structures can easily be converted to use MOSFETs, capacitors and op amps. An example will be given with the help of Figure 6. In (a), a Tow-Thomas biquad [16], is shown. The top op amp is only used for inversion. If balanced-output op amps were available, one would implement this filter as shown in (b). The circuit can be easily "doubled-up" to provide the structure in (c), which is simply a balanced version of (b). [It is easy to show that the op amp gain to each single-ended output should be halved in going from (b) to (c), if exact correspondence is to be maintained in the presence of finite gain effects.] The structure in Figure 6(c) can be seen to use building blocks of the form shown in Figure 5(b). Hence, all resistors in Figure 6(c) can be replaced by MOSFETs to produce the final circuit shown in Figure 6(d). The resistance values for each MOSFET will be given by (4), and for a given V_C, this resistance can be set by choosing the L/W ratios appropriately. The filter of Figure 6(d) can be represented compactly (for later use) by the circuit in Figure 6(e), where the rectangular boxes represent the circuit of Figure 3(b). If desired, the circuit of Figure 3(d) can instead be used for each box. More on this later.

Proceeding in a similar manner one can obtain a fifth-order elliptic filter as shown in Figure 7. The resulting circuit has been fabricated using CMOS technology, and is discussed in detail elsewhere [17]. This chip exhibited the high performance expected from computer simulation, the first time it was fabricated. No second iteration was needed and no unforseen effects were observed. This stems from the simplicity of the principle, the inherently "clean'" continuous-time operation and the readily available simulation tools for circuits in such operation. The reader can find other examples of MOSFET-C implementations in the literature, including ones for ISDN [18] and for video [14] applications.

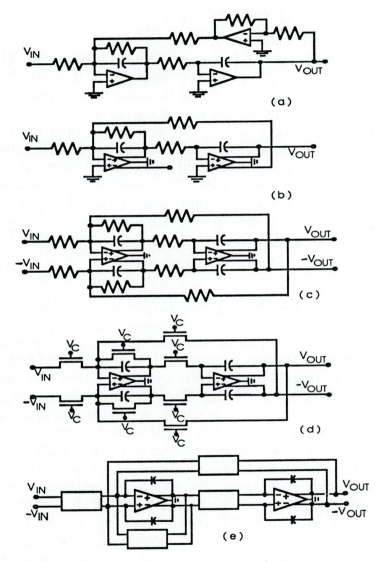

Fig. 6:   An example of filter synthesis using MOS transistors in lieu of resistors. (a) A Tow-Thomas biquad. (b) The circuit using balanced-output op amps. (c) Fully balanced version of the circuit. (d) The circuit in (c) with resistors replaced by MOS transistors. (e) Another representation of the circuit in (d), where each rectangular box represents the circuit of Fig. 3(b) or 3(d).

Although a balanced structure is inherently resistant to noise common to balanced paths, for critical applications it is desirable to provide control and substrate voltages (V_C and V_B) as free from extraneous signals as possible, to prevent such signals from varying the resistance values and creating modulation products with the input signal. [V_B affects the value of V_T through the body effect, and thus affects R in (4)]. This problem can be more severe if the extraneous signals contain frequency components in a range where the frequency response of the filter, or of one of its stages, is steep. Since V_B can carry large parasitic signals due to interference from other circuits on the chip, especially digital ones and analog drivers, the MOSFET "resistors" should be placed inside wells and fed from a "clean" V_B. For n-well CMOS processes, this implies the use of p-channel devices for implementing resistors.

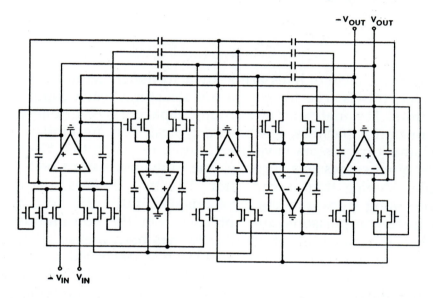

Fig. 7: A balanced fifth-order elliptic filter [17] (Copyright © 1985 by IEEE).

The V_C and V_B lines can be filtered rather easily since they do not carry DC current, or can be generated from reference voltages. On-chip crude filtering to rid the substrate of extraneous signals has also been used in switched-capacitor circuits.

As suggested by Czarnul [11], the integrators used in the filters described above can be modified by adding two transistors, essentially using the scheme of Figure 3(d) in lieu of the one in Figure 3(b). For this, one needs to replace each pair of MOSFETs in schematics like the one in Figure 5(a) by four devices. For the filter of Figure 6, one needs to use, in each box in (e), the circuit of Figure 3(d). The resulting circuits offer better immunity to substrate parasitic signals, since V_T does not appear in (7). Also, parasitic signals common to V_{C1} and V_{C2} in that equation cancel out. However, one should use caution since the scheme is based on signal substraction, and its internally generated noise can be significant. Note also that, because of the larger number of devices used, the effect of device mismatch is somewhat more significant. Finally, larger currents are needed to drive the four transistors of Figure 3(d); this should be taken into account when considering op amp output stages and power dissipation. We note also that, while the scheme in principle cancels odd-order nonlinearities, in practice the cancellation is not perfect due to the fact that the mobility μ in (1) depends on the gate potential. For the same reason, although assuming a constant μ shows that the technique can been applied even to non-balanced circuits [19], in practice such circuits do not behave as well as balanced ones [20]. Fully balanced configurations are desirable anyway for improved power-supply and substrate noise rejection.

2.3.3. Accuracy

Capacitors in MOS technologies are known to match within 0.1% "with extra care." However, this is true for composite structures, each composed of many "unit" capacitors. The matching of two unit capacitors, or of one unit capacitor to an array of unit capacitors, is worse; a typical working limit is around 0.5% [21]. Long-channel transistors have been known to match to 0.1%-0.3% for some time [1]. This often comes as a surprise, due to the transistor's being viewed as a more complex device than a capacitor. The two devices, though, undergo different processing steps and the mechanisms contributing to edge effects are different; direct comparisons are not easy. On the continuous-time filters reported [10, 17], a matching of about 0.2% has been measured for transistors next to each other. Although good matching is observed even for 4μ wide devices, such devices can show "narrow width effects" [9]; these effects increase the threshold voltage and are often not well modelled. Preferably somewhat larger widths should be used, unless the technology employed is well characterized. MOSFET "resistor" matching can be easily measured with a small drain-source voltage, along with appropriate gate-source and substrate-source bias. Such measurements are preferable to measurements in the saturation region, since the latter is characterized by mechanisms different from those in nonsaturation (where the above "resistors" operate); one can thus expect that the achievable degree of matching may also be different in the two regions. This is one of the reasons that op amp input offset numbers cannot be used to infer "resistor" matching, other reasons being that the offsets are determined by several devices, and that some of these devices have relatively short channels.

The high degree of MOSFET resistor matching can also be inferred indirectly from the reliability with which precision specs can be realized on chips (e.g., the fifth-order elliptic filter of Figure 7 exhibited a 0.06 dB ripple [17]).

2.3.4. Distortion and dynamic range

Distortion in the filters described above depends on such factors as transistor and capacitor matching, op amp output balancing, op amp input offset, and magnitude of V_C and V_B. Complete integrated filters such as the one in Figure 7 exhibit distortion of less than 0.1% for signal amplitudes of several volts for ± 5 V supplies [17], and of over 1 V for single 5 V supplies. The fact that a complete filter can have a distortion as low, or even lower than, a single integrator is due to additional nonlinearity cancellations that take place in the filter structure as a whole and which come more or less as an unexpected bonus to the designer. Such cancellations can eliminate both even and odd-order nonlinearities in certain cases [22]. Due to the low distortion and low noise, dynamic ranges of 95-100 dB for complete voice-band filters are easily achieved as demonstrated experimentally [17]; this number is based on a 1% THD overload level.

For low distortion, the chip internal signal levels should be kept low. To make the handling of a large external signal possible, a filter of the type described above can be placed between an on-chip attenuator and an on-chip amplifier, thus maintaining internal signal levels sufficiently low. With small attenuation and amplification factors (e.g., 2 to 3), this approach will not cause a large noise increase.

2.3.5. Parasitic capacitance

The parasitic capacitances of a MOSFET "resistor" with DC gate and substrate biases are shown in the small-signal equivalent circuit of Figure 8. Here C_s and C_d are the junction capacitances of the source and the drain; C_p is the distributed parasitic capacitance of the channel to the gate and to the substrate, consisting of the sum of the gate-channel and body-channel capacitances [1,9].

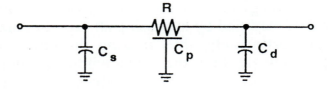

Fig. 8: A small signal model for a resistor implemented by a MOS transistor.

The effects of C_s and C_d are not significant in the structures presented, at frequencies where the op amp is practically ideal. These capacitances are either voltage driven, or are connected to one of the op amp input terminals in Figure 5(a). Due to the matched transistor pairs in this figure, the total junction capacitance between either op amp input terminal and ground is the same. As V_z varies with time, these capacitances add equal parasitic currents to I_z and I'_z, and thus cancel out in the difference in (10).

The effect of the distributed capacitance C_p can be very important, and can distort the frequency response if not taken into account at the design stage [23]. A simple empirical technique that has proven very successful is the following [10, 17]. A filter is first designed assuming ideal resistors with no distributed capacitance. Next, distributed effects are added and the filter is computer-simulated to determine whether such effects have significantly affected the frequency response. The distributed effects can be included in three ways. One can replace each transistor by the model of Figure 8, where the middle part is simulated using the transmission line model found today in some CAD programs. One can also use simple lumped RC circuits to approximate the distributed circuit in that figure [23]. Alternatively, one can use the approach illustrated in Figure 9. A transistor is split into several sections, with lengths adding up to the total transistor length. This results in a many-segment lumped approximation to the transistor. Since the individual devices in Figure 9 are fictitious and no source-drain junctions are meant to exist at the intermediate points, the model for these devices should of course not include any junction or gate overlap capacitances at these points; neither should it include short channel effects. Computer simulation including distributed effects usually predicts a certain effect on the frequency response (typically peaking at the band edge for a low-pass filter). Provided this peaking is relatively small (e.g., a fraction of 1 dB), it can be eliminated by empirically modifying slightly the value of one or two capacitor pairs in the filter, using computer simulation as a guide. The capacitors to which the response appears most sensitive are preferable. This process is fast, and has resulted in successful chips from the first run; design iterations were found unnecessary, as the measured response had been very well predicted during the above simulation stage. If the peaking produced by the distributed capacitance is large, the above procedure will not work well since elimination of the peaking might cause distortion of the response at other points (e.g., an increase of the passband ripple). In such cases, the filter should be first modified by increasing all capacitor sizes and decreasing all resistor sizes proportionately, thus making the effect of the distributed capacitances less pronounced; then the above empirical adjustment should be carried out. Increasing C and decreasing R, by the way, also makes the thermal noise lower. Another way is the use of passive compensation [23].

Another observation concerns the choice between p-channel and n-channel devices for implementing the MOSFET resistors. Since the mobility of p-channel devices is lower, a given resistance value can be implemented with a smaller L (see eq. 4), and thus the distributed parasitic effects will be smaller. In n-well CMOS processes, p-channel devices are a natural choice anyway since they are inside wells and can thus be protected from parasitic substrate voltages.

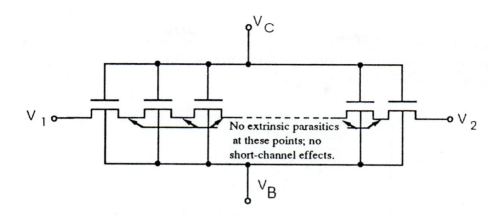

Fig, 9: A lumped approximation to a MOS transistor with distributed effects.

In passing, we note that the channel capacitance has been considered for implementing transistor-only filters, using "digital" fabrication processes which provide no capacitors. A 40 MHz experimental filter of this type has been reported [24].

2.4. Worst-case design

In the circuits described above, the value of V_C is assumed to be automatically changed with fabrication process and with temperature, to keep the RC products stable at a desired value. This is done by the automatic tuning system, to be described in section 5. Since drain and source voltages must be kept at least a threshold voltage below V_C to maintain operation in the nonsaturation region, one desires a V_C as high as possible to allow for large signal swings. A worst-case design strategy can be summarized as follows, considering a single RC product for simplicity. First, all process tolerances and temperature are combined in the worst-case direction resulting in a required V_C as high as possible for a given value of τ = RC. Curves of the required V_C versus might look as in Figure 10, where each curve is for a different nominal value of channel length for the MOSFET resistor (assuming L/W > 1 and that W has been set to the minimum acceptable value). If the maximum allowed value of V_C is set at the power supply voltage, V_{DD}, then by choosing L = L_2 from the figure we are guaranteeing that V_C will never need be higher than that value to adjust to the required value of τ; and it will only be equal to V_{DD} in the improbable case where all process tolerances and temperature happen to combine in a worst-case sense to require maximum V_C. Keeping now L at the chosen nominal value L_2, all process tolerances and temperature are combined in the opposite worst-case direction, resulting in the curve shown in Figure 10(b) where the middle curve from Figure 10(a) is also reproduced for comparison. From the new curve, the minimum V_C value that will be required is found. The corresponding minimum guaranteed peak swing will then be approximately one threshold voltage below $V_{C, min}$. A detailed numerical example is considered elsewhere [25].

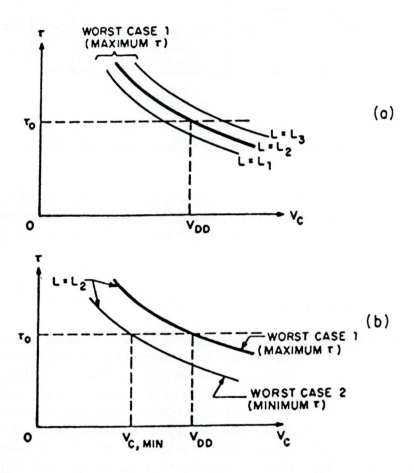

Fig. 10:   Ilustration of worst-case design procedure. (a) Choosing channel length for a maximum allowable V_C and worst-case corresponding to large RC. (b) Determining minimum V_C required for the choice in (a).

3. Transconductor-C Filters

3.1. Introduction

In conventional CMOS technology (as opposed to BiCMOS technology), operation of MOSFET-C filters at high frequencies is hampered by two effects: (1) Finite gain-bandwidth product of the op amp, and (2) non-zero output impedance of the op amp at high frequencies. These nonidealities perturb the frequency response of the filters, especially at the passband edge(s). One way to deal with the problem is to use passive compensation schemes that partially nullify the effect of these nonidealities [23]. However, these techniques are not effective when the filter has to operate in frequencies exceeding a few MHz, even when using op amps with gain-bandwidth products of 100 MHz. This is mainly due to two reasons: (1) it is difficult to estimate nonidealities quantitatively at high frequencies, and (2) compensation techniques are mostly first-order correction techniques, so they are ineffective when the non-ideality is relatively large. An alternative to the MOSFET-C technique is to use transconductor-capacitor (G_m-C) structures for implementing high frequency filters [26, 27, 28, 5]. The schematic of a typical G_m-C integrator is as shown in Figure 11. A current proportional to the input voltage is derived using a transconductor with a transconductance G_m. This current is then forced into a capacitance C. The transfer function of this scheme is given by

$$H_i(s) = \frac{V_o(s)}{V_i(s)} = \frac{G_m}{sC} \tag{12}$$

Since the transconductors are usually made with a single stage, they can operate at high frequencies.

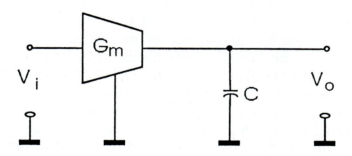

Fig. 11: G_m-C integrator.

In most high-frequency implementations, a fully-differential G_m-C topology is preferred. Although this topology constraint necessitates the use of additional circuits to provide common-mode feedback (CMFB), the improvement obtained in the dynamic range, power supply rejection, and immunity to noise coupled from other parts of the chip justifies the increase in circuitry.

3.2   Transconductor design

In most filter designs, the linearity of the G_m-C integrator is determined by the linearity of the transconductor. In this subsection, a classification of various linearization schemes is made.

3.2.1. *Designs that are based on MOSFET characteristics in the saturation region*

Here, transconductor designs that exploit the V-I characteristics of the MOSFET operating in the saturation region are considered. In all the design examples, these MOSFETs are assumed to be sufficiently long and wide such that short and narrow channel effects are absent. Although many different methods of obtaining a linear V-I relationship for the transconductor exist in the literature, we will examine only a few representative examples that are appropriate for high-frequency operation.

Perhaps the simplest of all schemes is the one in [26]. A simplified schematic of the transconductor configured as an integrator is as shown in Figure 12. M1 and M2 are the input transistors of a differential pair. When this transconductor is driven by a pair of balanced signal $+V_{in}/2$ and $-V_{in}/2$, the output voltage, V_{out} is devoid of even-order nonlinearities because of balanced operation. The transconductance of the cell can be varied by changing the bias current I_{bias}, thus implementing an integrator with a voltage-controlled gain constant. The simplicity of the structure makes this a good candidate for high-frequency operation. However, the strong odd-order nonlinearities of the simple differential pair severely limit the signal swing for moderate levels of distortion in the output voltage. Another drawback of the scheme is the following. To vary the G_m of the cell, the bias currents needs to be changed; in most VLSI processes, to tune out process tolerances, effects of aging, etc, the transconductance may need to be varied by as much as ± 75%. This tunability requirement implies a possible large variation in the bias current. Since sufficiently large signal swings are to be maintained over large variations of bias current, the design of this transconductor is difficult. Another problem with this implementation is the relatively low impedance at the output nodes. This introduces a leading phase error when the transconductor is configured as an integrator, and makes the circuit less suitable for high selectivity filters (filters that have a high selectivity generally need integrators with very low phase errors).

A topology that has good linearity is shown in Figure 13(a) [29]. For the terminal voltages as shown in the figure, the current I is approximately given by:

$$I = I_1 - I_2 = 4KV_x (V_1 - V_2) \tag{13}$$

where $K=(1/2) \mu C_{ox}W/L$ for M1 and M2. The floating voltage sources can be implemented using source followers and the corresponding implementation is as shown in Figure 13(b). The transconductance can be varied by changing the voltage V_x in Eq. 13. In the implementation shown in Figure 13(b), this is achieved by changing the current I_{bias}.

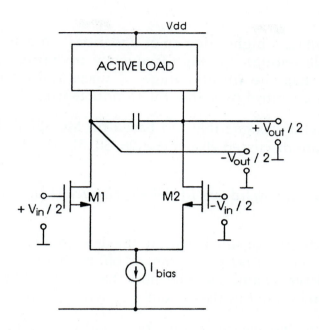

Fig. 12: Transconductance-C integrator implementation [26].

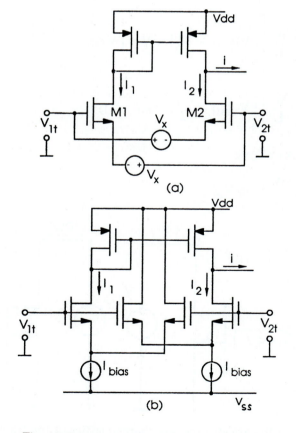

Fig. 13: Transconductor presented in [29].

This structure exhibits a higher degree of linearity than a simple source coupled pair and is simple enough for high-frequency operation. Just like the latter, though, this cell has the disadvantage of necessitating the change of bias currents to obtain a control over the transconductance.

Other transconductor designs that are based on MOSFET characteristics in the saturation region are discussed elsewhere [28, 30, 31, 32, 33].

3.2.2. *Designs that are based on MOSFET characteristics in the non-saturation region*

Several transconductor designs exploit the I-V characteristics of a MOSFET operating in the *non-saturation* region to obtain a linearised transconductor. Some representative examples will now be given. Wherever possible, the linearisation schemes used in these examples will be explained using Figure 3.

A scheme utilizing the triode region transconductance is shown in Figure 14 [34, 35, 36]. Two MOS transistors M1 and M2 are biased in the triode region. With the terminal voltages as shown in the figure, we have:

$$I_{DS1} = K' \frac{W}{L} \left[\left(V_{GS0} + \frac{V_{in}}{2} - V_T \right) V_{DS} - \frac{V_{DS}^2}{2} \right]$$

$$I_{DS2} = K' \frac{W}{L} \left[\left(V_{GS0} - \frac{V_{in}}{2} - V_T \right) V_{DS} - \frac{V_{DS}^2}{2} \right]$$

$$I_{out} = I_{DS1} - I_{DS2}$$

$$= K' \frac{W}{L} V_{DS} V_{in} \tag{14}$$

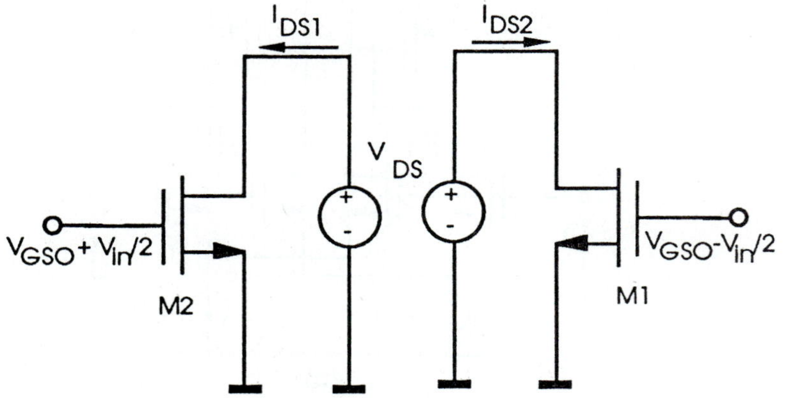

Fig. 14: Principle of linearisation of V-I characteristics of the transconductor in [34].

Thus the difference between the drain currents is a linear function of the input differential voltage. Several different circuit techniques can be employed to perform the difference operation. Measured results from filters that were implemented using this transconductor are described elsewhere [34,36].

All previous schemes used the transconductance of MOS transistors directly. The scheme to be described now uses instead the drain-source conductance as the key element of the transconductor. Figure 15(a) shows a simplified schematic of the scheme [37]. M1 and M2 are two source followers driving the transistor MW, operating in the triode region. Let the transconductor be driven by a pair of balanced voltages, $+V_{in}/2$ and $-V_{in}/2$. The voltages at the drain and source terminals of MW are $-V_Q + V_{in}/2$ and $-V_Q - V_{in}/2$, where $-V_Q$ is a level shift introduced by the source followers. If the W/L of M1 and M2 is fairly large, the value of V_Q remains fairly constant even for large swings in V_{in}. Thus the transistor MW sees an environment as shown in Figure 15(b). All the terminal voltages of MW can now be level-shifted up by V_Q (ignoring body effect). The correspondence of the result of this simplification to Figure 3(a) is obvious. Thus the current through MW is a linear function of V_{in}. This current is ultimately pumped into the capacitance C [Figure 15(a)] to implement an integrator. The transconductance of this cell can be controlled by the voltage V_G.

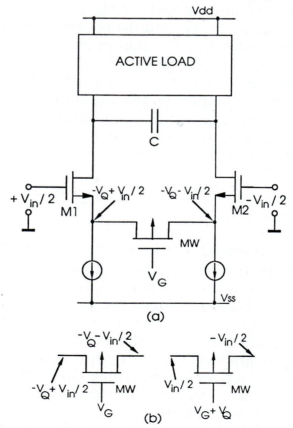

Fig. 15: (a) Transconductor-C integrator presented in [37]. (b) Principle of operation.

The simplicity of the scheme makes this design a good choice for high-frequency operation. Note that the transconductance can be varied through V_G without changing the current source bias, which makes it easier to maintain large swings over the tuning range, especially with limited power supply voltage. Also, although M1 and M2 have large widths, as shown in [26], the parasitic drain-gate overlap capacitances of the two devices do not affect circuit performance. The principal disadvantages of the scheme are the following: (1) The output impedance of the transconductor is fairly low, (2) the signal swing is limited due to odd-order nonlinearities and (3) the drain-source resistance of MW is controlled by a voltage dependent on the input common-mode voltage of the transconductor. Thus the G_m of this topology is dependent on the input common-mode voltage. Modifications, like using a folded-cascode can be used to improve the output impedance, at the cost of increased circuit complexity [5]. Also a cross-coupled input-stage [37] can be used for improved linearity. Filters using the above scheme are described elsewhere [5, 38].

The above transconductor used a device as in Figure 3(a) as a voltage-controlled resistor; if instead one uses the voltage controlled resistor of Figure 3(c) [1], one obtains the transconductor shown in Figure 16(a) [27]. Transistors MW1 and MW2 operate in the triode region. In Figure 16(b), these two transistors are shown with their corresponding terminal voltages. As in the previous case, all the voltages can be shifted up by V_Q and the operation of these two transistors simplifies to the case in Figure 3(c). The sum of the currents through MW1 and MW2 is now a linear function of the input voltage. As shown in [27], the linearity can be improved further when the ratio $[(W/L)M_{1,2}]/[(W/L)MW_{1,2}]$ is around 7. Unlike the previous topology, the G_m of this cell is not influenced by the input common-mode voltage, which is an advantage; the disadvantage of the scheme is that in order to tune the transconductance, one must vary the bias point of all the devices in the circuit through I_{bias}.

3.3. Gm-C filter synthesis

High-frequency, high-performance filter structures can be implemented using integrators and summers as building blocks. Figure 17 shows the two blocks constructed using transconductors. A number of well-known filter structures can be implemented using G_m-C topologies [39, 40, 41] A state-variable biquad is chosen as an example. The development of the G_m-C version is as follows. Figure 18(a) shows the active-RC Tow-Thomas filter. Figure 18(b) shows the corresponding signal-flowgraph (SFG). In Figure 18(c), the SFG is implemented using the basic building blocks of Figure 17 in a G_m-C topology. The SFGs can also be obtained directly from passive filter structures and converted to their corresponding active G_m-C equivalents.

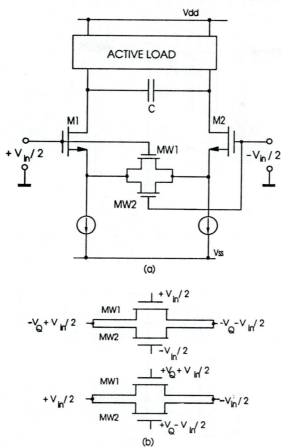

Fig. 16: (a) Transconductor topology in [27]. (b) Principle of operation.

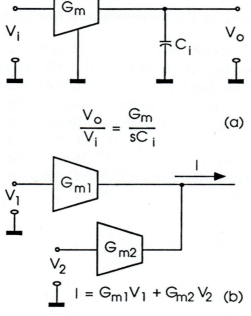

Fig. 17: (a) G_m-C integrator. (b) Summer.

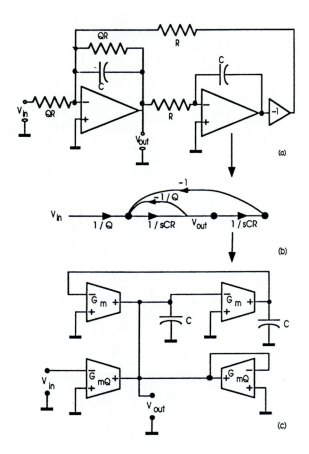

Fig. 18: Deriving the G_m-C equivalent of a Tow-Thomas biquad.

Along similar lines, a 7th-order elliptic filter can be synthesized as shown in Figure 19. (This structure was derived from the SFG obtained directly from the passive doubly-terminated ladder using the leapfrog synthesis technique.)To facilitate signal addition, a two input-stage, single output-stage transconductor was used. The resulting circuit was successfully implemented in a 1 μm CMOS technology with a cutoff frequency of 4.4 MHz. Further details of the implementation are discussed elsewhere [5]. Worst-case design or G_m-C filters follows along the lines presented for MOSFET-C filters in section 2.4.

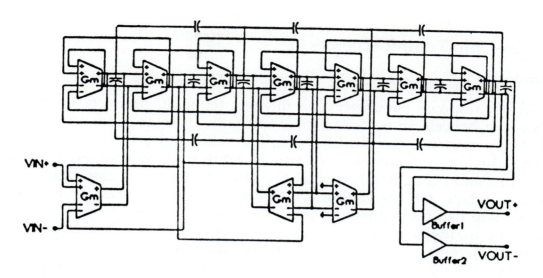

Fig. 19: Seventh-order elliptic filter (G_m-C realization) [5] (Copyright © 1990 by IEEE).

4. MOSFET-C vs Transconductor-C Filters

The main advantage of the MOSFET-C approach is simplicity. By using MOS resistors, capacitors, and op amps as basic building blocks the design is kept very simple and immune to parasitic capacitance, and a wealth of knowledge on active RC filters can be used to advantage. In addition, since the op amp is a generic element it continues to be improved for a variety of applications and is available as a building block in a typical design environment. The filter designer thus is not faced with the problem of designing special-purpose building blocks, and the design time is short. Transconductor-C filters use transconductors, which are not generic blocks and have to be designed at the device level for most filtering tasks. Parasitic capacitances in such filters should be taken into account and be "designed in." Nevertheless, transconductor-C filters have been initially favored for high-frequency applications, the argument being that transconductors, which do not need to drive resistors, can be designed with higher frequency performance than those of op amps that must drive resistors. This argument may be valid for standard CMOS processes, due to the limited amplifier design flexibility. However, with the advent of BiCMOS processes, in which excellent bipolar transistors are available with cut-off frequencies of the order of 5-10 GHz, this argument may no longer be valid. Indeed, professional video filters are now being designed using both MOSFET-C and transconductor-C techniques [14].

5.   On-Chip Tuning Schemes

Introduction

Frequency response parameters of a filter are set by RC products or C/G_m ratio. (They will henceforth be called "time constants.") An accurate and predictable frequency response for the filter implies stable and accurate time constants within the filter. However, because of fabrication tolerances, temperature variations and aging, absolute values of time constants in a chip can vary by as much as $\pm$ 50%. On the contrary, ratios of component values of like elements on the chip can be maintained to an accuracy of around 0.5%. An "on-chip" automatic tuning scheme implements time constants as a function of a stable external reference quantity and a ratio of component values of like elements [42]. Thus, it helps maintain an accurate frequency response for the filter.

Phase errors in the integrators of the filter[2] also cause large deviations in the frequency response of the filter. Using integrators that have a variable phase at unity-gain frequency (say using a second control voltage), it is possible to devise an on-chip tuning scheme that can correct phase errors also. This tuning loop that corrects both time constant and phase-errors is called a "vector-locked loop" [43].

Automatic tuning schemes are usually negative feedback systems. Like all other such systems, they observe a given parameter (in our case it is the time constant or integrator phase error), calculate the error in the given parameter, and using the error signal change the observed parameter such that the error is reduced. In most automatic tuning schemes, the external reference used is the frequency of a crystal oscillator. Also, instead of measuring the errors in the filter (slave) itself, a similar filter (master) is constructed on the same chip. The control circuitry monitors its performance continuously and makes the necessary corrections to maintain an accurate master filter frequency response. Since the master and slave are well matched, the slave also has an accurate and stable frequency response.

One general approach to the automatic tuning of the filters presented is a variation of schemes proposed elsewhere [4, 17, 27, 42, 44-53] and is illustrated in Figure 20. The parameter that is tuned is the integrator time constant (RC product or C/G_m ratio). The reference circuit (or master) can be an oscillator made by the same basic structures as are used in the filter itself. A phase comparator continuously compares the output of the oscillator to an external

---

[2]   The phase error in an integrator is the difference between the actual phase of the integrator at its unity-gain frequency and the ideal value of -90 degrees. Negative phase errors are said to be "lagging" errors and positive ones are said to be "leading" errors.

clock signal and adjusts V_C, a control voltage that can change the gain constant of the integrators, until the oscillator tracks the clock. At that point, the gain constants of the integrators within the oscillator attain the desired value. Since all relevant transistors and capacitors in the main filters are ratio-matched, respectively, to the transistors and capacitors of the oscillator, and since a common control voltage is used for all transistors (used for tuning the time constants), the various integrator gain constants within the main filter attain the desired value, and frequency response becomes stabilized.

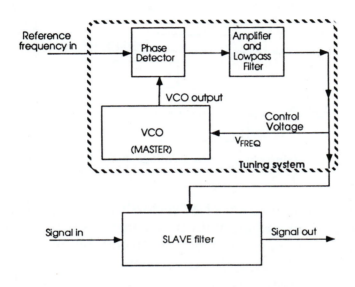

Fig. 20: PLL type tuning scheme.

Most of the tuning schemes are based on the principle explained above. However, different variations have been proposed in the literature with their relative merits and demerits.

One variation of the above tuning scheme uses a *filter* as the reference circuit [44, 45, 26, 5], again made of structures matched to those in the main filter. A block diagram of such a scheme is shown in Figure 21. The external reference signal is fed into the phase detector and the master filter, which is usually a second order section. The phase difference, measured by the phase detector, gives a measure of how much the center frequency of the 2nd-order section is different from the frequency of the reference signal. This "error" signal is amplified and fed back as the voltage that controls the time constants of the integrators in the master filter. At steady state, the error signal is made very small,which means that the center frequency of the 2nd-order section will be coincident with the frequency of the reference signal. This stabilizes the time constants of the integrators in the reference filter. As mentioned before, since the components in the main filter are ratio-matched to those in the reference filter, the frequency response of the main filter gets stabilized.

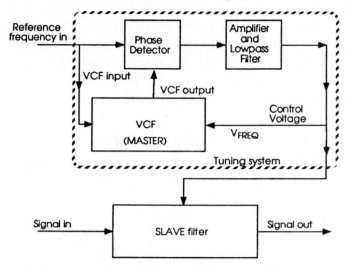

Fig. 21:   Automatic tuning circuit with a voltage-controlled filter as the master.

Figure 22 shows the block diagram of a "vector-locked loop." Both the slave and the master are made of integrators with time constants controlled by the voltage V_{freq}, and phase at unity-gain frequency controlled by another voltage V_{phase}. (The master can be a VCO, but in this discussion, we will restrict it to a voltage-controlled filter). The feedback system consisting of the phase detector and lowpass filter form a negative feedback loop, whose operation was described above. The network within the dotted lines along with the master form an amplitude-locked loop. The topology of the master filter is such that any phase errors in its constituent integrators will manifest themselves as a change in its gain at resonance frequency. The amplitude-locked loop detects any such changes in the resonance gain from the ideal value and tunes them out using the control voltage V_{phase}. Again, since the same control voltage controls the phase of the integrators in the slave, a stable and accurate frequency response is obtained from the slave too. The design of vector-locked loops can be very complicated. It is best to make their use unnecessary by starting from a robust design and limiting automatic tuning to a single control loop (Figures 20, 21).

The factors that introduce tuning errors are [54] (1) offsets in the phase detector or lowpass filter and (2) distortion in the output of the reference circuit (whether it is an oscillator or a filter). Keeping distortion low at the output of a reference filter is relatively simple, since the output level can be controlled by the reference input level. However in a scheme that has an oscillator as the reference circuit, special circuit techniques must be used to maintain small signal levels for low distortion. In [10] and [27] nonlinear elements were used across capacitors in the integrators to limit oscillation amplitude. In a vector-locked loop that has an oscillator as the reference circuit, the control of oscillation amplitude is performed using a different technique [48, 49]. As described before, in a vector-locked loop, the phase of the integrators at unity-gain frequency can be varied using the phase-control voltage (or current). When the phase at unity-gain frequency of the integrators is less than -90 degrees, the poles of the VCO, ignoring for the moment, nonlinear effects, move into the right-half plane and the oscillation amplitude increases. When the phase is greater than -90 degrees, the poles of the VCO move into the left half-plane and the amplitude of oscillation decreases. When the poles are on the jω axis, that is when the integrators have no phase errors, the oscillation amplitude will remain steady. This property can be used to control the oscillation amplitude to low values using the phase-control voltage. Thus a vector-locked loop with an oscillator as the reference circuit achieves both phase tuning and amplitude limiting.

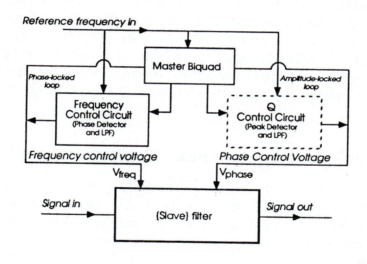

Fig. 22:   Vector-locked loop.

Some of the advantages of the tuning scheme that uses a VCO as the master are (1) *no* tuning error is introduced when the various blocks in the circuit have offsets [27] and (2) the reference input can be a square instead of a sine wave. In contrast, when the reference circuit is a filter, care has to be taken to reduce the offsets in the various subcircuits. The reference input to the tuning system which has a filter as the master should have low distortion. In most applications, conversion of square-waves to sine waves can be performed using simple non-critical circuits [55]. Some of the advantages in choosing a voltage-controlled filter instead of a VCO as the master are the following: (1) A non-linear amplitude limiting scheme is necessary to maintain low distortion in a tuning scheme that uses a VCO. This is unnecessary when the master is a voltage-controlled filter. (2) Design of a stable automatically tuned filter with a wide capture/lock range is easier than that for a PLL with similar characteristics. (3) Simulating automatically tuned filter systems is easier, more accurate, and faster than simulating for a VCO based tuning system.

Yet another scheme uses no clock signal at all, the external reference being a temperature-insensitive resistor [56]. The reference circuit is now simply a resistor or group of resistors, and a comparison circuit compares resistance value of the external resistance and the internal time constant determining resistance. The error signal so obtained is processed in such a way that the absolute value of the error is reduced. The external resistance must be initially adjusted until the prescribed frequency response is achieved. However, this type of control loop cannot tune out any subsequent variations in the capacitances due to temperature.

A direct "self-tuning" scheme, in which the filter itself is being periodically tuned (as opposed to being locked to a reference circuit), is described elsewhere [57] and has been used in commercial products [18]. If the service cannot be interrupted while the filter is being tuned, another filter can be placed in the signal path through a suitable switching arrangement that avoids transients [57]. This scheme may prove attractive for high-Q applications.

6. A Warning Concerning Computer Simulations

Although decent theory for correctly modeling MOSFETs is available, the implementation of MOSFET models in CAD programs is often totally inadequate for analog applications; the way model parameters are chosen by automatic "parameter extraction systems" often compounds this problem. Most of these problems can be traced to the "historical" view of the MOSFET as a "digital" device, and the fact that even to this day MOSFET technology is driven mostly by digital applications. The robustness of digital design against poor model implementations and/or bugs has helped the latter go unnoticed for many years. For analog work, though, sound models are a must. Numerous common modeling errors, and ways to rectify them, are described elsewhere [9].

7. Conclusion

Analog filters provide advantages in terms of power, chip area, or high frequency performance in applications requiring moderate dynamic range. Between switched-capacitor and continuous-time techniques, the latter offer advantages in certain cases, and are now used commercially in several applications; for example, in video signal processors, in pulse equalizers used in computer communication networks, and in disk drives. Further commercial applications are in the offing. Thus, continuous-time filters have been added to the choices available to the designer of high-performance integrated systems.

References

[1]　Y.P. Tsividis, M. Banu, and J. Khoury, "Continuous-time MOSFET-C filters in VLSI," *IEEE Journal of Solid-State Circuits*, vol. SC-21, pp. 15-28, 1986.

[2]　E. A. Vittoz, "Future of analog in the VLSI environment," *Proc. 1990 Int. Symp. Circuits Systems*, pp. 1372-1375.

[3]　J.O. Voorman, W. Bruls, and P. Barth, "Integration of analog filters in a bipolar process," *IEEE J. Solid-State Circuits*, vol. SC-17, pp. 713-722, August 1982.

[4]　K. Miura, Y. Okada, M. Shiomi, M. Masuda, E. Funaki, Y. Okada, and S. Ogura, "VCR signal processing LSIs with self-adjusted integrated filters," *Proc. Bipolar Circuits and Technology Meeting*, pp. 85-86, 1986.

[5]　V. Gopinathan, Y.P. Tsividis, K.S. Tan, and R.K. Hester, "Design considerations for high-frequency continuous-time filters and implementation of an antialiasing filter for digital video," *IEEE Journal of Solid-state Circuits*, vol. 25, pp. 1368-1378, December 1990.

[6]　G.A. De Veirman and R.G. Yamasaki, "Fully integrated 5 to 15 MHz programmable bipolar Bessel lowpass filter," *Proc. 1990 Int. Symp. Circuits Systems*, pp. 1155-1158.

[7]　F. Goodenough, "Voltage-tuneable linear filters move onto a chip," *Electronic Design*, pp. 43-54, February 1989.

[8]　J. Khoury, "A 15MHz CMOS continuous-time Bessel filter for disk drives," *Digest*, ISSCC, pp. 134-135, February 1991.

[9]　Y.P. Tsividis, *Operation and Modeling of the MOS Transistor*. New York: McGraw Hill, 1987.

[10]　M. Banu and Y.P. Tsividis, "Fully integrated active RC filters in MOS technology," *IEEE J. of Solid-State Circuits*, vol. SC-18, pp. 644-651, December 1983.

[11]　Z. Czarnul, "Modifications of Banu-Tsividis continuous-time integrator structure," *IEEE Transactions on Circuits and Systems*, vol. 33, pp. 714-716, July 1986.

[12] F.S. Shoucair and W.R. Patterson, "Harmonic suppression in unbalanced analog MOSFET circuit topologies using body signals," *Proc. ISCAS*, pp. 1151-1154, 1990.

[13] M. Banu, J.M. Khoury, and Y. Tsividis, "Fully differential operational amplifiers with accurate output balancing," *IEEE Journal of Solid-state Circuits*, vol. 23, pp. 1410-1414, December 1988.

[14] A. van Beooijen, N. Ramalho, and J.O. Voorman, "Balanced integrator filters at video frequencies," *Proc. European Solid-state Circuits Conference*, 1991.

[15] Y. Tsividis and J.O. Voorman, *Integrated Continuous-time Filters*. New York: IEEE Press, to be published.

[16] A.S. Sedra, and P.O. Brackett, *Filter Theory and Design: Active and Passive*. Portland, and OR: Matrix, 1978.

[17] M. Banu and Y. Tsividis, "An elliptic continuous-time CMOS filter with on-chip automatic tuning," *IEEE J. Solid-State Circuits*, vol. SC-20, pp. 1114-1121, Dec. 1985.

[18] G.J. Smolka, U. Reidl, U. Grehl, W. Veit, and G. Geiger, "A low noise trunk interface circuit," *Proc. European Solid-state Circuits Conference*, 1991.

[19] M. Ismail, S. V. Smith, and R.G. Beale, "A new MOSFET-C universal filter structure for VLSI," *IEEE Journal of Solid-state Circuits*, vol. 23, pp. 183-194, February 1988.

[20] Z. Czarnul, S.C. Fang, and Y. Tsividis, "Improving linearity in MOS fully-integrated continuous-time filters," *Proc. ISCAS*, 1986.

[21] P.R. Gray, "Private communication."

[22] Y. Tsividis and B. Shi, "Cancellation of distortion of any order in integrated active-RC filters," *Electronics Letters*, vol. 21, pp. 132-134, February 14, 1985.

[23] J.M. Khoury and Y.P. Tsividis, "Analysis and compensation of high-frequency effects in integrated MOSFET-C continuous-time filters," *IEEE T. Circuits and Systems*, vol. CAS-34, pp. 862-875, August 1987.

[24] Y. Tsividis, "Minimal transistor-only micropower integrated VHF active filter," *Electronics Letters*, vol. 23, pp. 777-778, July 16, 1987.

[25] M. Banu and Y. P. Tsividis, "Detailed analysis of nonidealities in MOS fully integrated active RC filters based on balanced networks," *IEEE Proc. Part G*, vol. 131, pp. 190-196, October 1984.

[26] H. Khorramabadi and P.R. Gray, "High-frequency CMOS continuous-time filters," *IEEE J. Solid-state Circuits*, vol. SC-19, pp. 939-948, December 1984.

[27] F. Krummenacher and N. Joel, "A 4 MHz CMOS continuous-time filter with on-chip automatic tuning," *IEEE J. Solid-state Circuits*, vol. SC-23, pp. 750-758, June 1988.

[28] C.S. Park and R. Schaumann, "Design of a 4 MHz analog integrated CMOS transconductance-C bandpass filter," *IEEE J. Solid-state Circuits*, vol. SC-23, pp. 987-996, August 1988.

[29] A. Nedungadi and T.R. Viswanathan, "Design of linear CMOS transconductance elements," *IEEE T. Circuits and Systems*, vol. CAS-31, pp. 891-894, October 1984.

[30] R. Torrance, T.R. Viswanathan, and J.V. Hanson, "CMOS voltage to current transducers," *IEEE T. Circuits and Systems*, vol. CAS-32, pp. 1097-1104, November 1985.

[31] E. Klumperinck, E. Zwan, and E. Seevinck, "CMOS variable transconductance circuit with constant bandwidth," *Electronics Letters*, vol. 25, pp. 675-676, May 1989.

[32] B. Nauta and E. Seevinck, "Linear transconductance element for VHF filters," *Electronics Letters*, vol. 25, pp. 448-450, March 1989.

[33] S.W. Kim and R.L. Geiger, "Design of a CMOS differential amplifier using a source-coupled backgate pair," *Proc. 1990 Int. Symp. Circuits Systems*, pp. 929-932.

[34] J. Pennock, P. Frith, and R.G. Barker, "CMOS triode transconductor continuous-time filters," *Proc. 1986 Custom Integrated Circuits Conference*, pp. 378-381.

[35] U. Gatti, F. Maloberti, and G. Torelli, "A novel CMOS linear transconductance cell for continuous-time filters," *Proc. ISCAS*, pp. 1173-1176, 1990.

[36] R. Alini, A. Baschirotto, and R. Castello, "8-32 MHz tunable continuous-time filter," *Proc. European Solid-state Circuits Conference*, 1991.

[37] Y. Tsividis, Z. Czarnul, and S.C. Fang, "MOS transconductors and integrators with highlinearity," *Electronics Letters*, vol. 22, pp. 245-246, February 1986.

[38] A. Kaiser, "A micropower CMOS continuous-time lowpass filter", *IEEE Journal of Solid-state Circuits*, vol. 24, pp. 736-743, June 1989.

[39] J.O. Voorman, "Analog integrated filters or continuous-time filters for LSI and VLSI," *Revue de Physique Appliqué*, vol. 22, pp. 3-14, 1987.

[40] R.L. Geiger and E. Sanchez-Sinencio, "Active filter design using operational transconductance amplifiers - a tutorial," *IEEE Circuits and Devices Magazine*, vol. 1, pp. 20-32, 1985.

[41] A.C.M. de Queiroz, L.P. Colaba, and E. Sanchez-Sinencio, "Signal-flow graph OTA-C integrated filters," *Proc. ISCAS*, pp. 2165-2168, 1988.

[42] J.R. Canning and G.A. Wilson, "Frequency discriminator circuit arrangement," *UK Patent no. 1421093*.

[43] D. Senderowicz, D. Hodges, and P.R. Gray, "An NMOS integrated vector-locked loop," *Proc. 1982 Int. Symp. Circuits Systems*, pp. 1164-1167.

[44] K.R. Rao, V. Sethuraman, and P.K. Neelakantan, "A novel 'follow the master' filter," *Proc. IEEE*, vol. 65, pp. 1725-1726, December 1977.

[45] J.R. Brand, R. Schaumann, and E. M. Skei, "Temperature stabilized active filters," *Proc. 1977 Midwest. Symp. Circuits Systems*, pp. 295-300.

[46] K.W. Moulding, J.R. Quartly, P.J. Rankin, R.S. Thompson, and G.A. Wilson, "Gyrator video filter IC with automatic tuning," *IEEE J. Solid-State Circuits*, vol. SC-15, pp. 963-968, December 1980.

[47] K.S. Tan and P.R. Gray, "Fully integrated analog filters using bipolar-JFET technology," *IEEE J. Solid-state*, vol. SC-13, pp. 814-821, December 1978.

[48] Y.T. Wang, F. Lu, and A.A. Abidi, "A 12.5 MHz CMOS continuous-time bandpass filter," *Digest, ISSCC*, pp. 198-199, February 1989.

[49] B. Nauta and E. Seevinck, "Automatic tuning of quality factors for VHF CMOS filters," *Proc. 1990 Int. Symp. Circuits Systems*, pp. 1147-1150.

[50] P.M. VanPeteghem and R. Song, "Tuning strategies in high-frequency integrated continuous-time filters," *IEEE Transactions on Circuits and Systems*, vol. 36, pp. 136-139, January 1989.

[51] K.A. Kozma, D.A. Johns, and A.S. Sedra, "An adaptive tuning circuit for integrated continuous-time filters," *Proc. ISCAS*, pp. 1164-1166, 1990.

[52] G. Venugopal and K.R. Rao, "A novel technique for the on-chip tuning of monolithic filters," *Proc. IEEE*, vol. 75, pp. 257-258, February 1987.

[53] P. Ashar and K.R. Rao, "Magnitude locked loop," *Proc. of IEEE*, vol. 76, pp. 201-203, February 1988.

[54] V. Gopinathan, *High-frequency transconductor-C continuous-time filters*. New York: Ph.D. Thesis, Columbia University, 1990.

[55] J. Fattaruso and R. G. Meyer, "MOS analog function synthesis," *IEEE J. of Solid-state Circuits*, vol. SC-22, pp. 1056-1063, December 1987.

[56] J.O. Voorman, W. Bruls, and P.J. Barth, "Integration of analog filters in a bipolar process," *IEEE J. of Solid-state Circuits*, vol. SC-17, pp. 713-722, August 1982.

[57] Y. Tsividis, "Self-tuned filters," *Electronics Letters*, vol. 17, pp. 406-407, June 1981.

Chapter 7

Switched-Capacitor Filter Synthesis

Adel S. Sedra and Martin Snelgrove

Department of Electrical Engineering, University of Toronto, Toronto, Ontario, Canada M5S 1A4

1. Introduction

There are currently two general approaches to the implementation of precision analog filtering functions in monolithic integrated-circuit form. Although both approaches use MOS technology, they differ in a fundamental way: while in one the analog signal is processed directly in its continuous-time form, the other (switched-capacitor or switched current) approach is based on processing samples of the analog signal and thus the resulting circuits are discrete-time or sampled-data systems. The continuous-time approach is described by Tsividis and Gopinathan in another chapter of this book. The object of the present chapter is to provide a concise exposition of some of the practical methods of designing switched-capacitor (SC) filters. The newer switched-current filters are very similar to SC filters, and their designs may be systematically derived from SC designs [1], so they are not considered separately.

Switched-capacitor filters have grown from active-RC filters. The latter utilize op amps together with resistors and capacitors and, depending on production volume, are implemented either on printed circuit boards using IC op amps, metal-film resistors and polystyrene capacitors, or as thick or thin film hybrid circuits. Active-RC is now a mature filter implementation technology with a comprehensive, well-understood design theory. These filters are extensively employed in telecommunication and instrumentation systems and excel in low-frequency (<100 kHz) applications.

Attempts to *directly* fabricate active-RC filters in monolithic form have not been successful for two reasons: (1) the need for large-valued resistors and capacitors (especially for low-frequency filters), and (2) the need for accurate RC time constants. It will be seen that the SC filter technique circumvents both problems. Switched-capacitor filters were originally developed on the principle that a capacitor C periodically-switched between two circuit nodes at a sufficiently high rate (f_c) is approximately equivalent to a resistor $R = 1/Cf_c$ connecting the two nodes. It is thus possible to realize filter functions using op amps, capacitors and periodically-operated switches. Since MOS technology provides high-quality capacitors, offset-free switches, and moderate-quality op amps it is eminently suited for the realization of SC filters.

Switched-capacitor filters have two other features that make them particularly suited for MOS Ic technology:

(1) Large resistors can be simulated using small-sized capacitors. This is a result of the inverse relationship between the value of the switched capacitor C and the realized equivalent resistance. As an example, a 1 pF capacitor, which occupies about 1000 μ^2 of chip area, when switched at the rate of 100 kHz simulates the operation of a 10 MΩ resistor.

(2) The precision of the realized frequency response is dependent on the precision of the clock frequency and the tolerance to which capacitor ratios are implemented. To see how this comes about consider a time constant C_2R_1 and let R_1 be realized as a capacitor C_1 switched at the rate of f_c. Thus the time constant realized is $1/f_c$ (C_2/C_1). Since f_c can be accurately controlled using a crystal resonator in the clock oscillator and since capacitor ratios can be realized to a high accuracy (as good as 0.1%) in MOS technology, MOS switched-capacitor filters can achieve very precise frequency responses.

Although the principle of generating frequency-selective responses using periodically-switched-capacitors has been known since at the least the mid 1960s, integrated circuit switched-capacitor filters became a reality only in the late 1970s [2, 3].

While SC filters were originally designed as approximations to RC circuits, more accurate methods are now available and are described in the later parts of this chapter. These analyze the SC circuit in discrete time, using the z transform, and are closely related to design techniques for digital filters. While SC and digital design share a mathematical base, the practicalities of the two areas lead to important differences; for example,

(1) common digital filter structures such as companion form and parallel form are almost never used in SC circuits, because they demand unattainable component accuracy and amplify internal noise.

(2) digital filters cannot implement "delay-free loops," which are routinely used in SC circuits for notch formation.

It is interesting to note that the best SC circuits still take their basic structures from active-RC circuits and from passive LC ladders, rather than from digital filter theory.

The focus of this chapter is on synthesis methods for SC filters. It is important, however, that the designer acquire an appreciation of MOS technology limitations, especially the effect of such limitations on the characteristics of SC circuit components (op amps, switches and capacitors). For such a study we refer the reader to [4, 5].

2. First-Order Building Blocks

One of the most popular methods for designing active-RC filters utilizes op amp-RC integrators as basic building blocks. A second-order filter section, known as a *biquad*, is realized using a feedback loop containing one inverting and one noninverting integrator. A high-order (i.e., >2) filter is realized either as a cascade of these two-integrator-loop biquads or, if a low-sensitivity circuit is desired, as a simulation of the operation of an LC ladder prototype. These ladder simulation circuits also utilize op amp-RC integrators for simulating the operation of the L and C elements of the ladder prototype.

As in the active-RC case, integrators are the basic building blocks of SC filters. Figure 1a shows the inverting and Figure 1b shows the noninverting SC integrator circuits that have become standard building blocks of SC networks [6, 7]. It will be seen that these integrator circuits are insensitive to the stray capacitances present between the source and drain diffusions of the MOS transistors and the substrate. This feature turns out to be the most important in the selection of a building block. For this reason we shall concentrate on SC filter design methods that employ only these integrators.

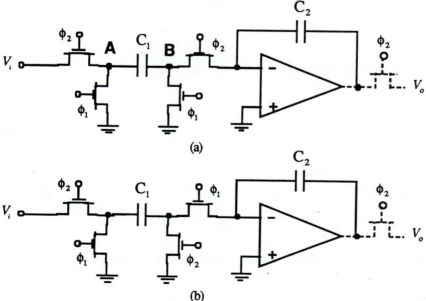

Fig. 1: Standard stray-intensive SC filter building blocks: (a) inverting integrator, (b) noninverting integrator.

2.1. Basic operation

Both integrator circuits utilize a two-phase nonoverlapping clock having the waveform sketched in Figure 2. Operation of the inverting integrator circuit of Figure 1a is illustrated in Figure 3. During clock phase ϕ_1 both plates of C1 are grounded (Figure 3a) and thus C_1 is fully discharged. Meanwhile the op amp is isolated and thus C_2 maintains its charge. Then, during clock phase ϕ_2 capacitor C_1 is connected between the source υ_1 and the virtual ground input terminal of the op amp (Figure 3b). Current flows in the direction indicated and C_1 charges to the instantaneous value of υ_I. Since the same current flows through C_2, the charge on C_2 will change by an amount equal to the charge deposited on C_1. It follows that during one clock period, T, a charge of $C_1\upsilon_1$ is transferred to C_2 in the direction indicated by the arrow in Figure 3b.

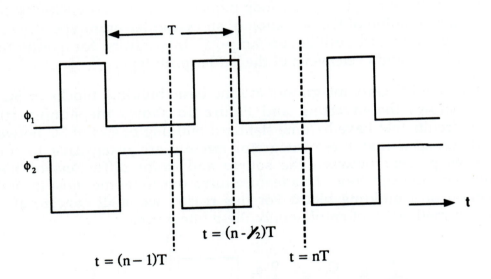

Fig. 2:   Waveform of the two-phase nonoverlapping clock used in SC filters.

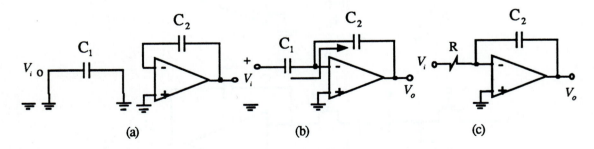

Fig. 3:   Operation of the inverting integrator circuit of Fig. 1a: (a) during ϕ_1, (b) during ϕ_2, and (c) continuous-time equivalent circuit.

Now, if clocking is performed at a rate ($f_c=1/T$) much higher than the rate of change of the input signal the charge transfer process appears almost continuous and thus can be modeled by a continuous current I flowing between v_I and the op amp input with $I=C_1 v_I/T$. This current can be obtained by connecting a resistance R between v_I and the op amp inputed as indicated in Figure 3c. The value of this equivalent resistance R is given by

$$R \equiv \frac{v_I}{I} = \frac{T}{C_1} \tag{1}$$

The continuous-time equivalent circuit in Figure 3c is obviously that of an inverting (Miller) integrator. Its transfer function is given by:

$$\frac{V_0(\omega)}{V_i(\omega)} = -\frac{1}{j\omega C_2 R} \tag{2}$$

Substituting for R from (1) gives the transfer function of the SC circuit of Figure 1a as

$$\frac{V_0(\omega)}{V_i(\omega)} = -\frac{1}{j\omega T}\frac{C_1}{C_2} \tag{3}$$

Thus, the circuit of Figure 1a when clocked at a frequency much higher than the signal frequencies implements an inverting integrator function with a time constant of $T(C_2/C_1)$. The SC circuit of Figure 1a will have the transfer function derived above irrespective of the value of parasitic capacitances between nodes A and B and ground. This follows because during ϕ_1 these stray capacitances are fully discharged, and during ϕ_2 one is connected across v_1 and the other is connected between the op amp input terminals. The end result is that neither of the two parasitic capacitors participates in the charge transfer process and the circuit is fully strays-intensive.

The operation of the circuit in Figure 1b is illustrated in Figure 4. During ϕ_2, C_1 is charged to the value of v_1 and C_2 maintains its previous charge. During ϕ_1, C_1 is fully discharged, thus transferring a charge of $C_1 v_1$ to C_2 in the direction indicated in Figure 4b. Following a reasoning process similar to that used above for the inverting circuit, one can show that if clocking is done at a sufficiently high rate the circuit has the continuous-time equivalent shown in Figure 4c. This latter circuit has a transfer function similar to that in (2) except with a positive sign.

Thus the SC circuit of Figure 1b when clocked at a high rate implements a noninverting integrator function with a time constant of $T(C_2/C_1)$. Note that unlike the active-RC case where an additional op amp is usually needed to implement a noninverting integrator, here a simple reversal of clock phases achieves the required inversion. Finally, the reader can easily verify that this circuit too is fully insensitive to stray capacitances.

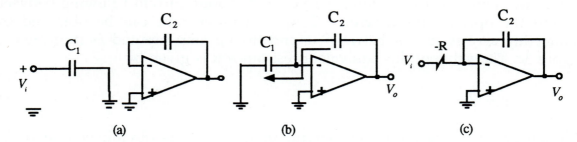

Fig. 4:   Operation of the noninverting integrator circuit of Fig. 1b: (a) during ϕ_2, (b) during ϕ_1, and (c) continuous-time equivalent circuit.

2.2. Exact transfer functions and the LDI variable

In deriving the transfer functions above it was assumed that the clocking frequency is much higher than the signal frequencies so that the discrete-time operation of the circuits appears almost continuous. Under this approximation, a switched capacitor C_1 is equivalent to a resistance (T/C_1).

The equivalent resistance is positive if the switch phasing of Figure 1a is used and is negative with the phasing of Figure 1b. This switched-capacitor-resistor equivalence has been used as the basis for the design of SC filters, as will be illustrated in the next section. In many cases, however, the approximation inherent in this equivalence leads to filter circuits whose responses deviate from the desired. In order to understand the reason for such performance deviation we shall derive the exact transfer functions of the basic building blocks of Figure 1.

Consider the inverting circuit of Figure 1a and refer to the clock waveform in Figure 2. The switch drawn in broken lines indicate that the output is sampled at the end of clock phase ϕ_2, that is, at the instant t=nT. The value of the output voltage at this instant, $v_o(n)$, can be expressed (using the circuit in Figure 3b) as the sum of the output voltage just before ϕ_2 goes high, $v_o(n-1/2)$, and the change in the output voltage that occurs as a result of the charge transferred to C_2 during ϕ_2. This charge is $C_1v_1(n)$ and the transfer is in the direction indicated in Figure 3b. Thus,

$$v_0(n) = v_0(n-1/2) - \frac{C_1v_1(n)}{C_2} \tag{4}$$

The value of $v_0(n-1/2)$ can be found from Figure 3a which shows the circuit configuration during ϕ_1. We see that during ϕ_1, C_2 maintains its charge, thus

$$v_0(n-1/2) = v_0(n-1) \tag{5}$$

Substituting in (4) results in

$$v_0(n) = v_0(n-1) - \frac{C_1}{C_2} v_1(n) \tag{6}$$

Since the operation of the circuit in Figure 1a is repeated identically every clock period, the first-order difference, eq. (6), provides a complete description of the operation. To obtain the frequency response of the circuit, the z transform is applied to (6) yielding

$$V_0(z) = z^{-1} V_0(z) - \frac{C_1}{C_2} V_i(z)$$

which after rearranging results in

$$\frac{V_0(z)}{V_i(z)} = -\frac{C_1}{C_2} \frac{1}{1-z^{-1}} \tag{7}$$

This transfer function can be written in the alternate form

$$\frac{V_0(z)}{V_i(z)} = -\frac{C_1}{C_2} \frac{z^{1/2}}{z^{1/2} - z^{-1/2}} \tag{8}$$

To obtain the response for physical frequencies ω we substitute $z=e^{j\omega T}$, which corresponds to evaluating the transfer function in (8) along the unit circle in the complex z plane (*see* Figure 5). This results in

$$\frac{V_0}{V_i}(\omega) = -\frac{C_1}{C_2} \frac{e^{j\omega T/2}}{j2 \sin(\omega T/2)} \tag{9}$$

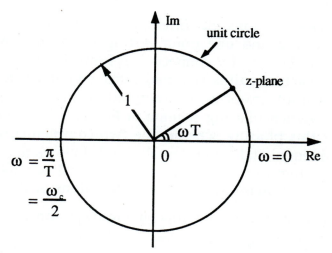

Fig. 5:   The frequency response of discrete-time circuits is found by evaluating the transfer function along the unit circle in the z plane.

This transfer function is obviously not that of an ideal integrator; apart from the excess phase lead represented by the factor $e^{j\omega T/2}$ the magnitude is inversely proportional to $\sin(\omega T/2)$ and not to ω as is the case for an ideal integrator. It will be seen shortly that the excess phase lead is not important. The dependence on $\sin(\omega T/2)$, however, unless taken into account in the synthesis process, causes deviations in the filter response. These deviations can be kept small by clocking the filter as a sufficiently high rate so that $\omega T/2 <<1$ and thus $\sin(\omega T/2) \cong \omega T/2$ which leads to

$$\frac{V_o}{V_i}(\omega) \cong -\frac{C_1}{C_2}\frac{e^{j\omega T/2}}{j\omega T} \tag{10}$$

Note that apart from the excess phase lead, this transfer function is identical to that derived above [eq. (3)] using the SC-resistor equivalence.

A similar analysis can be performed on the noninverting circuit of Figure 1b and results in the transfer function

$$\frac{V_o(z)}{V_i(z)} = \frac{C_1}{C_2}\frac{z^{1/2}}{z^{1/2} - z^{1/2}} \tag{11}$$

For physical frequencies, $z=e^{j\omega T}$,

$$\frac{V_o}{V_i}(\omega) = \frac{C_1}{C_2}\frac{e^{-j\omega T/2}}{j2\sin(\omega T/2)} \tag{12}$$

Except for the expected polarity reversal and the fact that the excess phase here is a lag, this transfer function is identical to that of the inverting circuit. Thus, the comments made above regarding the nonideality of the inverting circuit as an integrator apply equally well to the noninverting circuit.

As will be seen in later sections, SC filters are designed using feedback loops comprising one inverting and one noninverting integrator. We note that the circuits in Figure 1 can be directly connected to form these two-integrator loops because the input of each circuit is sampled during ϕ_2 and it has been assumed that the output of each integrator is also sampled during ϕ_2. We shall, therefore, consistently use the switch phasing scheme of Figure 1 in most of our designs. Furthermore, under this phasing scheme, the two complementary circuits have excess phase shifts of equal and opposite polarity. It follows that the excess phase around every two-integrator loop will be zero and we need not concern ourselves with the excess phase in the transfer functions of the integrators.

The transfer functions of the SC circuits of Figure 1 can be expressed in yet another useful form. Defining the complex variable γ as

$$\gamma \equiv 1/2\left(z^{1/2} - z^{1/2}\right)$$

Eqs. (8) and (11) can be written as

$$\frac{V_o}{V_i} = \frac{C_1}{2C_2} \frac{z^{1/2}}{\gamma} \tag{13}$$

and

$$\frac{V_o}{V_i} = \frac{C_1}{2C_2} \frac{z^{-1/2}}{\gamma} \tag{14}$$

respectively. Thus, apart from the unimportant numerator terms, the SC circuits behave as perfect integrators in terms of the γ variable. In the digital filter literature, γ is known as the Lossless Digital Integrator (LDI) variable [8]. Physical frequencies ω map to locations on the imaginary axis of the complex γ plane according to

$$Im(\gamma) = \sin(\omega T/2) \tag{15}$$

We note that $\omega=0$ maps to $\gamma=0$ and $\omega=\pi/T$ maps to $Im(\gamma)=1$. We shall make use of the γ variable in SC filter synthesis.

2.3. Damped integrators

Damping an SC integrator can be accomplished by connecting a switched-capacitor across the integrating capacitor C_2, as illustrated in Figure 6 for the noninverting integrator. Here it is important to note that the clock phasing of the switches in the damping branch results in an equivalent positive resistor. The transfer function of the damped integrator of Figure 6 can be derived directly in the z domain as follows. We consider the circuit as an integrator with two inputs: V_i through the switched capacitor C_1 and V_o through the switched capacitor C_3. Thus, we can express the output voltage $V_o(z)$ as the sum,

$$V_o(z) = \frac{C_1}{C_2} \frac{z^{-1/2}}{z^{1/2} - z^{-1/2}} V_i(z) - \frac{C_3}{C_2} \frac{z^{1/2}}{z^{1/2} - z^{-1/2}} V_o(z)$$

Collecting the terms in $V_o(z)$ results in the transfer function

$$\frac{V_o(z)}{V_i(z)} = \frac{C_1 z^{-1/2}}{C_2(z^{1/2} - z^{-1/2}) + C_3 z^{1/2}} \tag{16}$$

For physical frequencies,

$$\frac{V_o}{V_i}(\omega) = \frac{C_1 e^{-j\omega T/2}}{j2C_2\sin\left[\frac{\omega T}{2}\right] + C_3\left[\cos\frac{\omega T}{2} + j\sin\frac{\omega T}{2}\right]} \tag{17}$$

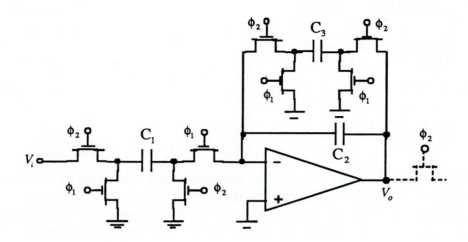

Fig. 6: Damped noninverting integrator.

Note that the damping term in the denominator is a function of frequency and, furthermore, includes an imaginary part. Thus, damping modifies the time constant of the SC integrator. This effect can, however, be accounted for in the design by simply combining the imaginary part of the damping term with the integrator time constant to obtain

$$\frac{V_o}{V_i}(\omega) = \frac{C_1 e^{-j\omega T/2}}{j2\left[C2 + \frac{C_3}{2}\right]\sin\left[\frac{\omega T}{2}\right] + C_3\cos\frac{\omega T}{2}} \tag{18}$$

Now the damping term though real is still frequency dependent. For $(\omega T/2) \ll 1$,

$$\frac{V_o}{V_i}(\omega) \approx \frac{C_1 e^{-j\omega T/2}}{j\omega T\left[C2 + \frac{C_3}{2}\right] + C_3} \tag{19}$$

Thus the integrator 3dB (corner) frequency is $\frac{1}{T}\left(\frac{C_3}{C_2 + C_3/2}\right)$.

The transfer function of the damped integrator [eq. (16)] can be expressed in terms of the LDI variable γ as follows. We define a new complex variable

$$\mu \equiv \frac{1}{2}\left(z^{1/2} + z^{-1/2}\right) \tag{20}$$

and express the $z^{1/2}$ term in the denominator of (16) as

$$z^{1/2} = \gamma + \mu$$

Thus (16) can be written as

$$\frac{V_o}{V_i} = \frac{C_1 z^{-1/2}}{2\gamma C_2 + (\gamma + \mu)\, C_3}$$

which can be manipulated to yield

$$\frac{V_o}{V_i} = \frac{C_1 z^{-1/2}}{\gamma(2C_2 + C_3) + \mu C_3} \tag{21}$$

This represents the transfer function of a damped integrator in the γ plane. Still an unusual feature of this integrator is the dependence of the damping term on frequency (through μ). Here we note that for physical frequencies, eq. (20) shows that μ is real and is given by

$$\mu = \cos(\omega T/2) \tag{22}$$

2.4. General form of the first-order building block

Figure 7 shows a general form of the strays-sensitive first-order building block. It has three feed-in branches: an inverting integrator branch through the switched capacitor C_1, an noninverting integrator branch through the switched capacitor C_2 and an inverting gain branch through the unswitched capacitor C_3. In addition, the circuit is damped via the switched capacitor C_f. To obtain the transfer function of the general block we express V_o as the sum

$$V_o = \frac{-C_1}{C}\, \frac{z^{1/2}}{z^{1/2} - z^{-1/2}}\, V_i + \frac{C_2}{C}\, \frac{z^{-1/2}}{z^{1/2} - z^{-1/2}}\, V_i - \frac{C_3}{C}\, V_i - \frac{C_f}{C}\, \frac{z^{1/2}}{z^{1/2} - z^{-1/2}}\, V_o$$

This leads to the transfer function

$$\frac{V_o}{V_i} = \frac{-C_1 z^{1/2} + C_2 z^{-1/2} - C_3\left(z^{1/2} - z^{1/2}\right)}{C\left(z^{1/2} - z^{1/2}\right) + C_f z^{1/2}} \tag{23}$$

which can be expressed in terms of γ and μ as

$$\frac{V_o}{V_i} = \frac{-C_1 z^{1/2} + C_2 z^{-1/2} - 2\gamma C_3}{\gamma(2C + C_f) + \mu C_f} \tag{24}$$

With a bit of practice one can write these transfer functions by inspection.

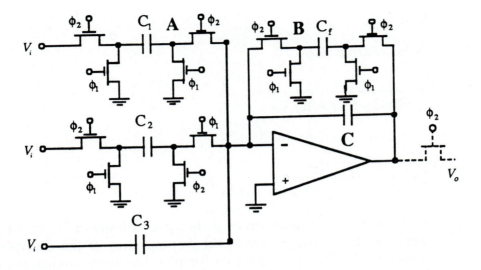

Fig. 7:   General form of the strays-insensitive first-order building block.

2.5. Switch design

Two practical circuit issues can be discussed at this point: switch sharing and charge injection.

Notice that the nodes marked A and B in Figure 7 are shorted together (through switches) in both clock phases. Circuit operation is therefore unaffected if they are simply wired together. This makes one pair of switches redundant, as A and B share a single switch pair. In more complicated circuits it is often possible to share many switches.

Secondly, when MOS transistors turn off, the inversion charge under their gates is injected into the circuit. This effect can cause DC offsets and clock feedthrough to the system output, often on the order of 100 mV. Because clocks switch to the power supply, the effect can also limit power-supply rejection. Further, when the injected charges depends on the signal, transfer function errors and nonlinear distortion result. Charge injection is difficult to predict or cancel, because it depends on device characteristics, "on" voltage V_{GS}-V_T, clock slew rates and the impedance levels on both sides of the switch [9, 10].

While many schemes have been reported for minimizing the charge injection problem, one that is now common involves the creation of a new clock ϕ'_2 which turns off slightly before ϕ_2. This clock is used on the grounded (or virtual ground) ends of input capacitors and minimizes the signal-dependent effects of charge injection.

2.6. Differential circuits

The most effective defense against offset and feedthrough from charge injection is to use differential circuits [11]. This is now a standard strategy, especially in low-voltage processes, and also reduces the influence of a wide range of other noise difficulties. Because signals are represented by the difference between voltages in two symmetrical circuits, any common-mode disturbance has no effect.

Figure 8 shows a simple differential version of the integrator of Figure 1a. It uses an amplifier with $V_0^+ = -V_0^- = A_v(V^+ - V^-)$, which has very high differential gain A_v. The transfer function, defined as $T(z) = (V_0^+ - V_0^-)/(V_i^+ - V_i^-)$ is $-C_1/C_2$ $z^{1/2}/(z^{1/2} - z^{-1/2})$ as before. By changing the output-sampling phasing, the delayed function $T(z) = (C_1/C_2) z^{-1/2}/(z^{1/2} - z^{-1/2})$ may be obtained.

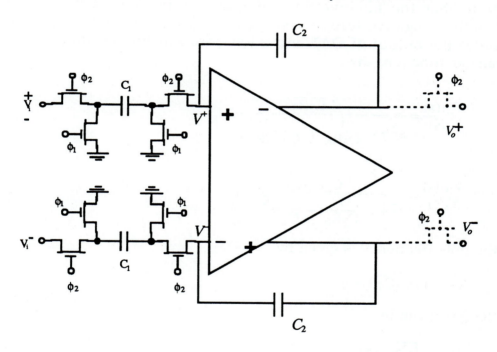

Fig. 8: Differential integrator modelled on Fig. 1a.

We have a new degree of freedom, however, with differential circuits: while in single-ended circuits a positive gain was always associated with a delayed output ($z^{1/2}$ numerator), in differential circuits the sign of the gain may be chosen arbitrarily merely by interchanging input or output terminals.

3. Biquadratic Sections and Cascade Design

In this section the first-order building blocks will be employed in the design of SC biquads, from which an arbitrary transfer function may be formed by cascading sections. These biquad circuits will be generated by simply replacing the resistors in active-RC biquads with their SC equivalent branches. Thus the response of the resulting SC biquads will approach that of their active-RC counterparts only if the clocking frequency is much greater than the biquad pole frequency.

3.1. Circuit generation via equivalent resistors

Figure 9a shows an active-RC biquad based on the two-integrator-loop topology [12]. Note that the noninverting integrator is implemented using a Miller circuit with a negative feed-in resistor ($-R_3$). This biquad realizes a bandpass function at the output of OA1 and a low-pass function at the output of OA2. The transfer functions are

$$\frac{V_{o1}(s)}{V_i(s)} = \frac{-s \, \dfrac{1}{C_2 R_1}}{s^2 + s \, \dfrac{1}{C_2 R_6} + \dfrac{1}{C_2 C_4 R_3 R_5}} \tag{25}$$

$$\frac{V_{o2}(s)}{V_i(s)} = \frac{-s \, \dfrac{1}{C_2 C_4 R_1 R_3}}{s^2 + s \, \dfrac{1}{C_2 R_6} + \dfrac{1}{C_2 C_4 R_3 R_5}} \tag{26}$$

Thus the pole frequency is given by

$$\omega_0 = 1/\sqrt{C_2 C_4 R_3 R_5}, \tag{27}$$

the pole-Q is given by

$$Q = \sqrt{\frac{C_2}{C_4}} \, \frac{R_6}{\sqrt{R_3 R_5}} \tag{28}$$

and the center-frequency gain of the bandpass function is given by

$$G_0 = -\frac{R_6}{R_1} \tag{29}$$

Usually the biquad is designed to have equal time-constants for the two integrators as follows

$$C_2 = C_4 = C, R_3 = R_5 = R$$

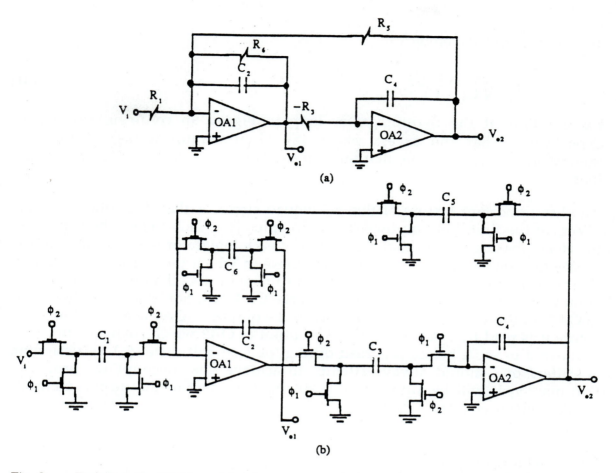

Fig. 9: Deriving the SC biquad from an active-RC circuit; (a) two-integrator-loop active-RC biquad; (b) the equivalent SC biquad obtained by replacing each resistor R_i with a switched capacitor $C_i = T/R_i$.

Thus,

$$CR = \frac{1}{\omega_0}$$

The damping resistor R_6 is given by $R_6 = QR$ and the feed-in resistor R1 determines the gain.

The equivalent SC biquad in Figure 9b is obtained by replacing each resistor with a switched capacitor. R_1 is replaced with $C_1 = T/R_1$, and so on. Note that positive resistors are replaced with inverting SC branches while the negative resistor is replaced with the noninverting SC branch. If the clocking frequency is much greater than the pole frequency, the SC biquad of Figure 9b realizes a second-order low-pass function at V_{o2}. The pole frequency is given by

$$\omega_0 = \frac{1}{T} \sqrt{\frac{C_5}{C_2} \frac{C_3}{C_4}} \tag{30}$$

and the pole-Q is given by

$$Q = \sqrt{\frac{C_2}{C_4}} \frac{\sqrt{C_3 C_5}}{C_6} \tag{31}$$

Examination of the two-integrator loop in Figure 9b confirms our earlier expectations that the excess phase ($z^{\pm 1/2}$ numerator terms) of the two integrators cancel out. Note that there is a delay of one clock period around the loop.

Another important feature of this SC biquad circuit is that it is tolerant of the effects of the finite settling time of the op amps [13]. This comes about because the clock phasing is such that the output of each integrator is not applied to the succeeding integrator during the clock phase in which the output is changing. To be specific, consider OA1. Its output will be changing during ϕ_2. Simultaneously, the output is used to charge C_3. It will not be applied, however, to OA2 until the next half cycle (ϕ_1). Thus the output of OA1 is allowed to settle before it is read by OA2. Similarly, the output of OA2 which changes during ϕ_1 is allowed to settle before it is read during the next half cycle (ϕ_2). In designing SC filters one should always attempt to achieve this "decoupling" property. Fortunately, it will be automatically obtained in most of the design methods studied in this chapter.

3.2. Exact transfer function: LDI design

The exact transfer function of the SC biquad in Figure 9b can be derived directly in the z domain. For the two output voltages we write by inspection

$$V_{o1}(z) = -\frac{C_1}{C_2} \frac{z^{1/2}}{z^{1/2} - z^{-1/2}} V_i(z) - \frac{C_6}{C_2} \frac{z^{1/2}}{z^{1/2} - z^{-1/2}} V_{o1}(z)$$

$$- \frac{C_5}{C_2} \frac{z^{1/2}}{z^{1/2} - z^{-1/2}} V_{o2}(z)$$

$$V_{o2}(z) = \frac{C_3}{C_4} \frac{z^{-1/2}}{z^{1/2} - z^{-1/2}} V_{o1}(z)$$

These two equations can be combined to obtain the two transfer functions V_{o1}/V_i and V_{o2}/V_i. As an example, for the first transfer function we obtain

$$\frac{V_{o1}(z)}{V_i(z)} = \frac{-\frac{C_1}{C_2} z^{1/2} \left(z^{1/2} - z^{-1/2}\right)}{\left(z^{1/2} - z^{-1/2}\right)^2 + \frac{C_6}{C_2} z^{1/2} \left(z^{1/2} - z^{-1/2}\right) + \frac{C_5}{C_2}\frac{C_3}{C_4}} \tag{32}$$

For physical frequencies we have

$$\frac{V_{o1}}{V_i}(\omega) = \frac{-j\frac{C_1}{C_2} 2\sin\left(\frac{\omega T}{2}\right) e^{j\omega T/2}}{\left[\frac{C_5}{C_2}\frac{C_3}{C_4} - 4\sin^2\left(\frac{\omega T}{2}\right)\left(1 + \frac{C_6}{2C_2}\right)\right] + j\frac{C_6}{2C_2}\sin\omega T} \tag{33}$$

which for $\omega T \ll 1$ reduces to

$$\frac{V_{o1}}{V_i}(\omega) \cong \frac{-j\omega T\frac{C_1}{C_2} e^{j\omega T/2}}{\left[\frac{C_5}{C_2}\frac{C_3}{C_4} - \omega^2 T^2\left(1 + \frac{C_6}{2C_2}\right)\right] + j\left(\frac{C_6}{C_2}\right)\omega T} \tag{34}$$

Apart from the unimportant numerator term $e^{j\omega T/2}$, this equation is in the form of the second-order bandpass function

$$T(\omega) = \frac{-j\omega k}{\left(\omega_0^2 - \omega^2\right) + j\frac{\omega\omega_0}{Q}}$$

It follows, that

$$\omega_0 = \frac{1}{T}\sqrt{\frac{C_5}{C_2}\frac{C_3}{C_4}\Big/\left(1 + \frac{C_6}{2C_2}\right)} \tag{35}$$

and,

$$Q = \sqrt{\frac{C_2}{C_4}}\frac{\sqrt{C_3 C_5}}{C_6}\sqrt{1 + \frac{C_6}{2C_2}} \tag{36}$$

These values are slightly different from those given by eqs. (30) and (31). The difference is due to the assumption, made in deriving (30) and (31), that damping does not affect the integrator time constant. That this is not the case for SC integrators has been demonstrated in the previous section. Thus eqs. (35) and (36) are more accurate than (30) and (31) and should be the ones used in design. Nevertheless, the resulting design will still be approximate since it is based on the assumption that $\omega T \ll 1$. Better results can be obtained if design is based on the exact transfer function in eq. (33). For instance, ω_0 can be found from

$$4\sin^2\frac{\omega_0 T}{2} = \left(\frac{C_3}{C_4}\frac{C_5}{C_2}\right)\Big/\left(1 + \frac{C_6}{2C_2}\right) \tag{37}$$

More details on this approach for designing SC biquads can be found in [6].

From the exact transfer function in eq. (32) it can be seen that the SC biquad of Figure 9b has one transmission zero at z=1 and the other at z=0. While the first zero corresponds to ω=0, the second has no corresponding physical frequency. Therefore it contributes little to the selectivity of the bandpass filter response. In other words, while the original continuous-time active-RC-circuit has a zero at ω=0 and another at ω=∞, both contributing to selectivity, the discrete-time SC filter has only one physical-frequency zero (at ω=0). A more selective bandpass response can be obtained by placing a transmission zero at half the clock frequency (ω=π/T).

3.3. Capacitive damping

The active-RC circuit of Figure 9a is damped by placing the resistor R_6 across the integrator capacitor C_2. The damping resistor can of course be placed across the other integrator capacitor, C_4. Alternatively, damping can be achieved by placing a capacitor across -R_3 or R_5. This capacitive damping is seldom used in active-RC filters because in that technology one attempts to use resistors in favor of reducing the number of capacitors. Capacitive damping, however, can be attractive in SC filters. As an example, it can be implemented in the circuit of Figure 9b by connecting an unswitched capacitor between the output of OA2 and the inverting output of OA1 (while, of course, removing the C_6 branch).

Both forms of damping biquads have been used in the literature [7, 14, 15]. Depending on the application, one form might lead to lower total capacitance and hence becomes preferred.

3.4. Biquads for exact design: the bilinear variable

Next we will present circuits for the *exact* synthesis of SC filters. Exact design methods result in an SC filter whose transfer function is identical to the desired one, without assuming that ωT<<1. Exact design for cascade-structure filters will require biquads with input feedforward components to provide zeros at more general locations than are provided by the basic loop.

All exact design methods [14, 16, 17] make use of the bilinear transformation [18, 12]

$$\lambda = \frac{z-1}{z+1} \tag{38}$$

which maps the unit circle in the z-plane (which is the contour of physical frequency points) onto the entire imaginary axis of the λ plane. The relationship between the frequency ω and the corresponding point Ω on the imaginary axis of the λ plane is given by

$$\Omega = \tan(\omega T/2) \tag{39}$$

We note that $\omega=0$ maps to $\Omega=0$ and $\omega = \dfrac{\pi}{T}$ (half the clocking frequency) maps to $\Omega=\infty$. This is a one-to-one mapping with half the unit circle ($\omega T=0$ to $\omega T=\pi$) being mapped onto the positive half of the $j\Omega$ axis ($\Omega=0$ to $\Omega=\infty$). In the continuous-time analog filter literature there exists a wealth of design data and synthesis methods for filters whose responses are specified along the imaginary axis of a complex plane. These tools can be applied to the design of discrete-time filters (of which SC filters is a special case) via the bilinear transformation.

Synthesis of SC filters using the bilinear transformation proceeds as follows. The given attenuation specifications versus frequency ω are recast versus Ω. This step, known as *prewarping*, uses the relationship in (39) and results in modifying the critical frequency points along the horizontal axis without affecting the vertical (attenuation) axis. For example, simple low-pass attenuation specifications transform to those shown in Figure 10.

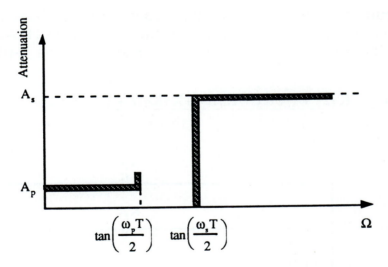

Fig. 10: The bilinear transform: specifications prewarped according to eq. (39).

We note that the attenuation specifications versus Ω look identical to those of continuous-time analog filters. Thus filter design tables and/or computer programs [19] can be used to obtain a transfer function $T(\lambda)$ whose attenuation function meets the prewarped specifications. If $T(\lambda)$ has a transmission zero at $\lambda=\infty$, the corresponding SC filter will have a zero at $\omega=\pi/T$. Thus, unlike the approximate design method of Section 3.2., transmission zeros at ∞ are not lost and their full effect on filter selectivity is realized.

Having performed filter approximation, it now remains to obtain an SC circuit implementation. In the following we present a number of methods for obtaining SC realizations starting from a transfer function $T(\lambda)$ or from an LC ladder realization (section 4) of $T(\lambda)$.

3.5. Biquad circuits for exact design

The SC filter can be realized as a cascade of biquads and, in the case of odd-order filters, a first-order section. Cascade design is performed by factoring the function $T(\lambda)$ obtained from filter approximation. We will address other details of cascade design at the end of this section, but first we must find circuits for the sections.

Second-order transfer functions obtained in the factorization of $T(\lambda)$ will be special cases of the general biquadratic function

$$t(\lambda) = \pm\frac{a_2\lambda^2 + a_1\lambda + a_0}{\lambda^2 + b_1\lambda + b_0} \tag{40}$$

SC biquad circuits capable of realizing this general function have been given in [7, 14, 15]. As an example, we show one such circuit in Figure 11. Analysis of this circuit, following the method of Section 3, results in the z-plane transfer function

$$\frac{V_0(z)}{V_i(z)} = -\frac{K_3z^2 + (-2K_3 + K_1K_5 + K_2K_5)z + (K_3-K_2K_5)}{z^2 + (-2 + K_4K_5 + K_5K_6)z + (1 - K_5K_6)} \tag{41}$$

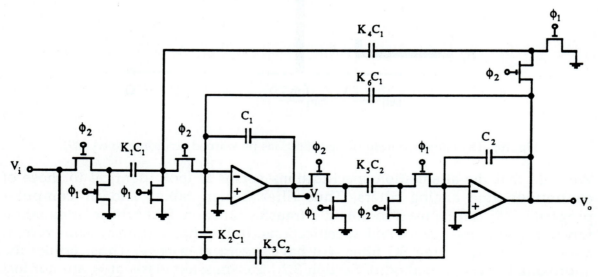

Fig. 11:   An SC biquad circuit capable of realizing the general bilinearly transformed function in eq. (40).

The λ-plane transfer function is obtained by substituting

$$z = \frac{1+\lambda}{1-\lambda} \qquad (42)$$

which is the inverse of the transformation in (38). Design equations for the SC biquad can be then obtained by equating the coefficients in the resulting transfer function to the corresponding coefficients in (40). Equating corresponding denominator coefficients results in two equations in the three capacitor ratios that determine the biquad poles,

$$K_4 K_5 = 4m b_0 \qquad (43)$$

$$K_5 K_6 = 2m b_1 \qquad (44)$$

where $\quad m = 1/(1 + b_1 + b_0) \qquad (45)$

Because we have only two equations in the three unknowns, K_4, K_5 and K_6, one of the capacitor ratios can be chosen arbitrarily. This choice is usually done with a view to maximizing the biquad dynamic range. For high-Q biquads, this is approximately obtained when the two loop time constants are made equal; that is, $K_4 = K_5$.

Design equations for the feed-in capacitor ratios, which determine the biquad transmission zeros, are obtained by equating coefficients of corresponding numerator terms. The result for some special cases are:

(a) Low-Pass ($a_1 = a_2 = 0$)

$K_2 = 0 \quad K_1 K_5 = 4m a_0 \quad K_3 = m a_0$

(b) Bandpass ($a_0 = a_2 = 0$)

$K_1 = 0 \quad K_2 K_5 = 2m a_1 \quad K_3 = m a_1$

(c) High-Pass ($a_0 = a_1 = 0$)

$K_1 = K_2 = 0 \quad K_3 = m a_2$

(d) Notch ($a_1 = 0$)

$K_2 = 0 \quad K_1 K_5 = 4m a_0 \quad K_3 = m(a_0 + a_2)$

The circuit of Figure 11 realizes a negative bandpass function. As will be seen below, a positive bandpass is frequently needed. It can be obtained by taking the output at V_1. While the pole design equations remain unchanged [eqs. (43) and (44)] the zeros' equations become

$$K_1 = 0 \quad K_3 K_4 = 2ma_1 \quad K_2 = ma_1 \left[1 + \frac{b_1}{b_0} \right]$$

The reader is urged to note that low-pass and bandpass biquad circuits designed using the bilinear transformation (such as the circuit in Figure 11) have frequency responses that are more selective than those studied in Section (3). This is a result of the transmission zero placed at $\omega = \pi/T$ (which is the mapping of $\lambda = \infty$).

3.6. Differential biquads

The single-ended loop used "integrators" with two different transfer functions (differing by a factor z in the numerator) for the positive-gain and negative-gain parts of the two-integrator loop at the heart of a biquad. The same thing can be done with a differential circuit.

Because the sign of the gain of a differential circuit can be reversed just by interchanging its outputs or inputs, many other configurations are possible [20, 21], of which three important examples appear in Figures 12a, 12b and 12c. These circuits all have the same design equations for capacitor values as the single-ended circuit, because they all have a total of one clock delay around the loop ("LDI phasing"). They differ in their sensitivity to op-amp settling time.

The circuit of Figure 12a has the best performance in the presence of finite-bandwidth op amps (assuming zero output impedance) [20, 21], because both amplifiers are given one full period in which to settle before their outputs are sampled. When simpler transconductance circuits are used as amplifiers, better results have been reported [20] for the "differencing-input" circuit of Figure 12b. This circuit requires input capacitors that are smaller by a factor of two than those for the other circuits, because they are switched between $\pm V_i$ rather than between an input and ground. In the (now prevalent) transconductance-amplifier case this smaller load can help compensate for the shorter settling times available.

A circuit built of two "non inverting type" integrators clocked on alternate phases only allows 1/2 period of settling time for each amplifier (because amplifier outputs are sampled on the phase in which they are created) and so performs poorly for slow amplifiers. On the other hand, because only one amplifier is active in each phase, this circuit allows the designer to use a single multiplexed op amp to implement the entire biquad [22] as in Figure 12c. This circuit is useful when clocking at moderate speeds, although the fact that it involves switching the integrator feedback capacitors (as well as the input capacitors) gives rise to difficulty with charge injection, with op-amp behaviour while open-loop, and can cause crosstalk between states.

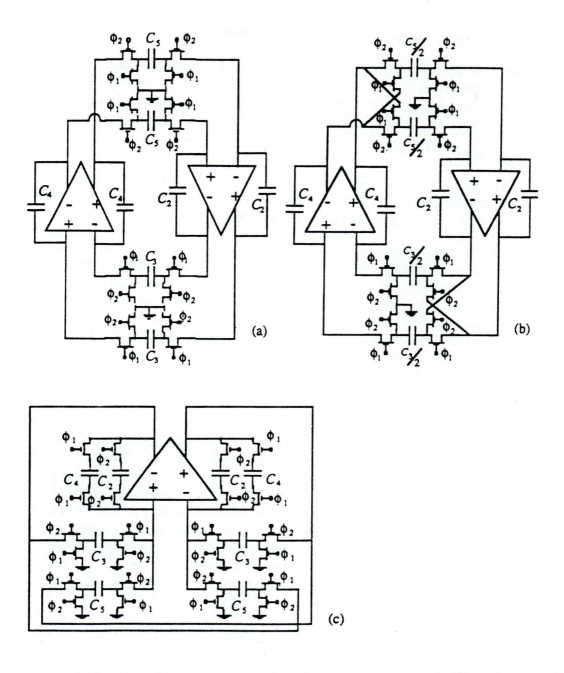

Fig. 12:   Differential biquads; (a) with improved settling time allowance, (b) with differencing inputs, (c) a multiplexed-amplifier biquad.

Many variants of the two-integrator loop can be considered. For example, one could construct a biquad with two "noninverting-type" integrators or two "inverting-type" integrators with identical (rather than LDI) phasing. While these can implement an arbitrary second-order function, they have very poor sensitivity to capacitors when realizing high-Q functions, and may even become unstable in the presence of small errors.

3.7. Cascade design

In principle, an arbitrary transfer function may be implemented by cascading second-order sections (and perhaps one degenerate first-order block to implement an odd-order filter). Several degrees of freedom remain, however, and careful choices are required to make good filters.

The freedoms available are:

(a) *pairing* of poles and zeros, which will affect both noise performance and total capacitance. A common rule of thumb is to pair high-Q poles with their nearest zeros.

(b) *ordering* of sections. If too many sections with high gain at one side of the passband are cascaded, later sections will have to favour the other side and will incidentally amplify noise in that band. Further, it is desirable to remove as much stopband energy as possible at the beginning of the cascade, so that later sections do not have to waste dynamic range responding to it. For example, bandpass filters will often begin with a bandpass (rather than, say, a notch) section so as to immediately attenuate DC and high-frequency signals.

(c) *scaling* of intermediate signals, which is often done conservatively to guarantee that the peak gain (over frequency) from the input to any output is 1. Choosing a higher gain leads to the possibility of clipping, while a lower gain to an internal node forces the remainder of the filter to have high gain *from* that node, and so to amplify noise unnecessarily. A less conservative scheme scales for root-mean square signal levels rather than peak gains, and yields lower output noise at the cost of occasional distortion for large sinusoidal inputs. This "L_2" scaling is convenient for signals with well-understood spectra such as occur in data communications applications.

Cascade designs are widespread in industry, primarily because of their generality, mathematical simplicity, and the direct correspondences between particular capacitors and particular poles or zeros. Since designers have to be able to make small adjustments to their designs to account for layout and circuit non idealities, they appreciate the easy debugging and trimming afforded by cascade design. A typical modern commercial filter would use the bilinear design equations of section 3.5 and the biquads of Figs. 11 and 12a.

4. Ladder Filters

Doubly-terminated LC ladder networks that are designed to effect maximum power transfer from source to load over the filter passband feature very low sensitivities to variations in their component values [23]. This fact has over the years spurred considerable interest in finding active-RC, digital and switched-capacitor filter structures which simulate the internal workings of LC ladder prototype networks. Essentially these filter structures consist of a connection of first-order blocks that implement the I-V integral relationship of the L and C elements in the prototype. That is, the basic building blocks are integrators.

These ladder-based circuits offer improved sensitivity and SNR over cascade designs, though at some cost in simplicity of design. They should be seriously considered for high-order and high-Q designs, where their advantages are most marked.

The straightforward approach to the design of SC filters based on the simulation of LC ladder prototypes, to be referred to as SC ladder filters, is to first design a suitable active-RC filter and then replace the integrators with the SC blocks in Figure 1. This approach works well only if the clock frequency is much higher than the signal frequencies. In this section we study a more refined approach to the design of SC ladder filters. A better approach takes into account the fact that the transfer functions of the SC building blocks are inversely proportional to $\sin(\omega T/2)$ and not ω. It also takes into account the fact that damping affects the time constants of SC integrators. The resulting designs, therefore, have frequency responses that are closer to the desired ideal than those obtained by replacing the resistors in an active-RC filter. Nevertheless, the resulting designs are still not exact because the method does not take into account the frequency dependence inherent in the damping of SC integrators. Exact design methods will then be studied in Section 4.3.

4.1. Circuit generation

LDI design makes use of the observation made in Section 2 that the SC blocks behave as perfect integrators in the γ plane. Therefore, we shall perform our design in that plane. The starting point in the design is a set of attenuation specifications posed versus frequency ω, such as those shown in Figure 13a for a low-pass filter. The specifications can be cast versus the γ-plane frequency by transforming the horizontal axis points according to $\text{Im}(\gamma) = \sin(\omega T/2)$. This process, known as *prewarping,* results in the attenuation specifications shown in Figure 13b. Note that half the clock frequency maps to $\text{Im}(\gamma)=1$.

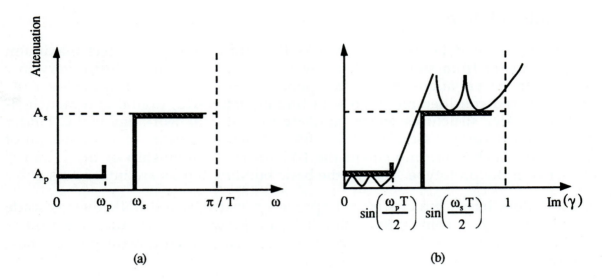

Fig. 13: Frequency warping for the LDI transformation; (a) attenuation specifications for a low-
          pass SC filter; (b) the specifications in (a) versus Im(γ) = sin($\omega T/2$).

Next, filter design tables or computer programs are used to find a suitable
transfer function and/or LC ladder network that meets the specifications in
Figure 13b. For illustration, we shall assume that a fifth-order elliptic filter
meets the given specifications. The attenuation response for such a filter is
shown in Figure 13b, and Figure 14a shows its LC ladder network realization.
Because this network is designed to meet specifications posed versus Im(γ), its
inductances have impedances γl_i and its capacitances have impedances $1/\gamma c_i$.
(This of course has no bearing on the process finding the element values of the
LC network.)

The next step in the design involves the use of Norton's theorem to replace the
bridging capacitors C_2 and C_4 by voltage-controlled current sources, resulting in
the modified network shown in Figure 14b. For this network we can write five
equations that describe the operation of its five reactive components, as follows:

$$V_1 = \frac{(V_i/r_s) - I_2 + \gamma c_2 V_3}{\gamma c'_1 + (1/r_s)} \tag{46}$$

$$I_2 = \frac{V_1 - V_3}{\gamma l_2} \tag{47}$$

$$V_3 = \frac{I_2 - I_4 + \gamma c_2 V_1 + \gamma c_4 V_5}{\gamma c'_3} \tag{48}$$

$$I_4 = \frac{V_3 - V_5}{\gamma l_4} \tag{49}$$

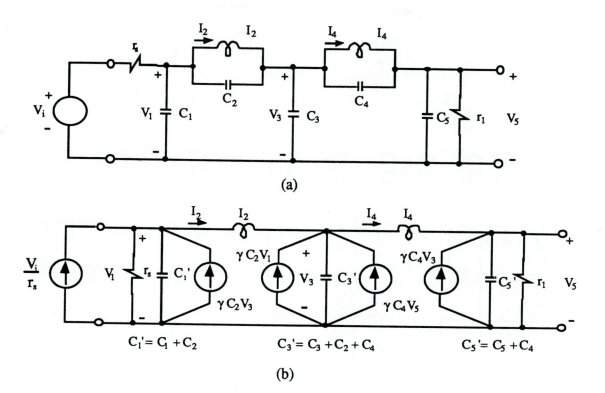

Fig. 14:   Passive prototype and notch formation. (a) γ-plane LC ladder network realization of a fifth-order elliptic filter that meets the specifications in Fig. 13b; (b) Norton's theorem is used to replace C_2 and C_4 with voltage-controlled current sources.

$$V_5 = \frac{I_4 + \gamma c_4 V_3}{\gamma c'_5 + (1/r_l)} \qquad (50)$$

These equations are in the form of the transfer function of the general first-order block of Figure 7 [eq. (24)]. From this analogy we can directly sketch the SC circuit realization shown in Figure 15. The transfer functions of the five blocks in Figure 15 can be written by inspection as

$$V'_1 = \frac{-C_1 z^{1/2}V'_i - C_2 z^{1/2}V'_2 - \gamma 2 C_{16}V'_3}{\gamma(2C_3 + C_s) + \mu C_s} \qquad (51)$$

$$V'_2 = \frac{C_4 z^{-1/2}V'_1 + C_5 z^{-1/2}V'_3}{\gamma 2 C_6} \qquad (52)$$

$$V'_3 = \frac{-C_7 z^{1/2}V'_2 - C_8 z^{1/2}V'_4 - \gamma 2 C_{15}V'_1 - \gamma 2 C_{17}V'_5}{\gamma 2 C_9} \qquad (53)$$

$$V'_4 = \frac{C_{10}z^{-1/2}V'_3 + C_{11}z^{-1/2}V'_5}{\gamma 2 C_{12}} \qquad (54)$$

$$V'_5 = \frac{-C_{13}z^{1/2}V'_4 - \gamma 2 C_{18}V'_3}{\gamma(2C_{14} + C_l) + \mu C_l} \qquad (55)$$

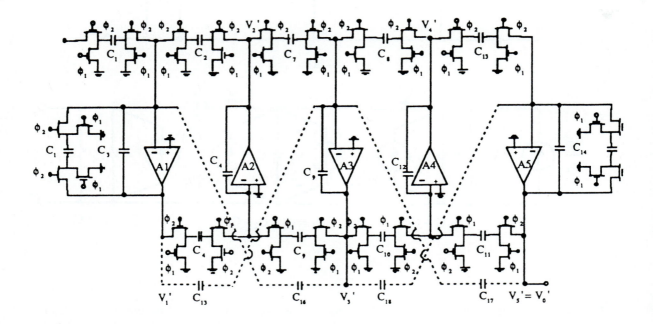

Fig. 15:   Switched-capacitor circuit that simulates the operation of the LC network of Fig. 14b.

Comparing corresponding equations between the set (46)-(50) and the set (51)-(55) we obtain the following correspondences between the ladder currents and voltages (Figure 14b) and the op am voltages (Figure 15):

$$(-V'_i) \Leftrightarrow V_i \qquad \left(z^{-1/2}V'_1\right) \Leftrightarrow V_1 \qquad V'_2 \Leftrightarrow I_2$$

$$\left(-z^{-1/2}V'_3\right) \Leftrightarrow V_3 \qquad (-V'_4) \Leftrightarrow I_4 \qquad \left(z^{-1/2}V'_5\right) \Leftrightarrow V_5$$

With these correspondences we see that the SC circuit in Figure 15 exactly simulates the operation of the LC network except for the two μ terms in the denominators of eqs. (51) and (55). For physical frequencies $\mu = \cos(\omega T/2)$ and thus the SC circuit simulates a ladder having frequency-dependent terminations. We should therefore expect the frequency response of the SC circuit to deviate from that of the ladder with the deviation being small at clock frequencies that are sufficiently high so that $\cos(\omega T/2) \cong 1$. A simple formula that predicts the expected deviation in attenuation response has been given in [24].

Note that the unswitched capacitive inputs used for notch formation produce delay-free loops through amplifiers (such as the loop A1-C15-A2-C16-A1). The good passive sensitivities of this design style are obtained at a slight cost in settling performance. Because of this delay-free loop, these filters cannot be directly implemented as digital filters.

4.2. Design

Initial values for the capacitors of the SC circuit in Figure 15 are obtained by equating the coefficients of corresponding terms between the set (46)-(50) and the set (51)-(55). The result is

$$C_1 = 1/r_s \qquad C_2 = 1 \qquad C_s = 1/r_s \qquad C_3 = \left(c'_1 - \frac{1}{r_s}\right)\Big/2$$

$$C_4 = C_5 = 1 \qquad C_6 = l_2/2 \qquad C_7 = C_8 = 1 \qquad C_9 = c'_3/2$$

$$C_{10} = C_{11} = 1 \qquad C_{12} = l_4/2 \qquad C_{13} = 1 \qquad C_l = 1/r_l$$

$$C_{14} = \left(c'_s - \frac{1}{r_l}\right)\Big/2 \qquad\qquad C_{15} = C_{16} = \frac{c_2}{2} \qquad C_{17} = C_{18} = \frac{c_4}{2}$$

Using these capacitor values gives the SC filter a dc gain equal to that of the LC ladder. If the gain of the SC filter is to be a factor K greater than that of the LC ladder then the value of C_1 must be changed to KC_1.

The initial capacitor values above must then be scaled so that the SC filter has the widest possible dynamic range. Dynamic range scaling is usually based on equalizing the peaks obtained at the outputs of the op amps as the frequency of an input sinusoid is swept over a desired band. To perform scaling one needs the value of the signal peaks, denoted $\hat{V}'_1$, $\hat{V}'_2$, $\hat{V}'_3$, $\hat{V}'_4$ and $\hat{V}'_5$. These can be found from a simulation of the SC circuit [25] with the capacitor values given above. Alternatively, the peak values can be obtained from an analysis of the LC ladder network.

Once the spectral peaks have been determined the capacitor values can be scaled as follows: capacitor C_{ij} which is connected between the output of op amp i and the input of op amp j is changed to $C_{ij}(\hat{V}'_i/\hat{V}'_j)$. As an example, capacitor C_2 changes from the initial value of unity to $(\hat{V}'_2\hat{V}'_1;)$. Note that this scaling process preserves the magnitude of loop transmission of every two-integrator loop, and hence the filter transfer function (except for a gain constant) remains unchanged. To keep the overall gain unchanged, C_1 must be scaled by the factor $(\hat{V}'_5/\hat{V}'_1)$.

The next and final step in the design involves scaling the capacitor values to minimize the total capacitance. This is achieved by considering the five blocks one at a time. For block #i, assume that the smallest capacitance connected to the virtual ground input of the op amp is of value C_i. We wish to scale so that this capacitor is equal to the minimum capacitor value possible with the given technology; call it C_{min}. (This may be 0.1 pF or so.) It follows that for block #i the scaling factor $k_i = C_{min}/C_i$. Every capacitor of block #i is multiplied by k_i. The transfer function of block #i obviously remains unchanged. The process is then repeated for the other blocks.

This completes the design procedure for SC ladder filters. As noted earlier, this design is not exact because of the frequency-dependent terminations realized. If the clock frequency is lowered to the point where the deviation in response is unacceptably high then one must consider using the exact design method of the next section. Another shortcoming of the method above is that transmission zeros at $\gamma = \infty$ do not map to physical frequencies and thus their effect on selectivity is lost.

4.3. A general exact design method

We now present a general method [16] for the synthesis of SC filters based on the exact simulation of LC ladder prototypes. Although the method will be demonstrated for the case of odd-order elliptic low-pass filters, it has been extended to other filter functions [28], to high-pass filters [29] and to bandpass filters. As will be seen shortly, the method presented removes the shortcomings of the LDI method studied in Section 4.2. The circuit structure is almost unchanged, but by using bilinear design and clever scaling we avoid the difficulties caused by the termination in LDI design.

The general method employs the three complex variables, γ, μ and λ. From the definitions of these variables above we can write

$$\gamma = 1/2\left(z^{1/2} - z^{-1/2}\right) = 1/2\left(e^{sT/2} - e^{-sT/2}\right) = \sinh(sT/2) \tag{56}$$

$$\mu = 1/2\left(z^{1/2} + z^{-1/2}\right) = 1/2\left(e^{sT/2} + e^{-sT/2}\right) = \cosh(sT/2) \tag{57}$$

$$\lambda = \frac{z-1}{z+1} = \frac{z^{1/2} - z^{-1/2}}{z^{1/2} - z^{-1/2}} = \tanh(sT/2) \tag{58}$$

Two important relationships that we will find useful are:

$$\lambda = \gamma/\mu \tag{59}$$

and $\qquad$ $$\mu^2 = 1 + \gamma^2 \tag{60}$$

To illustrate the method consider as an example the simple low-pass filter specifications whose prewarped version (Figure 10) can be met by a fifth-order elliptic filter. Figure 16a shows the LC ladder realization of such a filter, which is designed in the γ-plane. If we attempt to directly simulate the operation of this network we will need SC circuits having transfer functions of the form $1/\lambda$. This could be done by using a differential bilinear integrator, resulting in a very straightforward design process. Often, however, this is unacceptable either because the settling performance of that circuit is too poor or because a single-ended circuit is desired. We therefore seek an alternative to the direct simulation of the λ-plane ladder of Figure 16a.

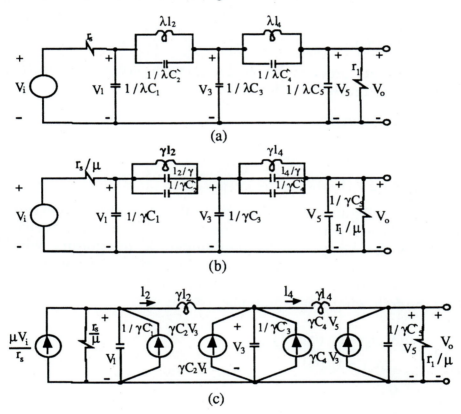

Fig. 16: (a) LC ladder realization of a fifth-order elliptic low-pass filter that meets the prewarped specification of Fig. 10. (b) The network of (a) with impedances scaled by μ. (c) The network of (a) after eliminating the bridging capacitors using Norton's theorem.

The alternative approach involves scaling the impedances of the ladder in Figure 16a by the factor μ. Specifically, if we divide all impedances by μ, the voltage transfer function remains unchanged and we obtain the ladder network in Figure 16b. Here we note that the termination resistors have changed to r_s/μ and r_l/μ. A capacitor c in the ladder of Figure 16a gives rise to an element with impedance $(1/\mu)(1/\lambda c)$ which, using eq. (59), is simply $1/\gamma c$. Thus a capacitor in the λ-plane ladder of Figure 16a transforms into a capacitor of equal value in the γ-plane ladder of Figure 16b.

An inductance l in the ladder of Figure 16a transforms to a branch with impedance $\lambda l/\mu$ which, using eq. (60), can be shown to be parallel equivalent of the two impedances γl and l/γ. In terms of the variable γ, this is an inductance l in parallel with a capacitance $1/l$.

Consider now the transformed network in Figure 16b. The parallel capacitors in each of the series arms of the ladder can be combined as follows

$$c_2 = c'_2 + (1/l_2)$$

$$c_4 = c'_4 + (1/l_4)$$

These bridging capacitors can be replaced with voltage-controlled current sources as shown in Figure 16c where

$$c'_1 = c_1 + c'_2 + (1/l_2)$$

$$c'_3 = c_3 + c'_2 + (1/l_2) + c'_4 + (1/l_4)$$

$$c'_5 = c_5 + c'_4 + (1/l_4)$$

Now the network of Figure 16c is identical to that of Figure 14b except that here the terminations are frequency dependent. This is a most welcome result because we know that strays-insensitive damped SC integrators, which have to be used to simulate the end elements, do in fact realize frequency-dependent terminations. To be specific, the operation of the ladder in Figure 16c is fully described by the following five equations:

$$V_1 = \frac{\mu(V_i/r_s) - I_2 + \gamma c_2 V_3}{\gamma c'_1 + (\mu/r_s)} \tag{61}$$

$$I_2 = \frac{V_1 - V_3}{\gamma l_2} \tag{62}$$

$$V_3 = \frac{I_2 - I_4 + \gamma c_2 V_1 + \gamma c_4 V_5}{\gamma c'_3} \tag{63}$$

$$I_4 = \frac{V_3 - V_5}{\gamma l_4} \tag{64}$$

$$V_5 = \frac{I_4 + \gamma c_4 V_3}{\gamma c'_5 + (\mu/r_l)} \tag{65}$$

These equations are identical to the set (46)-(50) except for the μ terms in the denominators of (61) and (65) and the μ term that multiplies V_i in the numerator of (61). The denominator μ terms are automatically realized in the damped SC integrators. To multiply V_i by μ we need to feed V_i through the input branch shown in Figure 17.

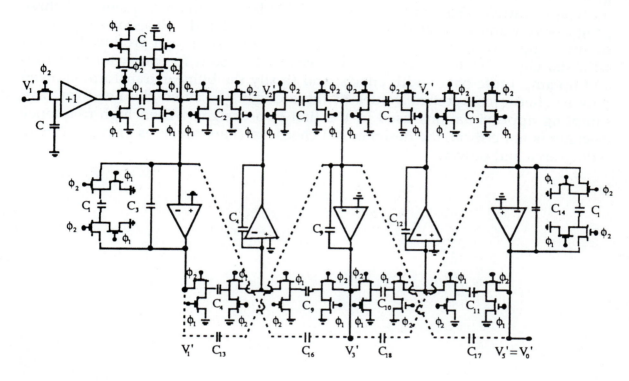

Fig. 17:   SC realization of a fifth-order elliptic low-pass filter designed using the general exact synthesis method of section 4.3.

Except for this modified feed-in branch the SC circuit configuration in Figure 17 is identical to that of Figure 15. Thus the resulting circuit is described by the set of five equations (51)-(55) except that (51) is replaced with

$$V'_1 = \frac{-2C_i \mu V'_i - C_2 z^{1/2} V'_2 - \gamma 2 C_{16} V'_3}{\gamma(2C_3 + C_s) + \mu C_s} \tag{66}$$

The remainder of the design procedure and indeed the equations giving the component values are identical to those in Section 4.2 with C_1 replaced by $(2C_i)$.

5.   Concluding Remarks

Beginning with the pair of strays-insensitive integrators we have presented a number of methods, approximate and exact, for the synthesis of SC filters. The general exact design method discussed above seems to yield the best results in most applications. For both cascade and ladder designs a sequence of three progressively more exact design styles was considered. "Resistor-equivalent" design suggested the basic topologies used; LDI-phased design produced circuits with the clocks arranged to give low sensitivities both for capacitance values and op-amp finite-bandwidth effects; and finally the bilinear transform showed how to choose component values in these circuits to avoid the need for fast-sampling approximations used in the first two designs. When high-frequency clocking is not critical, the designer can choose to share op amps or capacitors to reduce area and power.

6. References

[1] J.B. Hughes, I.C. MacBeth and D.M. Pattullo, "Switched Current Filters," *IEEE Proceedings* Part G, vol. 137, no. 2, pp. 156-162, April 1990.

[2] J.T. Caves, M.A. Copeland, C.F. Rahim and S.D. Rosenbaum, "Sampled analog filtering using switched capacitors as resistor equivalents," *IEEE J. Solid-State Circuits*, vol. SC-12, pp. 592-600, Dec. 1977.

[3] B.J. Hosticka, R.W. Brodersen and P.R. Gray, "MOS sampled-data recursive filters using switched capacitor integrators," *IEEE J. Solid-State Circuits*, vol. SC-12, pp. 600-608, Dec. 1977.

[4] R.W. Brodersen, P.R. Gray and D.A. Hodges, "MOS switched-capacitor filters," *Proc. IEEE*, vol. 67, no. 1, pp. 61-75, Jan. 1979.

[5] G.C. Temes and Y. Tsividis eds. "Special Section on Switched-Capacitor Circuits," *Proc. IEEE*, vol. 71, no. 8, pp. 926-1005, Aug. 1983.

[6] K. Martin, "Improved circuits for the realization of switched-capacitor filters," *IEEE Trans. Circuits Syst.*, vol. CAS-27, pp. 237-244, Apr. 1980.

[7] K. Martin and A.S. Sedra, "Strays-insensitive switched-capacitor filters based on the bilinear z transform," *Electron Lett.*, vol. 19, pp. 365-366, June 1979.

[8] L.T. Bruton, "Low-sensitivity digital ladder filters," *IEEE Trans. Circuits Syst.*, vol. CAS-22, pp. 168-176, Mar. 1987.

[9] W. Wilson, H. Massoud, E. Swanson, R. George and R. Fair, "Measurement and modeling of charge feedthrough in n-channel MOS analog switches," *IEEE J. Solid-State Circuits*, vol. SC-20, pp. 1206-1213, Dec. 1985.

[10] C. Eichenberger and W. Guggenbuhl, "On charge injection in analog MOS switches and dummy compensation techniques," *IEEE J. Solid-State Circuits*, vol. CAS-37, no. 2, pp. 256-264, Feb. 1990.

[11] K.C. Hsieh, P.R. Gray, D. Senderowicz and D.G. Messerschmitt, "A low-noise chopper-stabilized differential switched-capacitor filtering technique," *IEEE J. Solid-State Circuits*, vol. SC-16, pp. 708-715, Dec. 1981.

[12] A.S. Sedra and P.O. Brackett, *Filter Theory and Design: Active and Passive*, Matrix Publishers, Portland, Oregon, 1982.

[13] K. Martin and A.S. Sedra, "Effects of the operational amplifier gain and bandwidth on the performance of switched-capacitor filters," *IEEE Trans Circuits Syst.*, vol. CAS-78, no. 2, pp. 256-264, Feb. 1990.

[14] K. Martin and A.S. Sedra, "Exact design of switched-capacitor bandpass filters using coupled-biquad structures," *IEEE Trans Circuits Syst.*, vol. CAS-27, pp. 469-475, June 1980.

[15] P.E. Fleischer and K.R. Laker, "A family of active switched-capacitor biquad building blocks," *Bell. Syst. Tech. J.*, vol. 58, pp. 2235-2269, Dec. 1979.

[16] R.B. Datar and A.S. Sedra, Exact design of strays-insensitive switched-capacitor ladder filters," *IEEE Trans. Circuits Syst.*, vol. CAS-30, pp. 88-898, Dec. 1983.

[17] M.S. Lee, G.C. Temes, C. Chang and M.G. Ghaderi, "Bilinear switched-capacitor ladder filters," *IEEE Trans. Circuits Syst.*, vol. CAS-28, no. 8, pp. 811-822, Aug. 1981.

[18] A.V. Oppenheim and R.W. Schafer, *Digital Signal Processing,* Prentice Hall, Englewood Cliffs, N.J., 1975.

[19] W.M. Snelgrove, "FILTOR-2: A computer aided filter design program," Dept. of Electrical Engrg., Univ. of Toronto, Toronto, Canada 1981.

[20] D.B. Ribner and M.A. Copeland, "Biquad alternatives for high-frequency switched-capacitor filters," *IEEE Solid Sate Circuits*, vol. SC-20, no. 6, pp. 1085-1095, Dec. 1985.

[21] G.W. Roberts, D.G. Nairn and A.S. Sedra, "On the implementation of fully differential switched-capacitor filters," *IEEE Trans. Circuits Syst.*, vol. CAS-33, no. 4, pp. 452-455, April 1986.

[22] F. Montecchi, "Time-shared switched-capacitor ladder filters insensitive to parasitic effects," *IEEE Trans. Circuits Syst.*, vol. CAS-31, pp. 349-353, April 1984.

[23] H.J. Orchard, "Inductorless filters," *Electron. Lett.*, vol. 2, pp. 224-225, June 1966.

[24] K. Martin and A.S. Sedra, "Transfer function deviations due to resistor-SC equivalence assumption in switched-capacitor simulation of LC ladders," *Electron. Lett.*, vol. 16, no. 10, pp. 387-389, May 1980.

[25] S.C. Fang, Y. Tsividis and O. Wing, "SWITCAP; A switched-capacitor network analysis program," Parts I and II, *IEEE Circuits and Systems Magazine*, Sept. and Dec. 1983.

[26] D.G. Haigh, J.T. Taylor and B. Singh, "Continuous-time and switched-capacitor monolithic filters based on current and charge simulation," *IEEE Proc.* Part G, vol. 137, no. 2, pp. 147-155, April 1990.

[27] R. Gregorian and W.E. Nicholson Jr., "CMOS switched-capacitor filters for a PCM, voice CODEC," *IEEE J. Solid-State Circuits*, vol. SC-14, no. 6, pp. 970-980, Dec. 1979.

[28] R.B. Datar, "Exact design of strays-insensitive switched-capacitor ladder filters," *Ph.D. Thesis*, Dept. of Elect. Engrg., Univ. of Toronto, Canada, Oct. 1983.

[29] R.B. Datar and A.S. Sedra, "Exact design of stray-insensitive switched-capacitor high-pass ladder filters," *Electron. Lett.*, vol. 19, no. 29, pp. 1010-1012, 24th Nov. 1983.

Chapter 8

Multirate Switched-Capacitor Filters

José E. Franca and Rui P. Martins
Instituto Superior Técnico, Integrated Circuits and Systems Group, 1096 Lisboa Codex, Portugal

1. Introduction

In the early 80's multirate SC filters were firstly introduced as a means of relaxing the requirements of continuous-time pre- and post-filters in the context of a traditional SC filter system [1-4]. Then, the systematic application of the concepts of multirate discrete-time signal processing led to the development of new classes of multirate SC circuits [5-13] and their utilisation for the realisation of a variety of signal processing functions [14-17]. The most important among these include narrow and very narrow bandpass filtering [5, 15], high-frequency filtering [18, 19], linear phase filtering [6, 11], and frequency modulation and demodulation [17].

Traditional SC filters can also be employed in multirate signal processing systems, but it is usually desirable to design specialised multirate SC circuits which can take advantage of the processes of sampling rate reduction and sampling rate increase of discrete-time signals in order to reduce the speed requirements of the amplifiers. Such specialised multirate SC circuits are known as decimators and interpolators, respectively, for reducing and increasing the sampling rate, and their filtering functions are tailored to reject the unwanted alias and image frequency components associated with the signals sampled at the lower rate. Although SC decimator and interpolator circuits employ more complex digital circuits than traditional SC filters, and are required for generating multiphase switching waveforms, they may lead to considerable gains in the analogue circuitry, namely with respect to silicon and power consumption and also to the performance behaviour under non-ideal characteristics of the amplifiers. There are even some applications whose practical implementation in IC form can only be achieved using such multirate SC circuits. The need for analogue and non-trivial digital circuitry for implementing multirate SC signal processing is compatible with the technologies which are currently available to enable the strategic objective of realising on a single chip all aspects of signal processing, both analogue and digital.

The purpose of this chapter is to provide an overview of the most popular circuit techniques and design methodologies for multirate SC building blocks.

We start, in Section 2, by considering the design of finite impulse response (FIR) decimators and interpolators with single-amplifier architectures which are rather simple to design but which may have practical limitations considering the maximum speed of operation as well as the complexity of the IC realization. To overcome such limitations, we consider also SC decimator circuits based on multi-amplifier architectures which have much more favorable requirements with respect to component count, speed of the OA's, and complexity of the digital switching waveform generator. The design of multirate SC filters with infinite impulse response (IIR) discrete-time transfer functions is addressed in Section 3, for the case of decimating building blocks and, in Section 4, for the case of the complementary interpolating building blocks. The conclusions of this chapter are presented in Section 5.

2.   FIR SC Decimators and Interpolators

2.1.   Direct-form polyphase SC structure for decimation

The operation of decimation consists of reducing the sampling rate of the signals from a high value MF_s (M is an integer >1) to a lower value F_s, and is implemented by a decimator circuit [20] symbolically represented in Figure 1-a. The conceptual operation of the decimator is illustrated in Figure 1-b. The prototype filter at the heart of the decimator receives input samples at the higher rate MF_s, weights and combines those samples, and then produces output samples at the lower rate F_s. The prototype filter is a linear time-invariant filter with discrete-time transfer function $H(z)$ which, for FIR decimators, has the general form

$$H(z) = \sum_{n=0}^{N-1} h_n z^{-n} \tag{1}$$

where the unit delay period is $1/MF_s$. The length N and the coefficients h_n of the prototype filter are optimised such that its frequency response adequately attenuates the unwanted alias components associated with the output signal at lower rate F_s.

An SC decimator can be implemented using the direct-form polyphase structure shown in Figure 2 [6], in which the prototype filter is decomposed into a set of M sub-filters that operate at lower rate F_s. The discrete-time transfer function $H_m(z)$ of each sub-filter (m=0, 1,..., M-1) is given by

$$H_m(z) = \sum_{i=0}^{I_m} h_{m+iM} z^{-iM} \tag{2}$$

where the coefficients h_{m+iM} are equal to the coefficients h_n in (1) with $n=m+iM$; the length I_m of each sub-filter m is such that (I_m-1) is the integer part of $(N-1-m)/M$. The delay between equivalent samples of each sub-filter corresponds, in the frequency domain, to a phase shift and hence the sub-filters are designated *polyphase filters*.

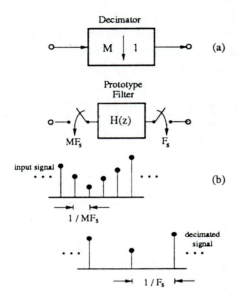

Fig. 1:   Sampling rate reduction by a factor M. (a) Symbol of the decimator; (b) illustration of decimation in time.

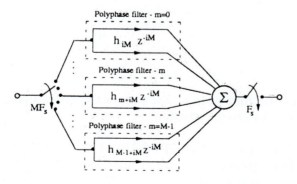

Fig. 2: Polyphase decimator structure with direct-form polyphase filters.

The implementation in SC form of the polyphase structure in Figure 2 can be divided into two parts. The first part corresponds to the implementation of the transmission factors using SC elements with only switches and capacitors. The simpler SC elements implement the terms with transmission factors h_m, and correspond to a situation where the input signal is sampled during the reference period. Such SC elements can be realised using conventional SC branches switched in the normal manner, which means that during the reference period the time slot for sampling precedes the time slot for charge transfer. The toggle-switch-inverter branch (TSI) in Figure 3-a realises positive coefficients whereas the open-floating-resistor branch (OFR) in Figure 3-b realises negative coefficients. An alternative realisation of the terms with negative coefficients uses the parasitic-compensated-toggle-switched-capacitor branch (PCTSC) in Figure 3-c. The delays of these SC branches refer to the reference period in Figure 3-d, and are determined with respect to the sampling slots. The SC elements in Figure 3 are canonic in the sense that they utilise only one SC branch with one capacitor in order to realize one term of the FIR impulse responses[1]. The SC elements that implement the terms with

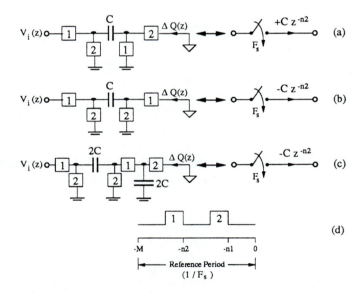

Fig. 3:   SC elements for the first term of the polyphase filters using a normal sequence of sampling followed by charge transfer.

transmission factors $h_{m+M} \cdot z^{-M}$ correspond to a situation where the input signal is sampled in the period preceding the reference period. Such SC elements can be realised by inverting the sequence of sampling and charge transfer, during the reference period. This operation can only be achieved using TSI and PCTSC branches because in such branches sampling and charge transfer take place in different time slots. The TSI in Figure 4-a realises positive coefficients, while

---

[1] We assume that the PCTSC branch with two capacitors 2C is equivalent to an SC branch with one capacitor C.

the PCTSC in Figure 4-b realises negative coefficients. It is easy to see that these SC elements are also canonic, as before. Finally, in order to implement terms with transmission factors $h_{m+2M} \cdot z^{-2M}$, $h_{m+3M} \cdot z^{-3M}$, and so forth, it is necessary to introduce SC branches with reduced sampling rate that can sample the input signal at two, three or more periods before the reference period. But because of the reduced sampling rate it is also necessary to have multiple SC branches to ensure that such terms are realised every reference period. Such a scheme is illustrated in Figure 5 for a transmission factor $h_{m+2M} \cdot z^{-2M}$. The SC elements consist of two parallel SC branches, with reduced sampling rate $F_s/2$ and

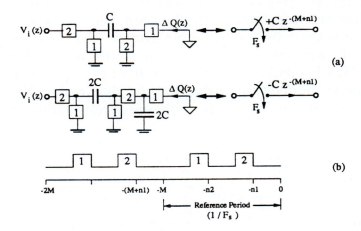

Fig. 4:   SC elements for the second term of the polyphase filters using an inverted sequence of charge transfer followed by sampling.

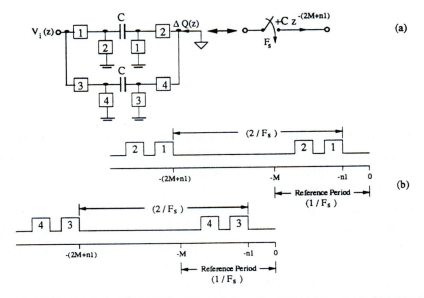

Fig. 5:   Example of a noncanonical SC element for the third term of the polyphase filters, using SC branches with reduced sampling rate.

with an inverted sequence of charge transfer followed by sampling. We should note that although this SC element is noncanonic with respect to the number of SC branches and capacitors, it is still a direct-form implementation since it realizes only one term of the FIR impulse response.

The second part of the polyphase structures consists of an SC accumulator that realises the addition of the packets of charge $\Delta Q_n(z)$ produced by the SC elements, and can be implemented using the structure shown in Figure 6. During the reference period of the decimator (i.e., one period of the output signal), the incoming packets of charge $\Delta Q_n(z)$ corresponding to all terms of the FIR prototype filter are transferred into the memoried feedback capacitor C. At the end of the reference period, in time slot 1, the new output voltage is sampled by the following SC circuit and also by the feedback toggle-switched-capacitor branch (TSC); then, in time slot 2 of the next reference period, the TSC transfers its charge to the feedback capacitor and thus resets the previous output voltage. At the same time, a new cycle of charge accumulation is executed.

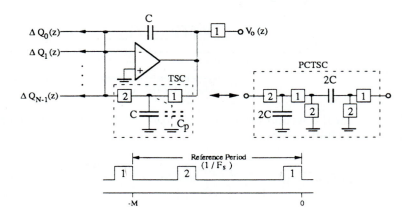

Fig. 6: SC accumulator for decimation.

As an example, we shall consider now the design of an SC decimator required to reduce the sampling rate from $4F_s=192kHz$ to $F_s=48kHz$ and whose impulse response is given in Figure 7-a. The prototype filter is decomposed into $M=4$ polyphase filters leading to the polyphase structure given in Figure 7-b. The SC implementation in direct form leads to the polyphase SC decimator shown in Figure 8-a operating with the time frame of Figure 8-b. It is easy to observe the correspondence between the SC decimator and the polyphase structure. For example, the input signal is sampled in time slot 0 by the polyphase filter m=0, in time slot 1 by the polyphase filter m=1, and so forth. The terms with transmission factors h_m are realised by SC branches switched in the normal manner, and thus the charge transfer slots appear after the sampling slots. But

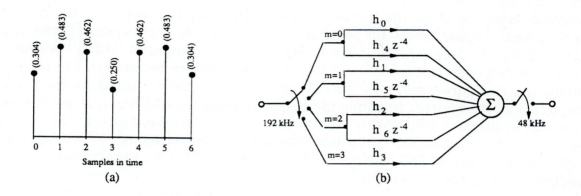

Fig. 7: FIR prototype filter for a decimator M=4. (a) Impulse response; (b) Polyphase structure.

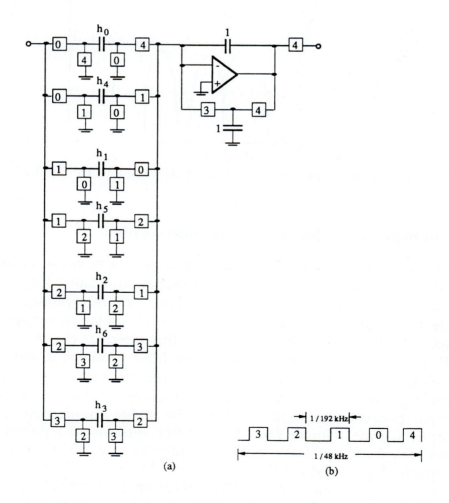

Fig. 8: Single-stage polyphase SC decimator M=4 with time frame.

the terms with transmission factors $h_{m+4}.z^{-4}$ are realised by SC branches with an inverted sequence of charge transfer followed by sampling, and thus the charge transfer slots appear before the sampling slots. The measured baseband response of the above SC decimator is given in Figure 9, and compared with a computer simulated response to demonstrate its good performance behaviour. In order to obtain the measured response in Figure 9 the detector is tuned to the same frequency as the source, and thus the response includes additional transmission zeroes at multiples of F_s=48kHz due to the sampled-and-hold effect. This effect was also simulated in the computer simulated response using SWITCAP. The slight attenuation at DC is due to the top-plate parasitic capacitance in the feedback TSC of the SC accumulator.

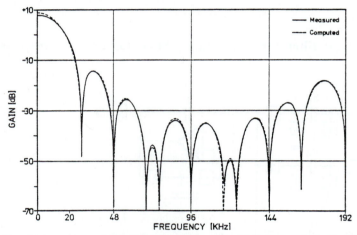

Fig. 9: Baseband response of the SC decimator M=4 [6] (copyright © 1985 by IEEE).

2.2. Direct-form polyphase SC structure for interpolation

The operation of increasing the sampling rate from a low value F_S to a higher value LF_S (L is an integer>1) is known as interpolation and complements the operation of decimation described above [20]. An interpolator circuit is symbolically represented as in Figure 10-a. The interpolation of sampled and held signals, which occur in SC circuits, is illustrated in Figure 10-b. Every period $1/F_S$ of the input signal, the prototype filter produces a sequence of L samples, each of which is determined by its discrete-time transfer function H(z), and held for a period $1/LF_S$. The frequency response of the prototype filter is also important because it should eliminate the image components associated with the input signal at lower rate. In SC filter systems, the decimators and interpolators may have the same factor of sampling rate alteration (i.e., M=L), and their original prototype filters may also have the same discrete-time transfer function H(z). But because the operation of interpolation refers to sampled-and-held signals, the original prototype filter for interpolation should be modified according to

$$H'(z) = H(z) \sum_{l=0}^{L-1} z^{-1}$$

(3)

where the unit delay period is $1/LF_S$.

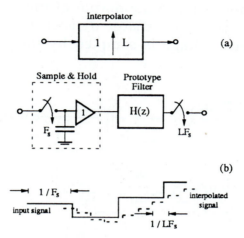

(a)

(b)

Fig. 10: Sampling rate increase by a factor L. (a) Symbol of the interpolator; (b) interpolation of a sample-and-held signal.

An analogue sampled-data interpolator with modified prototype filter can be implemented using the polyphase structure of Figure 11 [6]. Each one of the L polyphase filters at lower rate F_s produces one output sample determined by its transfer functions $H_l(z)$, and then the output signal of the interpolator is obtained by resampling the outputs of the polyphase filters at higher rate LF_s. The impulse responses of the polyphase filters are derived from the impulse response of the modified prototype filter using a decomposition process similar to that for the decimator. In this structure, the outputs of the polyphase filters only matter during the period of time when they are connected to the output of

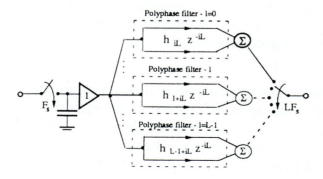

Fig. 11: Polyphase interpolator structure with direct-form polyphase filters using one time-shared accumulator.

the interpolator. It is possible to utilise only one time-shared accumulator, as shown in Figure 12, for the simple case of L=2. During one reference period of the interpolator (i.e., one period of the input signal), there are L=2 cycles of the operation of the SC accumulator, each of which produces one sample of the output interpolated signal. Firstly, in time slot 0, the charge packets of the polyphase filter l=0 are transferred to the output of the accumulator, while the previous output voltage is reset by the charge transfer from the feedback TSC. In time slot 2, the new output voltage is sampled by the feedback TSC. In the

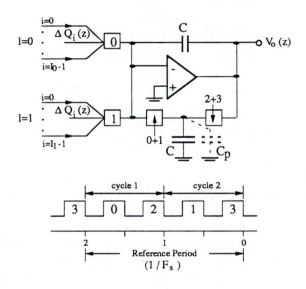

Fig. 12: A simple time-shared SC accumulator for interpolation (L=2).

second cycle, the packets of charge of the polyphase filter l=1 are transferred to the output of the accumulator in time slot 1, while the previous output voltage is reset by the charge transfer from the feedback TSC. This second cycle ends with time slot 3, when the new output voltage is sampled by the feedback TSC. With such an arrangement of the switch phasing the output of the interpolator is implicitly sampled and held, because the output voltage changes only during time slots 0 and 1.

For implementing the terms of the polyphase filters in the structure of Figure 11 we can also employ the basic SC elements described in Figures 3, 4, and 5. However, we have now to consider that the delay terms are defined with respect to the charge transfer slots whereas in the previous decimating structures they were defined with respect to the sampling slots.

As an example, we shall now consider the realisation of an SC interpolator that increases the sampling rate from $F_s=80KHz$ to $4F_s=320kHz$ and whose prototype filter leads to the modified impulse response shown in Figure 13-a. We can observe in the polyphase structure of Figure 13-b that, in order to implement the terms with transmission factors $h_{l+8}.z^{-8}$ in the polyphase filters $l=0$ and $l=1$, it is necessary to employ noncanonic SC elements with multiple SC branches. By employing the scheme given in Figure 5 we arrive at the polyphase SC interpolator and time frame of Figure 14, which produces the measured frequency response shown in Figure 15. The precision of this single-stage structure is affected by the increased sensitivity of the circuits using noncanonic SC elements as well as the narrower time slots that increase the effects of nonzero settling time and limited slew rate of the OA's. The noncanonic SC elements also introduce additional clock feedthrough at the lower rate $F_s/2$ but this can be reduced by employing well-known compensation techniques.

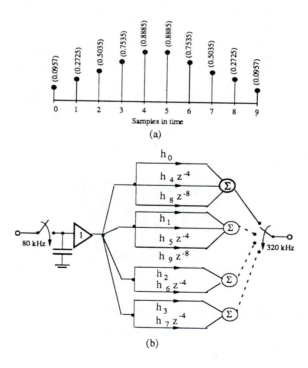

Fig. 13:   FIR prototype filter for an interpolator L=4. (a) Modified impulse response; (b) polyphase structure.

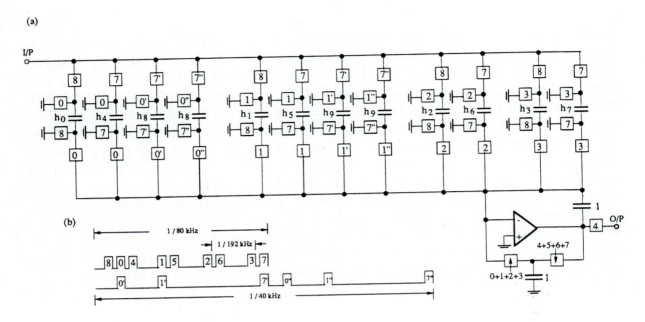

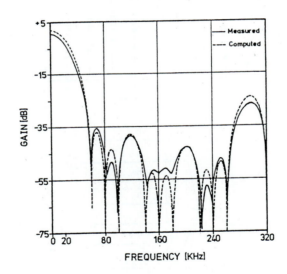

Fig. 14: Single-stage polyphase SC interpolator L=4 with time frame.

Fig. 15: Baseband response of the single-stage polyphase
SC interpolator L=4 [6] (copyright © 1985 by IEEE).

2.3. Multi-amplifier FIR SC decimators

We saw before that any type of FIR polyphase decimator can be characterised by
the length N of its impulse response and the decimation factor M determining
the number of polyphase filters. In the specific type of polyphase architectures to
be considered now, the polyphase filters share a number of common blocks to

realise the basic operations of coefficient multiplication, delay and accumulation using minimum hardware. They will, therefore, be referred to as active-delayed-block (ADB) polyphase architectures [11].

The ADB polyphase architecture schematically illustrated in Figure 16 consists of B common blocks [B is the smaller integer equal to or greater than (N/M)], each of which realises a *single term* of the associated transfer function of the polyphase filters. The first block - b_0, realises the terms h_m of the polyphase filters (m=0, 2, ..., M-1) which have an overall zero delay. The second block - b_1, realises the terms of the polyphase filters which have an overall delay of z^{-M} and whose coefficients are h_{m+M}. In general, the block b_j (j=0,1,..., [B-1]) realises the terms of the polyphase filters which have an overall delay of z^{-jM} and whose coefficients are h_{m+jM}, subject to the condition m+jM$\leq$N determined by the length N of the overall impulse response of the decimator. A total of [B-1] common blocks, each with an overall delay of z^{-M}, and an additional common block directly connected to the output accumulator are required to implement this type of polyphase architecture.

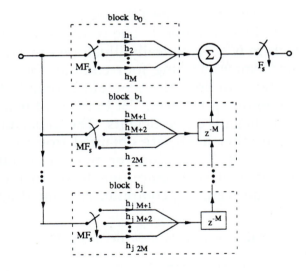

Fig. 16: Schematic representation of the architecture of an ADB polyphase decimator.

As in the previous decimating architectures, the input SC branches are usually implemented using either the TSI branch, for positive coefficients, or the PCTSC branch, for negative coefficients. Also as before, the typical SC implementation of an accumulator employs an OA with a feedback integrating capacitor (C_I) and reset network, such that the accumulation of the packets of charge produced by the input SC branches is scaled by the charge-to-voltage conversion factor ($-1/C_I$). By combining such an SC accumulator together with a simple output TSI branch, with equivalent capacitance value D, we obtain the typical implementation of an SC active-delayed block illustrated in Figure 17 whose charge-to-charge conversion factor becomes $(D/C_I)^2$. Hence, based on such circuits we can see that the first block - b_0 of the ADB architecture is formed by M SC branches with capacitors C_m (m=0, ..., M-1), and are directly connected to the output accumulator with integrating capacitor C_{I0}. Each remaining block b_j is also formed by M input SC branches with capacitors C_{m+jM} and one SC active-delayed block with integrating capacitor C_{Ij} and output TSI branch with capacitor D_j. The design of such an architecture is illustrated below.

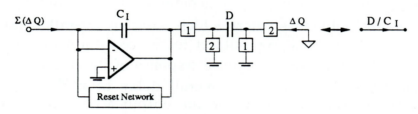

Fig. 17: Charge-to-charge conversion factor of an SC active-delayed block.

The impulse response coefficients of the FIR decimator to be considered here, with N=19 and M=5, are given in Table 1. The resulting SC circuit based on the ADB polyphase architecture is schematically illustrated in Figure 18, where five polyphase filters are implemented using four common blocks. The sign of the impulse response coefficients defines the type of SC branch to be employed at the input of the four common blocks, i.e., TSI branches for positive coefficients and PCTSC branches for negative coefficients. The capacitance values of the input SC branches, on the other hand, are obtained from the magnitude of the impulse response coefficients. Designing for maximum 0dB gain from the input to the output terminal of each amplifier, leads to the final capacitance values given in Table 2. The resulting computer simulated amplitude response is shown in Figure 19.

Table 1: Impulse response coefficients of the decimator FIR transfer function.

$h_1 = -0.0072681$	$h_6 = 0.060197$	$h_{11} = 0.11294$	$h_{16} = 0.015366$
$h_2 = -0.0029322$	$h_7 = 0.087489$	$h_{12} = 0.11215$	$h_{17} = 0.0028416$
$h_3 = 0.0028416$	$h_8 = 0.11215$	$h_{13} = 0.087489$	$h_{18} = -0.0029322$
$h_4 = 0.015366$	$h_9 = 0.11294$	$h_{14} = 0.060197$	$h_{19} = -0.0072681$
$h_5 = 0.035000$	$h_{10} = 0.13556$	$h_{15} = 0.035000$	

2 For simplicity, we disregard here the delay factors.

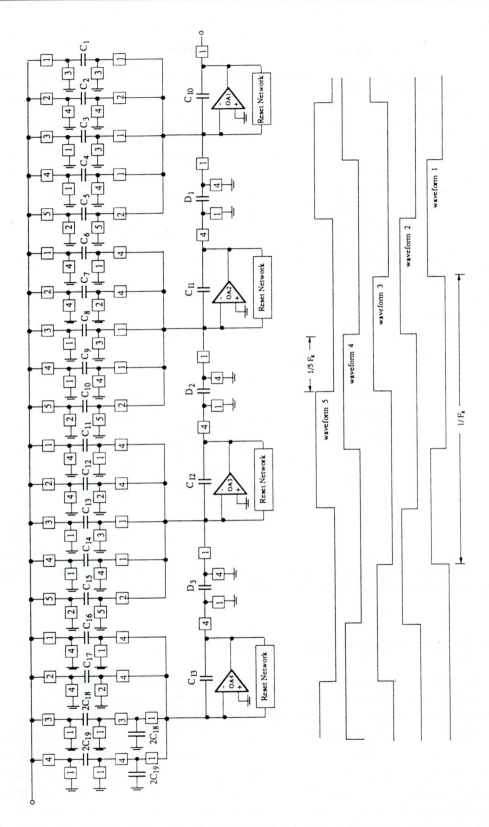

Fig. 18: SC decimator circuit with ADB polyphase architecture N=19; M=5.

Table 2:   Final normalised capacitance values of the circuit in Fig. 18 for absolute maximum signal handling capability.

Block b_0	Block b_1	Block b_2	Block b_3
$C_1 = 2.5578$	$C_6 = 1.0000$	$C_{11} = 5.0555$	$C_{16} = 5.4077$
$C_2 = 1.0319$	$C_7 = 1.4533$	$C_{12} = 4.3824$	$C_{17} = 1.0000$
$C_3 = 1.0000$	$C_8 = 1.8630$	$C_{13} = 3.4187$	$C_{18} = 1.0319$
$C_4 = 5.4077$	$C_9 = 2.1492$	$C_{14} = 2.3522$	$C_{19} = 2.5578$
$C_5 = 12.3173$	$C_{10} = 2.2518$	$C_{15} = 1.3676$	———
$C_{I0} = 351.9198$	$C_{I1} = 15.8976$	$C_{I2} = 16.8895$	$C_{I3} = 9.0060$
$D_1 = 336.7845$	$D_2 = 7.1801$	$D_3 = 1.0000$	———

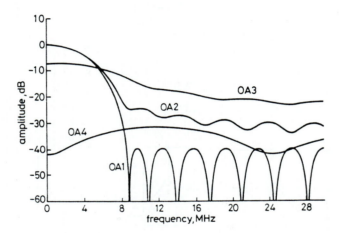

Fig. 19:   Optimised amplitude responses for absolute maximum signal handling capability in the SC decimator circuit of Fig. 18 [22] (copyright © 1991 by IEE).

3.   IIR SC Decimator Building Blocks

Infinite impulse response (IIR) SC decimators can be implemented based on *non-optimum* and *sub-optimum* classes of circuits, according to the resulting speed requirements of the OA's. The non-optimum class of IIR SC decimator circuits, schematically illustrated in Figure 20, consists of an SC filter operating at MF_s and whose output signal is sampled at the lower rate F_s. Although this offers a straightforward solution for the realisation of high selectivity amplitude responses using simple bi-phase SC filters, it does not take advantage of the sampling rate reduction inherent to the decimation process. Hence, the resulting speed of the OA's and capacitance spread are both determined by the

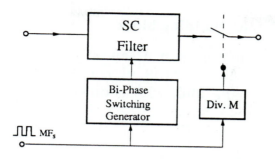

Fig. 20: Non-optimum implementation of IIR SC decimators.

higher input sampling rate MF_S. Figure 21 represents the sub-optimum class of IIR SC decimator circuits consisting of the cascading of an SC filter operating at MF_S and a non-recursive polyphase structure with a decimating factor of M [7]. This architecture gives greater design flexibility than the above non-optimum class of IIR SC decimators, and also allows slower OA's in the polyphase structure with output sampling rate F_S. However, faster OA's are still needed in the SC filter operating at the higher sampling rate MF_S.

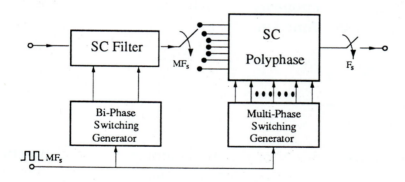

Fig. 21: Sub-optimum implementation of IIR SC decimators.

In order to overcome the speed limitations of the above architectures a new optimum class can be considered for the implementation of IIR SC decimators where the speed of the OA's is determined by the lower output sampling rate F_S. For this purpose we shall describe next a systematic methodology for the design of first-order, second-order, and Nth-order SC decimator building blocks with arbitrary numerator polynomial functions and integer decimating factors [8, 9, 12].

3.1. First-order SC decimator building block

An optimum IIR SC decimator with a sampling rate reduction factor of M, schematically illustrated in Figure 22 [8, 9], consists of a single network whose input sampling rate is MF_S and whose output sampling rate is F_S. The

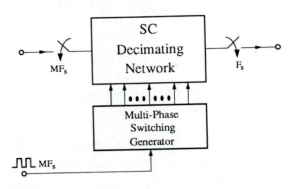

Fig. 22: Optimum implementation of IIR SC decimators.

amplitude response of such network is designed to meet the desired passband specifications and, at the same time, achieve the required rejection of the unwanted alias signal components associated with the signal sampled at F_S. The corresponding z-transfer function

$$H(z) = \frac{\sum_{j=0}^{N} a_j \cdot z^{-j}}{b_0 - z^{-1}} \tag{4}$$

can have an arbitrary order N of the numerator polynomial function. The unit delay period is relative to $1/MF_S$ corresponding to the sampling period at the input of the decimator. By using the equivalence

$$b_0 - z^{-1} = \frac{b_0^M - z^{-M}}{\sum_{I=0}^{M-1} b_0^{M-1} \cdot z^{-I}} \tag{5}$$

the above z-transfer function can also be expressed as

$$\overline{H}(z) = \frac{\displaystyle\sum_{m=0}^{N+M-1} \overline{a}_m \cdot z^{-m}}{b_0^M - z^{-M}} \tag{6}$$

with the following numerator coefficients

$$\sum_{m=0}^{N+M-1} \overline{a}_m \cdot z^{-m} = \sum_{j=0}^{N} a_j \cdot z^{-j} \sum_{I=0}^{M-1} b_0^{M-I} \cdot z^{-I} \tag{7}$$

For consistency, we still represent the delay terms in (6) referring to a unit delay period of $1/MF_s$. Hence, the denominator terms proportional to z^{-iM} can now be implemented by a recursive structure operating at the lower sampling rate F_s. This is illustrated next for a simple example of a first-order SC lowpass decimator with a decimating factor of M=4.

The optimum implementation of a first-order SC decimator described by the z-transfer function

$$\overline{H}(z) = \frac{\overline{a}_0 + \overline{a}_1 \cdot z^{-1} + \overline{a}_2 \cdot z^{-2} + \overline{a}_3 \cdot z^{-3}}{\overline{b}_0 - z^{-4}} \tag{8}$$

and with M=4, is illustrated in Figure 23-a. The OA with integrating capacitor C_I and feedback SC branch C_F is viewed as a first-order charge-to-voltage converter, while the input SC branches C_i (i=0 to 3) are viewed as voltage-to-charge converters and are organised as in a polyphase structure. The operation of this SC decimator refers to the time frames given in Figure 23-b. Time frame T, for charge transfer, has two wide time slots like a conventional SC integrator operating at F_s, which determine the operation of the switches controlling the flow of charge through the virtual ground of the amplifier. On the other hand, the four time slots in time frame S, for signal sampling, control the switching operation of the switches which are responsible for the sampling of the input signal. The width of such time slots is simply limited by the equivalent RC time constants associated with the charging of each one of the input switched-capacitors, and thus can be substantially narrower than the time slots in time frame T. Time slot 1, in time frame S, occurs immediately before time slot E in time frame T. Hence, by referring to the time reference given in Figure 23-b, we have an unimportant flat delay factor of $z^{-3/8}$ between the output sampling instant (at the end of time slot E) and the most recent sampling instant of the input signal (at the end of time slot 1). By employing

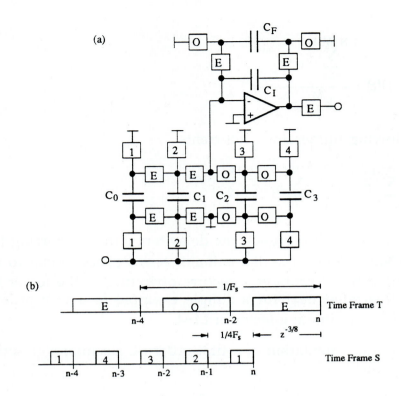

Fig. 23:   Example of an optimum first-order SC decimator with M=4. (a) Circuit; (b) time
            frames.

well-known signal flow graph (SFG) techniques for the analysis of SC networks,
we arrive at the z-transfer function

$$T(z) = z^{-3/8} \cdot \frac{C_0 + C_1 \cdot z^{-1} + C_2 \cdot z^{-2} + C_3 \cdot z^{-3}}{(C_I + C_F) - C_I \cdot z^{-4}} \tag{9}$$

In the above SC decimator building block, the proposed switching of the SC
branches is devised such that the sampling of the input signal is performed
during a short time interval, whereas the transfer of charge is performed during
a much wider time interval to maximise the time allowed for amplifier
settling. In comparison with the non-optimum implementation of this SC
decimator, shown in Figure 24 for clarity, the SC decimator building block of
Figure 23 employs a larger number of SC branches and switching waveforms.
However, we now obtain a significant M-fold increase (M=4 in this example) of
the time allowed for amplifier settling. In other words, for the same settling
time of the OA, the proposed circuit is capable of achieving an M-fold increase
of the input sampling rate compared with the traditional non-optimum
implementation. Next, we shall extend this technique to the design of second-
order SC decimator building blocks.

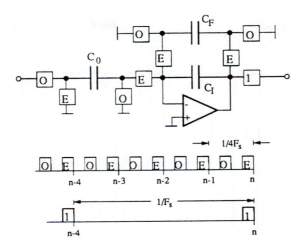

Fig. 24: Non-optimum implementation of a first order SC decimator with M=4.

3.2. Second-order SC decimator building block

The original z-transfer function of a second-order decimator, with arbitrary numerator polynomial function, is described by

$$H(z) = \frac{\sum_{j=0}^{N} a_j . z^{-j}}{1 - 2r_p . \cos\left(\theta_p\right) . z^{-1} + r_p^2 . z^{-2}} \tag{10}$$

which, according to (6), can also be written as

$$\overline{H}(z) = \frac{\sum_{j=0}^{N} a_j . z^{-j} \sum_{I=0}^{2(M-1)} \alpha_I . r_p^I . z^{-I}}{1 - 2r_p^M \cos\left(M\theta_p\right) . z^{-M} + r_p^{2M} . z^{-2M}} \tag{11}$$

where the coefficients α_I in the numerator function are given in Table 3 for practical decimating factors up to M=10.

The SC decimator building block implementing the z-transfer function (11) is described in Figure 25-a [8, 9], employing a basic TIL structure [21] which is responsible for the realisation of the denominator function. Capacitor E (capacitive damping), and switched-capacitor F (resistive damping), represent two alternative forms of damping that can be utilised in the loop. In a similar fashion as before, this TIL structure is viewed as a second-order charge-to-voltage converter whose input packets of charge are produced by the two sets of

Table 3:   Expressions giving the coefficients α_I in the transformation function for optimum second-order decimators [9] (copyright © 1990 by IEEE).

M	2	3	4	5	6	7	8	9	10
$\alpha 1$	$2x$	$2x$	$2x$	$2x$	$2x$	$2x$	$2x$	$2x$	$2x$
$\alpha 2$		$4x^2-1$	$4x^2-1$	$4x^2-1$	$4x^2-1$	$4x^2-1$	$4x^2-1$	$4x^2-1$	$4x^2-1$
$\alpha 3$			$8x^3-4x$	$8x^3-4x$	$8x^3-4x$	$8x^3-4x$	$8x^3-4x$	$8x^3-4x$	$8x^3-4x$
$\alpha 4$				$16x^4-12x^2+1$	$16x^4-12x^2+1$	$16x^4-12x^2+1$	$16x^4-12x^2+1$	$16x^4-12x^2+1$	$16x^4-12x^2+1$
$\alpha 5$					$32x^5-32x^3+6x$				
$\alpha 6$						$64x^6-80x^4+24x^2-1$			
$\alpha 7$							$128x^7-192x^5+80x^3-8x$		
$\alpha 8$								$256x^8-448x^6+240x^4-40x^2+1$	
$\alpha 9$									$512x^9-1024x^7+448x^5-192x^3+10x$

$$F(z) = \sum_{I=0}^{2(M-1)} \alpha_I \, r_p^I \, z^{-I}$$

$$\alpha_0 = 1, \; \alpha_{[2(M-1)-I]} = \alpha_I, \quad x = \cos(\theta_p)$$

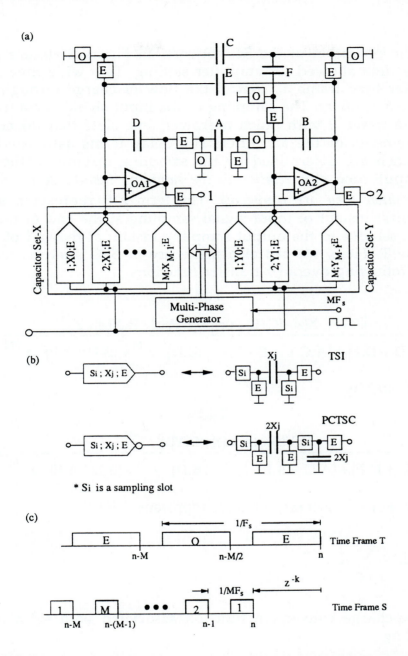

Fig. 25:   Optimum implementation of a second-order SC decimator. (a) General structure; (b) SC branches; (c) time frames.

SC branches, set-X and set-Y, respectively, connected to the virtual ground of the OA1 and OA2. These SC branches, organised in the form of polyphase structures, are represented by polygonal symbols where we indicate the time slot for input signal sampling, the time slot for output signal sampling, and the capacitance value of the branch as well as the polarity of charge transfer. For greater design flexibility we shall always use different input and output time slots. Hence, PCTSC branches are utilised for positive charge transfers whereas TSI branches are employed for negative charge transfers. The time frames controlling the operation of the SC network, in Figure 25-b, are designed to maximise the time allowed for amplifier settling. Two wide time slots, in the charge transfer time frame T, control the flow of charge through the virtual ground of the amplifiers. The sampling of the input signal is controlled by the sampling time frame S with much narrower time slots that do not affect the speed requirements of the amplifiers. The delay terms associated with each input SC branch are determined by the sampling instants at the end of the respective input time slot *relative to the sampling instant at the output of the amplifier to which they are connected*. The term z^{-k} represents a flat delay between the first instant of input signal sampling at the end of time slot 1, in time frame S, relative to the output sampling instant at the end of time slot E, in time frame T. Here, we can also make use of SFG analysis techniques to arrive at the following overall z-transfer functions

$$T_1(z) =$$

$$\frac{\left[(B+F).X(z) - (C+E).Y(z)\right] + \left[E.Y(z) - B.X(z)\right].z^{-M}}{(B.D + D.F) + (A.C + A.E - D.F - 2B.D).z^{-M} + (B.D - A.E).z^{-2M}} z^{-k} \qquad (12\text{-}a)$$

for terminal 1, and by

$$T_2(z) =$$

$$\frac{D.Y(z) + \left[A.X(z) - D.Y(z)\right].z^{-M}}{(B.D + D.F) + (A.C + A.E - D.F - 2B.D).z^{-M} + (B.D - A.E).z^{-2M}} z^{-k} \qquad (12\text{-}b)$$

for terminal 2. For conciseness, we have represented by

$$X(z) = \sum_{u=0}^{M-1} X_u . z^{-u} \qquad (13\text{-}a)$$

the voltage-to-charge conversion function associated with set-X of input SC branches, and by

$$Y(z) = \sum_{v=0}^{M-1} Y_v . z^{-v}$$

$$(13\text{-}b)$$

the corresponding function associated with set-Y of input SC branches.

For each one of the above z-transfer functions (12) of the SC decimator building block, the coefficients of the z-transfer function (11) are determined as functions of the input switched-capacitors X_u and Y_v, respectively, in capacitor set-X and capacitor set-Y, as well as the capacitors in the loop of the decimator. The adoption of a particular topology, including the selection of the output terminal and type of damping, as well as the design of capacitor sets X and Y, is carried out on a case-by-case basis to minimise the capacitance spread and total overall capacitor area and also to obtain reduced sensitivity of the amplitude response with respect to capacitance ratio errors. Designing for maximum signal handling capability is achieved by using the conventional technique of scaling the voltages at the output of the OA's.

For illustration purposes, we shall consider now a design example corresponding to the optimum implementation of a second order lowpass decimator, with Tchebyshev approximation and maximum ripple of 0.05dB in the passband with cut-off frequency of 6KHz. The SC decimator reduces the sampling rate from $4F_S=1MHz$ at the input to $F_S=250KHz$ at the output, giving a minimum rejection of 50dB of the alias signal components associated with the lower sampling rate F_S (i.e., in the frequency band from F_S-6KHz to F_S+6KHz), and their corresponding repetitions around $2F_S$ and $3F_S$. To meet these specifications we investigated four alternative design solutions for the SC decimator circuit shown in Figure 26-a, which operates with time frames of Figure 26-b. Such solutions correspond to the implementation of z-transfer

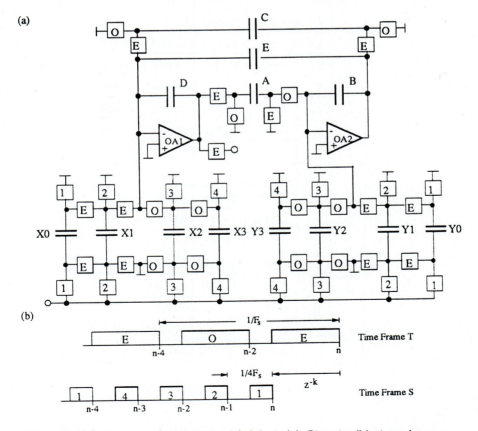

Fig. 26: SC lowpass decimator with M=4. (a) Circuit; (b) time frames.

functions (12-a), for output terminal from OA1, and (12-b), for output terminal from OA2, both of which can have either resistive-damping (E=0) or capacitive-damping (F=0). After scaling for maximum signal handling capability these solutions led to the normalised capacitance values indicated in Table 4. Solution 1 is clearly the most economical both in terms of capacitance spread (19.7) and total capacitor area (68.3). The nominal passband and overall computer simulated impulse sampled baseband amplitude responses are shown, respectively, in Figure 27-a and Figure 27-b. Overall, the measured frequency-translated amplitude response of the SC decimator is in agreement with the theoretical response except for some local deviations around F_S=250KHz, and its multiples $2F_S$ and $3F_S$, which are due to well-known clock feedthrough effects in the operation of the circuit.

Table 4: SC lowpass decimator with M=4 capacitance values.

Solution		1	2	3	4
Output Terminal		1		2	
Damping		E	F	E	F
Amplifier Capacitor Set-1	Y0	1.00957	1.38411	1	1
	Y1	1.01232	1.24985	1.89417	1.89417
	Y2	1.00892	1.12178	2.68790	2.68789
	Y3	1	1	3.38663	3.38663
	A	4.07695	4.07695	31.4045	55.6255
	B	19.6998	9.2309	174.292	114.338
	F	0	4.84026	0	59.9537
Amplifier Capacitor Set-2	X0	1.17406	3.31861	1.00956	1.00956
	X1	1.11664	2.10469	1.01232	1.01232
	X2	1.05849	1	1.00891	1.00891
	X3	1	0	1	1
	C	4.53678	42.8870	4.07694	4.07693
	D	11.8521	156.858	9.27309	16.4250
	E	19.6998	0	17.7031	0
C_{spread}		19.6998	156.858	174.292	114.338
ΣC		68.246	229.072	249.749	263.419

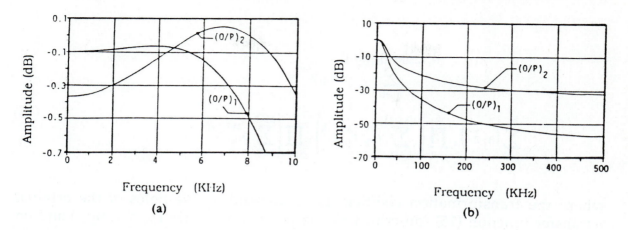

Fig. 27: Nominal computer simulated amplitude responses of the SC decimator in Fig. 26 [9] (copyright © 1990 by IEEE). (a) Passband; (b) overall.

3.3. Nth-order SC decimator building block

The original z-transfer function of an Nth-order SC decimator building block can be expressed as

$$H(z) = \frac{\displaystyle\sum_{j=0}^{N_p} a_j \cdot z^{-j}}{\displaystyle\prod_{i=1}^{S}\left(1 - 2r_{p_i}\cos\left(\theta_p\right) \cdot z^{-1} + r_{p_i}^2 \cdot z^{-2}\right) \cdot \prod_{i=1}^{F}\left(b_{0_i} - z^{-1}\right)} \tag{14}$$

where the unit delay period corresponds to the sampling period $1/MF_S$ at the input of the decimator. The numerator polynomial function can have an arbitrary order N_P+1. The order of the denominator polynomial function is $N=2S+F$, where S and F represent, respectively, the number of second and first order sections. The modification of the original z-transfer function (14) leads to

$$\overline{H}(z) = \frac{\displaystyle\sum_{m=0}^{N_p+2S(M-1)+F(M-1)} \overline{a}_m \cdot z^{-m}}{\displaystyle\prod_{i=1}^{S}\left(1 - 2r_{p_i}^M\cos\left(M\theta_p\right) \cdot z^{-M} + r_{p_i}^{2M} \cdot z^{-2M}\right) \cdot \prod_{i=1}^{F}\left(b_{0_i}^M - z^{-M}\right)} \tag{15}$$

The above modified numerator polynomial function is expressed by

$$\sum_{m=0}^{N_p+2S(M-1)+F(M-1)} \bar{a}_m z^{-m} =$$

$$\sum_{j=0}^{N_p} \left(a_j z^{-j} \right) \cdot \prod_{i=1}^{S} \left(\sum_{k=0}^{2(M-1)} \alpha_{k_i} r_{p_i}^k z^{-k} \right) \cdot \prod_{i=1}^{F} \left(\sum_{l=0}^{M-1} b_{0_i}^{M-1} z^{-1} \right) \tag{16}$$

where the transformation coefficients α_k depend on the poles of the original z-transfer function (15) (described by the polar coordinates r_{p_i} and θ_{p_i}) and on the decimating factor M.

The block diagram of Figure 28 illustrates the general architecture of the Nth-order IIR SC decimator building block [12]. This can be divided into two major parts, namely one high selectivity recursive network primarily responsible for the implementation of the denominator polynomial function and one low selectivity polyphase network determining the numerator polynomial function. Such polyphase network is formed by a varying number of simple SC branches depending both on the decimation factor M and on the complexity of the numerator polynomial function. The recursive network, on the other hand, can realise an arbitrary Nth order combination of real and complex

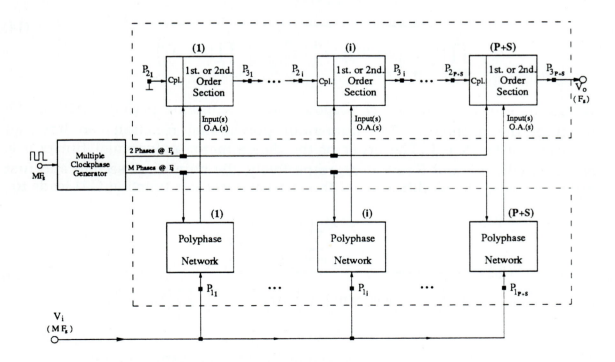

Fig. 28: General architecture of an optimum Nth order IIR SC decimator building block.

conjugated poles determining the total number of OA's in the circuit, *all of which operate at the low output sampling rate F$_S$*, and is formed by the cascade of three types of basic sections, namely a first-order section, a second order section and a coupling section. The first- and second-order sections are implemented using the type of building blocks previously described. For the coupling between sections we can select either a TSI branch (positive coupling) or an OFR branch (negative coupling). Given a specific topology, selected depending on the resulting capacitor area, capacitance spread and also the performance robustness under non-ideal characteristics of the amplifiers, SFG analysis techniques can again be utilised to derive the overall z-transfer function of the circuit and obtain the resulting design equations.

The application of the above building block can be demonstrated considering an SC lowpass video decimator [19] designed to reduce the sampling rate from $3F_S$=40.5MHz to F_S=13.5MHz (M=3), and possessing a fifth-order elliptic lowpass frequency response with passband ripple of 0.2dB, cut-off frequency F_C=3.6MHz, and minimum 35dB rejection above 4.44MHz. Based on a computer aided filter synthesis tool we obtained first the bilinear discrete-time coefficients for the original z-transfer function of the decimator prototype filter. Then, we derived the corresponding modified z-transfer function for optimum implementation yielding the SC decimator circuit shown in Figure 29-a which operates with the switching waveforms of Figure 29-b. The topology of this circuit has been selected in order to obtain low capacitance spread and total capacitor area, and also reduced response degradation under non-ideal characteristics of the OA's. The final normalised capacitance values obtained after scaling for maximum signal handling capability are also indicated in Figure 29-a. An optimising procedure has been employed to minimise the capacitance spread and total capacitor area in the circuit yielding, respectively, 8 and less than 100 units. Figure 30 gives the computer simulated overall amplitude response of this video decimator obtained using the ELDO electrical simulator and based on practical real modelling of the devices.

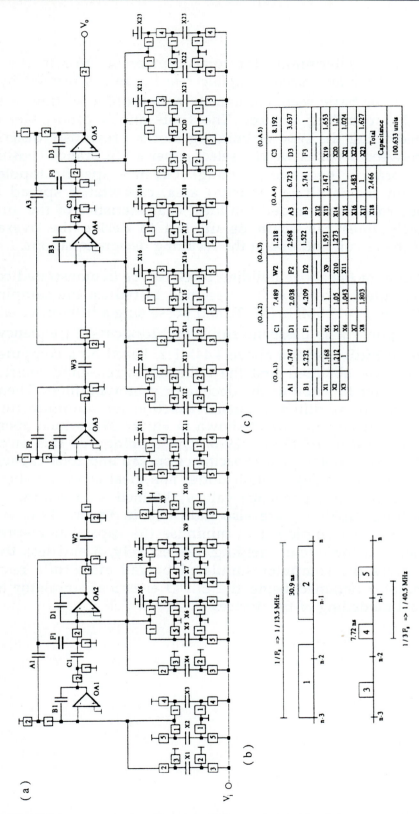

Fig. 29:   Fifth-order SC lowpass video decimator (M=3, Fs=13.5 MHz). (a) Circuit; (b) time
frames; (c) normalized capacitance values.

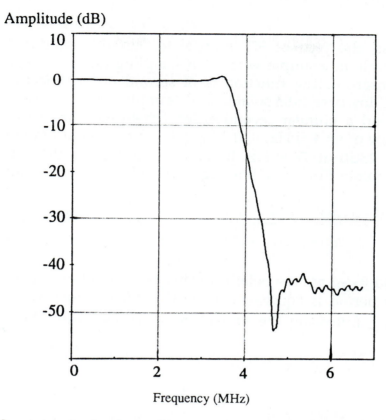

Amplitude (dB)

Frequency (MHz)

Fig. 30: Computer simulated overall amplitude response of the circuit in Fig. 29.

4. IIR SC Interpolator Building Blocks

IIR SC interpolators have been traditionally implemented based on the non-optimum and sub-optimum classes of circuits defined according to the resulting speed requirements of the operational amplifiers (OA's), in a similar form as for the case of the decimating building blocks. We can recall from the previous section that non-optimum IIR SC interpolator circuits are usually implemented using classical bi-phase SC filters where the resulting speed of the OA's and capacitance spread are both determined by the higher output sampling frequency LF_S. The sub-optimum type of IIR SC interpolator circuits consist of the cascading of a non-recursive polyphase structure with an interpolating factor of L and an SC filter operating at LF_S [7]. Although the polyphase structure can employ OA's operating at the low sampling rate F_S, faster OA's are still needed in the SC filter operating at LF_S. In the alternative type of SC interpolator circuit, with optimum implementation, all OA's can operate at the lower sampling rate F_S and the realization of arbitrary interpolating functions can be achieved employing a reduced number of switching waveforms and requiring low capacitance spread and total capacitor area.

For conciseness, the type of SC interpolator circuit considered here will be described through an example where the sampling rate increases by a factor of L=3 and the interpolating function is of second order. Specifically, we shall consider an SC lowpass interpolator with sampling rate increase from 384kHz to 1.152MHz and a bilinear second-order Tchebyshev lowpass response with cut-off frequency of f_c=4.8kHz, 0.01dB passband ripple, and minimum stopband attenuation of 25dB at 76.8kHz. Based on a computer-aided filter synthesis procedure we obtain first the following original bilinear z-transfer function

$$H(z) = k \frac{(1 - 2r_z \cos(\theta_z) z^{-1} + z^{-2})(1+z^{-1})}{1 - 2r_p \cos(\theta_p) z^{-1} + r_p^2 z^{-2}} = 0.0062 \frac{(1 - 0.9074 z^{-1} + z^{-2})(1+z^{-1})}{1 - 1.8835 z^{-1} + 0.8902 z^{-2}} \quad (17)$$

where $r_{z(p)}$ and $\theta_{z(p)}$ are the polar coordinates of both the zeroes and poles, and the unit delay period is equivalent to $1/3F_s$. After modification for optimum implementation, following the procedure previously described, this function leads to

$$H'(z) = \frac{\sum_{i=0}^{2L+1} a'_i z^{-i}}{1 - 2 r_p^L \cos(L\theta_p) z^{-L} + r_p^{2L} z^{-2L}} =$$

$$= 10^{-3} \frac{6.165 + 12.183 z^{-1} + 18.028 z^{-2} + 19.094 z^{-3} + 18.971 z^{-4} + 17.792 z^{-5} + 10.789 z^{-6} + 4.886 z^{-7}}{1 - 0.7055 z^{-3} + 1.6515 z^{-6}} \quad (18)$$

where the unit delay period still refers to the higher sampling rate $3F_s$.

The above z-transfer function can be implemented using the SC interpolator shown in Figure 31 [13]. The circuit architecture of Figure 31-a consists of a TIL network whose amplifiers *operate at the low input sampling frequency F_s*, together with two output polyphase networks whose characteristics depend on the interpolation factor L and on the complexity of the numerator polynomial function. The interpolated signal is obtained at the output of an SC accumulator which linearly combines the packets of charge transferred through the branches of the polyphase networks. Such accumulator employs a unity gain buffer for reduced power and silicon consumption and high frequency of operation. The operation of the interpolator presented in Figure 31-a is referred to the switch timing indicated in Figure 31-b, which is divided into three sets of time slots corresponding to time frames S, T and R. Time slots E and O of time frame S define the input sampling instants (as shown in Figure 32) and control the operation of the recursive part of the circuit in a similar way as in a conventional biquad. Such time slots are also utilised to define the input sampling of the polyphase networks and whose operation additionally requires

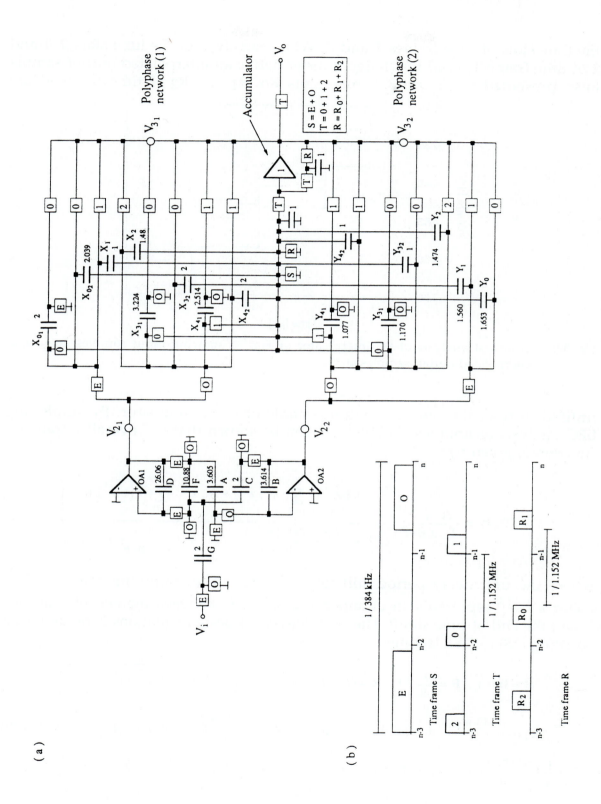

Fig. 31: Second-order SC interpolator with L=3. (a) Circuit architecture; (b) time frames.

the time slots of time frames T and R. While each type of the time slots 0, 1 and 2 of time frame T, used for charge transfer define an interpolated output sample (also presented in Figure 32), time slots R_0, R_1 and R_2 of time frame R are

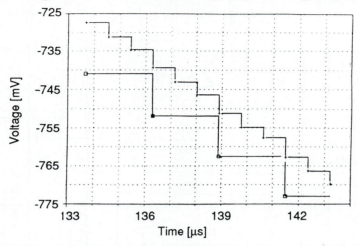

Fig. 32:   Transient analysis of SC interpolator of Fig. 31 showing input samples at F_s and interpolated output samples at $3F_s$.

utilised to reset the previous output sample of the accumulator. By employing SFG analysis techniques, as before, it can be shown that the overall z-transfer, function is given by

$$T(z) = -\frac{G \cdot z^{-1/16}}{D \cdot (B+F)} \cdot \frac{[(1-z^{-3}) \cdot \sum_{i=0}^{4}(X_i \cdot z^{-i}) + A \cdot z^{-3} \cdot \sum_{i=0}^{4}(Y_i \cdot z^{-i})]}{[1 + (\frac{A.C-D.F-2B.D}{D \cdot (B+F)}) \cdot z^{-L} + (\frac{B.D}{D \cdot (B+F)}) \cdot z^{-2L}]} \tag{19}$$

where the unit delay period still refers to the output sampling rate $3F_s$. The factor $z^{-1/16}$ represents an unimportant flat delay between the end of time slot E and the end of time slot 0. The complete set of design equations is obtained by equating (18) to (19), leading to

$$B = D = 1, \quad F = \frac{1}{r_p^{2L}} - 1, \quad A = C = \sqrt{[2 + F - (2 r_p^L \cos(L\theta_p))(1+F)]}$$

$$X_i = \frac{-(1+F) a'_i}{G}, \text{ if } i = 0, 1, 2 \quad \text{and} \quad X_i = \frac{-(1+F) a'_{i+3}}{G}, \text{ if } i = 3, 4 \tag{20}$$

$$Y_i = -\frac{1}{A}[\frac{-(1+F) a'_{i+3}}{G} + X_{i+3} - X_i], \text{ if } i = 0, 1, \quad Y_i = -\frac{1}{A}[\frac{-(1+F) a'_{i+3}}{G} - X_i], \text{ if } i = 2$$

$$\text{and } Y_i = X_i, \text{ if } i = 3, 4$$

Designing for reduced capacitance spread is achieved by making, in some equations, $X_i = X_{i1} - X_{i2}$ or $Y_i = Y_{i1} - Y_{i2}$. The final normalized capacitance values obtained after scaling for maximum signal handling capability are indicated in Figure 31-a yielding a total capacitor value of only 88.57 units. Figure 33 shows the nominal computer simulated amplitude responses of the proposed interpolator, in the passband (Figure 33-a) and from DC to $3F_s = 1.152MHz$ (Figure 33-b). The notch at $3F_s$ is due to the sample-and-hold-effect at the output of the accumulator.

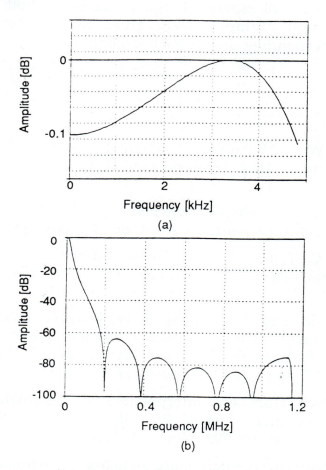

Fig. 33: Amplitude response of SC interpolator of Fig. 31. (a) Passband (DC-4·8 kHz); (b) from DC to 1·152 MHz.

5. Conclusions

This chapter presented an overview of the most popular circuit techniques and design methodologies for SC decimators and interpolators, both with FIR and IIR impulse responses. Although such multirate SC building blocks employ more complex digital circuits than traditional SC filters, which are required for generating multiphase switching waveforms, they usually can lead to considerable gains in the analogue circuitry, namely with respect to silicon and power consumption and also to the performance behaviour under non-ideal characteristics of the amplifiers. There are even some applications whose practical implementation in IC form can only be achieved using such multirate SC circuits. The need for analogue and non-trivial digital circuitry for implementing multirate SC signal processing is compatible with the technologies which are currently available to enable the strategic objective of realising on a single chip all aspects of signal processing, both analogue and digital. Plentiful design examples were given to illustrate the various architectures and corresponding design methodologies.

References

[1]   R. Gregorian, W.E. Nicholson, "Switched-capacitor decimation and interpolation circuits," *IEEE Transactions CAS*, Vol. CAS-27, No.6, pp. 509-514, June 1980.

[2]   D.C. Grunigen, et al., "Simple switched-capacitor decimation circuits," *Electronics Letters*, Vol. 17, No.1, pp. 30-31, 8th. January 1981.

[3]   M.B.Ghaderi, G.C. Temes, S. Law, "Linear interpolation using CCD's or switched-capacitor filters," *Proc. Inst. Elect. Eng.*, Vol. 128-G, No.4, pp. 213-215, August 1981.

[4]   D.G. Grunigen, R. Sigg, M. Ludwig, U.W. Brugger, G.S. Moschytz, H. Melchior, "Integrated switched-capacitor low-pass filter with combined anti-aliasing decimation filter for low frequencies," *IEEE J. Solid-State Circuits*, Vol. SC-17, pp. 1024-1029, December 1982.

[5]   J.E. Franca, "Decimators and interpolators for narrowband switched-capacitor bandpass filter systems," *Proc. IEEE International Symposium on Circuits and Systems*, pp. 789-793, Montreal, Canada, May 1984.

[6]   J.E. Franca, "Non-recursive polyphase switched-capacitor decimators and interpolators," *IEEE Transactions CAS*, Vol. CAS-32, No.9, pp. 877-887, September 1985.

[7]   J.E. Franca, "Switched-capacitor decimators and interpolators with biquad-polyphase structures," *Proc. Midwest Symp. Circuits and Systems*, pp. 1181-1184, Syracuse, New York, August 1987.

[8]   J.E. Franca, D.G. Haigh, "Optimum implementation of IIR SC decimators," *Proc. IEEE International Symposium on Circuits and Systems*, pp. 76-79, Philadelphia, USA, May 1987.

[9] J.E. Franca, R.P. Martins, "IIR Switched-capacitor decimator building blocks with optimum implementation," *IEEE Trans. CAS*, Vol. CAS-37, No. 1, pp. 81-90, January.1990.

[10] R.P. Martins, J.E. Franca, "Optimum implementation of a multistage IIR SC bandpass decimator for a voiceband analogue interface system," *Proc. IEEE International Symposium on Circuits and Systems*, pp. 1661-1664, Helsinki, Finland, June 1988.

[11] J.E. Franca, S. Santos, "FIR switched-capacitor decimators with active-delayed block polyphase structures," *IEEE Trans. CAS*, Vol. CAS-35, No. 8, pp. 1033-1037, August 1988.

[12] R.P. Martins, J.E. Franca, "A novel Nth-order IIR switched-capacitor decimator building block with optimum implementation," *Proc. IEEE International Symposium on Circuits and Systems*, pp. 1471-1474, Portland, Oregon, May 1989.

[13] R.P. Martins, J.E. Franca, "Infinite impulse response switched-capacitor interpolators with optimum implementation," *Proc. IEEE International Symposium on Circuits and Systems*, pp. 2193-2197, New Orleans, Luisiana, May 1990.

[14] J.E. Franca, "A single-path frequency-translated switched-capacitor system for filtering and single sideband generation and detection," *Proc. IEEE International Symposium on Circuits and Systems*, pp. 1625-1628, Kyoto, Japan, June 1985.

[15] J.E. Franca, "A single-path frequency-translated switched-capacitor bandpass filter system," *IEEE Transactions on Circuits and Systems*, Vol. CAS-32, No. 9, pp. 938-944, September 1985.

[16] J.E. Franca, D.G. Haigh, "Design and applications of single-path frequency-translated switched-capacitor Ssystems," *IEEE Transactions on Circuits and Systems*, Vol. CAS-35, No. 4, pp. 394-408, April 1988.

[17] J.E. Franca, "Multirate switched-capacitor system approach to frequency division multiplexing", *Electronics Letters*, Vol. 24, No. 8, pp. 501-503, April 14, 1988.

[18] R.P. Martins, J.E. Franca, "An experimental 1.8μm CMOS anti-aliasing switched-capacitor decimator with high input sampling frequency," *Proc. European Solid State Circuits Conference - ESSCIRC'91*, pp. 13-16, Milan, Italy, September 1991.

[19] R.P. Martins, J.E. Franca, "A 2.4μm CMOS switched-capacitor video decimator with sampling rate reduction from 40.5MHz to 13.5MHz," *Proc. 1989 IEEE Custom Integrated Circuit Conference*, San Diego, California, May 1989.

[20] R. Crochiere, L. Rabiner, *"Multirate Digital Signal Processing,"* Prentice Hall, Inc., Englewood Cliffs, NJ. 1983.

[21] P.E. Fleisher, K.R. Laker, "A Family of active switched-capacitor biquad building blocks," *Bell System Technical Journal*, Vol. 58, pp. 2235-2269, December 1979.

[22] J.E. Franca, V. Dias, "Systematic method for the design of multiamplifier switched-capacitor FIR decimator circuits," *IEE Proceedings-G*, Vol. 138, No. 3, pp. 307-314, June 1991.

Chapter 9

Analog-Digital Conversion Techniques for Telecommunications Applications

Paul R. Gray and Robert R. Neff
Department of Electrical Engineering and Computer Sciences, University of California, Berkeley, California, U.S.A.

1. Introduction

A key element of the continuing progress in improving the performance capabilities of voice and data communications systems has been the application of digital VLSI logic and memory technology to the switching, transmission, storage, and processing of voice, data, and video information. The representation of data in digital form is fundamental to the operation of these systems. However, such systems must often interface with analog signal sources, the most important example being voice information. Other examples are voiceband data signals which have complex phase, frequency, and/or amplitude modulation applied as well as phase and amplitude distortion introduced by the transmission medium; data signals transmitted on wire pairs which have experienced severe phase and frequency distortion and thus must be extensively processed prior to reinterpretation as digital information; and finally video information, including both image information and data which has been phase, frequency, and/or amplitude modulated on a video channel in a broadband local area network system.

The field of monolithic analog-digital converter (ADC) design is a very broad one. This chapter presents a brief tutorial overview of some of the most important circuit techniques used for the implementation of the A/D converter function in a mixed signal IC environment. In Section 2, the role of A/D conversion in telecommunication systems is discussed. In Section 3, ADC implementations are classified and compared. In Section 4, some aspects of the implementation of important building blocks in monolithic ADCs is described. In Section 5, some of the practical and fundamental limitations to ADC performance are described. A very important ADC technique for telecommunications, sigma-delta modulation, is discussed in detail in a later chapter.

2.   The Role of A/D Converters in Telecommunications Systems

Because of the large number of analog-digital interfaces in modern telecommunications systems, the development of low-cost monolithic analog-digital conversion techniques has been an important element in reducing the cost of digital transmission and switching systems, particularly in telephony where a very large number of interfaces are required. The role of these interfaces in VLSI digital systems in general is depicted in Figure 1. Important types of interfaces include the interface to transducers and actuators as used in industrial control, instrumentation and robotics, the interface to storage media such as optical disk and magnetic tape, and the interfaces to transmission media. In this chapter we will focus on those interfaces of particular importance in telecommunications.

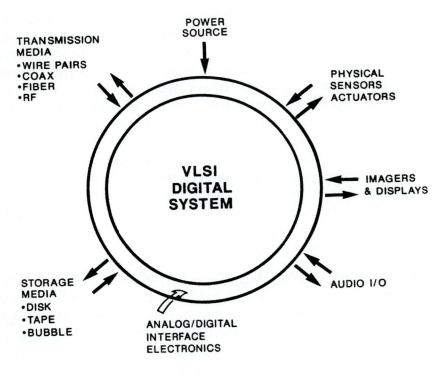

Fig. 1: Role of A/D interfaces in VLSI systems.

One of the most important analog-digital interfaces from a commercial viewpoint is the interface to the analog subscriber loop in digital switching systems. Presently, most new telephone digital switching systems are implemented with per-line codec/filter chips on the line card, the chip containing A/D, D/A, anti-alias filtering, line driver, and miscellaneous digital control, sequencing, and data storage functions. Presently most such codec chips utilize companding charge-redistribution coders with switched capacitor filters, but as MOS technology feature sizes continue to fall oversampling coders with digital filters will become more widely used, particularly where programmable features are important.

In voiceband data modems, half duplex modems at data rates above 1200bps typically utilize adaptive equalization to compensate for the phase and amplitude distortion of the line, and as a result are best implemented with at least the adaptive portion of the required signal processing realized digitally. An analog-digital converter with resolution in the 8-10 bit range is thus required. For full duplex modems, the echo problem requires either that extensive analog bandsplit filtering be included or, if the filtering is to be done digitally, that a 12-14 bit linear A/D converter be incorporated in front of the digital filter.

In digital subscriber loops under ISDN, the digital transmission of baseband data over long twisted pairs introduces phase distortion, which in turn can introduce large intersymbol interference lasting over many symbol intervals. For half duplex transmission, the equalization required may be most easily implemented in the analog domain, but for duplex transmission the combination of the echo cancellation function together with the equalization required dictates that extensive digital processing be used. A good example is the U-interface transceiver in the North American ISDN system, which requires A/D and D/A functions of high resolution sampling at 80K or 160K samples/second.

In digital audio, compact disc players require 44 Khz D/A converters with a 16-bit dynamic range for audio playback from a digital source. With the emergence of recording to digital tape and disc formats, there will be a need for low cost A/D converters with similar specifications.

Digital processing, storage, and transmission of video information requires analog-digital converters operating at sample rates of 20Mhz and above and resolution levels of 8 bits. This class of converter has become progressively more important as the transition to digital-based television has occurred. The increasing use of digital video signal processing in studio equipment, in multimedia computer systems, in NTSC/PAL/SECAm broadcast receivers, and in advanced television will make A/D converters with sample rates in the 20-100Mhz range and resolution in the 8-10 bit range much more commercially important in the coming years.

Each of these applications requires a conversion function with a unique combination of sampling rate, quantization noise as a function of signal level (i.e., companding or non-companding), and linearity error. In the next section circuit techniques for implementing several classes of such converters will be described.

3. Classification of A/D Conversion Techniques

Analog-digital converters can be classified in a number of ways, but one approach is to classify on the basis of how the decision is made about the value of the input signal. The basic function of a quantizer is to electronically define a range of input values, subdivide that range into a set of subregions, and then decide within which subregion the input sample lies. The simplest approach to doing this is to examine one subregion at a time, going from one end of the range to the other, as in serial or integrating converters. These approaches are widely used in instrumentation because of their simplicity and insensitivity to hardware imperfections, but are too low in throughput to be useful for most telecommunications applications. Obtaining a result requires on the order of 2^n clock periods where n is the number of bits.

In successive approximation, a binary search of possible subregions is made, instead of a linear search. The advantage is that a result can be obtained in on the order of n clock periods, but the cost is that a high-speed DAC with precision on the order of the conversion itself is required. Successive approximation is widely used in telecommunications ADCs, including many voiceband coder/decoder applications. A block diagram of a typical successive approximation converter is shown in Figure 2.

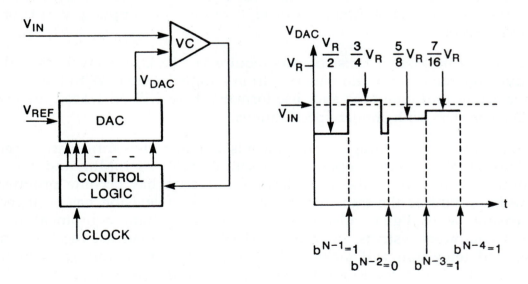

Fig. 2: Typical successive approximation A/D converter.

In parallel conversion all subregions are examined simultaneously using one comparator per subregion. A result can be obtained in on the order of one clock cycle, but the cost in hardware is much greater than successive approximation, especially for higher resolutions. These techniques and variations thereon are widely used in video applications. A typical example is shown in Figure 3.

Two other important variations on these techniques are quantized feedback and quantized feedforward converters. Quantized feedback converters reduce overall quantization error by subtracting from the input a reconstructed analog estimate of the input based on past history, and then quantizing the error between that and the actual input. When the sampling can be made far enough above the nyquist rate of the signal that the samples are highly correlated, large reductions in the required resolution of the DAC and ADC blocks can be realized. By inserting a suitable analog filter in the forward loop prior to the quantizer, much of the quantization noise can be moved away from the baseband, further improving SNR. Techniques in this class include interpolative converters, delta modulators, sigma-delta modulators, and others. Sigma-delta modulators have become widely used in voiceband applications. A general block diagram of a quantized feedback converter is shown in Figure 4.

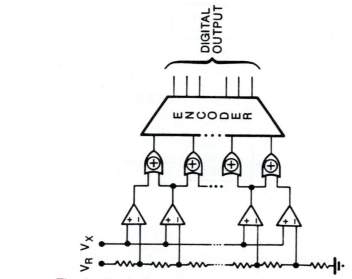

Fig. 3: Typical parallel, or flash, A/D converter.

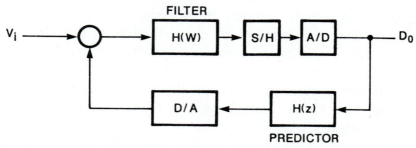

Fig. 4: Block diagram, quantized feedback converter.

Quantized feedforward converters utilize a series of cascaded blocks in which low-resolution A/D converters are used to estimate the signal, and a D/A converter and analog subtraction block removes this estimate from the signal. A remainder voltage is passed on to the next stage where the process is repeated. The most common example is the pipelined converter. These converters can potentially achieve a throughput rate comparable with parallel converters but at much less hardware cost at large resolutions. A block diagram is shown in Figure 5.

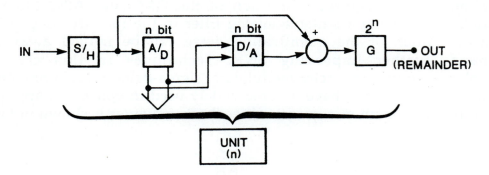

Fig. 5:   Block diagram of quantized feedforward converter.

Many techniques exist which embody more than one of the ADC approaches outlined above, so that the method of classification is not unique or always useful. However, the classification provides qualitative guidelines to expected performance. One way of comparing the approaches is illustrated in Figure 6. Here, the methods are compared based on the number of "analog clock cycles" required to obtain one signal sample at the ADC output after any required processing or averaging. An analog clock cycle includes such delays as comparator decision time, DAC settling time, and sample/hold aperture time. It is generally characteristic of the technology being used. In Figure 6, the horizontal axis is analog clock cycles, and the vertical axis is the resolution in bits. At the 12-bit level, several thousand clock cycles are required for a serial converter. For second order sigma delta, a form of quantized feedback coder, about 100 cycles are required (higher order versions can get down to 30-50 cycles). For successive approximation, about 12 cycles are required, and for parallel and pipeline converters about 1. A key point from this figure is that the various approaches complement each other and the optimum approach for a given application depends directly on the required sample rate, resolution, and speed capability of the technology. Perhaps the most significant development in telecommunications ADCs in recent years has been that with faster technologies the sigma-delta approach can now achieve sample rates up to near 1 Mhz at 12 bits, encompassing many telecommunications applications. This technique is described in more detail in a later chapter.

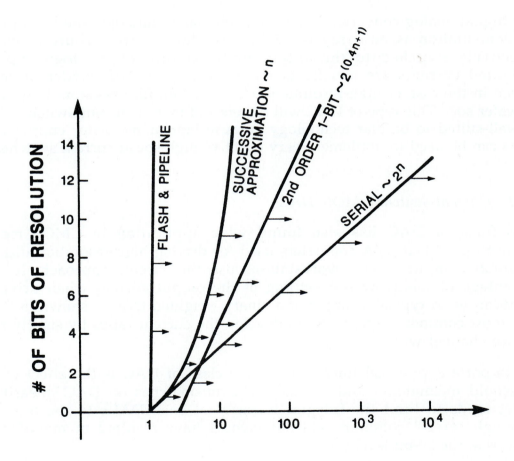

Fig. 6:   Comparison of A/D conversion approaches.

4.   Circuit Building Blocks for A/D Converters

All of the A/D conversion techniques described above require the same basic set of building blocks: sample/hold circuits for taking a sample of the signal, voltage comparators for quantizing the signal, in many cases high speed DACs for signal reconstruction, and switched capacitor integrator/gain blocks for implementing functions like sigma-delta modulators and interstage elements of pipeline ADCs. In the rest of this chapter the design of several of these blocks is discussed.

4.1.   High-speed DACs in MOS technology

For digital-analog converters realized in bipolar technology, the most prevalent implementation is an array of binary weighted current sources which are selectively switched to the output under control of the logic input. The weighted currents are usually generated using an R-2R ladder of resistors, which in the case of high-resolution DACs are thin film resistors laser trimmed at wafer sort. This type of DAC will be referred to as a current-switched DAC. It is well-suited to bipolar technology because bipolar transistor emitter-coupled pairs can be used to implement very effective high-speed current switches.

4.1.1   Current-switched MOS DACs

This form of DAC has also found wide application in CMOS for video applications. Most D/A converters used for driving high resolution displays in personal computers and workstations use this circuit approach. It has the advantage of relatively fast settling of the output current when driving low impedances. A typical example of a binary weighted DAC is shown in Figure 7. The most common approach is to develop the current ratios by simply ratioing device channel widths.

An important practical limitation of this class of DAC is the effect of device threshold mismatches and channel W/L mismatches on DAC linearity. This effect tends to limit the integral linearity of such DACs to the 8-bit range, although careful common centroid layouts have resulted in manufacturable devices at the 10-bit level (1).

For video applications, differential linearity and glitch energy are more important considerations, and have led to the widespread use of segmented approaches to these DACs in which the two or three most significant bits are implemented with an array of equal current sources rather than a binary array (2,3).

4.1.2 Resistor-string DACs

MOS transistors are particularly useful as zero-offset analog switches, and the resistor-string DAC makes good use of this property. A typical resistor string DAC is shown in Figure 8. The tree of analog switches is used to connect the output to one of the taps on the string. For an n-bit DAC, 2n resistor elements and 2n+1 transistor switches are required. However, the resistor string itself can be laid out in an x-y addressed array which is very economical in area (4,5). For example, a typical 8-bit resistor string DAC occupies about 0.2 square mm in a 1-micron technology.

The principal advantages of the resistor string are small size for moderate resolution levels, high speed, and inherent monotonicity (i.e., each output level is inherently higher than the previous one). An important disadvantage is that the linearity of the transfer characteristic can be strongly affected by the differential temperature coefficient and the voltage coefficient of the resistors in the string, which are usually implemented with a diffused region. This type of DAC is particularly useful in technologies which contain no capacitor, and is used extensively in NMOS microprocessors with on-board analog-digital converters (4,5).

4.1.3 Charge-redistribution DACs

A second useful property of MOS transistors is the fact that the input resistance of the device is virtually infinite. As a result, the transistor can be used to sense the voltage on a small capacitor continuously for a period of milliseconds without discharging the capacitor and destroying the stored information. This property allows the implementation of dynamic MOS memory, switched capacitor filters, and charge-redistribution analog-digital converters.

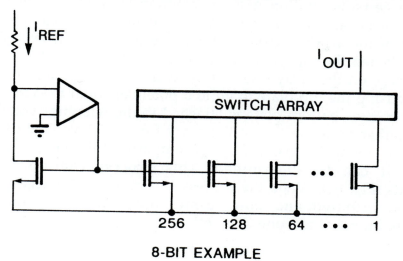

8-BIT EXAMPLE

Fig. 7: Typical current-switched DAC.

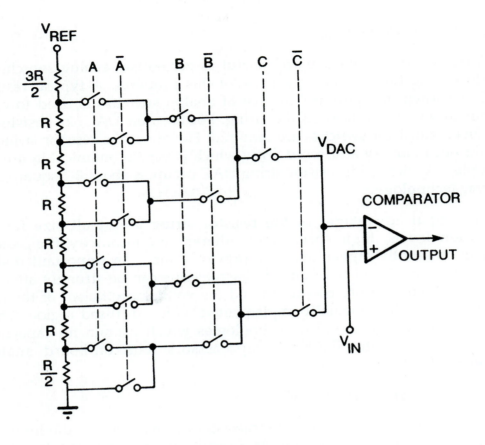

Fig. 8:   Typical resistor-string DAC. V_{DAC} is the DAC output. With the addition of the comparator this circuit can be used as an ADC.

A typical charge-redistribution ADC is shown in Figure 9. It consists of a binary weighted set of capacitors and associated switches, and a comparator. Initially the top plate of the array is grounded and the bottom plate is connected to the input signal. When the top plate switch is opened, the signal is sampled in the form of a charge on the array (Fig. 9a). Next the bottom plates are all connected to ground, causing the top plate to go to a potential equal to -Vin (Fig. 9b). Next the largest capacitor bottom plate is attached to the reference voltage. Since this capacitor together with the rest of the array form a 2:1 voltage divider, the top plate goes to a voltage -Vin+Vref/2. By comparing this bit to ground the most significant bit can be determined. This process continues through the remaining bits (Fig. 9c). When the process is completed, the charge on the array has been redistributed and now resides only on those capacitors whose corresponding bit value is a one (Fig. 9d).

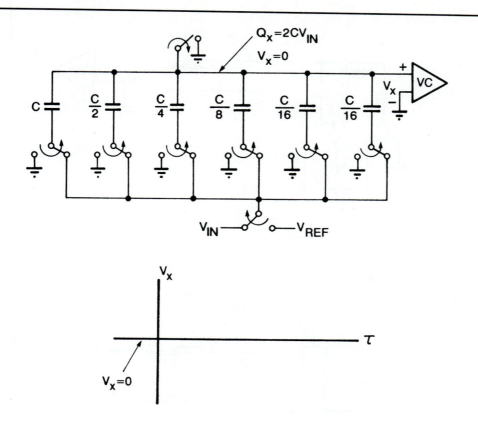

Fig. 9a: Charge redistribution ADC, sample mode.

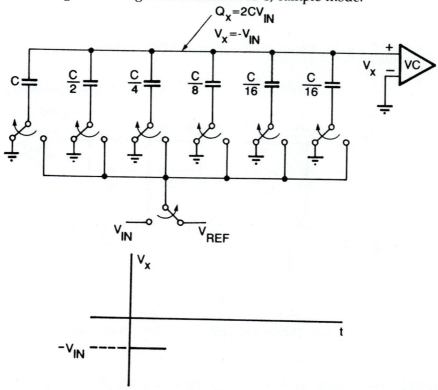

Fig. 9b: Charge redistribution ADC, hold mode.

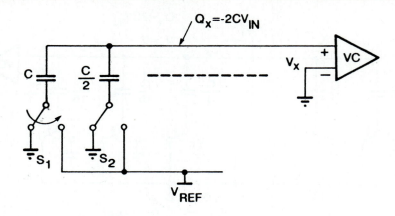

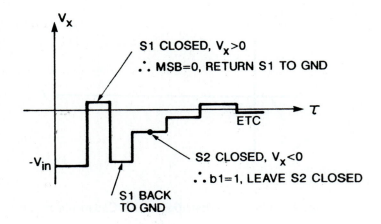

Fig. 9c: Charge redistribution ADC, first bit trial.

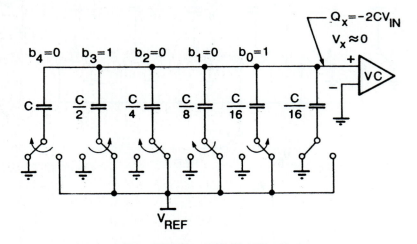

Fig. 9d: Charge redistribution ADC, final result.

The principal advantage of the charge-redistribution ADC is the fact that it incorporates the sample/hold function, which would otherwise have to be implemented separately. Another important advantage, discussed later, is the fact that the memory of the array can be used to carry out an auto-calibration cycle to remove the effects of component ratio errors. Also, the temperature coefficient long-term stability of the linearity of the ADC is primarily dependent on the stability of the S_iO_2 capacitors. Since this dielectric is a form of quartz, the long-term stability and temperature stability is very good.

An important disadvantage of charge-redistribution converters is the requirement that the technology produce a S_iO_2 capacitor with two conducting electrodes. Such capacitors occur naturally in metal gate MOS technologies and in silicon gate technologies with two layers of polysilicon. However, in technologies with one layer of polysilicon, an additional process step must usually be added to produce a capacitor of suitable quality.

As in the case of current switched DACs, segmentation can be used to improve the differential linearity of capacitive DACs in a number of ways. One is to use the capacitor array to interpolate between taps in a resistor string as illustrated in Figure 10. Here, a resistor string is used as the primary DAC, and a capacitor array is used to interpolate between the taps on the resistor string (6). This configuration has the advantage that the ADC is monotonic as long as the low-resolution capacitor array is monotonic, and has been used to implement a 16-bit data converter system for signal processing applications (7). In Figure 11, a segmented array of equal capacitors is used as the main DAC and a resistor string is used to drive the bottom plate of the capacitors. This configuration is inherently monotonic.

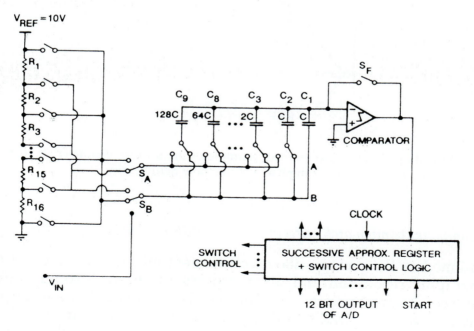

Fig. 10: DAC in which a capacitor array is used to interpolate between taps on a resistor string.

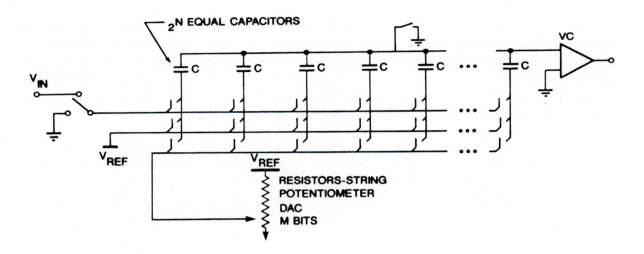

Fig. 11a: DAC in which an array of equal capacitors is used as the main DAC, and a resistor string is used to step the lower terminal of the capacitors one at a time to construct the transfer characteristic.

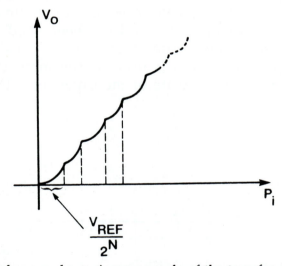

Fig. 11b: The scalloped curve shown is an example of the transfer characteristic that would be observed if the resistor string had a strong bow nonlinearity.

4.1.4 Component mismatch effects in monolithic DACs

As mentioned in the sections above, component mismatches due to limitations in lithographic resolution and other process variations strongly affect DAC performance. As was pointed out, the use of segmentation can improve the differential linearity of DACs in the presence of component mismatches, but in all of the DACs mentioned integral linearity is no better than the accuracy with which components (resistors, capacitors, transistors) can be matched to each other. A binary weighted capacitor array is shown as an example in Figure 12. Here, a mismatch between the largest capacitor and the rest of the capacitor array results in a nonlinearity in the transfer characteristic at the major carry point. In order to achieve 1/2 lsb integral linearity at the major carry, approximately 0.4% matching is required for 8-bit resolution, 0.1% for 10 bits, and 0.025% for 12 bits. In the fabricated state, with no trimming of ratios, a typical capacitor array would display an average mismatch of largest capacitor to the rest of the array of perhaps 0.1 to 0.2% due to processing gradients, random processing variations, and limited lithographic resolution (8,9). Thus the fabrication of DACs at resolutions of 10 bits or above usually requires some kind of trimming or calibration.

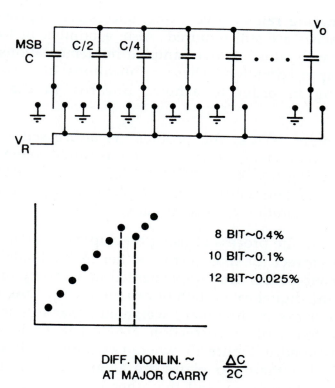

Fig. 12:   Effect of capacitor mismatches on linearity of capacitive DAC. In this example the largest capacitor is too small with respect to the rest of the array; all other ratios are exactly binary.

For many telecommunications applications, integral linearity on the order of 8 bits is adequate. Thus the fabrication of PCM companding codecs, for example, does not require trimming of DAC linearity. However, applications in full duplex data communications over wire pairs, both in voiceband modems and in subscriber loops, may require substantially higher resolution and linearity, particularly if the echo cancellation or band separation filtering is to be done digitally.

For ADCs with sampling rates in the several hundred kilohertz and below, quantized feedback coders such as sigma delta offer perhaps the best approach to removing the dependence of ADC linearity on component matching. By oversampling enough that the resolution of the internal DAC and ADC is reduced to 1 bit, all dependence of integral linearity on component matching is removed. This approach has become widely used for applications such a digital audio, full duplex voiceband data modems, and ISDN U-interface transceivers where integral linearity of 12 bits and above is required at moderate sampling rates and at low cost.

For ADCs with sampling rates above 1Mhz, successive approximation, flash, and pipeline approaches are still required. Also, a number of lower frequency applications are not well suited to oversampled implementations, requiring the use of successive approximation. These applications in turn require close component matching in order to achieve linearity in the DAC transfer characteristic.

In conventional bipolar ADCs, the problem is addressed with laser trim technology to adjust components on the IC to the required tolerance. In a mixed signal environment using CMOS or BICMOS VLSI, it is not desirable to incur this additional process and fabrication complexity, hence the evolution of various forms of self-calibration of DAC circuits.

Self-calibration refers to the process of using on-chip circuitry to measure errors in components and correct for those errors either in the analog or the digital domain. The basic concept has been implemented in a variety of ways. Perhaps the best example is the digital calibration of capacitor array DACs (10,11). In this technique, the ratio errors in the binary weighted capacitors is measured and subsequently corrected with a correction DAC. The basic concept of the measurement is illustrated in Figure 13, where two supposedly equal capacitors are connected such that their top plates are at ground, one bottom plate is connected to a reference voltage, and the other bottom plate is connected to ground. If the top plate switch is opened and the bottom plate voltages are interchanged, the top plate voltage should be unchanged if the two capacitors are equal. If there is a ratio error, the residual top plate voltage will be proportional to this error, and this voltage can be measured and used to quantize the ratio error. In a binary weighted capacitor array DAC, the ratio error in the largest capacitor is measured by doing the above experiment using the largest capacitor as one element and the rest of the array as the other. The residual error remaining on the top plate of the array is encoded by carrying out a normal A/D conversion following the switching sequence described above. The result is stored in a RAM on chip. The same process is repeated for the remaining bits as required. Subsequently, for a given bit pattern supplied to the DAC, the digital correction values for the non-zero bits are summed and applied to an auxiliary correction DAC to correct the linearity of the output. Experimental versions of such converters have achieved integral linearity of 1/2 lsb at 15 bits. The block diagram of such an experimental converter is shown in Figure 14.

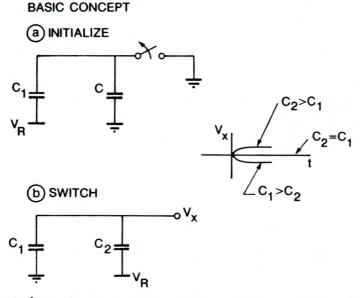

Fig. 13: Concept of measurement of ratio error between two nominally identical capacitors.

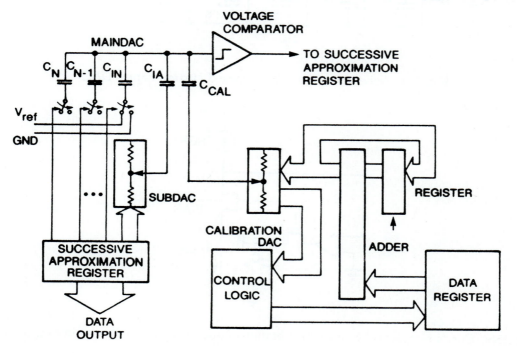

Fig. 14: Block diagram of an example self-calibrating charge-redistribution ADC.

Another example of self-calibration is the current-copier DAC. The basic concept is illustrated in Figure 15. This type of self-calibration is most often implemented using a segmented DAC where a number of equal segments are used. In order to calibrate each segment current, the current is compared to a reference current and the difference used through a feedback path to set the value of the gate voltage so that the drain current is equal to the reference value. The gate voltage is then held on the gate capacitance to preserve that current value. The process is repeated for the other currents. This approach has been use in DACs for digital audio reconstruction applications (12).

The use of self-calibration in DACs has major impact on the achievable performance in high-speed, high-resolution ADCs. It has been applied to pipeline converters where it has the potential to realize video sampling rates at 12-bit resolution levels (13).

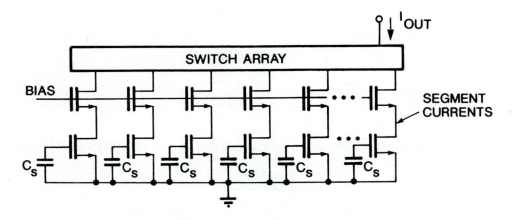

Fig. 15a: Schematic of a self-calibrated current-copier DAC.

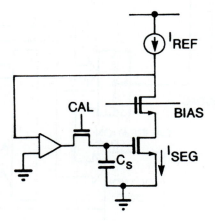

Fig. 15b: Current-copier DAC during calibration of one segment. Feedback forces $I_{seg} = I_{ref}$.

4.2. Monolithic sample/hold amplifiers

Analog-digital conversion requires two fundamental operations: sampling the signal and quantizing the signal. Quantization is carried out using comparators and level generators (DACs). The sampling process involves storage of an analog quantity, usually charge, that is captured and whose value represents the signal at the sample instant. The sample/hold function and its performance at high signal frequencies limits the achievable overall sampling rate in a given technology, since in principal the quantizer throughput can always be increased by paralleling more quantizers once the signal samples are captured.

The simplest implementation of the S/H function, a switch and capacitor, is shown in Figure 16. When a voltage is applied to the gate of an MOS transistor it turns on, and the input voltage passes through the transistor and is sampled on the capacitor. When the voltage is removed and the transistor is turned off, the voltage is held on the capacitor.

SIMPLEST S/H CIRCUIT

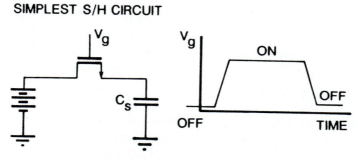

Fig. 16: Simple MOS sample/hold.

The thermal noise in the switch resistance is sampled onto the capacitor and appears as a random component of the sampled value with a variance equal to kT/C. The only factor affecting the value of this noise is the value of the sampling capacitor. The value of the signal energy stored on the capacitor is $CV^2\backslash2$ while the value of the noise energy stored on the capacitor is kT/2. This thermal noise phenomenon represents a fundamental limit on the per-sample dynamic range achievable in monolithic sample/hold amplifiers as the technology is scaled down and capacitor size is decreased. The value of kT/C for various capacitor sizes is shown in Figure 17, along with a second parameter, the "equivalent bits," corresponding to the resolution of a converter with quantization noise that is equal to this kT/C noise, assuming a 1-volt RMS signal.

In most actual ADC implementations there are multiple sampling operations, such as signal sampling or offset nulling, and each sampling operation makes a kT/C noise contribution. This total kT/C noise is a very significant limitation at the higher resolution levels. Ultimately, the ability to scale the physical size of the S/H function to smaller dimensions is limited by this phenomena.

C	$\sqrt{kT/C}$	EQUIV. BITS
100pF	6.4μV	15.7
10pF	20.2μV	14
1pF	64μV	12.3
0.1pF	202μV	10.6

Fig. 17:   Values of kT/C noise, expressed as a voltage, and the equivalent bits of resolution that would produce that value of quantization noise for a 1V RMS signal.

Practical S/H functions suffer from many other nonidealities including finite sample mode bandwidth, sample/hold offset, and aperture delay and jitter. The minimization of offsets and signal-dependent charge injection during the sampling process as the switch turns off is perhaps the single greatest challenge in implementing S/H amplifiers at high sampling rates and signal bandwidths. Many circuit approaches have been devised for reducing these effects. One example is shown in Figure 18 (14).

The circuit is implemented in a differential configuration so that charge injection offsets cancel to first order. The sampling of the signal is carried out by turning off the bottom plate switch first, (called bottom plate sampling), so that the charge injection that does occur is first-order signal independent. In this configuration the ultimate limit to the accuracy of the sampling process is due to the fact that for high-frequency signals, a large displacement current flows in the sampling capacitor and switch in the sampling mode due to large rates of change on the incoming signal. This causes an ohmic drain-source drop in the switch which is signal slope dependent. The linear portion of this drop simply limits sample-mode bandwidth, but the drop also changes the distribution of channel charge that occurs upon switch turn off. The nonlinear components of this effect on the final equilibrium voltage result in a high-frequency nonlinearity that limits the ultimate linearity performance at high signal frequencies. This effect is manifested as a reduction in the overall SNR of the ADC when the signal frequency approaches the nyquist rate and the signal amplitude approaches full scale.

Many other circuit techniques have been proposed and used to reduce signal-dependent charge injection. These include bootstrapping gate drive to the signal voltage, and connecting the sampling capacitor as an integrating capacitor in an operational amplifier with the sampling switch at the summing node (15).

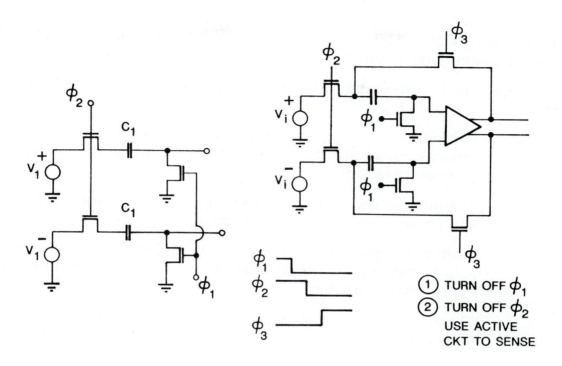

Fig. 18: Example S/H using passive bottom plate sampling. The bottom plate switches controlled by ϕ_1 turn off first. An auxiliary amplifier is needed to recover a low-impedance voltage corresponding to the signal following sample acquisition. (a) Sampling capacitor and switches. (b) Complete circuit, including amplifier and amplifier switch.

4.3. Monolithic voltage comparators

Once the signal is sampled, it must be quantized. The most basic element of a quantizer is a decision circuit, or comparator. The time of conversion is limited by many factors but can be no shorter than some multiple of the comparator delay.

The design of a practical comparator is a complex process involving many tradeoffs. The number of gain stages must be selected, and in the MOS case an offset cancellation strategy must be selected. The response of the comparator when initially precharged to a condition far from the threshold and subsequently driven by a small overdrive must be optimized. All these factors will have an effect on the comparator delay actually achieved.

A complete discussion of all the aspects of comparator design is beyond the scope of this chapter. However, it is useful to try to determine a technological limit to the minimum obtainable delay time in a comparator built with a given technology. This can be determined under a set of simplifying assumptions.

Consider the greatly simplified multistage comparator shown in Figure 19. The following simplifying assumptions are made:

The comparator is considered to be a series of identical MOS source-coupled pairs.

- Parasitic capacitances are neglected as are transistor drain-gate capacitances so that the only capacitances present are the transistor gate-source capacitances.

- The internal nodes are initialized to the balanced condition at t=0, and an input voltage Vin is then applied.

- During the portion of the transient of interest the differential stages remain in their linear region of operation with approximately constant g_m.

- The load impedances are high enough that the time constants at all drain nodes is much longer than the delay time of the comparator.

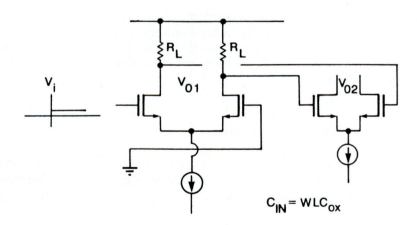

Fig. 19: Simplified multistage comparator.

Under this last assumption the individual stages can be considered to behave as an integrator for the time period of interest, as illustrated in Figure 20. It is then possible to obtain a closed-form expression for the output voltage as a function of time, the number of stages n, and the input drive voltage, and from this determine the optimum number of stages for minimum delay with a given drive level at the input and desired logic level at the output (16). This optimum number of stages is illustrated in Figure 21.

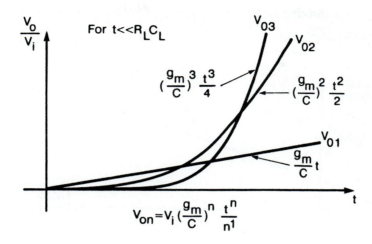

Fig. 20: Time response to step input at the output of successive stages.

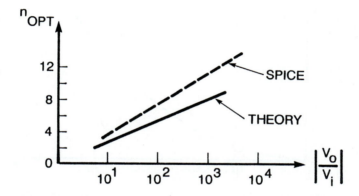

Fig. 21: Optimum number of stages as a function of input level and output logic level.

Using this, the minimum comparator delay achievable under these idealized conditions is illustrated in Figure 22. It is approximately equal to the transit time of the devices used multiplied by the natural log of the desired gain. It is interesting to note that this result is the same as would be obtained if the comparator were implemented using a regenerative latch in which the latch was initialized at an initial differential voltage of V_{in} and then allowed to regenerate to the same final value. Thus the result is first-order independent of the number of gain stages used before the regenerative latch.

The results of this simple analysis indicate that a comparator delay of a small multiple of the device transit time should be achievable. The same result applies for other devices such as bipolar transistors and MESFETs. For a 1-micron effective channel length nmos transistor operated at a V_{dsat} of 0.5 volt, the transit time is approximately 25ps, giving a comparator delay on the order of 0.2ns for a gain of 1000.

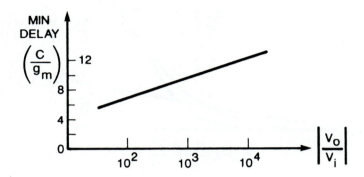

Fig. 22a: Minimum comparator delay assuming optimum number of stages is used.

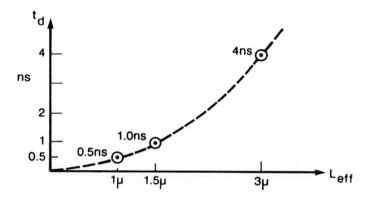

Fig. 22b: The actual delay as a function of $L_{effective}$ for an MOS implementation, using $\dfrac{V_{out}}{V_{in}} = 1000$ and $C_p \ll C_{gs}$. This is computed from $t_{d_{min}} \approx 1.2\ \tau\ \ln\left[\dfrac{V_{out}}{V_{in}}\right]$, where $\tau = \dfrac{C_{TOT}}{g_m} = \dfrac{C_{gs} + C_p}{g_m} \approx \dfrac{1}{2\pi f_t}\left[1 + \dfrac{C_p}{C_{gs}}\right]$.

Actual comparators used in A/D converters require much longer times than this to make decisions because of practical design issues (16). The most important of these is the overload recovery behavior of the comparator. In most applications during the period immediately prior to the decision period the comparator has some large voltage applied to its input. As a result of this the internal nodes do not start in the balanced state, but start with some large initial offset in one direction or the other. Overcoming this initial condition occupies most of the comparator delay time. Means of minimizing the required recovery time include the use of low-impedance load elements in the individual stages, initialization of the comparator internal nodes with a switch, and others. The actual delay achieved tends to be about an order of magnitude longer than that predicted by idealized theory (16). A typical relationship between initial overdrive and delay time is shown in Figure 23.

An important design choice in practical comparators is the number of stages of gain before the regenerative latch. The least hardware is used when the number of stages is minimized. However, the input-referred offset due to charge injection when the latch mode is initiated usually dictates the use of some gain in the comparator if offset is an issue in the application. The number of stages of gain is set primarily by the amount of offset allowable, and the details of the implementation of the latch.

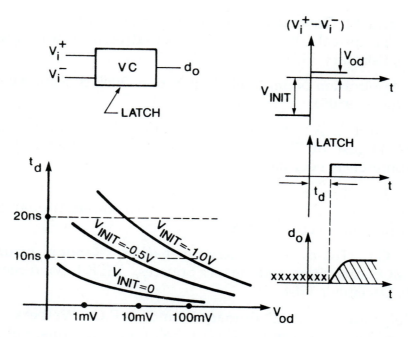

Fig. 23: Typical relationship between overdrive and comparator delay with initial input voltage as a parameter.

Summary

As more types of electronic systems come to rely on digital implementations, the role of the A/D and D/A converters will grow in importance. The increasing use of digital signal processing in electronic systems tends to push the ADC closer to the analog signal source, which usually increases the performance requirement on the speed and resolution of the ADC. For example, advanced video systems of the future will require ADCs with rates in the 100Mhz range and resolutions in the 10-12 bit range. These performance levels will require continued advances in the application of concepts such as self-calibration and pipelining.

References

1. M. Pelgrom, "A 50Mhz 10b CMOS D/A converter," *Digest of Technical Papers*, 1990 International Solid-State Circuits Conference, February 1990.

2. C.A.A. Bastiaansen, D.W.J. Groeneveld, H.J. Schouwenaars, and H.A.H. Termeer, "A 10-b 40-MHz, 0.8-um CMOS current-output D/A Converter," IEEE J. Solid-State Circuits, vol. SC-26, pp. 917-921, July 1991.

3. J.M. Fournier and P. Senn, "A 130 MHz 8-bit CMOS video DAC for HDTV applications," IEEE J. Solid-State Circuits, vol. SC-26, pp. 1073-1076, July 1991.

4. M.E. Hoff and M. Townsend, "An analog input-output microprocessor," Digest of Technical Papers, 1979 International Solid-State Circuits Conference, February 1979.

5. Intel 8022 Application Note, Intel Corporation, 3065 Bowers Avenue, Santa Clara, California 95051.

6. B. Fotuchi and D.A. Hodges, "An MOS 12b monotonic 25us A/D converter," Digest of Technical Papers, 1979 International Solid-State Circuits Conference, February 1979.

7. D. Hester, K.S. Tan, and C.R. Hewes, "A Monolithic Data Acquisition Channel," Digest of Technical Papers, 1983 International Solid-State Circuits Conference, February 1983.

8. J.-B. Shyu, G.C. Temes, and F. Krummenacher, "Random error effects in matched MOS capacitors and current Sources," IEEE J. Solid-State Circuits, vol. SC-19, December 1984.

9. M.J.M. Pelgrom, A.C.J. Duinmaijer, and A.P.G. Welbers, "Matching properties of MOS transistors," IEEE J. Solid-State Circuits, vol. SC-24, pp. 1433-1440, October, 1989.

10. H.S. Lee, D.A. Hodges, and P.R. Gray, "A self-calibrating 15b CMOS A/D Converter," IEEE J. Solid-State Circuits, vol. SC-19, pp. 813-819, December 1984.

11. K.S. Tan et al., "A 5Volt, 16bit, 10usec differential CMOS ADC," Digest of Technical Papers, 1990 International Solid-State Circuits Conference, February 1990.

12. D.W.J. Groeneveld, H.J. Schouwenaars, H.A. H. Termeer and C.A.A. Bastiaansen, "A self-calibration technique for monolithic high-resolution D/A converters," IEEE J. Solid-State Circuits, vol. SC-24, pp. 1517-1522, December 1989.

13. Y.M. Lin, B. Kim, and P.R. Gray, "A 13-bit, 2.5 MHz self-calibrated pipelined A/D converter in 3-um CMOS," Digest of Technical Papers 1990 Symposium on VLSI Circuits, Honolulu, Hawai, June 1990, pp. 33-34.

14. H. Ohara, H.X. Ngo, M.J. Armstrong, C.F. Rahim and P.R. Gray, "A CMOS Programmable self-calibrating 13-bit eight-channel data acquisition peripheral," IEEE J. Solid-State Circuits, vol. SC-22, December 1987.

15. M. Nayabi and B.A. Wooley, "A 10-bit biCMOS track-and-hold amplifier," IEEE J. Solid-State Circuits, vol. SC-24, pp. 1507-1516, December 1989.

16. D.C. Soo, "High-frequency voltage amplification and comparison in a one-micron technology," Ph.D. Thesis, Memorandum No. UCB/ERL 85/96, University of California, Berkeley, December 1985.

17. D.C. Soo et al. "A 750 MS/s NMOS latched comparator," Digest of Technical Papers, 1985 International Solid-State Circuits Conference, February 1985.

18. M. Yeung P. Cheung, R. Araluanantham, "A 5V front-end chip for a universal voiceband modem," Digest of Technical Papers, 1990 International Solid-State Circuits Conference, February 1990.

19. R. Batruni et al., "Mixed digital/analog signal processing for a single chip 2B1Q U interface transceiver," Digest of Technical Papers, 1990 International Solid-State Circuits Conference, February 1990.

20. H.J. Schouwenaars, D.W. J. Groeneveld and H.A. H. Termeer, "A low-power stereo 16bit CMOS D/A converter for digital audio," IEEE J. Solid-State Circuits, vol. SC-23, pp. 1290-1297, December 1988.

21. P.W. Li, M.J. Chin P.R. Gray, and R. Castello, "A ratio-independent algorithmic A/D conversion technique," IEEE J. Solid-State Circuits, vol. SC-19, December 1984.

22. S.H. Lewis and P.R. Gray, "A pipelined 5-Msample/s 9-bit analog-to-digital converter," IEEE J. Solid-State Circuits, vol. SC-22, December 1987.

23. S.H. Lewis et al., "A pipelined 9-stage video-rate A/D converter," IEEE 1991 Custom Integrated Circuit Conference, 26.4.1, May 1991.

24. A.G.F. Dingwall and V. Zazzu, "An 8-Mhz CMOS subranging 8-bit A/D converter," IEEE J. Solid-State Circuits, vol. SC-20, December 1985.

25. K. Tsuji, H. Sugiyama, and N. Sugawa, "A CMOS 20Mhz 8-bit 50mW ADC for mixed A/D ASICs", IEEE 1991 Custom Integrated Circuit Conference, 26.3.1, May 1991.

Chapter 10

Delta-Sigma Data Converters

Gabor C. Temes

Department of Electrical and Computer Engineering, Oregon State University, Corvallis, OR 97331, U.S.A.

1. Introduction

Analog-to-digital interfaces are becoming increasingly more important as translators between state-of-the-art digital signal processing (DSP) systems and the stubbornly analog outside world. With strict demands for higher accuracy in signal processing, these interface stages must also become more precise. In particular, the data converters, both analog-to-digital (A/D) and digital-to-analog (D/A), must be at least as accurate as the overall precision of the DSP systems which they connect to analog systems. At the same time, as both the feature sizes and bias voltages of very-large-scale integrated (VLSI) systems decrease, the accuracy and dynamic range of analog components are reduced, making the fabrication of high-resolution data converters more difficult. Thus, unless trimming and/or external (i.e., off-chip) components are used, the realization of high-accuracy converters becomes a very difficult task in newer technologies.

Since the state-of-the-art technology offers high-density and high-speed realization of both analog and (especially) digital circuitry, a tradeoff in which enhanced analog resolution is obtained in exchange for increased temporal resolution and digital complexity is desirable. This can be accomplished by using oversampling data converters. In these circuits, the sampling and processing of the input signal is performed at a clock rate f_s which is much higher than the Nyquist rate needed to avoid aliasing. If the bandwidth of the signal is f_0, then the Nyquist rate is $2f_0$; the ratio $f_s/(2f_0)$ is therefore called the oversampling ratio (OSR). In most cases, OSR values range from 64 to 1024; for practical convenience, they are usually chosen as integer powers of 2.

The basic categories of oversampling converters include the noise-shaping, predictive, and combined predictive/noise-shaping converters. All contain a feedback loop in which the input x of the converter is compared with the output y, or with a filtered version of y. The difference is then used (either directly or in a filtered form) to provide the output y. Figure 1 illustrates the three systems, as used for A/D conversion. In the noise-shaping system of Figure 1a, as will be shown later, the output y contains the converted but otherwise unmodified input x, plus a filtered replica n of the quantization error e introduced by the internal A/D converter. The filtering suppresses e in the signal band, and thus allows a digital filter to separate the signal x subsequently from the noise n contained in y.

In the predictive system shown in Figure 1b the task of the digital predictor $H_d(z)$ is to generate an accurate prediction w of the current value of x, and thus to reduce the value of the analog input v of the internal A/D converter. This allows much larger values for x and thus increases the dynamic range. Now the quantization error e and the input x are both filtered by the inverse of $1+H_d(z)$.

In the combined system of Figure 1c both x and e appear in y filtered, as $\hat{x}$ and n.

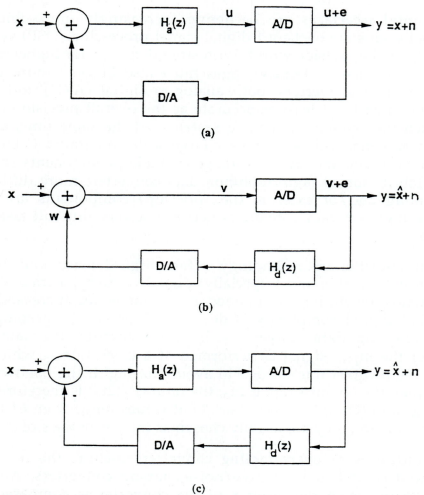

Fig. 1:   Feedback loops for oversampling A/D converters: (a) noise-shaping converter; (b) predictive converter; (c) predictive/noise-shaping converter.

In the following, only noise-shaping data converters will be discussed. The reason is that these systems appear to have important practical advantages over the competing ones in terms of insensitivity to the inevitable imperfections of the analog circuitry, and hence are used in most practical applications of oversampling systems.

Since the required sampling rate used in oversampling converters is typically two orders of magnitude higher than in conventional (Nyquist-rate) ones, their conversion rate tends to be relatively low for a given maximum clock frequency. In fact, they are second only to the counting-type converters in both conversion accuracy and in required conversion time. If state-of-art technology is used in the realization, oversampling converters can provide conversion rates up to 10^6 samples/s, with an accuracy of around 14 bits. For lower (say, audio-frequency) conversion rates, even 20-bit accuracy can be achieved.

It should be noted that there are important basic differences between Nyquist-rate and oversampling converters. In the Nyquist-rate A/D converter, the sampling rate f_s of the analog input signal is the same as the output data rate. Hence, a one-to-one correspondence exists between analog input samples and digital output words. By comparing their values, the accuracy of the conversion can be evaluated for each sample, and expressed in terms of (say) integral and differential nonlinearity. Spectral considerations (other than harmonic distortion and maximum frequency of operation) seldom enter into the design of Nyquist-rate converters. By contrast, as will be discussed later, every output word of the oversampled data converter is found as a weighted average of many consecutive analog input samples, and thus no one-to-one correspondence can be established. Hence, the performance is evaluated by comparing either the input and output spectra in the frequency domain, and/or the complete input and output waveforms in the time domain. From these comparisons, the rms values of the signal and noise can be found, and the signal-to-noise ratio SNR calculated. Finally, the resolution R can be estimated from the approximate relation R(bits) $\approx$ [SNR (dB)]/6. The linearity of the oversampling converter can also be characterized by the ratio of the signal power S and the total harmonic distortion power THD.

Typical applications of Nyquist-rate converters include digital voltmeters, instrumentation interfaces, etc. Typical application examples of the oversampling converter are digital audio, digital TV, and digital radar signal processing.

For large OSRs, only very simple (or no) antialiasing filtering is needed before the signal is entered into the A/D converter. Similarly, the postfilter following an oversampling DAC may be fairly simple if the OSR is large. Most of the necessary filtering is therefore performed by the adjacent digital circuitry, in the spirit of the tradeoff mentioned earlier. Thus, the analog front end of the oversampling ADC is followed by a digital low-pass filter which suppresses the out-of-band noise and reduces the data rate (usually all the way down to the Nyquist rate) for the converter output signal. In a complementary way, the digital filter in the oversampling DAC increases the data rate, and eliminates spectral replicas centered at multiples of the slower clock frequency prior to truncation and the actual A/D conversion.

2.   Noise-Shaping A/D Converters

In this section, the operation of the noise-shaping A/D converter schematically illustrated in Figure 1a will be discussed and alternative configurations will also be introduced and analyzed.

2.1. Quantization noise

To understand the operation of noise-shaping converters, a basic understanding of quantization noise is required. Figure 2 illustrates the input-output characteristics of a 6-level quantizer [1], with a continuous-amplitude input x and a quantized output y. The error $e = y - x$ is also shown as a function of x. As the diagram shows, $|e| \leq 1$ as long as $|x| \leq 6$. The range $|x| \leq 6$ is thus the linear (unsaturated) range of the converter. In practice, the bound on $|e|$ is valid only if the input x is nearly constant between quantization operations. Figure 3 illustrates the situation when x changes with time. If the quantization is performed at $t = T, 2T, \ldots$, in principle it is possible that between some kT and $(k+1)T$ it will occur that $|e(t)| = |y(t) - x(t)| > \Delta/2$, where Δ is the *level spacing on quantization step* of the quantizer. (In Figure 2, $\Delta = 2$.)

If $x(t)$ varies randomly and sufficiently rapidly, the following commonly used approximation may give reasonably realistic results. Instead of evaluating $e(t) = y(t) - x(t)$ directly using the quantization characteristics such as shown in Figure 2, we assume that $e(t)$ is a random additive white noise with a uniform amplitude distribution between $-\Delta/2$ and $\Delta/2$. To illustrate the validity of this assumption, Figure 4 shows the output spectrum of an 8-bit (256-level) quantizer with a sine wave input $x(t)$. The line around $\omega = 0.24 \, \omega_s$ is the spectral replica of $x(t)$, while the random-looking noise floor surrounding it is the spectrum of the quantization error. Clearly, $e(t)$ resembles a white noise, even though it is in reality a causal function of $x(t)$. The amplitude and rms value of the observed noise are also consistent with the uniform $-\Delta/2 \leq |e| \leq \Delta/2$ assumption.

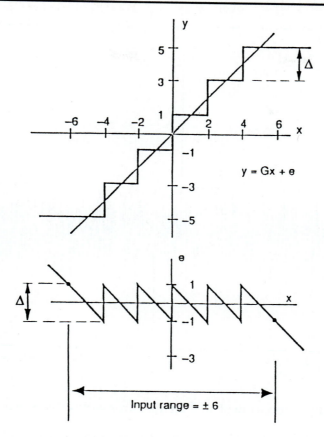

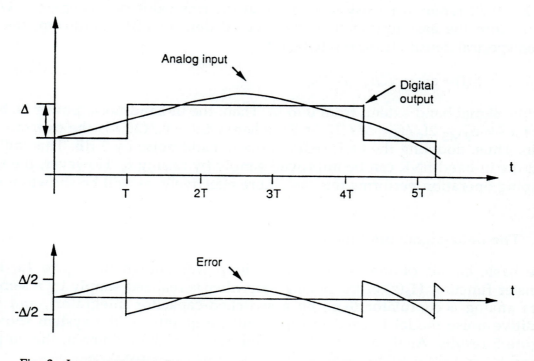

Fig. 2: Input-output characteristics for a 6-level quantizer: x is the input, y is the output, and e=y=-x the quantization error.

Fig. 3: Input, output and error waveforms of an ideal rounding A/D converter.

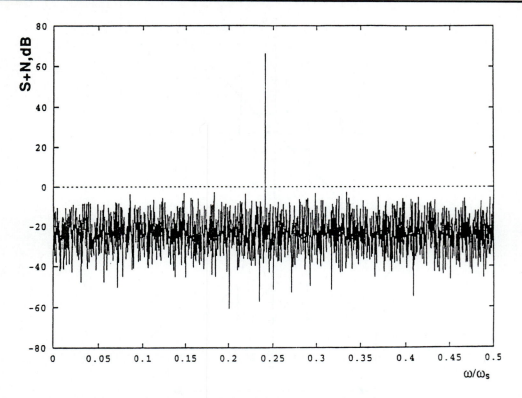

Fig. 4: The output spectrum of an 8-bit quantizer with a sine-wave input.

From the assumption that the value of e(t) lies with equal probability in the $-\Delta/2$ to $\Delta/2$ range, it is easy to verify that the rms value of e is $e_{RMS} = \Delta/\sqrt{12}$. Also, from the assumption that the spectral density E(f) is uniform, the one-sided spectral density turns out to be [1]

$$E(f) = e_{RMS} \sqrt{2/f_s} = \Delta/\sqrt{6f_s}. \tag{1}$$

Let the signal band extend from 0 to f_0. Then, the in-band noise power is, by eq. (1), $f_0 E^2 = e_{RMS}^2 \, 2f_0/f_s = \Delta^2/(12 \, OSR)$, where $OSR = f_s/(2f_0)$ is the oversampling ratio. Thus, doubling the OSR reduces the in-band noise by 3 dB. This indicates that the in-band SNR can be improved simply by raising f_s. However, the noise-shaping operation performs this task more effectively, as will be shown next.

2.2. The delta-sigma modulator

The basic circuit of the noise-shaping loop was shown in Figure 1a. If the transfer function $H_a(z)$ of the analog loop filter is realized simply by a sampled-data analog accumulator (e.g., by a switched-capacitor integrator), and if the additive noise model is used to represent the quantizer, the system shown in Figure 5 results. Analyzing this linearized circuit in the z-domain, the output is found to be

$$Y(z) = H_S(z)X(z) + H_N(z)E(z), \tag{2}$$

where H_S and H_N are the signal and noise transfer functions, respectively. They are given by

$$H_S(z) = \frac{H_a(z)}{H_a(z) + 1} = z^{-1}$$

$$H_N(z) = \frac{1}{H_a(z) + 1} = 1 - z^{-1}, \tag{3}$$

respectively, where the last parts of the equations follow since here the loop filter transfer function is $H_a(z) = z^{-1}/(1 - z^{-1})$. Thus, the digital output contains a delayed replica of the analog input signal, plus a noise n whose spectrum is that of the quantization noise e shaped by the noise filtering function due to H_N

$$|H_N(e^{j\omega T})| = 2\sin(\omega T/2), \tag{4}$$

where $T = 1/f_s$ is the sampling period. For low frequencies, where $\omega T \ll 1$, $|H_N| \approx \omega T$.

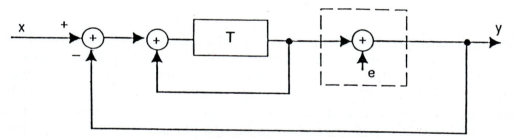

Fig. 5: Delta-sigma modulator loop.

H_N is clearly a high-pass filter function, and hence it tends to suppress the quantization noise at low frequencies including the baseband, but enhances it near $f = f_s/2$. Thus, the noise power is swept out of the baseband (where it would have overlapped with the signal) and into the high-frequency range where it can be eliminated by a low-pass digital filter following the modulator stage.

The A/D and D/A converters in the loop of the system of Figure 1a can be single-bit circuits. Then the A/D function can be performed by a comparator, and the D/A by a few switches and passive elements. In this case, the conversions are inherently linear since there is only one step in the quantization characteristics (cf. Figure 2). Nonlinearity results from a mismatch of steps, which cannot occur for the single-bit converter.

For historical reasons, this simplest noise-shaping converter loop (which uses an integrator as a loop filter and only a single-bit quantizer) is called a delta-sigma (or, in some instances, sigma-delta) modulator [2]. Often, these names are also used for any noise-shaping converter.

The in-band noise power P_N for the delta-sigma modulator can be estimated from its linearized model (Figure 5). From eqs. (1)-(3), assuming $\omega T \ll 1$,

$$P_N = \int_0^{f_o} |H_N E|^2 \, df \approx \frac{\pi^2}{36} \Delta^2 \left[\frac{2f_o}{f_s}\right]^3 = \left(\frac{\pi\Delta}{6}\right)^2 (\text{OSR})^{-3} \qquad (5)$$

Now the doubling of the OSR improves the SNR by 9 dB, rather than by only 3 dB as for the A/D converter without noise shaping.

From the description of the basic operation of the delta-sigma modulator it is clear that the largest signal value which a system using a single-bit quantizer can convert is $|x| = \Delta/2$, where it is assumed that the DAC outputs are $\pm\Delta/2$. Thus, for a sine-wave input the largest permissible signal power is $P_S = \Delta^2/8$. Therefore, the signal-to-noise power ratio is, from (5), approximately given by $(4.5/\pi^2)(\text{OSR})^3$.

A major shortcoming of the simple delta-sigma system described above is that it may generate low-frequency (and thus in-band) tones for special values of the input x. To see how this may occur, assume for the moment that the two output levels of the 1-bit D/A converter in the system of Figure 1a are 0 V and +1 V, and that the input x is a constant voltage k/m volt where k and m are integers and k < m. Then, it is possible to have a periodic D/A converter output pattern with each period m pulses long and containing k pulses of value 1 V and (m - k) pulses of 0 V. The average value of the DAC output will then equal x, and hence the loop can settle into a steady state oscillation under these conditions. If m is sufficiently large so that $m > f_s/f_o = 2\text{OSR}$, then the fundamental component f_s/m of the oscillation wave will fall into the signal band, and it may therefore appear as a sine-wave tone of considerable amplitude in the final data output. This tone (often called pattern noise) appears as one (or more) sharp peaks in the output spectrum, and as a sharp dip in the SNR vs. $|x|$ characteristics.

The amplitude and frequency of the pattern noise tones can be predicted from simple formulas [3].

The generation of pattern noise may be avoided by injecting an extra signal (called *dither*) into the loop as an additive noise. If the amplitude and frequency of the dither are both sufficiently large, then it will prevent the conditions for pattern noise. Of course, the dither will appear in the ADC output, and hence it must be subsequently removed. This is easy if the dither frequency falls into the stopband (and, preferably, coincides with a transmission zero) of the digital decimation filter following the loop. Unfortunately, a large dither signal will restrict the maximum value of x allowed for linear operation, and hence it will reduce the dynamic range of the system.

Another way to avoid pattern noise is to reduce the correlation between the input x and the quantization noise e by using a more complex loop filter. As will be shown in the next section, this will also result in more selective noise shaping, and thus in an improved SNR. In this case, the internal signals earlier stored in the accumulators act as dither, reducing the correlation between the current values of e and x.

2.3. Higher-order single-stage converters

The noise-shaping function $H_N(f)$ can be made more selective by using a higher-order loop filter. Two cascaded integrators are needed to realize a second-order loop filter, three to realize a third-order one, etc. To maintain stability, the open-loop transfer function of loop (which has all its poles at s = 0 due to the cascaded integrators) must also have some finite zeros so that at the unit-gain frequency a proper phase margin can be maintained. The system shown in Figure 6 (where, as in Figure 5, the ADC is represented by an additive noise source corresponding to the quantization noise e) satisfies this condition. Note that both feedback loops contain one delay. Delay-free loops are usually impractical, and two or more delays in a loop tend to impair the stability and hence also the SNR performance of the system. Linear analysis shows that the z-transform of the output y can again be described as in eq. (2), and the signal transfer function $H_S(z)$ is again given by z^{-1}, corresponding to a one-clock-period delay; the noise transfer function, on the other hand, now becomes $H_N(z) = (1 - z^{-1})^2$. Thus, the noise-shaping function becomes the *square* of that given in eq. (4), so that

$$|H_N(e^{j\omega T})| = 4\sin^2(\omega T/2).$$
(6)

Accordingly, the noise is now better suppressed at low frequencies, and there is more noise enhancement at higher ones. For $\omega T \ll 1$, $|H_N| \approx (\omega T)^2$ and integration of the modulation noise power density $|H_N E|^2$ over the signal band results in the inband noise power

$$P_N = \frac{(\pi^2 \Delta)^2}{60}(OSR)^{-5}.$$
(7)

The maximum signal-to-noise power ratio for a sine-wave input is therefore given by $(7.5/\pi^4)(OSR)^5$. For every doubling of the OSR, the inband noise is reduced (and thus the SNR is enhanced) by 15 dB. Also, the generation of inband noise tones is largely prevented by the added feedback path which reduces the dependence of the quantization noise e on the recent values of x, and thus decreases their correlation. In fact, the feedback signal added to the input of the second quantizer acts as a dither signal. As an external dither signal would, it unfortunately also reduces the maximum permissible input signal range [4].

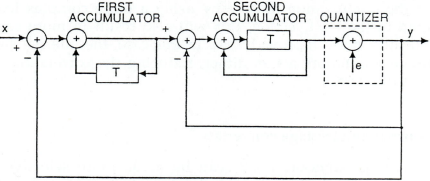

Fig. 6: Second-order noise-shaping loop.

In principle, even higher-order loop filters can be used, resulting in even more selective noise shaping. The resulting reduction of the inband noise for increased values of the order L is illustrated in Figure 7 [4]. However, the stability of the loop becomes precarious if L ≥ 3 is used. Linearized analysis is not a reliable predictor of the stability, since the single-bit quantizer is a grossly nonlinear element whose equivalent gain varies abruptly with the value of its input. A consideration of the Bode plot of the open-loop gain of the high-order loop reveals that for guaranteed stability the equivalent quantizer gain must be high. Since the output amplitude of the quantizer is fixed, this will be achieved only if the quantizer input is small. To insure this, the maximum amplitude of

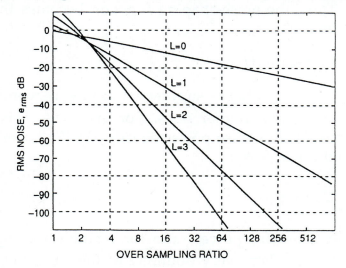

Fig. 7: Inband quantization noise for Lth noise-shaping modulators.

x may be restricted, and/or limiters may be used to reduce the signal swing at the outputs of the integrators. Also, reset switches may be used to discharge the integrator capacitors whenever the internal signals reach some critical level.

An alternative structure for realizing higher-order noise-shaping converters, which is free of the stability problems associated with the high-order single-stage converters described above, is the multi-stage or cascade architecture discussed in the next section.

2.4. Multi-stage (cascade) converters

In a basic two-stage converter (Figure 8), an analog signal x_2 containing the quantization noise e_1 of the first stage is fed to a second noise-shaping converter where it is translated into digital data [5]. Next, the output data streams from the two stages are digitally filtered and then combined in such a way that e_1 is canceled and the only remaining modulation noise is the second-order high-pass filtered quantization noise e_2 of the second stage. Next, we shall show that the system of Figure 8 accomplishes this. We shall assume that each ADC incorporates a one-clock-period delay.

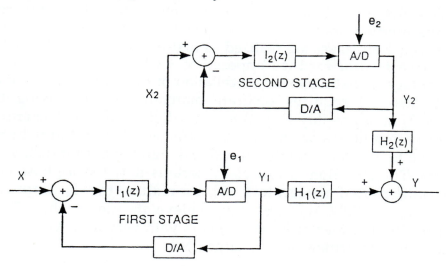

Fig. 8: Second-order cascade modulator.

Linear analysis of the system of Figure 8 shows [1] that the z-transform $Y(z)$ of the overall output signal is given by

$$Y = H_S X + H_{N1} E_1 + H_{N2} E_2 \tag{8}$$

where the noise transfer function H_{N1} can be written in the form $[(1+z^{-1}I_2)H_1 - z^{-1}I_1I_2H_2]/[(1+z^{-1}I_1)(1+z^{-1}I_2)]$.

Here, the I_k and H_m are the transfer functions of the various blocks as indicated in Figure 8. If H_2/H_1 is chosen to equal $(1+z^{-1}I_2)/(z^{-1}I_1I_2)$, then $H_{N1}(z) \equiv 0$ results and e_1 will not affect the output y. This choice will also result in H_5 becoming equal to H_1. Thus, if H_1 is chosen as a delay function z^{-k}, then the output y will contain only the delayed input x and the shaped noise e_2. The transfer function of the latter is given by $H_{N2} = z^{-k+1}/(I_1I_2)$. If $I_1 = I_2 = 1/(1-z^{-1})$, then the cancellation of e_1 is achieved if $H_1 = z^{-1}$ and $H_2 = 1-z^{-1}$. Then, $H_{N2} = (1-z^{-1})^2$ results. Thus, the output contains only the delayed input and the second-order filtered quantization noise of the second stage, as desired.

The principle of cascading can be extended to realize a third-order A/D converter [6]. Now a third first-order delta-sigma loop is added to the previously described system. The added stage converts the second-stage noise e_2 into digital form. The three data streams generated by the three stages are digitally filtered and combined in such a way that e_1 and e_2 are both cancelled, and y contains only a delayed replica of x and a third-order-shaped form of e_3. A detailed analysis of the third-order cascade system reveals that now the transfer functions of the component filters should satisfy $H_1 = z^{-1}$, $H_2 = z^{-1}/I_1 = z^{-1}(1-z^{-1})$, and $H_3 = 1/(I_1I_2) = (1-z^{-1})^2$ if all I_i are delay-free integrators.

An obvious advantage of higher-order cascade modulators (as compared with the single-stage modulators of the same order) is that no stability problems arise. This is because all stages are realized as first-order loops, and their outputs are combined in a feedforward (as opposed to feedback) configuration. Their largest disadvantage is that the exact cancellation of the error e_1 (and, in a third-order system, also e_2) requires accurate matching of the analog transfer functions I_1 and I_2 (and possibly also I_3) to some digital functions determined by the $H_i(z)$. If these conditions are not exactly satisfied, then unfiltered or poorly filtered noise due to e_1 (and possibly e_2) will leak into the output data y, and the SNR plummets. The situation is especially serious if I_1 has a phase error [7]. Using gain-compensated integrators [8] to realize the I_k improves the performance, but the SNR will still not achieve its ideal value. The noise leakage is especially detrimental for modulators with very large OSR which rely on noise cancellation to achieve a very large SNR.

A major improvement results for the third-order modulator if it is realized as a cascade of a second-order and a first-order stage (Figure 9). Now a mismatch between the analog and digital transfer functions causes only first- and second-order filtered quantization noise to appear in the output [9,10]. Hence, the matching accuracy need not be so extreme. The condition for noise cancellation is now satisfied if H_1 is a delay and $H_2 \propto 1/(I_1I_2)$.

In most applications, cascade modulators require no external dither signal, since the multipath structure cancels the tones generated for special input levels.

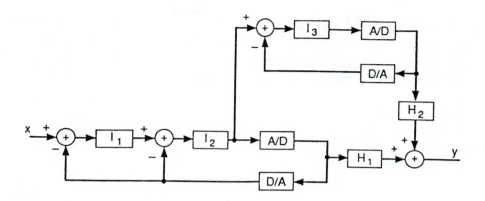

Fig. 9: Third-order cascade modulator with second-order first stage and first-order second stage.

2.5. Modulators with multibit quantizers

The permissible range of the input signal x is limited by the output range of the DAC in the feedback path. The latter is usually determined by the fabrication technology, and hence can be regarded as a fixed quantity. However, the maximum SNR of the modulator can be increased by reducing the RMS value $\Delta/\sqrt{12}$ of the quantization noise, where Δ is the level spacing in the quantizer characteristic (Figure 2). This is achieved by using multibit quantization, i.e., employing M-bit (M > 1) rather than single-bit internal A/D and D/A converters. It is easy to see that if the input range of the quantizer remains the same, then an (M-1)6 dB SNR increase, and hence an (M-1)-bit resolution improvement, results, since Δ is now reduced by a factor of 2^{M-1}. Also, the range of x nearly doubles and hence the SNR improvement is close to Mbits. The reduced high-frequency noise power also decreases the necessary complexity of the digital filter used to suppress it. Finally, the multibit quantizer is a better approximation to a linear amplifier than a single-bit one; hence, the stability properties are better and the agreement between the behavior predicted by linear theory and the actual performance is improved.

The multibit internal A/D converter must be a parallel (flash) type circuit, since stability and noise cancellation allow only one clock period for conversion. On the other hand, the ADC nonlinearity (manifested by unequal level spacings) merely increases the quantization noise somewhat, and will be suppressed by the noise shaping process. By contrast, any nonlinearity of the DAC will directly affect the output signal. This can be seen by noting from Figure 1a that at low frequencies where the loop gain is sufficiently high, the output signal of the DAC is forced by the feedback loop to follow the input x of the modulator, while its input is the digital modulator output y. Thus, if the input-output characteristics of the DAC is nonlinear, then so is the y = f(x) relationship. In fact, it is clear from this argument that the DAC linearity must be at least as accurate as the desired overall linearity of the complete conversion process. A direct realization of such high-precision converter is sometimes possible (using, e.g., trimmed resistors [11]) but it is expensive and defeats the very purpose of using oversampling as discussed in the introduction of this chapter.

There are several alternative techniques for reducing the effects of DAC nonlinearity. One is based on converting the error introduced into y by this nonlinearity into a random noise, minimally correlated with x [12]. For a sine-wave input x(t), this converts the spectral peaks representing harmonic distortion in the output spectrum into a small increase of the noise floor. This is often preferable to nonlinear distortion. To achieve the randomization, the DAC may be constructed from equal-valued elements (resistors, capacitors, or current sources), and for every conversion a new element array can be selected randomly from the set of all elements [12]. In an alternative technique [13], which is considerably more complex but also much more effective, the harmonics generated by the nonlinear DAC are modulated by a carrier frequency and thus shifted out of the signal band. To achieve this effect, the past history of each possible DAC output signal level must be memorized and taken into account in reconfiguring the DAC from its components.

A completely different approach was proposed by Leslie and Singh [14]. Their approach is conceptually similar to the cascade technique. As before, a delta-sigma modulator (with a single-bit quantizer) acts as the first stage, and the analog input voltage to its quantizer is fed to the second stage (Figure 10a). Here, however, the second stage is a conventional M-bit A/D converter, rather than a second sigma-delta loop. As before, the digital outputs of the two stages are filtered and then combined in such a way that the quantization error e_1 of the first stage is cancelled and replaced by the high-pass-filtered quantization error e_2 of the M-bit ADC which is 2^{M-1} times smaller.

Replacing each A/D converter by a delay followed by the addition of the quantization noise e_1 or e_2, linear analysis gives

$$Y = H_1Y_1 + H_2Y_2 = \left(H_1z^{-1} + H_2\right)\frac{IX}{1+z^{-1}I} + \frac{H_1-H_2I}{1+z^{-1}I}\ E_1+H_2E_2 \qquad (9)$$

Hence, e_1 can be eliminated from y if $H_1 = H_2I$ is chosen. Then, the factor of X in Y becomes H_1, so that $H_1=z^{-k}$ may be used. Therefore, we obtain $H_2 = H_1/I = z^{-k+1}(1-z^{-1})$ if I is a delaying integrator, and $H_2 = z^{-k}(1-z^{-1})$ if I is delay free. In the former case, $k = 1$ can be used; in the latter, $k = 0$ is permissible.

In practice, the most significant bit (MSB) of the M-bit ADC may be used as the 1-bit ADC output of the delta-sigma loop. The resulting system (assuming a delay-free I) is shown in Figure 10b.

Since the M-bit ADC of the modulator of Figure 10a is not in the feedback loop, it need not perform the conversion in a single clock period. Thus, a slower converter (e.g., a pipelined ADC) may be used. In this case, $H_1(z)$ must provide the same delay as the M-bit A/D converter.

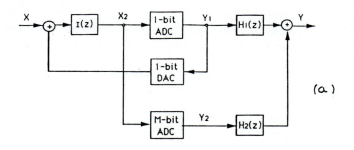

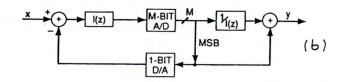

Fig. 10: Noise-shaping modulator with dual quantizer: (a) basic concept; (b) a possible implementation.

Note that by eliminating the M-bit DAC, the inherent nonlinearity of the M-bit loop disappears. The nonlinearity noise of the M-bit ADC is spectrally shaped (along with its quantization noise) by the digital high-pass filter H_2, and hence it is largely suppressed in the signal band. It is also possible to extend the principle to higher-order modulators. The generalized scheme is shown in Figure 11a and the special case of a second-order modulator in Figure 11b. In general, assuming a delay in each ADC we have the linearized relations

$$Y_1 = H_S X + H_N E_1$$

$$X_2 = z(Y_1 - E_1) = z H_S X - z(1 - H_N) E_1 \tag{10}$$

$$Y_2 = z^{-1} X_2 + E_2 = H_S X - (1 - H_N) E_1 + E_2$$

and the overall output is hence

$$Y = H_1 Y_1 + H_2 Y_2 = H_N(H_1 + H_2) X + (H_1 H_N + H_2 H_N - H_2) E_1 + H_2 E_2 \tag{11}$$

To cancel e_1 in y, we must make $H_2 = (H_1 + H_2) H_N$. Also, the factor of X should be a delay, so $H_S(H_1 + H_2) = z^{-k}$. Then

$$H_2 = (H_1 + H_2) \, H_N = z^{-k} \, H_N/H_S$$

(12)

$$H_1 = z^{-k}/H_S - H_2 = z^{-k}(1-H_N)/H_S$$

are the design equations. Here k can be chosen as the number of delays in $H_S(z)$. If $H_S(z) = z^{-k}$, then under ideal conditions $H_2 = H_N$, $H_1 = 1 - H_N$ and $Y = z^{-k}X + H_N E_2$ results. Thus, the system replaces e_1 by the much smaller quantization error e_2 in the output y. For the second-order system of Figure 11b, we get $H_S = z^{-1}$ and $H_N = (1-z^{-1})^2$. Hence,

$$H_1 = 2z^{-1} - z^{-2}$$

(13)

$$H_2 = (1-z^{-1})^2$$

should be used, making Y equal to $z^{-1}X + (1-z^{-1})^2 \, E_2$.

The reduction of the quantization noise can also be obtained simply by replacing the single-bit ADC and DAC by multibit ones in the second stage of a cascade modulator. This will be effective for either the second-order system of Figure 8 or for the third-order one of Figure 9 [15]. In either case, after the cancellation of e_1, only the shaped noise e_2 remains which is now subject to a second- (third-) order high-pass filtering. The nonlinearity error of the second-stage DAC does enter the output; however, it is typically much smaller than an M-bit LSB, and is first- (second-) order shaped. Hence, it is usually not critical. Note, however, that this realization is more complex than that of Figure 10, and hence it is advantageous only for third- or higher-order modulation, where the cascade structure avoids the stability problems which may arise for the single-stage realization.

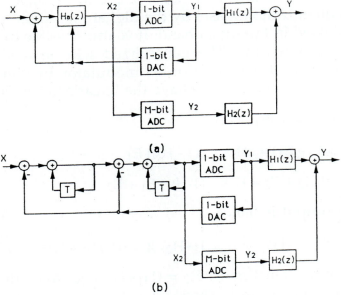

Fig. 11: Higher-order dual-quantizer modulators: (a) general scheme; (b) second-order loop.

A single-stage system which is also based on the separate feedback of the MSB of the M-bit ADC [16] is shown in Figure 12. Here, as in the schemes of references [14] and [15], the use of the 1-bit DAC in the feedback path to the first-stage input insures that x will appear in the output y without substantial nonlinear distortion. At the same time, the feedback signal from the M-bit DAC makes the cancellation of e_1 possible, and contributes to the linearization of the system. The nonlinearity error of the M-bit DAC (denoted by δ in Figure 12) appears in the output y filtered by $1/H_2(z)$, where H_2 is the transfer function of part of the loop filter preceding the M-bit DAC. For the system of Figure 12,

$$Y(z) = z^{-1}X + 2(1-z^{-1})^3 E_2 - 2z^{-1}(1-z^{-1})^2 \Delta, \tag{14}$$

where Δ is the z-transform of δ. Thus, the DAC error is second-order high-pass filtered as in the scheme of ref. [15].

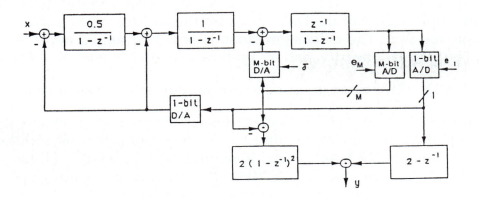

Fig. 12: An alternative dual-quantizer modulator scheme.

An alternative to the randomization or filtering is the digital cancellation of the DAC nonlinearity, as illustrated in Figure 13 [17]. Here, the RAM is programmed to provide the digital equivalent of all DAC output levels. Thus, for any M-bit digital output y, the resulting analog DAC output y_1 and the digital RAM output y_2 have equivalent values. Since in the baseband the feedback loop causes $x \approx y_1$ with a high accuracy, we have the desired $y_2 \doteq y_1 \approx x$ equivalence.

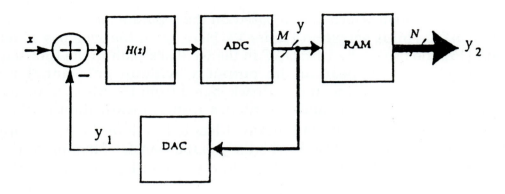

Fig. 13: Digitally corrected noise-shaping modulator.

The programming of the RAM can be achieved by a rearrangement of the original converter system (Figure 14). Here, an M-bit counter (not shown) provides the input signal of the DAC, and also the address of the corresponding RAM storage location. The delta-sigma loop is now rearranged into a single-bit front end, insuring its linearity, and an up-down counter is used to calculate the average value $\sum_{n=0}^{n_a-1} y(n)/n_a$ of the comparator output data. To achieve a 1/2 LSB accuracy, we must use $n_a \geq 2^{N+2}$ clock cycles for the averaging, where N is the required resolution of the overall converter system. Since this procedure must be performed for all 2^M DAC levels, it requires 2^{M+N+1} clock pulses. If, for example, M = 4, N = 16, and f_s = 1 MHz, the calibration requires about two seconds and hence can be performed at power-up times.

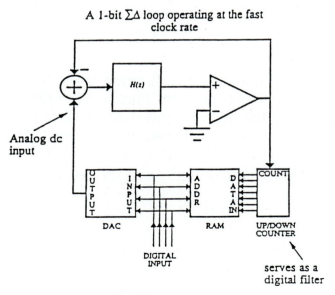

Fig. 14: Calibration scheme for digital correction.

3. Noise-Shaping D/A Converters

The block diagram of a noise-shaping D/A converter is shown in Figure 15. The input x is an N-bit data stream occurring at a slow clock rate f_s/OSR, where f_s is the data rate at the D/A block. The interpolator filter raises the data rate to f_s, and suppresses all spectral replicas of the signal except those centered at kf_s, k=0, ±1, ±2, etc. The digital truncator is a noise-shaping loop, similar in its operation but not necessarily in its structure, to the analog noise shaping loops discussed in the preceding sections. Its role is to reduce the word length of the oversampled data stream to M << N without significantly raising the in-band noise level. Usually, M = 1 is used, since this allows the use of a single-bit (and thus inherently linear) D/A converter.

Fig. 15: Oversampling D/A system.

Two alternative systems for accomplishing this task are shown in Figures 16 and 17. The general system of Figure 16 is an exact equivalent of the analog noise-shaping loop of Figure 1a, with a high-gain digital low-pass filter and a truncator or rounder in its forward path. The structure of Figure 17 is a first-order truncation loop based on error prediction, somewhat similar to the system of Figure 1b. Simple analysis reveals that $Y(z) = X - (1-z^{-1})E$, where E(x) is the z-transform of the (N-M+1)-bit truncation error e(n). Thus, e(n) is first-order high-pass filtered as required. Figure 18 illustrates how a second-order filtering of the truncation noise can be achieved using the prediction principle [18]. It can easily be shown that if the limiter is inactive the z-transform of the output is given by $Y = X + (1-H)E = X + (1-z^{-1})^2E$, where $H = z^{-1}(2-z^{-1})$ is the transfer function of the feedback path and E the truncation error. The purpose of the digital limiter is to prevent the input of the quantizer (truncator) from becoming very large for large input x. This prevents instability and overflow. An alternative system [19] which does not need a limiter is shown in Figure 19.

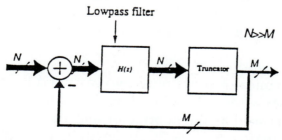

Fig. 16: Digital truncation loop using truncation noise shaping.

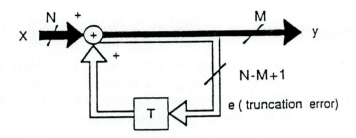

Fig. 17: Digital truncation loop based on prediction.

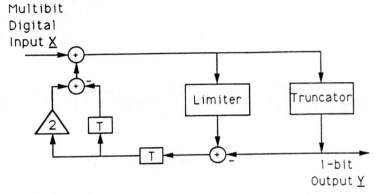

Fig. 18: Digital truncation loop using second-order predictor filter.

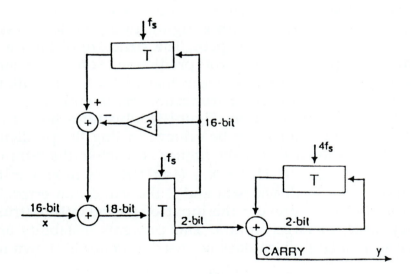

Fig. 19: Two-stage truncation system.

A second-order digital truncator containing two cascaded first-order stages [20] is shown in Figure 20. If $M_1 = M_2 = 1$ and each H_i block is preceded by a single-bit DAC, then H_1 and H_2 as well as the subtractor and the following low-pass filter can be realized using analog (active-RC or switched-capacitor) circuitry. The analysis of this system is very similar to that used for the analog cascade loops, and is left to the reader as an exercise. A third-order cascade truncator is described in [21].

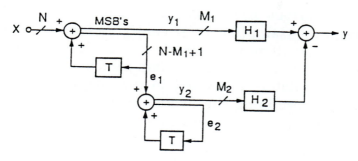

Fig. 20: Cascade truncation system.

As discussed in Sec. 2.5, it is possible to use a multibit internal DAC, as long as special steps are taken to suppress or cancel the noise and harmonics caused by its nonlinearity. Thus, the techniques described in refs. [12] and [13] remain applicable to oversampling DACs. Figure 21 shows how digital error correction can be used to cancel the nonlinearity error of the DAC [17]. As before, the RAM is filled with the digital equivalents of the DAC levels, obtained by the same calibration process as was described in connection with Figure 14. Since the DAC and RAM have the same inputs, their outputs will be equivalent signals and thus the feedback action will force the digital output and analog input to be equivalent in the high-gain frequency region of the loop.

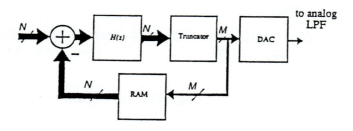

Fig. 21: Noise-shaping truncation loop incorporating digital correction of the DAC.

4.   Conclusions

Oversampling data converters use increased temporal resolution and digital complexity to obtain a high amplitude resolution and signal-to-noise ratio which would otherwise not be achievable in integrated data converters. This chapter examined the structures available for the realization of the most practical version of these oversampling data converters, the delta-sigma converters. These circuits use noise shaping (i.e., high-pass filtering of the quantization noise), to achieve a large inband signal-to-noise ratio.

The properties of the various A/D converter architectures were examined and compared, and some oversampling D/A configurations were also discussed. Due to a lack of space, no attempt was made to include the design of the digital filters and the continuous-time analog filters which precede and follow the modulators and demodulators described in the chapter. They are, however, important factors in determining the overall performance of the system, and should be designed carefully to achieve a functional device.

Acknowledgement

The author is grateful to Dr. J.C. Candy of AT&T Bell Labs and to Dr. R. Schreir of Oregon State University for helpful discussions.

5. References

[1] J.C. Candy and G.C. Temes, "*Oversampling methods for A/D and D/A conversion,*" in Oversampling Delta-Sigma Data Converters, J.C. Candy and G.C. Temes, eds., IEEE Press, New York, NY, 1991.

[2] H. Inose, Y. Yasuda, and J. Murakami, "A telemetering system code modulation Δ-Σ modulation," IRE Trans. Space Elect. Telemetry, vol. SET-8, pp. 24-209, Sept. 1962.

[3] J.C. Candy and O.J. Benjamin, "The structure of quantization noise from sigma-delta modulation," IEEE Trans. Commun., vol. COM-29 pp. 1316-1323, Sept. 1981.

[4] J.C. Candy, "A use of double integration in sigma-delta modulation," IEEE Trans. Commun., vol. COM-33, pp. 249-258, Mar. 1985.

[5] T. Hayashi, Y. Inabe, K. Uchimura, and T. Kimura, "A multi-stage delta-sigma modulator without double integration loop," ISSCC Dig. Tech. Papers, Feb. 1986, pp. 182-183.

[6] Y. Matsuya, K. Uchimura, A. Iwata et al., "A 16-bit oversampling A-to-D conversion technology using triple-integration noise shaping," IEEE J. Solid-State Circuits, vol. SC-22, pp. 921-929, Dec. 1987.

[7] M. Rebeschini, N. van Bavel et al., "A high resolution CMOS sigma-delta A/D converter with 320 kHz output rate," IEEE Proc. ISCAS'89, pp. 246-249, May 1989.

[8] W.H. Ki and G.C. Temes, "Offset-compensated switched-capacitor integrators," ISCAS Proc., May 1990, pp. 2829-2832.

[9] L. Longo and M. Copeland, "A 13 bit ISDN-band oversampled ADC using two-stage third order noise shaping," IEEE Proc. Custom IC Conf., pp. 21.2.1-21.2.4, Jan. 1988.

[10] D.B. Ribner, "A comparison of modulator networks for high-order oversampling sigma-delta analog-to-digital converters," IEEE Trans. Circuits Sys., vol. CAS-38, pp. 145-159, Feb. 1991.

[11] K. Matsumoto and R.W. Adams, "An 18b oversampling A/D converter for digital audio," ISSCC Dig. Tech. Pap., pp. 202-203, Feb. 1988.

[12] L.R. Carley, "A noise-shaping coder topology for 15+ bit converters," IEEE J. Solid-State Circuits, vol. SC-24, pp. 267-273, April 1989.

[13] B.H. Leung, "Architectures for multi-bit oversampled A/D converter employing dynamic element matching techniques," ISCAS Proc., June 1991, pp. 1657-1664.

[14] T.C. Leslie and B. Singh, "An improved sigma-delta modulator architecture," IEEE Proc. ISCAS'90, pp. 372-375, May 1990.

[15] B. Brandt and B. Wooley, "A CMOS oversampling A/D converter with 12b resolution at conversion rates above 1 MHz," ISSCC Dig. Tech. Pap., pp. 64-65, Feb. 1991.

[16] A. Hairapetian, G.C. Temes, and Z.X. Zhang, "Multibit sigma-delta modulator with reduced sensitivity to DAC nonlinearity," El. Letters, vol. 27, pp. 990-991, May 1991.

[17] T. Cataltepe, G.C. Temes, and L.E. Larson, "Digitally corrected multi-bit Δ-Σ data converters," IEEE Proc. ISCAS'89, pp. 647-650, May 1989.

[18] P.J. Naus, E.C. Dijkmans et al., "A CMOS stereo 16-bit D/A converter for digital audio," IEEE J. Solid-State Circuits, vol. SC-22, pp. 390-395, June 1987.

[19] H.G. Musmann and W. Korte, "Generalized interpolative method for digital/analog conversion of PCM signals," 1984 U.S. Patent, No. 4,467,316 (filed 1981).

[20] J.C. Candy and An-Ni Huynh, "Double interpolation for digital-to-analog conversions," IEEE Trans. Commun., vol. COM-34, pp. 77-81, Jan. 1986.

[21] Y. Matsuya, K. Uchimura, A. Iwata, and T. Kaneko, "A 17-bit oversampling D-to-A conversion technology using multistage noise shaping," IEEE J. Solid-State Circuits, vol. 24, pp. 969-975, Aug. 1989.

Chapter 11

Layout of Analog and Mixed Analog-Digital Circuits

Franco Maloberti

Department of Electronics, University of Pavia - 27100 Pavia, Italy

1. Introduction

In many of the chapters of this book, theoretical considerations of the design of analog circuits, data converters and mixed analog-digital systems have been discussed. The design flow consists of the following steps after theoretical analysis: logic, electrical simulation and verification and finally the translation of results into the physical description (and physical verification) of the system its layout.

It is well known that designing a layout is a tedious and error prone task. Thus, in order to avoid trivial work and to reduce the possibility of (human) error, convienient CAD tools for automatic, error-free layout generation have been introduced. For the design of digital circuits, where an extremely large number of transistors is very common, the use of such CAD tools is mandatory. However, the principles on which these tools are based do not coincide with the design strategies that should be used for analog or mixed analog-digital systems. For digital circuits compactness and speed are the main criteria driving electrical and physical implementation; to guarantee analog circuit accuracy, signal-to-noise ratios (SNR) and bandwidth (for operation with small signals) are important design issues. They must be carefully taken into account in the layout design phase.

To date, economic and technical aspects (simplicity and better codificability) have favoured the development of CAD tools oriented towards dealing with digital design problems. However, because of the increasing importance of integrated mixed analog-digital systems, a new generation of CAD tools will be made available soon. They will be capable of addressing the design problems (including, of course, layout) involved with entire signal processing systems.

It is important for designers to be capable of critically analysing automatically generated solutions, in order to identify any weak points and suggest suitable modifications. Moreover, critical analysis of the results favours the technical growth of the designer himself. The purpose of this chapter is to provide the basic background necessary in order to carry out the functions of the designer mentioned above.

2.   Differences Between Layout and Circuit

Silicon technology (and its limitations) has been extensively covered in another chapter of this book. Here, we shall only recall those results which are relevant to the correct design of circuit layouts. In particular we will analyse why a designed layout and its corresponding real circuit are different. There are several reasons for this. Among them, the most relevant are:

- lateral diffusion

- etching under protection

- boundary dependent etching

- three-dimensional effects

and, obviously, the errors and limitations associated with mask production and mask alignment.

The diffusion of doping atoms is intrinsically a three-dimensional effect. Thus, the edge of a metallurgical junction does not proceed only in the direction perpendicular to the silicon surface but also laterally under the mask-defined protection. If the depth of the junction is x the lateral diffusion is in the range *0.6-0.8 x*. (Fig. 1.a) [1]. This effect must be predicted and taken into account at the circuit design level. However, fluctuations in lateral diffusion contribute to a certain amount of random error.

In order to be sure that unwanted patterns are completely eliminated, materials are over-etched and the etching also proceeds slightly undermask-defined protection even if unisotropic (plasma assisted) etching is used. It turns out that the resulting pattern is smaller than the one defined by the layout (Fig. 1.b) (undercut effect). Moreover, the undercut effect is boundary dependent: etching is less active when small strips of material must be removed (Fig. 1.c). So, not only undercut but also its boundary dependence must be taken into account.

The surface of an integrated circuit, in particular when small linewidths are used, is not flat. The selective oxidation of silicon determines a silicon consumption and, at the same time, the growth of an oxide layer whose thickness is larger than that of the consumed silicon. Selective growth of chemical vapox deposed layers also contribute to non-flat structures. So it results (Fig. 2) that the geometrical dimensions (L1) of patterns grown on the top of non-flat surfaces are different from those of their designed layout (L).

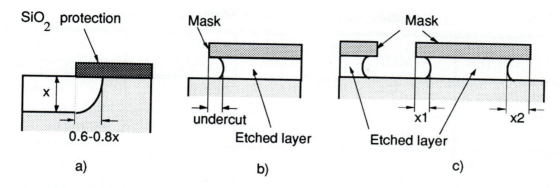

Fig. 1 : a) Lateral diffusion; b) etching under the protection; c) boundary dependent etching.

The above described effects are not very substantial for digital systems; by contrast, they may have a significant impact on the accuracy of analog circuits and must be avoided or compensated.

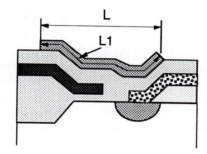

Fig. 2: Error in the pattern size due to tri-dimensional effects.

3. Absolute and Relative Accuracies

The global performance of an analog circuit is strongly dependent on the absolute and the relative (matching) accuracy of its basic components. In turn, these accuracies depend on the relevant properties of materials and on the geometries of the components. The absolute accuracy of the properties of materials can be controlled only at the technological level; by contrast, relative inaccuracies, due to gradients and local variations, which are also controllable at the technological level, can be compensated with suitable layout techniques.

The factors affecting the static performances of MOS transistors, capacitors and resistors are summarized in Tables 1, 2 and 3, [2]. The parameters are influenced in process controlled and layout controlled effects. The relative importance can be deduced by their typical absolute and relative accuracies.

Table 1: Parameters relevant for transistor performance.

PARAMETER	SYMBOL	TYPICAL VALUE	ABSOLUTE ACCURACY	MATCHING ACCURACY
Threshold	V_{th}	0.8 V	0.2 V	5 mV at 100 μ dist.
mobility (n)	μ_n	700 cm^2/V sec	15%	1% at 100 μ dist.
mobility (p)	μ_p	300 cm^2/V sec	15%	1% at 100 μ dist.
oxide capacit.	C_{OX}	30 nm	6%	0.1% at 100 μ dist.
length	L	-	0.3 μm	0.04 μm at 20μ
width	W	-	0.3 μm	0.04 μm at 20μ

Table 2: Parameters relevant for capacitor performance.

PARAMETER	SYMBOL	TYPICAL VALUE	ABSOLUTE ACCURACY	MATCHING ACCURACY
Diel. constant	ε_R (SiO$_2$)	3.8	2%	0.02%
oxide thickness	t_{ox}	40 nm	10%	0.03% at 100μ dist.
poly-oxide thick.	t_{ox}	50 nm	14%	0.05% at 100μ dist.
length	L	-	0.3 μm	0.04 μm at 20μ
width	W	-	0.3 μm	0.04 μm at 20μ

Table 3: Parameters relevant for resistor performance.

PARAMETER	SYMBOL	TYPICAL VALUE	ABSOLUTE ACCURACY	MATCHING ACCURACY
diff. resistivity	ρ_{diff}	30 ÷ 50 Ω/sq	25 %	2%
poly resistivity	ρ_{poly}	20 ÷ 30Ω/sq	35%	4%
diff. thickness	$x_{j,diff}$	0.3 μ diff 3 μ well	15%	0.5%
poly thickness	$x_{j,poly}$	300-400 nm	15%	1%
length	L	-	0.3 μm	0.04 μm
width	W	-	0.3 μm	0.04 μm

Since the various parameters depend on different technological steps, they can be assumed to be statistically independent. It results that the inaccuracy affecting the considered element can be evaluated by summing up quadratically the various error contributions. From the tables it follows that geometrical parameters (W and L) exibit a 10% absolute inaccuracy if their value is 3 μm and a 1% mismatch if their value is 4 μm. In the case in which other parameters contribute with lower error or mismatch, it is necessary to use dimensions larger than the ones indicated, in order to have inaccuracies not dominated by geometrical parameters.

From the tables it also appears that the mismatch is, at least in some of the parameters, dominated by gradient effects: as the distance of matched elements increases the matching accuracy worsens. In this case it is necessary to compensate gradient effects with interdigitized and common centroid layout arrangements. These arrangements will be considered in a following section.

4. Layout of MOS Transistors

Typically MOS transistors in analog circuits have a relatively large W/L aspect ratio (in the range 10-100). In special cases (for example, transistors in the output stage of a line driver) the aspect ratio W/L can be as high as 1000 or more. The layout of such elements must face two main problems: the large aspect ratio of the resulting physical structure and the series resistance at the source and at the drain. It is quite common to use non-straight gates for digital application, in order to optimize the aspect ratio, as shown in Fig. 3. Given that the matching of orthogonal elements is very poor and the contribution to the gate length of elements at the corners is not well defined, this is not used much in analog precision applications. An additional big disadvantage of the layout in Fig. 3 is that the drain and the source connections are taken only in two localized points. Since the active area has a specific resistance that is as large as something like 10 Ω per square ($\Omega / \square$), and since the source and the drain span several squares, the actual equivalent circuit is not that of a single transistor but the more complex configuration shown in Fig. 3 where many elementary transistors are interconnected to one another through drain and source resistances. The voltages at the source and the drain of the elementary transistor are not the same. Even for a relatively low current (100 μA) a voltage drop of several mV can result across the parasitic resistor of the chain. This drop voltage is equivalent to an input-referred offset which, for precision applications, cannot be neglected.

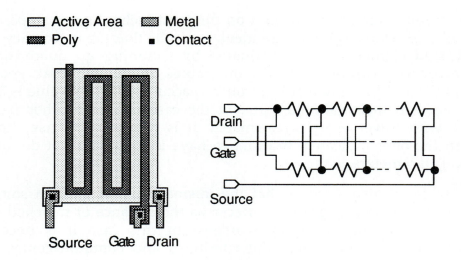

Fig. 3: Layout of a digital MOS transistor and its equivalent circuit.

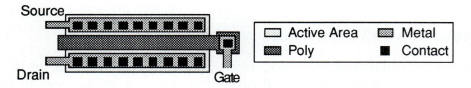

Fig. 4: Layout of an analog MOS transistor.

This drawback is overcome by using straight transistors with multiple source and drain contacts (Fig. 4). The structure is again equivalent to distributed elementary transistors connected in parallel; however, the metal line on the top of the drain and source contacts the underneath area at many points, thus shorting the drain and source diffused resistances. In some styles of layout only one big contact ranging over the entire source or drain area is used instead of multi-contacts placed at the minimum distance allowed by the design rules. However, even if the two solutions are circuitally equivalent, the approach shown in Fig. 4 is usually preferred because it guarantees a better circuit reliability. Many contacts, with minimum spacing, lead to a reduced curvature of the surface of the metal, thus reducing the risk of micro-fractures (potential sources of failure) in the body of the metal connections (Fig. 5).

In the case of transistors with a large W/L aspect ratio a wide straight transistor is not very manageable. To get a more suitable shape it is common to split such a transistors into a number of parallel equal parts arranged in a stack [3]. Fig. 6 shows an example of layout of a wide transistor split into eight equal parts. All the sources and drains of each element are connected in parallel by suitable metal connections. The gates are also interconnected by using polysilicon. It should be observed that some of the drain and source connections serve two different elements. This corresponds to a reduction of the silicon area and to an associated reduction of the parasitic capacitances of the two junctions source-substrate and drain-substrate. Metal lines run outside the active area of the transistor in order to ensure the best contact of source and drain. As a second-order effect, it should be noted that while the inner gates of the stack see the same boundary (two gates at the two sides), the gates at the end elements have a different periphery: unmatched undercut will determine slightly different lengths. This minor mismatch is not very important if it concerns elements of a stand-alone big transistor but, it should be taken into account when different transistors, which could be placed on the same stack, must match one another.

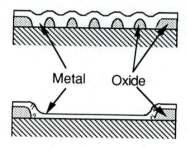

Fig. 5: Metal profile with multi-contacts and only one contact.

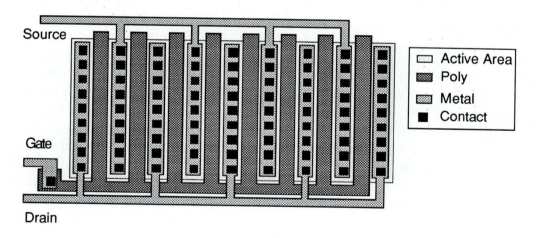

Fig. 6: Layout of a "stacked" wide transistor.

The layout structure in Fig. 6 that contains only one transistor can be generalized by putting different transistors or parts of them on the same stack. A typical situation would be a differential matched pair shown in Fig. 7.

Two transistors, M1 and M2, which are made up of a differential pair, are split in five equal parts and are arranged into an interdigitized fashion. The connection to the drains (DM1 and DM2) is taken with metal combs placed at the two sides of the stack. The common sources (SM1-SM2) are contacted with a serpentine metal path. It should be noted that crossings (with resulting parasitic resistances) with poly underpass are strictly avoided.

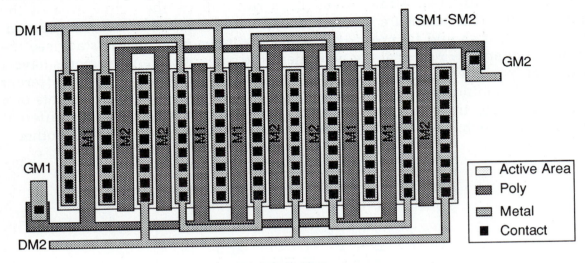

Fig. 7: Layout of interdigitized stacked differential pair.

5.   Layout of Resistors

Integrated resistors can be obtained by the suitable use of resistive layers [4]. The specific resistance of a typical resistive layer can range from a few tens of $\Omega/\square$ (ohm per square) to $K\Omega/\square$; hence, in integrated resistors the number of squares, and thus the L/W ratio, is often quite large. For this, it is common to make the layout with a serpentine arrangement. As in the case of transistors, the main practical problems concern the estimation of the contribution of the corners of the serpentine and the control of absolute and matching accuracies. For resistors where a limited accuracy is permitted it is enough to assume that the contribution of a corner element is equivalent to one-half square; for precise realizations this approximation is not suitable and it is suggested to use rounded corners (Fig. 8 a) or straight elements interconnected by metal. In this case controlling and taking into account the contact resistances resistive layer-metal is very important.

The boundary-dependent undercut effect determines any mismatch. It can be avoided by the use of dummy elements placed around the resistor's layout (Fig. 8a, 8b). Gradients in the properties of the materials also can create problems. They are usually compensated for with interdigitized structures [5]. Fig. 8b shows a possible implemetation of the technique: equal elements of two resistors R_1 and R_2 that should match are interleaved and connected in series by metal (low resistive) connections. The centroids of the two structures are close to each other so that the mismatch due to gradients is limited in such a small distance. As already mentioned, it is important to carefully design the end points of the resistive elements. If the current is disturbed from its laminar flow a localized resistance at the end points can result, which can be as high as 1 square of material. In Fig. 8c examples of bad and good end points are shown along with a very rough estimation of the localized resistances (measured in squares) [6].

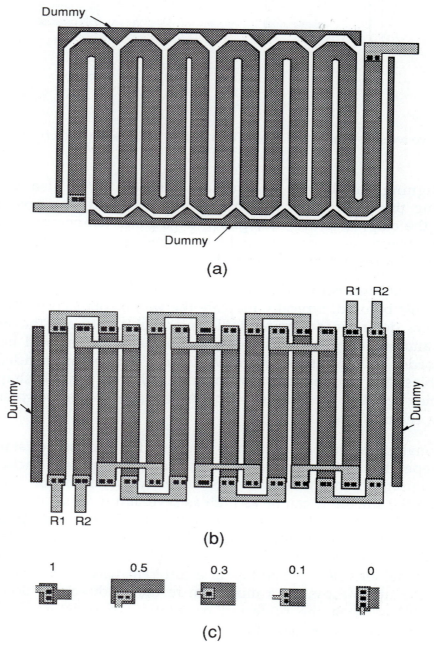

Fig. 8: Examples of layout of resistances and their terminations.

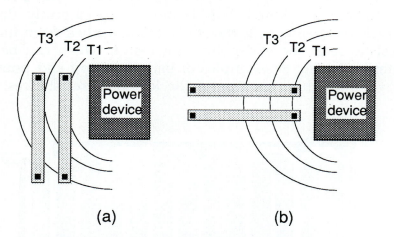

Fig. 9: Two resistors unmatched (a) and matched (b) with respect to thermal gradients.

When designing the layout of accurate resistors it is also worth while remembering that their value is temperature-dependent. If a power device is present on the chip, its power dissipation determines temperature gradients. In this case matched resistors should be arranged with their centroids placed symmetrical with respect to the power devices (Fig. 9).

With resistors of high values (hundreds of KΩ or MΩ), layers with high specific resistance must be used; typically they are the well or pinched wells. The use of such a layers determines a remarkable interaction between the body of the resistance and the surroundings (coupling, noise, etc,..., as will be discussed later). In order to limit these negative effects it is recommended to put a substrate bias ring around the resistor and, with serpentine shapes, to use substrate bias between the parallel strips. A better insulation between the strips is therefore ensured (Fig. 10).

An additional point to be taken into account for precise resistances is the pressure dependence of the resistive layer used. With plastic packages a fluid material is injected into a mould and its successive solidification gives rise to a big stress on the silicon. It can be as high as several hundreds of atmospheres (a value that is close to the fracture limit of silicon). Such a high pressure gives rise to an unpredictable resistor variation. Moreover, for monocrystalline layers realized in <100> silicon the effect is also unisotropic: the pressure induces variations of the electrical properties in a different way in the two orthogonal directions; the unisotropy has a minimum in the 45°direction. Because of this minimum, 45° is often the preferred orientation of resistors designed with monocristalline layers in <100> silicon.

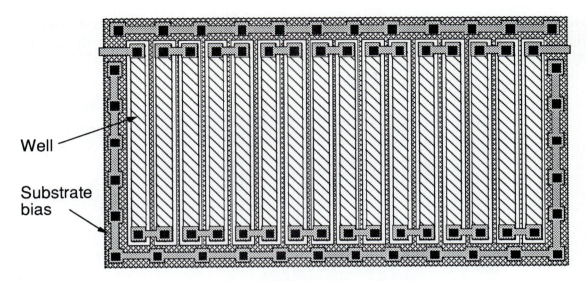

Well

Substrate
bias

Fig. 10: Layout of resistor realized by well diffusion.

6.   Layout of Capacitors

Capacitors are achieved in MOS technology by using diffusion, polysilicon or metal as plates and silicon oxide or polysilicon oxide as dielectric. The structures correspond to a parallel-plate element with capacitance given by:

$$C = \varepsilon_0 \, \varepsilon_r \frac{A}{t_{ox}} \tag{1}$$

ε_0 is, as known, the absolute dielectric constant. Relative dielectric constant (ε_R), the area of the plates (A) and the oxide thickness (t_{ox}) are the main parameters controlling the capacitor value. However, second-order effects must be taken into account when precise capacitors have to be designed. These effects are related to fabrication inaccuracies and to the fringing effects at the boundary of the structures.

Fabrication inaccuracies can give rise to errors in oxide thickness. When this error corresponds to a gradient, its first-order effect is cancelled by matched elements that are arranged in a common centroid fashion. Fig. 11 shows the layout of two capacitors (C_1 and C_2) that need to be equal. The two capacitors are split into eight equal parts that are connected in parallel. This arrangement guarantees that a gradient in the oxide thickness either in the x or y direction does not affect the capacitor matching [7].

Another significant limititation which is due to fabrication steps is the undercut effect. A capacitor is a parallel-plate structure with the upper one normally smaller than the lower one. The area of the smaller plate is assumed to be that of the capacitor. However, because of the undercut effect the actual area is smaller than the designed one. If the plate is a rectangle with designed sides a and b, with an undercut x, the actual area A' is smaller than the designed A:

$$A' = A - 2x(a + b) = A - xP \qquad (2)$$

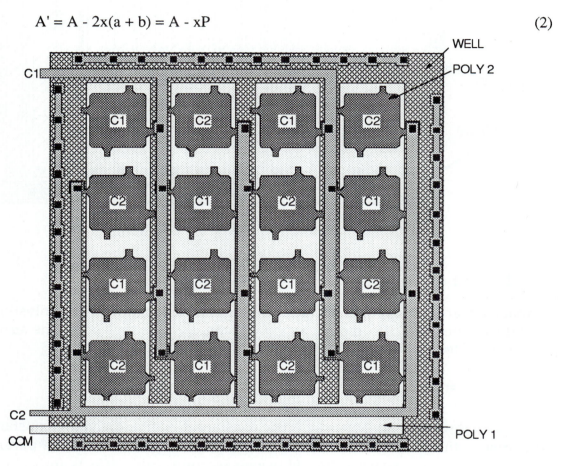

Fig. 11: Layout of two matched capacitors (common centroid structure).

The resulting reduction is proportional to the perimeters of the plate. It follows that in order to get the same proportional reduction in matched capacitors they should have the same area-perimeter ratio. This condition can easily be fulfilled in matched capacitances whose ratio must be a rational number. Equal "unit" capacitances connected in parallel can do the job; equal elements, of course have the same area-perimeter ratio. For non-integer ratios, the following strategy is usually used: a number of unit capacitors are connected in parallel with the addition of a rectangular element, which value ranges from 1 to 2 unit capacitances (Fig. 12). Some technologies allow use of a contact poly 2-metal even in the thin oxide region where poly 1 overlaps poly 2. In this case the layout is more efficient, as shown in Fig. 12 b.

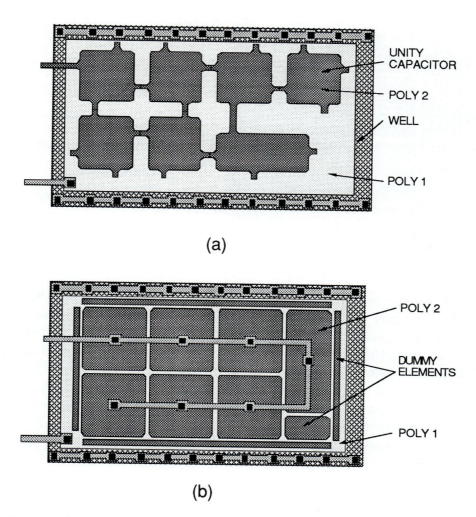

(a)

(b)

Fig. 12: Layout of a non-integer multiple of unity capacitance; a) poly interconnection of the top plates; b) metal interconnection of the top plates.

In order to limit the boundary-dependent errors, rounded or 45° corners are used. When the contact on the thin oxide area is not allowed by the technology (there is the risk of faliure), plate endings for exiting the thin oxide area must be used. Furthermore these poly endings should match. For precise applications even the contribution of parasitics of the metal lines used for interconnections is important. The capacitors should be designed in such a way that the parasitic capacitances are minimized and matched. It is not possible to concentrate the above considerations, illustrated in Fig. 13, into standard design rules. However, the awareness of the multiplicity of these practical problems will stimulate one to think about an optimum layout solution.

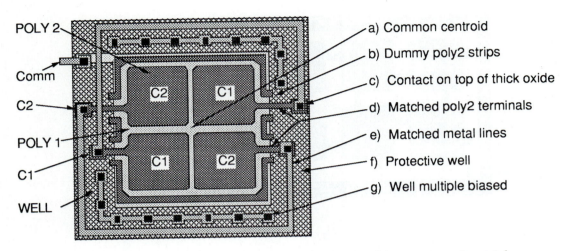

Fig. 13: Two matched capacitors laid out accordingly to some analog tricks.

Finally, it is worth mentioning a fundamental limitation on the capacitor accuracy. This comes from fringing effects. Equation (1) is valid under the assumption that the electric field between the two plates of the capacitor is uniform. In reality, at the boundary of the plates the electric field is not uniform and its fringing causes an intrinsic error in the use of equation (1). Since the fringing depends not only on the voltage of the two plates but also on the voltage of nearby conductors, its effect cannot be quantified. However, this contribuion to capacitor inaccuracy is proportional to the ratios t_{ox}/W and t_{ox}/L. If these ratios are smaller than 1/500 the effect becomes negligible compared to those resulting from fabrication-associated errors.

7.   Stacked Layout for Analog Cells

The advantages of using stacks of transistors have already been discussed. They involve reducing the parasitic capacitance of the source-substrate and drain-substrate junctions and have the advantage of saving area because they allow the sharing of the same contact space (for the source or drain). Moreover, the use of elements with the same orientation improves the matching and also reduces the effects of physical parameter gradients by allowing the use of interdigitized structures.

This section considers a generalization of the stacked technique applied to analog building blocks.

It is evident that the most efficient situation is the design of a stack with elements that have the same width, considering that the length of the stacked transistors is not very important. To get an idea of the topological degrees of freedom, it is worthwhile noting that, if a transistor is made of an even number of parts, its stacked representation will have its source connection (or the drain) on both sides of the stack; by contrast, when an odd number of elementary components are used, the source is on one side of the stack and the drain is on the other. Thus, depending on the number of parts in which a transistor is divided both source and drain or same terminal can be made available at the ending of the stack. If different transistors have sources or drains connected to the same electrical node they can be combined in the same stack in order to share the area corresponding to the common node. For this it is necessary that elements with equal width, whether they are single transistors or part of them, are combined into a single stack. Advantages in active area and junction capacitance reduction will result. Fig. 14 shows an example where four transistors (part of a telescopic folded-cascode op-amp) are designed on the same stack. Since the width of the transistors is an integer multiple of 40 μm, they are split into a number of parts 40 μm wide. The two transistors M_1 and M_2 are also interdigitized in order to get good matching. The sources of M_3 and M_4 partially share the same contact area (node 1 and 2) with the drains of M_1 and M_2.

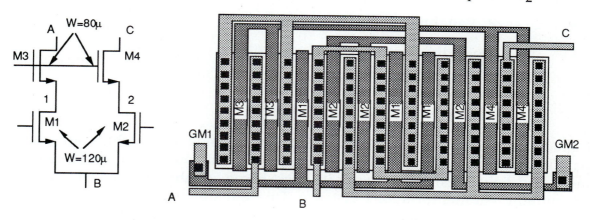

Fig. 14: Schematic and stacked layout of four n-channel transistors.

When designing an analog cell, simply changing the dimensions of a few transistors can lead to a very critical situation, whereas by contrast many other elements can, within limits, have a range of dimensions without significantly affecting the performance of the circuit. In order to achieve a well-designed layout, it is often worth changing the dimension of non-critical transistors (layout oriented design technique) in such a way that the dimensions fit the requirements of the stacked arrangement.

The layout of an analog cell can be organized as the interconnection of superimposed or side by side stacks; all the stacks should be made of elements which have equal width and a topological symmetry that corresponds to the electrical symmetry [8].

Two examples of a fully stacked layout of a two-stage op-amp and of a folded cascode operational amplifier are given in Fig. 15 and Fig. 16. The arrangement of the transistors is given in the associated schematics. It is interesting to observe that the layout drawn with the suggested approach is compact and regular with a close correspondence between the electrical and physical symmetries. Moreover, all the transistors have the same orientation, ensuring good matching. For the two stage op-amp the compensation capacitor occupies all the empty areas. Only one level of metal is used in the interconnections. If the technology makes more than one level of metal available the extra layers can be used for interconnecting at system level.

8.   Digital Noise Coupling

In analog circuits that are integrated with digital sections the problem of digital noise coupling and the techniques for its avoidance are very important issues.

The sources of digital noise coupling are:

- capacitive couplings

- couplings through the power supply

- couplings through the substrate

Capacitive coupling is mainly originated by analog lines which are routed parallel to the digital buses. Even if the transversal coupling of metal lines (or metal and poly lines) is very weak its effect becomes substantial when two paths are running parallel for a long distance. This situation is typical when reference voltages and clock signals must be delivered to a distant analog block, or, when for shielding purposes the analog V_{DD} (or V_{SS}) is used to bias a shielding ring that is placed close to another shielding ring which in turn is biased by a digital power supply. In such cases the noise protection given by the shielding can be eliminated by the capacitive coupling between the analog and the digital lines.

Another source of capacitive coupling is crossing. Generally digital signals are run as a bus with a lot of lines. It follows therefore that a crossing can often mean interaction with a lot of noisy sources. This kind of coupling is critical when the analog line is connected to the virtual ground of an operational amplifier. The collected noise is directly integrated into the feedback element and is made available at a low impedance node.

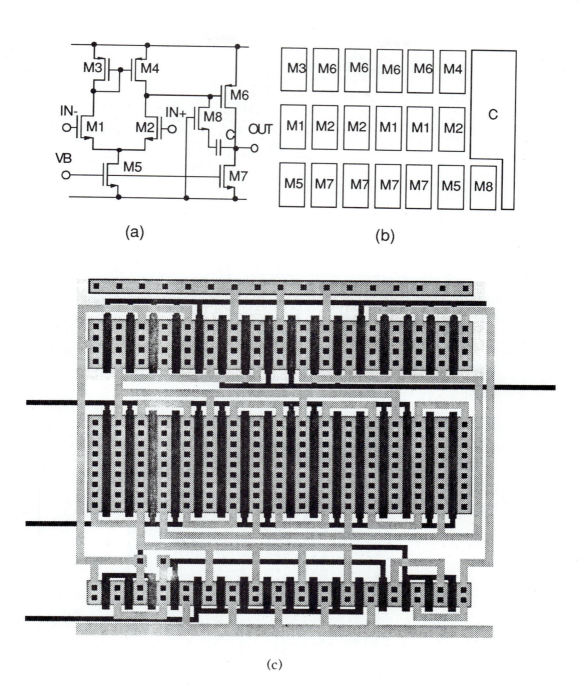

Fig. 15: a) Schematic of a two stages op-amp; b) transistor's arrangement; c) stacked layout.

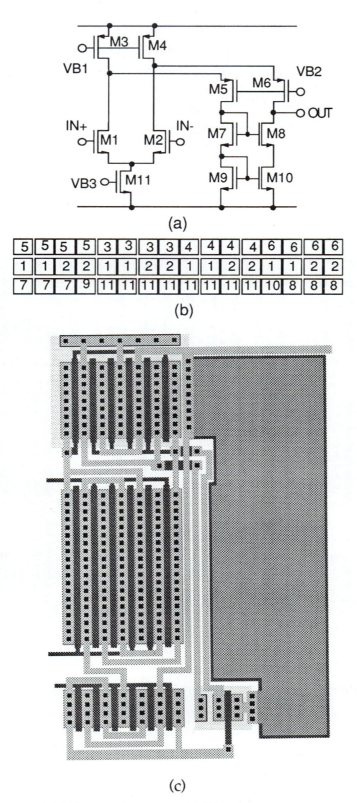

Fig. 16: a) Schematic of a folded cascode op-amp; b) Transistor's arrangement; c) Stacked layout.

In switched capacitor circuits, crossing between analog lines and clocks is often unavoidable. A typical layout of complementary switches is shown in Fig. 17 [9]. It should be noted that the metal used to interconnect the n-channel and the p-channel transistor must cross the digital phase driving the gate of the transistors that are placed in the well.

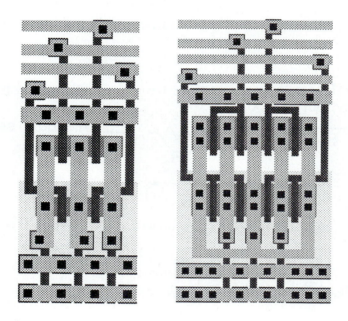

Fig. 17:   a) Layout of a minimum area complementary toggle switch; b) layout for large W/L ratio.

The reduction of noise injection due to capacitive coupling often is achieved from a simple awareness of the problem. A suitable placement and routing allows us to limit its effect. When parallel analog and digital lines (which should be kept as short as possible) are really necessary, they can be decoupled by using an appropiately large distance in between and, if necessary, by putting a grounded line (horizontal shielding) between the two critical paths.

Another important source of digital noise are power supply connections. An integrated circuit usually employs the same pins for both the analog and the digital biasing. Moreover the circuit is connected to the external world by a pad, a wire bonding and a pin. After the pad, a common analog and digital biasing line runs for a while before there is a definite splitting of the analog biasing network and the digital biasing network. This situation is described by the equivalent circuit shown in Fig. 18. In general, the current in the analog part of the circuit is constant or slightly varying in time; by contrast, the current flowing in the digital section is made up of sharp pulses, almost synchronous with the clock. The current pulses are necessary to charge or discharge the parasitic capacitance of the nodes driven from one logic state to the complementary one. The amplitude of the pulses is dominated by the switching of digital output drivers, since they are required to actuate external loads that are typically a capacitance of 100 pF (in series with the bonding inductance).

The bias current (analog and digital) determines a voltage drop ΔV due to the resistance of the connection, and more importantly to the inductance of the bonding.

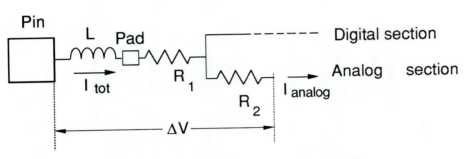

Fig. 18: Equivalent circuit of the bias connection in a mixed A/D circuit.

$$\Delta V = R_1 I_{tot} + R_2 I_{analog} + L \frac{dI_{tot}}{dt} \tag{3}$$

As a rule of thumb, the typical inductance of a bonding connection is around *1 nH/mm*. Thus, for usual bondings, it can range from *3* to *10 nH*. Moreover, the additional contribution of capacitance and inductance of frame and pin should also be taken into account. The resulting inductance is such that the drop voltage can be as large as tens or hundreds of mV. Table 4 shows typical bonding loads for dual-in-line (DIL) and chip-carrier (CC) 40-pin packages.

Table 4: Parasitic inductances and capacitances in packages.

Element	C (DIL)	H (DIL)	C (CC)	H (CC)
pin 1, pin 40	2.5 pF	15 nH	1.0 pF	3 nH
pin 5, pin 36	1.5 pF	8 nH	1.3 pF	6 nH
pin 10, pin 31	0.7 pF	4 nH	1.0 pF	3 nH
pin 20, pin 21	2.5 pF	15 nH	1.0 pF	3 nH

The drop voltage expressed by (3) (usually referred to as ground or V_{DD} bouncing) gives rise to an effect equivalent to a power supply noise. This noise has high frequency components; in the analog section it must be rejected by proper power supply rejection ratios (PSRR). Unfortunately PSRR is quite poor at high frequencies. Therefore it is very important to reduce the generation of the power supply noise. In general this is achieved by suitable bonding and biasing strategies that refer to the following general rules:

-   Firstly, it is advisable to keep as separate as possible the analog from the digital biasing networks. They should merge only very close to the pad.

-   Secondly, when possible, it is recommended to use separate pads for the analog and digital section with relative bonding (eventually multiple-ones) to the same pin (Fig. 19). In such a case the inductance of the analog bonding acts as a filtering element.

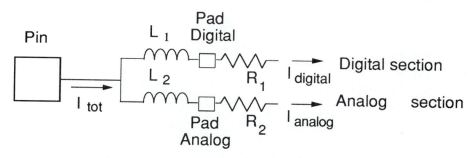

Fig. 19: Equivalent circuit of bias connection with separate bonding.

-   Thirdly, when extra pins are available, separate pins for the analog and the digital bias should be used. Such a complete separation in the biasing of the two sections gives evident advantages in terms of noise limitation, even if special care must be paid to substrate biasing. A delay in the substrate biasing (because of a delayed biasing of the analog or digital supply) can determine latch-up.

-   Fourthly, it is always suggested to choose for bias connections the pins that are in the middle of the frame. These result in the minimum parasitic inductance.

Another important source of noise is the coupling through the substrate. Output digital drivers very often employ transistors with a huge W/L aspect ratio since they must actuate large external capacitances. The drain diffused areas are consequently large and big capacitive couplings with the substrate results (because of the drain-substrate junction). When the driver switches from one logic state to the complementary state the output node exhibits a fast excursion with a resulting capacitive current. Moreover, during switching, high current flow can result in impact ionization in the drain area (hot carrier effects) with a resulting substrate current. The relatively high specific resistance of the substrate establishes a considerable amount of noise voltage even for an extremely low avalanche current.

The noise coming from the substrate can be limited either by reducing the noise injection or by using shielding strategies. In turn, the noise injected can be reduced by limiting the couplings with the substrate and by special care in the design of digital output drivers. They, in particular, should control the derivative of the output current in order to limit the inductance-dependent drop voltage component.

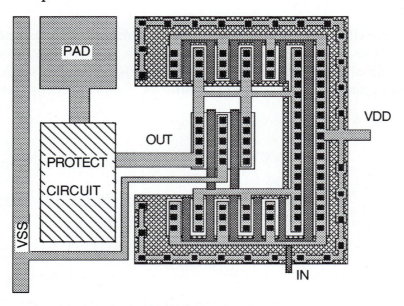

Fig. 20: Example of layout with well protection strategy.

Another method of reducing the substrate noise is to use suitable shields for intercepting the substrate noise and draining it towards non-critical nodes. Shielding can be achieved by plates or by rings. A typical plate shield is a well diffusion put under a capacitor array or under a critical metal line. In order to have real shielding (and non-noise collecting) the plate must be carefully biased and connected to a quiet, low-impedance node. Shielding can even be achieved by using the circuit itself or the wells inside which part of the circuit is integrated. For example, in a digital output driver the layout can be arranged in such a way that results in self-shielding: the transistor inside the well is designed surrounding its complementary transistor (Fig. 20).

Shielding rings are typically well rings or substrate-bias rings. Substrate bias offers a low impedance path for the noise currents that are going around the substrate. Wells create a surface barrier along the path of those noisy currents.

Very often a substrate bias ring and a concentric well placed on the side of the analog circuit are used in order to create a double protection for the analog sections.

A very much debated point concerns which voltage is better to bias the shielding rings: analog or digital V_{DD} or V_{SS}. The use of an analog bias determines a corruption of quiet lines (if a non-negligible current must be collected); by contrast, the use of a digital bias reduces the efficiency of the shielding. The decision is often decided on by the specific situation. However very often the digital biases are preferred.

9. Floor Planning of Mixed Analog-Digital Blocks

A typical mixed analog-digital block typically contains in the analog section an input signal conditioner, a continuous time or a switched capacitor filter and, eventually, data converters. These analog blocks are made up by active analog cells (operational amplifiers or comparators), passive components (resistors or capacitors) and switches. Moreover, very often a specific digital logic constitutes an essential part of the block (for example, the generator of the disoverlapped phases). The design of the layout of active and passive components must be oriented by analog system requirement. For example, if operational amplifiers must be placed side to side, it is worthwhile to use only one bias block. In this case, the biasing lines in the layout of the op-amp should cross the op-amp in a fashion that the connection is automatically established when the cells are placed side to side (Fig. 21). More in general, the input-output connection in the cell must be located in the proper side, in order to minimize the path and crossing of the inter-block routing. Thus, before designing the layout of the components, it is necessary to define the floor plan of the analog block.

The general of reference are:

- put the analog critical components as far as possible from the digital elements
- make the connections to the critical nodes as short as possible
- avoid crossing between the analog biasing lines and digital busses

Fig. 22 shows the typical floorplan of a switched capacitor filter. It can be noted that the switches that are the components closer to the digital world are placed on one side of the layout. The operational amplifiers that, by contrast, are the more critical analog elements, are placed on the other side of the layout. The capacitors that are in the middle are usually protected by a well shielding. The crossing of analog bias and signal lines and the digital bus bringing in the switches command is strictly avoided.

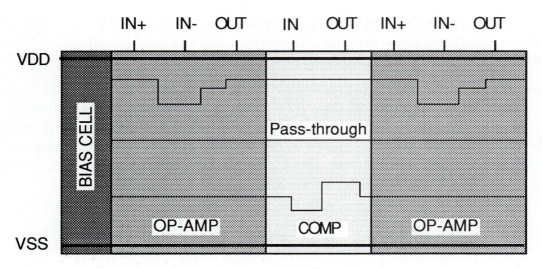

Fig. 21: Path of bias lines and V_{DD} - V_{SS} for basic analog cells.

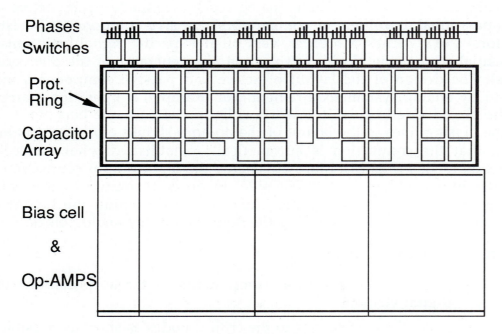

Fig. 22: Typical floorplan of an SC filter.

For a fully differential structure, the floor plan shown in Fig. 23 is normally used. The arrangement utilized for a single ended circuit is made symmetrical around the operational amplifiers. The switches are now at the two sides of the floor plan and the digital busses used to drive them are never crossing and far away from the analog ones.

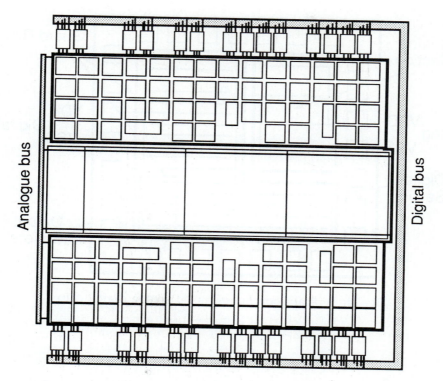

Fig. 23: Typical floorplan of a fully differential SC filter.

At a more complex level, when many analog and digital sections must be arranged on the same chip, guidelines similar to the ones already mentioned for floor planning of switched capacitor circuits should be used. In addition, it is important to introduce a well-defined physical separation between the analog and the digital circuitry with suitable protections and decoupling, as discussed in Section 8. A special care must be put in the power supply distribution: networks as separated as possible, for the analog and digital section, must be used. These recommendations are resumed in Fig. 24 where a possible floorplan of a mixed analog-digital circuit with protection and power supply distribution is shown.

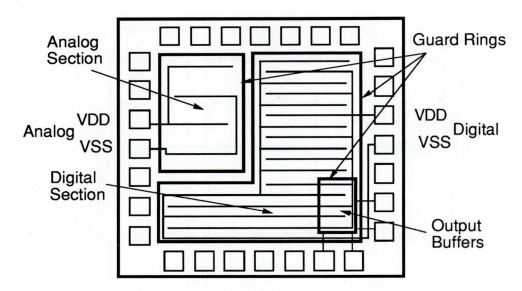

Fig. 24: Typical floorplan of mixed analog digital chip.

10. Concluding Remarks

The layout of analog and mixed analog/digital circuits is an important issue for integrated systems. While digital layouts are realized with automatic tools, the physical description of analog blocks is still mainly manual and, generally speaking, requires a specific expertise. The realization of mixed analog/digital circuit must combine results obtained from automatic tools with the results of manual activity. New computer aids for this kind of task will become available in the near future; however, it is necessary to have the knowledge to critically analyze and interactively optimize the solutions proposed by existing and coming CAD tools.

11. References

[1]   S.M. Sze (Ed.), *VLSI Technology*, McGraw-Hill, New York, 1983.

[2]   Austria Mikro Systeme: *AMS Design Rules CMOS Manuals.*

[3]   K.C. Hsieh, P.R. Gray, D. Senderowicz, D.G. Messerschmid, "A low noise chopper stabilized differential switched-capacitor filtering technique," *IEEE J. Solid-State Circuits*, SC-16, 708-715, 1981.

[4] P.R. Gray, R.G. Mayer, *Analysis and Design of Analog Integrated Circuits*, John Wiley & Sons, New York.

[5] D.J. Allstot, W.C. Black, "Technological design considerations for monolithic MOS switched capacitor filtering systems," *Proceedings of the IEEE*, 967-986, 1983.

[6] A.B. Grebene, *Bipolar and MOS Analog Integrated Circuit Design*, John Wiley & Sons, New York, 1984.

[7] J.L. McCreary, P.R. Gray, "All-MOS charge redistribution analog-to-digital conversion techniques," *IEEE J. Solid-State Circuits*, SC-10, 371-379, 1975.

[8] U. Gatti, F. Maloberti, V. Liberali, "Full stacked layout of analogue cells," *Proc. IEEE Int. Symp. on Circ. and Syst.*, 1123-1126, 1989.

[9] R. Gregorian, G. Temes: *Analog MOS Integrated Circuits*, John Wiley & Sons, New York, 1986.

Chapter 12

System Architectures and VLSI Circuits for Telecommunications

Guenter Weinberger
Siemens AG, Semiconductor Group, Balanstr. 73, 800 Munich 80, Germany

1. Introduction

The evolution of the worldwide telephone network from the analog "plain old telephone" - net towards a digital international communication network during the last decade has been invisible to residential customers. Only business users have taken advantage of enhanced quality and services like Group 3 Fax and others.

However, this digital network is building the backbone for the introduction of the revolutionary integrated services digital network (ISDN), offering end-to-end digital communication, thus representing a universal international communication network.

The differences between the last decades' evolution and the present decade's revolution are obvious. The digitalization of the analog telephone network was cost driven and replaced electromechanical parts by electronics as far as possible. Using the pulse code modulation (PCM) technique within the network while maintaining analog subscriber lines was a tremendous effort in system innovation based on the latest semiconductor technologies. The investment's benefits were reduced cost to both system manufacturer and network operator and a superior quality of voice and services.

Introducing digital end-to-end communication is logically the next innovative step; however, it will revolutionize our communication possibilities and therefore our lives. All of a sudden, voice and data are handled in the same way, and are conveyed by "bit pipes," of rate equal to a multiple of 8 kb/s. Whether voice or data services are used depends only on the type of terminal equipment the user has connected to the telephone line. Simultaneous voice and data communication as well as services changing on the fly are nothing special any more (e.g., three party video conferencing).

However, introduction of ISDN is only possible using the most advanced semiconductor technology to manage the special requirements and complexity of digital data transmission and communication. Moreover, moving the more than 500 million telephone subscribers from analog to digital will be a smooth evolution communication systems have to be prepared to accomplish.

It is the intention of this chapter to give an overview on architectures of both digital exchange systems and ISDN terminals, focusing afterwards on key functions for those systems, i.e., very large scale integrated (VLSI) circuits for digital telecommunications.

2. System Structure

According to the so-called OSI seven layer model end-to-end communication can be represented as given in Fig. 1. The communication network provides for physical data transmission (layer 1), link establishment (layer 2), and networking facilities (layer 3), thereby providing peer-to-peer communication capability. The end users implement the application-specific upper layers [1].

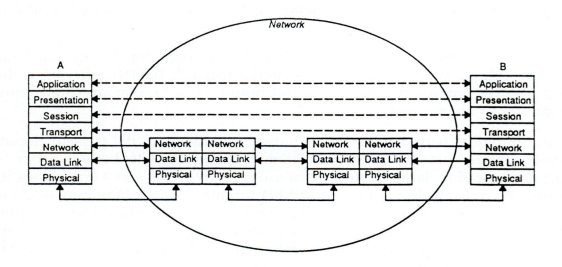

Fig. 1: OSI seven layer model representing a communication network.

A digital exchange system, representing layer one to three functionality, has to be able to handle different transmission schemes (layer 1) and associated link protocols (layer 2) while PCM channel switching functions can be implemented in the same way in any system. The associated architecture of a digital switching system is shown in Fig. 2. The core system comprises a switching network for 64/128 kb/s channels interfacing to subscriber and trunk line units as well as special function and service units (e.g., tone generation for analog phones or voice processing) via PCM highways, time division multiplexed (TDM) serial links running at 1.544 MHz (24 time slots per 8 kHz frame), 2.048 MHz (32 time slots), or multiples of these.

Supervision and control of the entire system, including connection setup, maintenance, and test is performed by a hierarchical structure of decentralized intelligence ranging from line card and group control level to system control level in a powerful central processing unit. The communication interfaces between the various control levels are manufacturer dependent and range from a standard parallel to various serial interface protocols, e.g., HDLC.

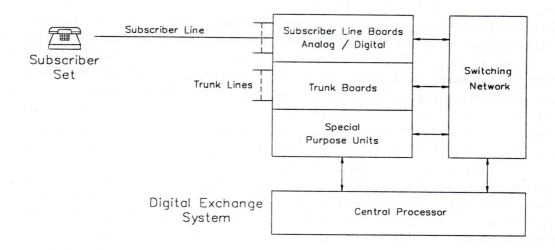

Fig. 2: Basic structure of a digital exchange system.

3. Communication ICs for Digital Exchange Systems

Telecommunication specific VLSI circuits are required throughout the entire exchange system. A distinction in switching, signalling, and line interface ICs is suitable to achieve a first order classification.

3.1. PCM switching network components

The objective of the switching network in a communications system is to establish connections between two or more subscribers. In fully analog configurations, this is performed by electromechanical or electronic physical links in a space division switch, which for n subscribers in its straightforward form requires n crosspoints of excellent analog transmission capability. Highly suited for memory oriented implementation, the basic principle of PCM time division multiplex (TDM) circuit switches is to transfer digital voice/data samples from the PCM-highway time slot attached to the "transmit" subscriber to that of the "receive" subscriber.

This merely requires an 8-bit memory location for each time slot of a PCM-highway, the contents of which are clocked to the outgoing highway according to another timing and, if required, a different clock frequency. The ability of connecting incoming time slots to one of several outgoing PCM highways results in a mixture of TDM and space division switches. Traditionally this type of switching network has been realized using standard LSI/VLSI circuits including shift registers, counters and memories. Dedicated VLSI devices implementing an 8x8 (which means 8 incoming and 8 outgoing serial PCM lines) and a 16x8 switching matrix were introduced by Mitel and Siemens in the early eighties. A block diagram of the Siemens PEB 2040 Memory Time Switch is shown in Fig. 3. The 16 incoming PCM lines are converted from serial to parallel and stored in the 4-kbit speech memory with 16x32=512 8-bit words. For this write procedure, the address is generated by an input counter. Interleaved with the writing function, a speech memory location is addressed by one of 256 words of the connection memory (just in time with the outgoing time slot) and the information is then clocked to one of the eight output lines. This is because the read address of each speech memory word is the input time slot, which itself is given by the contents of the connection memory word. This connection memory word is for its part sequentially addressed by the output time slot counter driven by the output clock. The connection memory is controlled by a microprocessor working asynchronously. Two of the devices allow the realization of a nonblocking switch for 512 subscribers, replacing quite a number of hitherto separate components.

In recent times Siemens has introduced a complete family of switching devices covering different sizes as well as offering additional functions such as conferencing. Fig. 4 gives an overview of up-to-date PCM switching devices based on modern CMOS technology [15].

3.2 Serial communication controllers

It is not only since the advent of ISDN that serial communication protocols have been a key element of telecommunication systems. With the introduction of PCM switching a lot of system manufacturers chose the HDLC protocol as their system standard to connect the decentralized control parts to the central processing unit.

A wide range of applications for the HDLC protocol, of which Fig. 5 shows only the telecommunication specific area, led to the development of various kinds of HDLC controller ICs. Depending on the IC, from one to 32 physically separate HDLC channels can be handled by one device. Their applications range from system internal communication, ISDN layer-2 signalling functions on inter-office trunk lines and subscriber lines, to layer-2 processing of user packet data.

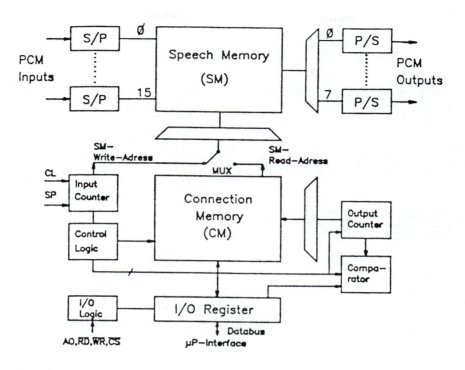

Fig. 3: Block diagram of a typical switching device.

Complete Switching and Conferencing IC Family

	MTSC PEB 2045	MTSS PEB 2046	MTSL PEB 2047	MUSAC PEB 2245
Switching capacity (time slots)	512 x 256	256 x 256	1024 x 512 (x 1024)	512 x 256
Clock rate (MHz)	8.192 or 4.096	8.192 or 4.096	8.192 or (16.384)	8.192 or 4.096
PCM data rate (Mbit/s)	2/4/8 + mixed mode	2	4/8 + mixed mode	2/4/8 or mixed mode
Conferencing	–	–	–	for up to 64 channels
Programmable attenuation input output	–	–	–	for up to 64 channels 0/3/6/9 dB 0/3 dB
Power consumption (max.) (mW)	50	50	90 (125)	100
Package	P–DIP–40 PL–CC–44	P–DIP–40 PL–CC–44	PL–CC–44	PL–CC–44
Compatible pinning	X	X		X

Fig. 4: Bandwidth in size and functionality of switching ICs.

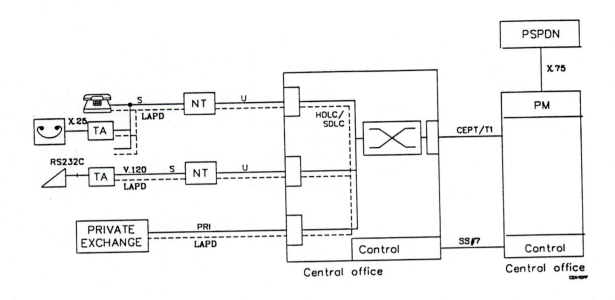

Fig. 5: Use of HDLC based protocol in telecommunication systems.

A typical IC to cover most of the above applications is the Siemens SAB 82525 HSCX (high-level serial communication controller extended), a dual channel device with 64-byte FIFO's per channel and direction to interface data to the host system. Two fully independent channels with time-slot assignment, collision detection, statistical multiplexing, PLL and baud rate generator, different coding schemes, and DMA support render this device a powerful and flexible communication device to be used for system control, layer-2 signalling, as well as packet data applications (e.g., V.120 rate adaption scheme).

3.3 Line card functions and architectures

Due to the fact that the hardware cost for line cards amounts to up to more than 70% of the total hardware cost for a digital switching system this area has always been the major area of cost reduction efforts via the use of the most advanced semiconductor technologies. For the semiconductor industry on the other hand this specific application is a very attractive business segment simply because of the number of lines being installed every year. Complete system solutions for both analog and ISDN line cards require highly complex filter functions as well as tricky analog front ends and high voltage parts.

Finally the smooth shift from analog to ISDN subscriber lines requires a well-prepared hardware architecture. It was a major goal of the telecom industry activities in the mid-eighties to globally redefine their line card architectures to meet the requirements of both analog and ISDN subscribers [7].

3.3.1. *Line card functions*

The line card interfaces to the subscriber line (or to an inter-office trunk line) on the one hand and to the core exchange system on the other hand. The latter is comprised of the PCM interface and a logically separate control interface.

The line interface functions are performed either by a codec filter and a SLIC in the case of analog lines (see 3.5) or a transceiver, a transformer, and a power controller in the case of ISDN lines (see 3.6). Serial transmission of the 64-kbit/s PCM data to the switching network via the so-called PCM highways is performed by selecting one receive and one transmit 8-bit timeslot within the 125-us (corresponding to 8 kHz) frame for each subscriber channel.

Line card control is achieved via a signalling link to the central control unit. Although there is no standard protocol HDLC based serial data transmission is widely accepted.

Fundamental differences between analog and ISDN subscribers exist in layer-2 signalling. While analog signalling is restricted to on/off hook, ground, and DTMF detection, the ISDN 16-kbit/s D channel with its layer-2 LAP-D protocol increases signalling capabilities tremendously. Moreover, in the case where, in addition to out-band signalling, packet data is also conveyed in the D channel, the peak traffic load to be handled by the line card processor drastically increases, if all D channel information is completely processed at the line card level. In order to avoid such situations alternatives must be considered to achieve a cost/performance optimum.

3.3.2. *Line card architectures*

There are basically two architectures supported by the semiconductor industry. They can be roughly characterized by vertical and horizontal integration, respectively, as indicated in Fig. 6.

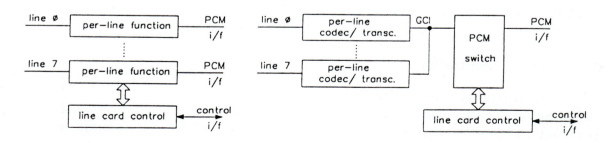

Fig. 6: Horizontal and vertical (GCI) line card architectures.

Horizontal integration emphasizes the per-subscriber approach ending up in subscriber line interface devices with direct access to the PCM highway. Programming of time-slot assignment (TSA) as well as subscriber line interface functions is performed via a direct (parallel or serial) interface to the microprocessor. The system internal signalling interface is implemented by an additional link controller or, in the case of simple protocols, a microprocessor's on-chip serial interface.

Vertical integration concentrates system internal interface functions for 8, 16, or 32 subscribers in a PCM interface device which connects also to the subscriber line devices via an internal serial line card interface. Since this architecture is very attractive for many reasons the internal line card interface has been of great interest to system manufacturers. In Europe this led to a joint definition activity of the largest system suppliers Alcatel, Italtel, Plessey, and Siemens in order to standardize the so-called general circuit interface (GCI). As a consequence, every semiconductor supplier with major telecommunication business is supporting the GCI by their devices.

Being the original promotor of vertical line card architecture together with Intel, Siemens offers a complete device spectrum for both analog and ISDN line cards. Due to the early in-house standardization Siemens refers to the GCI as the IOM^R-2 (ISDN oriented modular) architecture [16]. National Semiconductor as the protagonist of horizontal integration also offers a complete spectrum; their ISDN transceivers, however, support both architectures.

Fig. 7 shows the frame structure of the four wire GCI, connecting up to 8 ISDN or 16 analog subscribers to the appropriate PCM interface controller via a 4.096 MHz TDM line per direction. Besides the PCM channels and the ISDN D-channels the GCI offers complete control and monitoring of the line interface circuits.

Although vertical integration of line interface circuits (i.e., concentration of several subscriber channels in one device) has hardly happened so far, dual channel codec filter devices available since 1989 from AMD and Siemens indicate the feasibility based on device-internal DSP architectures (see 3.5.2).

3.4 Line card controller

With the introduction of the Siemens Peripheral Board Controller (PBC) in 1983 the line card was pushed towards state-of-the-art technology including complete software control and monitoring, HDLC based system internal signalling, flexible TSA to switch the PCM channels, and simpler line card wiring [2] (see Fig. 8).

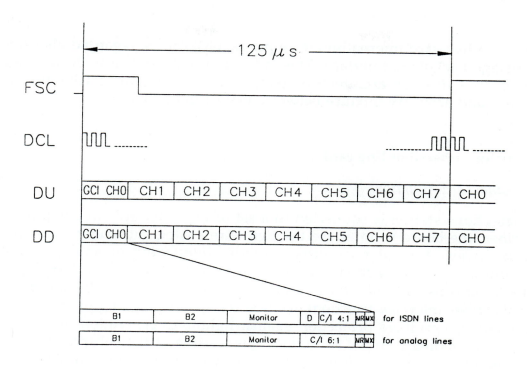

Fig. 7: GCI frame structure.

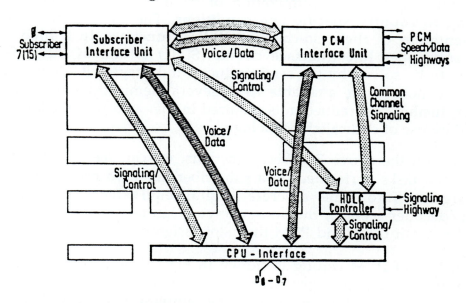

Fig. 8: Possible data streams within the PBC.

With the definition of the GCI, a second generation PCM interface controller family has been developed giving access to up to 32 ISDN or 64 analog subscribers.

Although a line card internal interface per se, GCI can be regarded also as being a first-stage backplane interface. This is in perfect coherence with the tendency towards line drawer architecture (single-subscriber line cards) for higher reliability and arbitrary mixture between analog and ISDN lines.

3.5. Analog subscriber line card devices

3.5.1. *Line card functions*

Each telephone station is connected to a set of line interface circuits placed in a subscriber line board to interface the two wire analog subscriber loop with the PCM highway. A similar circuitry provides interfacing of analog interoffice trunk lines. The two wire connection can be considered as a "floating" high voltage loop consisting of a wire "a", the tip wire, and a wire "b", the ring wire. The subscriber interface functions include the battery feeding of the subscriber equipment, overvoltage protection against lightning, generation of the ringing voltage and feeding it to the subscriber, supervision of the subscriber loop (e.g., "on-hook/off-hook" detection), hybrid 2/4-wire conversion, and test access. High transmission quality is achieved by careful matching of the complex termination impedance, optimization of balance network and level control according to country specific line conditions. The cost effective and reliable implementation of these so-called BORSHT functions in a high voltage all silicon subscriber line interface circuit (SLIC) is one of the most difficult problems to be solved by the semiconductor industry. SLIC devices realized in bipolar technology are available from several semiconductor manufacturers; however, further cost reduction is required in that area which could possibly be achieved by use of dedicated technology.

Together with the above BORSHT functions, we can see in Fig. 9 that further fundamental tasks performed on a subscriber line card are A/D conversion of voice signals into time discrete digital equivalents in a companding encoder/decoder (codec), the accompanying limitation of the voice band in a receive/transmit filter, and PCM time slot assignment.

The principle of PCM coding is shown in Fig. 10. According to the sampling theorem, in any sampling procedure the input frequency spectrum is mirrored around the frequency $f_s/2$ where f_s is the sampling frequency. Thus, to prevent aliasing (i.e., folding of frequency components above $f_s/2$ back into the voice band), the input signal must be first filtered to suppress frequencies above $f_s/2$. For PCM systems using a sampling frequency rate of 8 kHz, representing one voice sample every 125 us, a PCM filter therefore should provide attenuation of frequencies above 4 kHz; in practice, a 3.4 kHz stop-band limit frequency is used.

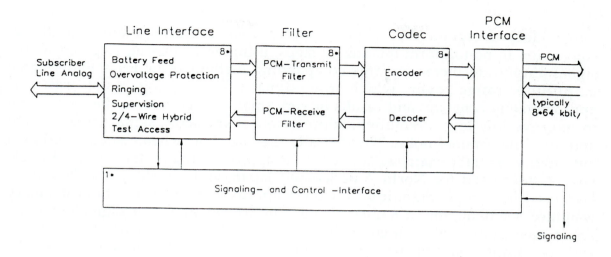

Fig. 9: Basic functions of an analog subscriber line card.

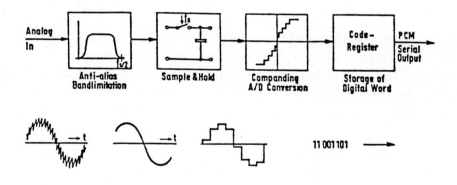

Fig. 10: Sequence of PCM coding.

In addition to this anti-aliasing function, a 300 Hz high pass prevents low-frequency distortion. After passing the anti-aliasing filter in the encode direction, the analog signal is encoded. It is subsequently loaded in an output buffer and clocked to the PCM system. In the receive direction the PCM-filter acts as a reconstruction low-pass filter. Since the decoder output sample-and-hold function has a sinx/x amplitude roll-off as frequency increases, this must be taken account of in the filter design.

Traditionally, anti-aliasing and reconstruction filters have been implemented as analog circuits, and the codec samples the analog voice at the same 8-kHz rate used for digital data transmission in PCM systems. However, advances in large scale integration enable digital instead of analog filtering methods, resulting in a different sequence of PCM coding.

The codecs designed for telecom applications perform analog to digital conversion in nonuniform quantizing steps, using an increasing step size as the signal amplitude increases. This companding method provides a wide signal amplitude range over which the signal-to-quantizing noise ratio is approximately constant, and results in very small step sizes near the zero origin to improve low level resolution, suppression of crosstalk, signal enhancement and low idle channel noise. Theoretically the encoding scheme should follow a logarithmic transfer characteristic, which is approximated by the so-called u-225 law, standardized for North America and Japan, and the slightly different A-law, used in Europe. Significant for these approximations is a nonlinear binary weighted segmentation of 8 intervals per sign (called chords or segments) and a linear quantization of each chord into 16 equal steps. This means that a chord and its companion stepsizes double in length as they move away from the origin. In total, the PCM word code format is described by 1 sign bit, 3 bits to specify one out of 8 chords, and 4 bits to specify one out of the 16 linear steps of the chord. The maximum resolution possible using an 8-bit PCM word is thus equivalent to a linear code of 12 bits for the A-law and 13 bits for u-law, respectively.

Optimized integration of the described functions in a SLIC, advanced codec filter and PCM interface controller in a cost-effective way has been one of the great challenges for the semiconductor industry. The implementation of new features has a decisive influence on the performance capabilities of the system as a whole, as well as on its structure, operating software and cost.

3.5.2 Codec and filter devices

Essential factors which describe the performance of a monolithic filter combo-device are

- stop band characteristic
- passband ripple
- inband signal-to-noise ratio
- harmonic out-of-band distortion
- idle channel noise contribution
- gain tracking
- dynamic range
- power supply rejection and crosstalk coupling.

In an analog solution encoding is performed in a successive-approximation sequence for the definition of sign, chord and segment bits. In an iterative algorithm consisting of setting bits in a successive approximation register, performing the D/A conversion of its contents and comparing the result with the input signal in a high resolution comparator, the register contents are successively matched to represent a digital equivalent of the input voltage.

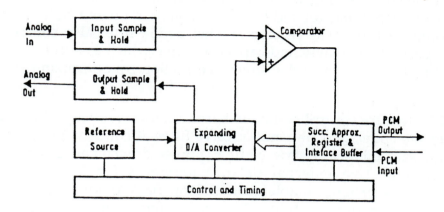

Fig. 11: Simplified block diagram of a successive approximation codec.

One of the key problems which had to be overcome for codec integration was to realize complex digital logic and high precision analog circuitry, such as operational input/output amplifiers, comparators, voltage reference and precision matched components, on one chip. While first types with analog capability were preferably realized in bipolar I^2L or mixed technology, advances in MOS analog capability also made monolithic codecs in NMOS and CMOS technology possible. Regarding analog circuitry, MOS is highly suited to voltage- and charge-switching, allows simple sample/hold realization and, furthermore, inherently produces capacitors of high linearity and precision. A key feature is its compatibility with switched capacitor filter technology. Therefore second generation combo devices are exclusively realized in MOS. Due to its low power consumption and better analog performance, the trend clearly is towards CMOS technology.

One solution is to use switched capacitor techniques to meet the high precision requirements for the integration of an active filter design by replacing resistors with switched MOS capacitors of high switching rate, where the switches are realized by MOS-transistors. Following this principle the band edge of a simple RC network is determined basically by the matching accuracy of capacitors on the order of 0.1%.

A comparison of performance characteristics shows that the switched capacitor approach has advantages in response accuracy, noise sensitivity, dynamic range and process sensitivity. Due to this, switched capacitor filters are the most commonly used filters world wide.

Typically, the transmit filter section consists of a third order-RC active anti-aliasing filter with a nominal 3-dB frequency of 30 kHz to prevent distortion due to high frequencies, a second- or third-order monotonic high-pass switched capacitor filter suppressing frequencies below 300 Hz, and a fifth-order elliptic low-pass switched capacitor ladder filter. This is followed by a simple second-order smoothing filter. The receive reconstruction filter also consists of a fifth-order low-pass elliptic switched-capacitor filter which is similar to that in the transmit direction, but provides sinx/x correction by optimizing pole and zero locations.

Intel's 29C51 feature control combo is based on switched capacitor technique. In addition to the codec and transmit/receive filter, it contains programmable gain adjustment, selectable u/A-law coding, power down, and selection of one of three on-chip or two external balance networks for 2/4 wire conversion. Moreover, it has programmable gain, a loop back mode, a secondary analog input and conferencing capability. The entire control of signaling to the SLIC is handled via the device by selectable interface leads.

A codec filter approach that is quite different from conventional solutions results from a digital filtering concept and digital signal processing techniques implemented in AMD's Am 7901 subscriber line audio processing circuit (SLAC) and Siemens's PEB 2060 digital signal-processing codec filter (SICOFI) [3, 4]. The basic idea of this DSP approach is to provide simple anti-aliasing filters and A/D conversion circuitry for speech digitizing by using some sort of oversampling noise shaping coder, followed by digital signal processing to produce PCM and band limitation. Fine resolution is achieved by averaging over a great number of coarse quantizations utilizing the fact that there is a correlation between sampling rate and necessary accuracy of the A/D-conversion. The noise shaping coder produces noise in a wide frequency range, which afterwards is limited to its voice band part by digital filters. In this way, the A/D converter can provide an excellent dynamic range and the resulting signal-to-noise ratio exceeds CCITT requirements considerably.

The DSP approach potentially enables the control of the device's analog behaviour by digital signal processing in a wide range of programmability. It takes advantage of typical DSP features such as highly predictable performance, tolerance to parameter fluctuations, low crosstalk and flexibility. Furthermore, the new approach also has significant impact on device-testing philosophy. By testing the signal processor, all digital filters and the programmable functions are verified, provided the A/D converter reaches its full performance. Therefore, fast testing can be reduced to the separate test of A/D-, D/A-converters, voltage reference and the digital part. Also, system test under operation can be supported by different loop back functions implemented in the microprogram.

The functional concept of both devices, the SLAC and SICOFI illustrated in Fig. 12, is quite similar. The encoder basically consists of a low order anti-aliasing filter, a predictive oversampling A/D converter, and a digital filter section performing downsampling to the 8-kHz PCM sampling rate and PCM band limitation. For decimation to a lower sampling rate, possible foldover distortion has to be taken into account. The result is an equivalent linear digital sample of the analog signal in each 125-us frame, which is then compressed in an additional step. The decoding direction includes linearization, band limitation and interpolation, D/A conversion at a high sample rate, and reconstruction post filtering.

Incorporation of programmable filters enables software controlled adjustment of the effective termination impedance and hybrid balancing for different applications, which traditionally have been achieved by hardware trimming discrete networks. Moreover, level control and frequency response correction can be easily implemented.

The impedance matching (z) filter implements a feedback loop, which modifies the effective termination impedance of either the SLIC or transformer hybrid towards optimized return loss. The hybrid balance (b) filter is a filter path from the receive to the transmit side, which, when properly programmed, cancels the receive echo signals and thus provides a high degree of balancing. In fact, the b-filter matches the full echo loop and generates a compensation signal, which is then subtracted from the transmitted signal.

The digital filtering is performed by a microprogrammed digital signal processor having a highly optimized architecture with CSD (canonic sign digit) code arithmethic. This method provides multiplication with simple shift and add operations.

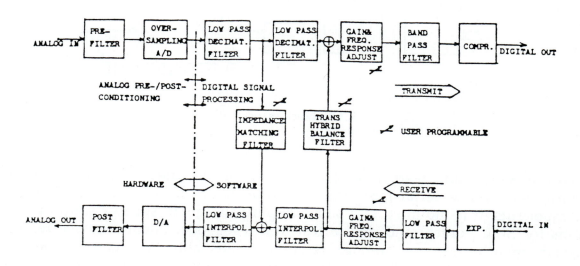

Fig. 12: SLAC/SICOFI signal flow (AMD/Siemens, respectively).

In spite of the similar functional structure used internally, the SLAC and SICOFI are completely different concerning device architecture and implementation. While the SLAC implements a 512-kHz delta modulation and separate signal processors for receive and transmit in a low voltage n-well MOS technology, the SICOFI design (Fig. 13) is based on a 2-um p-well double poly CMOS technology, optimized for telecom applications. This results in dense digital parts in spite of 10-V analog capability. The noise shaping, predictive A/D converter used provides high linearity at a sampling rate of 128-kHz resulting in a signal-to-noise ratio which does not suffer from intermodulation signals. Using an internal 4-MHz clock generated by an on chip phase locked loop from an external 512-kHz system clock, a single DSP performs all digital filtering including decimation and interpolation. To implement the recursive infinite impulse response band limitation filters, the device uses so-called wave digital filters developed in close cooperation with the University of Bochum, Germany. These filters are proved to be stable under all conditions. All filter algorithms are contained in a 10-kbit microprogram.

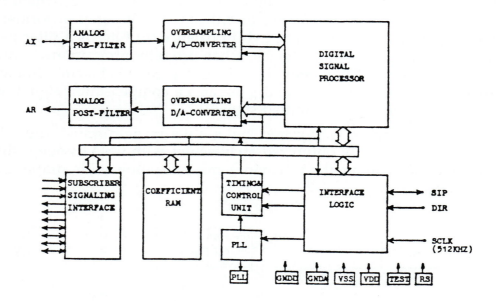

Fig. 13: SICOFI block diagram (Siemens).

Feature control and read back of coefficients are handled by a PCM interface controller via the serial interface. Since the device uses no external discrete components, it is mounted in a small 22-pin package, in spite of supporting 10 pins for SLIC control and device test in a test mode. Further device features worth mentioning are on-chip reference, selectable A/u-law, three-party conference support, hard- and software reset, supply voltage supervision, and power down mode.

3.5.3. *Subscriber line interface circuit (SLIC)*

The BORSHT functions previously described are traditionally realized with discrete components by the use of a transformer, relays, passive components and operational amplifiers. Depending on subscriber line termination resistance, battery feeding of the subscriber telephone requires voltage levels of up to and above 60 V. In the case of short lines, a high voltage must be dropped across a series on-board resistor. To avoid high power consumption, electronic current limitation or constant current feeding is to be used. Overvoltage protection has the objective of preventing damage to equipment by overvoltages, (e.g., lightning) which are reduced by protective devices to about 1000 V for 700 us. Furthermore, longitudinal common mode line voltages such as 50-Hz signals induced from sources adjacent to the line have to be suppressed. Longitudinal signal rejection is currently performed by a transformer, which simultaneously provides isolation from foreign potentials. Ringing with an ac-voltage of up to 70 Vrms and test access to the subscriber line is generally performed by interrupting the line with a relay; supervision of loop state is done by monitoring the subscriber loop current flow to detect the "off-hook" state or periodic dialing interrupts. Since the subscriber is connected to a 2-wire line, 2-wire/4-wire conversion is achieved by a hybrid which also provides high attenuation between the transmit and receive channels of the 4-wire port. The solution is basically a Wheatstone bridge, which is balanced by matching a balancing network to the line impedance. Traditionally, the hybrid is implemented by a transformer with multiple windings.

While an all silicon replacement of the interface functions discussed so far has in principle been shown to be possible, the implementation for public exchange systems suffers from the combination of extremely stringent requirements, including:

- capability of handling voltages in the range of 150 V
- high driving capability
- large on-chip power dissipation of up to 4 W
- high longitudinal balance exceeding 52 dB
- excellent reliability.

In any case, protection of low voltage electronics against high voltage transients (e.g., 1000 V, 700 us) must be provided externally, e.g., by series elements and clamp diodes placed at the tip and ring leads.

As a first generation several integrated versions of a SLIC were developed that perform some or all of the BORSHT functions for central office or PABX applications in conjunction with some external elements such as relays, power devices or transformers. In close cooperation with system manufacturers extensive work on fully electronic approaches was done by a variety of semiconductor companies such as AMD, Harris Corp., Motorola, Texas Instruments, RIFA (Sweden) and SGS-Thomson (Italy). It was shown that the optimized implementation of central switch demands can best by achieved by a two-chip approach using a high voltage bipolar technology for the realization of the high voltage functions, and standard bipolar or MOS technique to implement the low voltage parts, in conjunction with on-chip or external realization of the precision resistors. Based on its dielectrically isolated, complementary high-voltage bipolar technology and its conventional bipolar junction isolated process, for example, Harris came up with a two-chip solution which also integrates the ringing function on the chip.

The HC 5502 monolithic SLIC from Harris Corp. provides -48 V battery feed, switch hook and ground key detection, on-chip ring relay drivers, 2/4 wire conversion, and overvoltage protection with some external devices for PABX applications. This device competes with products from Motorola and others. Quite similar to Harris, Italy's SGS-Thomson presented its own relayless two-chip solution for public exchange systems which uses the company's multiwatt package [5].

The SLIC solutions introduced so far are primarily designed with standard codec filters in mind. Optimization with respect to the extended functions of the described third generation devices will reduce complexity. Moreover, integration of all low voltage functions together with the codec filter as a next step, eventually also for public exchange applications, will result in a two-chip subscriber line integrated solution.

3.6. Digital subscriber line card devices

3.6.1. Line card functions

Implementing digital transmission technology on the old telephone subscriber loops at 144-kb/s user rate as opposed to the old 4-kHz analog telephone bandwidth requires sophisticated concepts and circuit techniques.

Using an echo cancellation transmission scheme together with block codes has been proven most suitable to cover loop lengths of 6 km (18 kft) and more. Given the appropriate transceiver function, subscriber power feeding and overvoltage protection make the subscriber line interface function complete, as indicated in Fig. 14.

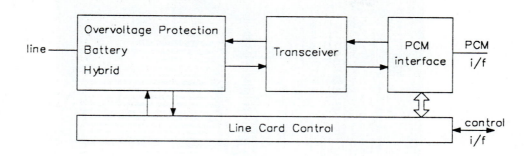

Fig. 14: Basic functions of an ISDN subscriber line card.

From the early stages of ISDN most attention has been paid to select the line code and frame structure including overhead bits for layer-1 maintenance and test facilities. Since CCITT could not find a worldwide accepted solution different standards were set. Today, however, the official American standard ANSI T1.601 has the (market) power to be adopted world wide during the nineties [11]. It uses a quaternary line code offering as much as 4-kb/s maintenance channel capacity in addition to the 144-kb/s 2B+D user bit rate while restricting line signal power spectrum below 100-kHz. For the basic technical aspects of the transmission technique please refer to the chapter "Integrated Circuits for the ISDN" by Peter F. Adams, contained in this book.

3.6.2. ISDN transceiver devices

Given the system requirements such as the ANSI specification VLSI solutions can be quite different, however. This became evident soon after the first 2B1Q transceiver samples were delivered by AT&T, Motorola, National Semiconductor, and Siemens Semiconductor. Provided the devices cover all loops required by ANSI they widely differ in die size and power consumption. This is due to different device architectures and circuit technologies as well as semiconductor technology generations.

As an example the Siemens PEB 2091 is designed in a two-micron CMOS technology (Fig. 15). Full custom DSP and analog design combined with a cell approach for layer-1 protocol and serial interface functions result in 75 mm^2 die size and less than 300 mW power consumption.

Fig. 15: Microphotograph of the Siemens PEB 2091 ANSI U-interface transceiver.

4.   Analog Telephone Sets

The conventional analog voice transmission telephone set can simply be divided into a voice module (including 2/4-wire hybrid), a dialing section and a ringing module. After electronic replacement of these functions, the trend clearly is towards high feature sets which incorporate extended intelligence and features like number storing, automatic last number redialing, repertory dialing, speakerphone, use of optical display and cordless operation. Fundamental functions to be included for the electronic circuits are polarity guard, overvoltage protection and voltage regulation. In general, low power consumption and low voltage operation with high dynamic range are main requirements.

The first step in converting subscriber set functions to electronic operation was the replacement of the rotary dial by a pushbutton keyboard operating in conjunction with either a pulse generator to originate the so-called loop disconnect dialing, or a tone generator for dual tone multifrequency (DTMF) dialing.

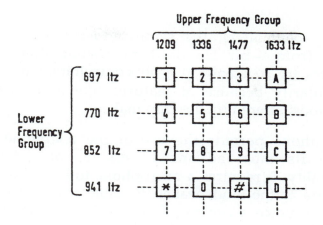

Fig. 16: Telephone pushbutton keyboard and associated DTMF dialling frequencies.

In the DTMF technique each digit is represented by a different pair of frequencies within the voice band. The various frequencies are derived from a crystal controlled oscillator, followed by a sine wave synthesizer consisting of a programmable divider for the upper and lower frequency groups, and a D/A-converter followed by smoothing filter.

Present IC solutions of the analog telephone are comprised of a set of two to five devices, depending on the manufacturer (Fig. 17). Speech circuits and tone ringer circuits in bipolar technology are available from numerous semiconductor manufacturers. Powered from the telephone line current the typical speech circuit device contains receive and transmit amplifiers and means of switching inputs, voltage regulator, gain control facilities, exchange power supply voltage corrections, microphone input, loudspeaker output drive and power supply to external circuits. These components still require quite an amount of external circuitry for gain adjustment, stabilization and side tone suppression.

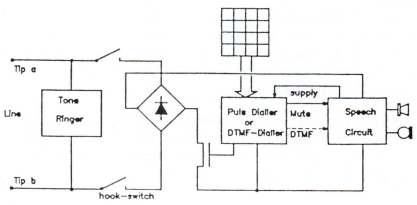

Fig. 17: Electronic telephone set.

In terms of dialing controllers, the semiconductor industry offers CMOS solutions ranging from 4-bit standard microcontrollers over dedicated telephone controllers up to standard 8-bit controllers. The latter have been gaining more popularity because of features like on-board E^2PROM and comfortable design tools offered by the chip makers.

The handsfree telephone permits the use of the telephone in the on-hook condition and, in its most advanced version, provides full two-way speakerphone capability. It requires a microphone to be connected, or mounted to the same enclosure as the loudspeaker. Possible acoustic feedback problems may be overcome by voice switching facilities implemented in an extended speech circuit. The simplest voice switching approach compares the signal from the microphone with the received signal and allows the one with the highest level to be amplified, while the other is attenuated.

In order to achieve further cost reduction in that specific mass consumer market, semiconductor suppliers are urged to introduce the next innovation step. One solution is the combination of the bipolar speech circuit and the CMOS controller onto one BiCMOS device like SGS-Thomson has done. However, an alternative all-CMOS solution could bring an even greater cost advantage. Siemens Semiconductor has worked very intensively in this area [8, 9].

Another approach to reduce system cost is the almost complete integration of any programming or trimming which leads to customized versions of standard base products. Especially in the analog area of speech and handsfree circuits sophisticated designs are required to reach that goal.

5.   ISDN Subscriber Equipment

According to CCITT's ISDN reference configuration illustrated in Fig. 18 there are different types of subscriber equipment to be connected to the ISDN. ISDN-compatible terminal equipment (TE1) is directly connected to the S or T reference point which are identical interfaces. Older, non-ISDN terminal equipment (TE2) representing voice and data terminals have to connect to the ISDN via a terminal adaptor (TA).

The conversion between the 2-wire U interface and the 4-wire S or T interface is managed by a non-intelligent network termination (NT1), which can connect directly to terminal equipment or to an intelligent network termination (NT2) called private exchange system [6, 10].

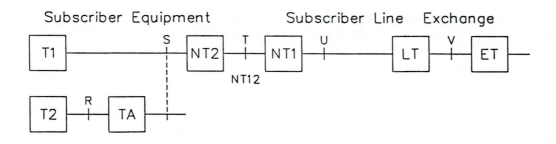

Fig. 18: CCITT's ISDN reference configuration.

5.1. ISDN terminals and terminal adaptors

As an end user equipment the terminal must cover OSI layers 1 through 7 resulting in a very complex functionality in comparison to the analog telephone, for instance.

The spectrum of ISDN specific terminal devices includes layer-1 transceivers to connect to the 4 wire S/T interface, HDLC controller implementing the major part of layer-2 signalling, dedicated ISDN codec filter for A/D conversion of PCM voice data and direct connection of handset microphone and earpiece, and various types of data rate adaption devices for the main rate adaption standards such as V.110 and V.120. A microcomputer system is used to implement layer-3 protocol, user interface functions and low level control functions.

Similar to the line card, different types of interfaces are used to connect the various IC's necessary for the ISDN TE. Again two mainstream architectures exist: the IOM^R-2 architecture supported by AMD and Siemens and the Microwire^{TM}/SCP (serial control port) architecture supported by National Semiconductor and Motorola.

Whereas the Microwire^{TM}/SCP architecture uses different interfaces for data (2B+D) and control the IOM^R-2 interface carries all on one serial interface. Also the Microwire^{TM}/SCP interface has to be supported by the microcomputer.

Due to the early introduction of the IOM architecture and chip set as well as worldwide support by many semiconductor companies such as TI, SGS-Thomson, Hitachi, Fujitsu, among others, the IOM^R-2 architecture is the most widespread approach for the ISDN terminal and offers the most comprehensive device family.

Specific attention has to be paid to IC solutions covering layer-2 control and voice codec because of their tremendous impact on system cost and performance.

5.1.1. *ISDN layer-2 terminal controller*

When the first standard ISDN LAP-D controller was sampled in 1986 by Siemens a standard was set concerning integration, cost and performance. The ISDN Communication Controller (PEB 2070, ICC) offers a complete bit oriented HDLC controller as well as advanced LAP-D protocol support. Using 64 byte FIFOs per direction to interface D channel data to the microcomputer reduces the dynamic requirements to the micro tremendously. This allows the use of cheap 8-bit controllers in ISDN terminal applications.

In the meantime a lot of other layer-2 implementations have been brought into market, ranging from low level bit-oriented HDLC controllers up to a fully integrated LAP-D controller implemented by NEC by combination of a standard CPU with on-board program ROM.

Further generations expected will probably put LAP-D controller and a standard microcomputer system onto one piece of silicon simply because at this stage reducing chip count is the most effective means to save system cost.

5.1.2. *ISDN codec filter*

At first sight the analog-to-digital conversion of PCM coded voice in the terminal can be understood as being the same function as on the analog line card, but on the subscriber end. However a closer look reveals reasons to follow another approach. Besides the core function codec and filter, extensive tone generation capabilities are required in the terminal. Ringing tones with flexible frequency and cadencing are a must in an ISDN phone from the user's point of view; dialing tones to be sent to the potentially analog communication partner are required from the network.

At the analog interface some driving capability is required to connect a handset rather than a SLIC. Last but not least implementation of handsfree operation may need some support by the codec function.

Various semiconductor suppliers therefore decided to bring special ISDN Codecs into the market, among them Siemens, SGS-Thomson, National

Semiconductor, and AMD, the latter integrating all ISDN-specific terminal function into one chip, the 79C30 DSC (digital subscriber controller) [14, 15].

Driven by ISDN market requirements, a second generation ISDN codec is being launched by AT&T and Siemens that offers cost-effective on-chip handsfree operation. While AT&T's T 7540 is still a traditional codec requiring extensive software support to realize the speakerphone, the PSB 2165 from Siemens has implemented the complete handsfree algorithm. Basically the same approach as in analog phones, volume detection and comparison of receive and transmit channels can be done very cost-effectively by additional computing steps in its internal DSP.

5.1.3. ISDN terminal adaptors

Aimed at the possibility of connecting existing DTE to the ISDN, terminal adaptors (TA) have to provide data rate adaption for any kind of subrate synchronous or asynchronous interfaces. In Europe a standard rate adaption scheme was defined by CEPT whereas in the US AT&T set an industry standard using HDLC packaging. Today CCITT recommends two rate adaption standards called V.110 (ECMA) and V.120 (US). A universal ISDN TA would then be able to connect to DTE's with synchronous and asynchronous interfaces like X.21, X.21 bis, V.25 bis, V.24, RS 232C with bit rates from 300 bit/s up to 64 kbit/s, applying either V.110 or V.120 rate adaption scheme to be able to communicate with any other DTE through the ISDN.

First generation rate adaptor devices introduced in the late eighties by companies like AMD, Siemens, Philips, and Mietec took into account these requirements. The most recent semiconductor solution in this area is the MC 68302 from Motorola.

This 68000 family based product incorporates a standard CPU (68000) together with three serial communication channels. Although it can be used in universal TA it is primarily aimed for workstations and PC cards with V.120 rate adaption [17].

However, an optimal solution for ISDN PC cards is required for the early 90's. It is up to the semiconductor industry to promote ISDN also by a highly integrated TA solution.

5.2. ISDN network terminations (NT1)

Although the functional characteristics of the NT1 are well defined administrational and political differences between Europe and the U.S. have led to different functional partitionings, as indicated in Fig. 19.

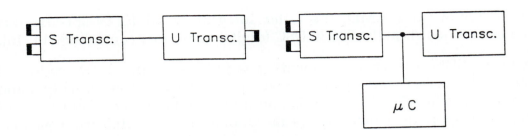

Fig. 19: U.S. (left) and European (right) NT1 Architecture.

The European Operating Companies as owner of the NT1 call for a simple layer 1 network termination to convert the 2-wire U interface to the 4-wire subscriber's S/T bus. Power feeding is provided to the terminals drawn from the mains; in emergency conditions restricted power is provided from the U line.

Since the NT1 owners in the U.S. are the residential users, additional maintenance facilities are required to allow both the owner and the network operating company to monitor functioning and performance of the NT1, each one from his network access point. A complete set of remote maintenance functions has therefore been defined in ANSI T1 standards which eventually led to an extended American S/T interface specification. The main difference in NT1 implementation turns out to be the use of an 8-bit microcontroller in the US NT1 whereas the European NT1 does not need anything like that [11, 12, 13].

Besides the little higher cost of a U.S. NT1 the pure fact that U.S. subscribers have to invest into NT1 equipment that does not offer any practical use generates a large potential market for terminals with a U interface instead of S/T interface. However, in order to be able to use the multipoint facility of the S/T bus, combined NT1/TE equipment is likely to be brought into the market (Fig. 20). Since semiconductor devices are well prepared to fit into this kind of equipment the market will develop according to the user's choice.

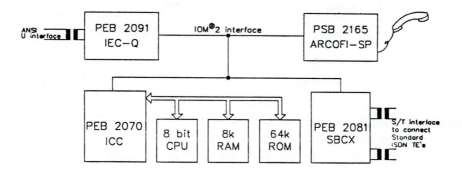

Fig. 20: Combined NT1/TE equipment based on IOMR-2 architecture.

6. DSP in Telecommunications

6.1. Introduction

The emerging use of digital signal processing (DSP) techniques in telecommunication systems has only come about with the advances in microelectronics and the development of appropriate DSP methodologies. But it should be emphasized that the feasibility of a new technology (e.g., DSP) is no argument per se for the replacement of an old one (e.g., analog signal processing). Only if manufacturing costs can be decreased or if better performance can be achieved at attractive costs will such a replacement be accepted by the system designers and equipment manufacturers.

Codecs (ref. chap. 3.5.2) were the first examples of a succesful transition from analog to digital solutions and was possible for the following reasons:

- Introduction of high perfomance and low cost A/D and D/A converters (e.g., sigma-delta).

- Advances in CMOS technology:

  * Decreasing power consumption and physical dimensions
  * Increasing speed (multitasking/channel realizations)

- Sophisticated VLSI architectures and DSP algorithms.

However the breakthrough of DSP technology in telecom application was achieved by the consequent evolution of existing systems and the definition of new fully digital end-to-end communication systems which unite a wide variety of voice and non-voice services. These new system features could only be implemented using digital signal processing techniques.

One example is ISDN in which the conventional 4kHz-twisted pair wires are used for the transmission of 144 kbit/s (2B+D) over cable lengths in the range of 4 to 8 km and with an error rate below 10^{-7}. There are a variety of solutions for the realization of such so-called U-transceivers in principle but the accepted methods are adaptive echo-cancelling in conjunction with adaptive line equalization in order to eliminate hybrid mismatches and line dispersions. The first VLSI components were available as multi-chip solutions in the early 80s but with the refinement of both DSP-algorithms and -architectures in conjunction with the advances in CMOS technologies one chip solutions have been launched in the late 80s by several Semiconductor companies such AT&T, Motorola, National semiconductor and Siemens.

A further example of the use of DSP techniques in transmission networks are modems. They enable the transmission of data via the existing telephone network with its PCM bandwidth (3.4kHz) and PCM resolution (8 bit). Like ISDN U-transceivers, high rate modems use complex adaptive echo-cancellation and line equalization techniques in conjunction with sophisticated coding principles. For example, V.32 modems operate at 9.6 kbit/s and standards are under definition that will enable data rates of 19.2 kbit/s and higher. This evolution is one of the reasons for the delayed introduction of ISDN.

Further applications for DSP are demonstrated by the extension of telecom systems by video and digital mobile communication services. The challenges here are the compression and decompression of video and speech signals in order to meet the constraints of available channel capacity (e.g., 64 or 2x64 kbit/s in ISDN) or to increase the number of transmission channels in a given radio frequency band. Furthermore, in mobile communications, channel coding/decoding and adaptive channel equalization is performed by dedicated DSP-algorithms (e.g., Viterbi).

Especially in terminals with speech and video interfaces many of the required features are only feasible using DSP techniques. Examples are:

- Voice activated dialing (voice synthesis and recognition)

- Handsfree telephoning (cancellation of acoustic echos)

- Face-to-face communication (compression and decompression of video signals)

- Voice mail (compression and decompression of speech signals for storage in RAM)

A measure of the complexity of such systems is the number of operations per second that are needed to implement the according DSP algorithm. The processing of speech signals in the aforementioned examples needs between 10 and 100 million operations per second (MOPS) whereby the processing of video signals needs about 1000 MOPS.

6.2. DSP design methodology

From a semiconductor point of view the designers can nowadays choose between a variety of different approaches, as indicated in Fig. 21. They can use programmable general purpose DSPs (GPDSP) and/or application specific ICs (ASICs). In the meantime there is a large variety of GPDSPs available on the market, with architectures ranging from high-speed, lower cost fixed-point to high-precision, higher cost floating-point devices. Although the term ASIC is commonly used to describe any non-standard IC (and what is standard ?), we define ASICs as ICs which are dedicated or optimized solutions for a given DSP-(sub-) system.

Fig. 21: Today's triangle of possible approaches.

Concerning the design methodology we differentiate between semi- and full-custom design methods, the latter generally performed by semiconductor houses whereas semi-custom designs can be carried out by both semiconductor and/or system houses.

The challenge for DSP-system designers is the choice of the appropriate approach which has been enhanced by the fact that since 1990 nearly all GPDSP vendors offer DSP-cores as macro-cells in their CAD libraries with which customers can design their own ASIC (or better CSIC, Customer Specific IC).

Although no strict differentiation is possible between the various (system-) design approaches depicted in Fig. 21, Table 1 summarizes the key features of alternative solutions.

Table 1: Comparison of different DSP design approaches.

	GPDSP	Semi-custom	Full-custom
Algorithm Development in Realtime	yes	no	no
Speed	< 10-40 MIPS	> 10-40 MIPS	"no limit"
Power Consumption	high	medium	minimum
Who can do the design	System Houses (SH)	SH/SC	Semiconductor Houses
Flexibility	high	medium	worse
Time-to-Market	1 month	1 year	> 1 year
System Integration	no (yes with core concept)	restricted	yes
Volume for cost advantage	< 10k	< 100k	> 200k (units)

7.   Conclusion

The revolutionary change from analog to digital communication is based on well-defined architectures and state-of-the-art technology. However, the digitalization of subscriber equipment is expected to last until the end of this century. This is due to the big share of "voice-only" users within the telecommunication community which have to be convinced by service, feature or cost advantages of digital over analog telephony.

Eventually it will be all together: flexible audio and video communication services for a price sensitive mass consumer market and comprehensive data communication facilities for the professional down to the laptop level. The technologies required are available. The success will depend to a high degree on the cooperation between system houses, semiconductor manufacturers, operating companies and international standardization organizations.

8.   References

[1]   H. Abramowicz and A. Lindberg, "OSI for telecommunications applications," *Ericsson Review*, Vol. 66, No. 1, pp. 2-12, 1989.

[2]   L. Lerach, G. Geiger and M. Strafner, "Peripheral board controller for digital exchange systems," *ISSCC*, New York, February 1983, p. 76.

[3] R. Apfel, H. Ibrahim and R. Ruebush: "Signalprocessing chips enrich telephone line card architecture," *Electronics*, Vol. 5, pp. 113-118, May 1982.

[4] D. Vogel and W. Pribyl, "CMOS digital signalprocessing codec-filter with high performance and flexibility," *ESSCIRC*, Edinburgh, Sept. 1984.

[5] L. Lerac,h "LSI/VLSI for Telephony," in *Design of MOS-VLSI Circuits for Telecommunications*. Prentice Hall, Inc., N,J,.

[6] G. Geiger and L. Lerach, "ISDN-oriented modular VLSI chip set for central office and PABX Applications," *IEEE J. Sel. Areas in Comm.*, Vol. SAC-4, pp. 1268-1274, Nov. 1986.

[7] G. Geiger and G. Weinberger, "Integrated circuits for ISDN - status and future," *IEEE Conference on Telecommunications*, York, UK, pp. 190-195, Apr. 1989.

[8] F. Marti, M. Robbe and J. Le Corre, "A single chip BIMOS telephone set," *ISSCC*, pp. 254-255, February 1989.

[9] F. Dielacher, A. Wurmitzer et. al., "A programmable CMOS telephone circuit", *ESSCIRC*, 1990.

[10] *CCITT Blue Book*; I series recommendations.

[11] American National Standards Institute, Inc., *ISDN-Basic Access Interface for Use on Metallic Loops for Application on the Network Side of the NT (Layer 1 Specification)*, T1.601, 1990.

[12] American National Standards Institute, Inc., *ISDN-Basic Access Interface for S and T Reference Points (Layer 1 Specification)*, T1.605, 1990.

[13] European Telecommunications Standards Institute, *ISDN-Basic user-network interface. Layer 1 specification and test principles*, ETS 300-012, 1990.

[14] Advanced Micro Devices, *ISDN Data Book*, 1989.

[15] Siemens Semiconductor, *ICs for Communications*, Data Book, 1989.

[16] Siemens Semiconductor, *ICs for Communications*, IOM[R]-2 Interface Reference Guide.

[17] Motorola Inc., *MC 68302, Integrated Multiprotocol Processor User's Manual*, 1990.

Chapter 13

Integrated Circuits for the ISDN

Peter F. Adams

British Telecom Laboratories, Martlesham Heath, Ipswich, Suffolk, England

1. Introduction

Integrated services digital networks (ISDNs) are a logical progression from the integrated digital networks (IDNs) that provided modern telephony communications. IDNs are digital as far as subscriber's line interface circuits in the central office. By extending digital transmission out to subscribers to support 64 kbit/s channels and a powerful access signalling system, a variety of services can be provided to a common subscriber interface — the essence of an ISDN.

The cost-effective wide-spread deployment of ISDNs relies heavily on the development of low-cost complex integrated circuits (ICs), especially for providing digital subscriber loops (DSLs) over existing networks of copper pair cables. This chapter is concerned with the ICs required for ISDNs; the circuits required for IDNs which, naturally, are also required for ISDNs are dealt with in another chapter.

2. ISDN Fundamentals

The functionality, interfaces and protocols for ISDNs are defined in the I Series of CCITT Recommendations. Fig. 1 illustrates the fundamental parts of the access portion of an ISDN. A number of reference points are defined — R, S, T, and U as shown. In many countries the T reference point marks the boundary between the public telecommunications network and the subscriber's equipment. Therefore, an interface is defined at that point to allow a competitive market in terminal equipment (TE) while ensuring interworking with the network. The interface allows a variety of wiring configurations for point-to-point or bus working. Exactly the same interface is defined for the S reference point so it is often referred to as the S/T-bus.

The regulatory environment for telecommunications in the USA has required the definition of a 2-wire subscriber interface at the U reference point. To allow interworking between line terminations (LTs) and network terminations (NTs) (which, because they can be supplied competitively, may be obtained from a variety of different manufacturers from the supplier of the LT) the interface specification must precisely define the format of all digital signals passing across it and their electrical parameters.

The R reference point is included to take account of the fact that there are a great many terminals designed to work to existing interfaces, and subscribers may want to connect them to an ISDN. A terminal adaptor (TA) is required to convert existing interfaces to the S/T-bus. In the longer term the need for terminal adaptors will disappear as existing terminal equipments (TE2s) are replaced by new terminal equipments (TE1s) designed specifically for working on ISDNs. The NT2 allows the implementation of a distribution function if required, e.g., a switch.

ISDN access is defined to support primary rate access of 30 (or 23) 64 kbit/s channels plus a 64-kbit/s channel for signalling and packet data, or basic rate access of two 64-kbit/s channels plus a 16-kbit/s signalling or packet data channel. The 64-kbit/s channels are called "B-channels" and the signalling/packet data channels "D-channels." Primary rate access is provided by a similar technology to that used to support such rates in IDNs and usually requires specially provided cables; consequently it will not be considered here. The widespread deployment of ISDNs depends critically on the ability of ICs to exploit installed copper pair networks for low-cost basic rate access.

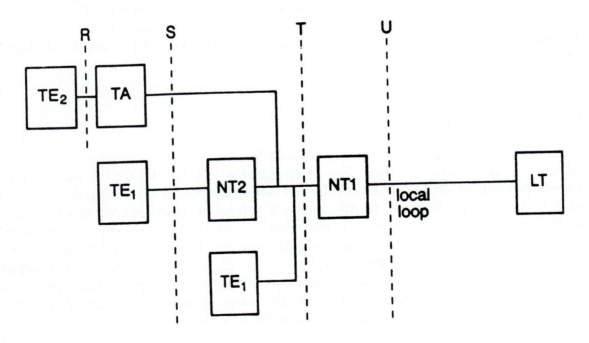

Fig. 1: CCITT ISDN subscriber access terminology.

2.1. ISDN access functional blocks

The functionality required for ISDN access could be provided in a variety of different ways by ICs. A breakdown of the main functional blocks is illustrated in Fig. 2.

Starting at the central office a D-channel controller is required. This terminates and controls the D-channel signalling known as the link access protocol for the D-channel (LAPD), which is an HDLC based protocol. The exact position of this function will vary depending on the internal architecture of the central office and at what point it is required to convert the D-channel signalling into the internal signalling format used in the central office. On the line card in the central office is a DSL transceiver which communicates with a similar device in the NT1 at the far end of the subscriber loop. Also in the NT1 is an S/T-bus transceiver. This interworks with similar transceivers on the T-bus. Finally, in the terminal equipment, is a D-channel controller. The D-channel controller and the S/T-bus transceiver interwork with the rest of the TE which contains application-specific functions. If an NT2 function is included then on either side of it are S/T-bus transceivers, working to the T-interface and S-interface, respectively.

The exact mapping of these functions on to ICs varies from manufacturer to manufacturer. For convenience it will be assumed here that the functions map on to specific devices, with deviations from this mentioned where appropriate.

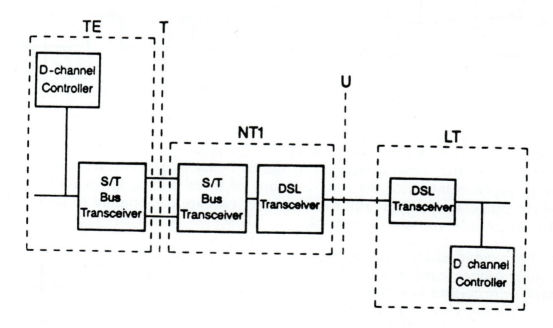

Fig. 2: ISDN subscriber access functional blocks.

3. S/T-Bus Transceivers

The S/T-bus is a 4-wire bus operating at a bit rate of 192 kbit/s in each direction. It carries the 144 kbit/s required for basic rate access with a further 48 kbit/s used for control, framing and DC balancing. Alternate mark inversion (AMI) line code, pulse amplitude modulation (PAM) is used to allow transformer coupling without DC wander occurring. A detailed explanation of the frame structure of the S/T-bus given in reference [1]. The functions of an S/T-bus is transceiver are illustrated in Fig. 3.

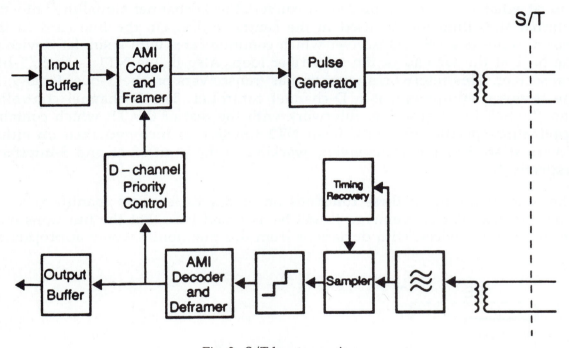

Fig. 3: S/T-bus transceiver.

On the transmit side the incoming 144-kbit/s data stream is buffered and the frame formed for transmission on to the S/T-bus. A transmitter converts the ternary representation of the data into one of three pulses, i.e., a null pulse, a positive pulse, or a negative pulse. The latter two are constrained to lie within a pulse amplitude voltage/time mask.

On the receive side, after appropriate filtering and sampling, a three-level decision device converts the line signal into a ternary code from which the binary information is recovered. When the transceiver is in a TE it also has to have a timing recovery circuit that recovers bit timing from the line signal to control the sampler. The limited transmission range of the S/T-bus means that there is no need for any reach dependent equalization.

When the transceiver is located in an NT1 the received D-channel data are echoed into a 16-kbit/s channel in the transmit data stream. This is to enable the TE to know that it has successfully communicated with the NT1. In T-bus configurations using multiple terminals contention for the D-channel occurs. TEs wait until the echo channel contains its idle pattern (continuous "1"s) before sending D-channel data. If a sending TE fails to see the correct echo it assumes that another TE has attempted to transmit D-channel data simultaneously, and so it stops sending and waits to see a number of idle state "1"s again in the echo channel before retrying. The number of "1"s it waits to see are determined by a priority mechanism that is designed to reduce the priority of terminals that have previously successfully sent an HDLC frame. This function requires a D-channel priority controller. The priority controller communicates to the D-channel controller the state of correspondence of the D-channel data and the echo channel data and the presence of a valid idle pattern. In return its priority control is set by the D-channel controller.

The S/T-bus transceiver is not particularly challenging for IC implementation, the main design issues being achieving the required pulse shaping and timing recovery circuit performance. A typical IC [2] has the following characteristics when implemented in a 2-micron CMOS technology:

- 10-k transistors
- 15-sq mm die size
- 7.68-MHz clock rate
- 60-mW power consumption

Such a modest die size allows a combined architecture where most of the D-channel controller functionality is also included on the S/T-bus transceiver IC. One such design [3] results in approximately double the die size in a 2-micron process, but with approximately half the clock rate and two-thirds the power consumption. This approach allows a reduced IC count in TEs, albeit the D-channel controller function is redundant in the NT1.

4. D-Channel Controllers

A typical D-channel controller is illustrated in Fig. 4. It is required to both capture and process incoming D-channel data and prepare outgoing D-channel data for insertion into the D-channel. D-channel data arrive on the serial interface which has a throughput rate of at least 16 kbit/s (64 kbit/s for primary rate access). An HDLC receiver processes the incoming information and an HDLC transmitter prepares outgoing information before it is transmitted via the serial interface. first-in-first-out (FIFO) registers provide data buffers for both transmit and receive paths. This function is managed by a controller and an interface enables interworking with a microprocessor.

Generally, D-channel controllers implement all the Layer-1 functions to support LAPD and some of its Layer-2 functions; the remainder are implemented by the microprocessor.

D-channel controllers are digital ICs which are straightforward to implement. To enable reduced IC counts in line card implementation dual controllers have been developed [4].

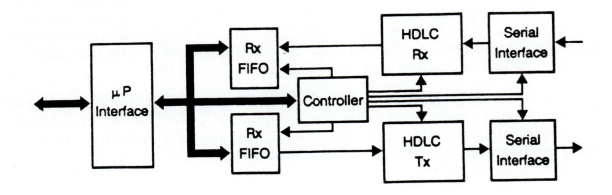

Fig. 4: D-Channel controller.

5. Digital Subscriber Loop Transceivers

In terms of functionality and design by far the most challenging of the ICs required for ISDN basic rate access are DSL transceivers. They are required to have the following capabilities:

- 2-wire duplex working,

- 100% loop coverage with acceptable error performance,

- 144-kbit/s transparent data rate,

- bit and frame timing recovery,

- automatic start-up on demand from the network,

- low power consumption to allow line powering if required, and

- maintenance features to allow fault detection and diagnosis.

Generally copper pair local loop plant is dimensioned to allow enough pairs to meet forecasted demand for telephone lines. If ISDNs are to be as ubiquitous as existing telephony networks then DSLs must operate over a single pair. Thus 2-wire duplex transmission to provide communication to and from the network is required.

For unrestricted application, DSL transceivers must allow as close to 100% coverage of local loop populations as possible. If the cost of installation is to be low then this must be achieved without special engineering in the outside plant. Thus, transceivers must be designed which deliver the required performance over extremes of the loop plant and transmission impairments.

Basic rate access requires the transport of 144 kbit/s in a transparent manner; i.e. there should be no constraints on the data transported. Any capacity required for control or other purposes must be provided additionally.

The transceiver in the LT is slaved to the network clock and to the frame marker that allows the individual channels to be distinguished from each other. In the NTI both clock and frame timing must be recovered from the signal on the pair.

As we will see later transceivers contain various signal processing functions that have to adapt to the characteristics of the loop plant when the DSL is started up. Start-up must be automatic, but under control of the network. There are two types of start-up: cold and warm. A cold start is required when the DSL is first installed, or when it is restarted again after normal operation has failed. A warm start refers to a fast re-establishing of normal operation after a cold start and subsequent deactivation. A cold start can take tens of seconds if necessary because speed is not important on initial start-up. In some countries the DSL is kept activated even when access is not required so only a cold start is necessary. A warm start, however, is used in countries where the DSL is deactivated when not being used by the subscriber, and it is important to re-establish access quickly when a call is made by, or to, the subscriber.

The choice of deactivating DSLs when service is not required is linked to the issue of powering of the NT1. In some countries it is intended to power the NT1 locally so power is not fed from the central office apart from that required to provide sealing current for the copper pair. In other countries line powering is used, especially where there is a requirement to power an emergency telephone. The amount of power available at the end of a long loop for a safe line feed voltage is not very large, typically of the order of one Watt. If power is provided by the central office then even this amount of power adds up to a great increase in the central office power demand when the number of ISDN lines is large. Deactivating loops when not in use can significantly reduce the average power demand. In addition deactivation can reduce the average pollution of the local loop plant with crosstalk coupled noise.

The available power at the end of a long loop affects the design of DSLs. Allowing power to feed an emergency telephone and the rest of the NT1 only a few hundred milliwatts is left for the DSL transceiver. Even in CMOS technology this is a significant constraint which places a limit on the amount of signal processing that the DSL transceiver can perform. It turns out, however, that it is possible to design transceivers that meet all the other requirements within this power budget.

The presence of an NT1 at the end of the loop allows, unlike on analogue loops, the possibility of the provision of maintenance functionality. This can be to monitor the performance of the loop (e.g., by error detection capability,) or provide diagnostic capability (e.g., by loop-backs.) The degree to which such capability is provided is dependent on the way in which operating companies intend to organise the maintenance of ISDN service. The provision of maintenance functionality in the NT1 implies that there must be extra capacity added to the DSL to convey maintenance instructions and information; this will, however, require only a modest bit rate.

5.1. Key loop characteristics

Local loops were designed to support analogue telephony which requires approximately a 4-kHz bandwidth. Fortuitously, they have sufficient useful bandwidth to allow transmission at ISDN basic rate over almost 100% of installed loops. Loops vary widely in length and make-up with several gauges of twisted pair concatenated in each connection. An additional problem in some countries is the use of bridge taps to provide network flexibility. The key loop characteristics that affect the performance of 2-wire local loop transceivers are:

- pulse attenuation
- crosstalk
- characteristic impedance
- dispersion

Transmission is usually by PAM. The attenuation of pulses ultimately determines the distance over which a transmission system will function. The pulse peak attenuation (in decibels) introduced by various gauges of copper pair after appropriate equalization varies approximately linearly with distance and increases with decreasing conductor diameter. The pulse attenuation also decreases with increasing pulse width suggesting that the signalling rate of the DSL should be kept as low as possible [5].

The main interference experienced by basic rate access systems is near end crosstalk (NEXT) from other basic rate systems operating in the same cable. Extensive measurements on cables have established that the NEXT power sum attenuation follows an $f^{1.5}$ law, where f is frequency.

The characteristic impedance of twisted pairs varies dramatically below about 100 kHz which is the band of interest for DSLs. It also varies for different cable gauges. The importance of this variability is that it makes it very difficult to achieve 2-wire duplex working by the use of simple hybrid balance impedances.

Dispersion of pulses occurs because of the rising attenuation with frequency characteristics of twisted pairs. The phase characteristics of cables is mostly linear with a small non-linear region at low frequencies. The combined effect is to produce dispersion that predominantly trails the peak of the pulse. Dispersion gives rise to inter-symbol interference (ISI) between successive symbols (modulated pulses). A typical pulse response of a pair is shown in Fig. 5.

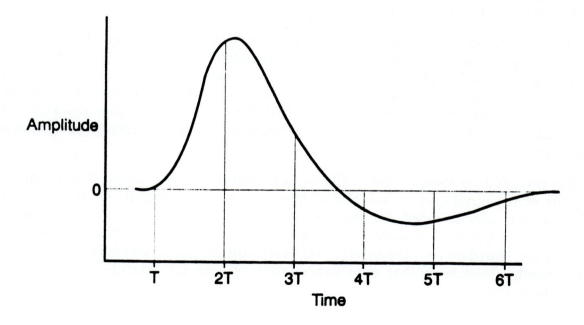

Fig. 5: Typical pulse response.

The response is that obtained by sending a dipulse, i.e., two contiguous pulses of opposite polarity. Such a response is representative of that obtained after suitable transceiver filtering. The key points to note are that the rise time of the pulse is only slightly greater than T, the pulse is balanced (i.e., integration over all time gives zero), and there is a decaying tail. As the tail is composed of mainly low frequency energy and the impedance of the pair is high at low frequencies, the decay rate of the tail is determined mainly by the inductance of the line coupling transformer in parallel with the transceiver output impedance. For high inductance transformers (say above 100 mH) the tail starts at a low amplitude but decays very slowly; low inductance transformers (say less than 10 mH) give a larger amplitude tail that decays rapidly.

5.2. 2-wire duplex operation

As mentioned above it is not possible to gain sufficient trans-hybrid loss to enable 2-wire duplex transmission. There are three techniques that have been considered to circumvent this problem: frequency division, burst mode and echo cancellation.

Frequency division simply uses different frequency bands for the two directions of transmission. The rising attenuation with increased frequency makes one direction of transmission much more susceptible to interference. Consequently, the technique has not found favour.

Burst mode uses time separation to prevent echo interference. It is a very simple technique to implement and a number of commercially available transceivers use it. However, it requires that the instantaneous bit rate for transmission in each direction is more than doubled. The increased pulse attenuation and noise bandwidth mean that the reach of burst mode systems is limited. Their main use has been in providing digital extensions for private switches.

Echo cancellation uses adaptive filters to model the echo path so that a replica of the echo signal can be subtracted from the received signal. Thus, the same bandwidth can be used for each direction of transmission simultaneously, resulting in the best potential performance, albeit at the cost of complex echo canceller hardware. However, IC technology has provided low-cost implementations such that echo cancellation is now the established technique for 2-wire DSLs.

5.3. Choice of line code

A critical choice in the design of a DSL transceiver is that of line code and modulation rate. It is well established [5] that for the longest reach in the NEXT limited environment of the local loop it is best to use low-redundancy multi-level baseband codes with a modulation rate as low as possible. The reason for this is that the reduced pulse attenuation on long loops and the smaller noise bandwidth outweigh the power penalty of multi-level codes. Therefore, modern transceivers have been designed using line codes such as 4B3T, 3B2T and 2B1Q. The modulation rates associated with these line codes are usually 120, 108 and 80 kBaud, respectively, each allowing an overhead for frame alignment and maintenance and control bits. The NEXT margins (the amounts in dBs by which a standard NEXT noise source can be amplified before an error rate of 1 in 10^7 occurs) achieved by these codes are illustrated in Fig. 6 compared to AMI, another popular choice for earlier transceiver designs. In the USA where the regulatory situation has forced the definition of an interface at the U reference point, 2B1Q has been chosen for the American National Standards Institute (ANSI) standard [6].

The choice of line code also has a significant effect on the signal processing required to perform echo cancellation, equalization and timing recovery as will be discussed later.

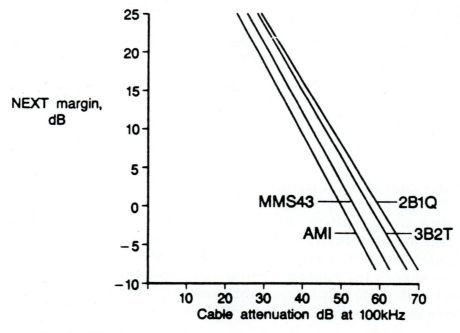

Fig. 6: NEXT margin of linearly equalized 144 kbit/s transmission.

5.4. Transceiver architecture

There are significant variations in the architectures of DSL transceivers. These mainly relate to the different types of echo canceller and equalizer that are used, different filter designs and the choice of timing recovery scheme. The choice of Digital Signal Processing (DSP) or analogue implementation also affects the architecture, especially in the positioning of an analogue-to-digital converter

(ADC), or digital-to-analogue converters (DACs). These variations will be described where appropriate. A general transceiver architecture, which encompasses most options, is shown in Fig. 7.

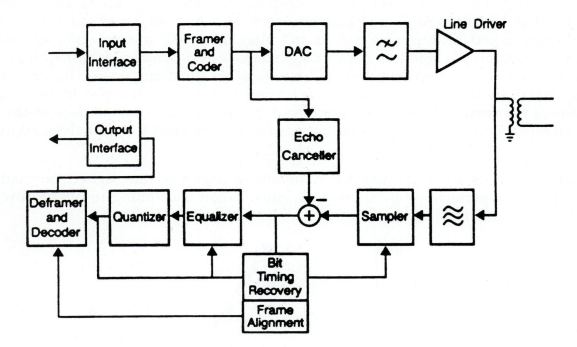

Fig. 7: DSL transceiver architecture.

The transmitter consists of a framer and coder, a DAC, a transmit filter and a line driver. The framer takes in the 144-kbit/s data and formats into a transmit frame structure, along with frame alignment information and additional bits for control and maintenance. The coder converts groups of bits into the words of the line code. The DAC converter line code symbols into voltage levels. The transmit filter shapes the transmit pulse as desired and provides attenuation against unwanted high-frequency components. The line driver provides the power to drive the twisted pair line, usually via a transformer.

On the receive side the signal from the twisted pair is low-pass filtered to remove unwanted out-of-band interference and sampled at least once per baud interval. Echo is removed by substracting off the output of the echo canceller. The signal is equalized and a quantiser decides which received symbol is present. A decoder and deframer provide complementary functionality to the coder and framer, respectively. Bit and frame timing recovery processes establish the correct phase for the sampling to take place and correct frame alignment for the deframer, respectively.

It is implicit in Fig. 7 that analogue-to-digital conversion is performed at the quantiser and the point of digital-to-analogue conversion in the echo canceller is left unspecified. However, in most transceiver designs DSP is used to implement the echo canceller and equalization functions, and sometimes the receive filtering. This requires an ADC at the front end of the transceiver. The exact balance of analogue and digital processing is a matter of debate and depends on the transceiver architecture.

5.5. Transmitter implementation

The coder and framer functions are usually straightforward digital processes to implement. The DAC is somewhat more difficult as there are, dependent on the type of echo canceller used, some stringent requirements on linearity. The linearity requirements are determined by the level of echo cancellation which is required to be sufficiently small compared to the received signal level. The use of non-linearity correction in the echo canceller may allow high levels of specific forms of non-linearity to exist. An M-level non-return-to-zero signal potentially has up to M^2 different pulse shapes if the M levels and the transitions between adjacent elements are non-ideal. These variations in pulse shape result in non-linearity, making it more difficult to obtain a given level of linear distortion than with a binary signal. In practice symmetries in tranmitter construction reduce the number of independent pulse shapes to less than M^2. Also, the element-to-element variation does not exist if return-to-zero pulses are used. It is possible to generate a signal with less than 2^b levels as the sum of b binary sources with their amplitudes in the ratios 2^{b-1} . . . 2:1. As accurately ratioed components are easily implemented in silicon technology it is, therefore, only slightly more difficult to construct an M-level transmitter, when M is 3 or 4, than a binary transmitter.

The Transmit Filter

The transmit filter is required to suppress out-of-band energy, partly to avoid unnecessary interference into the twisted pair, but also so that the line driver does not have to deliver energy into the pair that is not required. In order that the performance of transceiver is consistent it is necessary for the transmitted pulse shape to be accurately controlled. It is quite difficult to control the absolute accuracy of components in silicon technology without special trimming and so continuous filter designs are not appropriate. Instead switched capacitor or digital filter designs are employed for pulse shaping with a final non-critical low-order anti-aliasing continuous filter before the line driver. In these designs the question of linearity control is more complex but the elementary discussion above is still relevant.

Line Driver

The peak current required to be delivered to the line transformer and the tolerable level of non-linear distortion are key parameters affecting line driver design. Normally the power of the signal at the input to the copper pair is constrained to prevent unacceptable crosstalk interference into other pairs. The peak current will, therefore, be determined by the peak to rms ratio of the line code. The permissible level of non-linear distortion is proportional to the pulse attenuation which is related to the signalling rate. An M-level line code with zero redundancy has a peak to rms ratio which is the square root of $3(M-1)^2/(M^2-1)$. Thus, the peak currents for 2-level, 3-level, and 4-level line codes are in the approximate ratios 1:1.24:1.35. On long copper pairs, however, the reduction in pulse gain with increasing M outweighs this increase and so generally a slightly smaller peak current to rms distortion ratio is required for multi-level codes.

5.6. Receiver implementation

As mentioned earlier most receivers perform the complex signal processing for the receiver functions digitally. The exact position of the boundary between analogue and digital affects the type of conversion required and its accuracy. Most modern designs opt for the use of a front-end ADC with varying degrees of filtering in the analogue domain. The design of such an ADC is one of the key issues in transceiver implementation.

Analogue-to-Digital Converter

The ADC is a challenging design problem. Better than 70 dB signal-to-noise ratio is required with bandwidths up to 160 kHz or more, depending on the signalling rate and whether fractional tap adaptive equalization (see next section) is used in the transceiver. The signal-to-noise ratio required does not vary significantly with different line codes; however, the reduced signalling rate associated with multi-level codes lowers the bandwidth and so simplifies ADC design.

The type of ADC used depends more on the design experience of the manufacturer than on any inherent technical features. However, 1-bit over-sampled coders have achieved a degree of dominance. They are simple to realise and the digital decimating filter required to filter the out-of-band quantisation noise can also provide accurate pulse shaping and fixed equalization.

Equalization

The most general equalization scheme considered for use in DSL transceivers, which is shown in Fig. 8, consists of a decision feedback equalizer (DFE) preceded by a fractional-tap (FT) linear equalizer, both being adaptive. Adaptation is usually by the least mean squares (LMS) algorithm which will be described later for echo cancellers. The received pulses are sampled at m times the baud rate, where m/n is greater than twice the highest frequency of the line signal, n is the spacing between the taps on the linear equalizer, and m and n are integers. The FT linear equalizer produces an output at each baud interval. For simplicity usually m=2 and n=1. The linear equalizer reduces pre-cursor ISI, and the pulse tail outside the span of the DFE, while the DFE cancels post-cursor ISI. The linear equalizer is capable of equalizing the loop pulse response on its own (provided the line code spectrum has a null at zero frequency), but the use of a DFE reduces the noise enhancement caused by the linear equalizer attempting to equalize amplitude loss in the pair frequency response. In addition the FT equalizer can, at the same time, provide optimal filtering of the noise present with the received signal. A further advantage of the FT equalizer is that the sampling process at the front end of the receiver can, if the equalizer has enough taps, be at any arbitrary stable phase provided any frequency difference between the transceivers at each end of the loop is tracked.

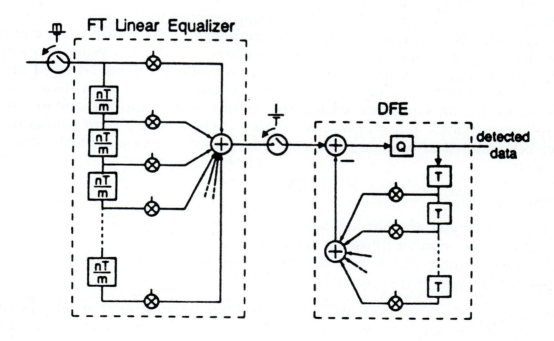

Fig. 8: General equalizer.

The advantages of the FT equalizer, however, are obtained at a high cost in signal processing resulting from multiple taps per baud interval and the higher front-end sampling rate. Consequently, most current designs for DSL transceivers which incorporate a linear equalizer use equalizers where m=1 — the T-spaced linear equalizer. The T-spaced equalizer cannot provide optimal noise filtering and it requires the front-end sampling to be close to an optimum if excessive noise enhancement is to be avoided. However, the equalizer still deals with ISI that falls outside the span of the DFE. The cost in performance terms is small, typically 1-2 dB in signal-to-noise ratio. A further simplification is to constrain the adaptive T-spaced equalizer to deal only with pre-cursor ISI using either a longer DFE or some fixed high-pass filtering (at a small cost in noise enhancement) to reduce the pulse tail.

The DFE and linear equalizer are both transversal filters. However, the DFE, because it is driven by data symbols, does not require any complex multiplications. The linear equalizer, however, involves multiplications of tap values by received signal samples and so is more complex to implement digitally.

With a further small loss of performance (typically of the order of 1 dB in signal-to-noise ratio) it is possible to sample the received pulses just prior to their peak such that the pre-cursor ISI is negligible. Combined with judicious high-pass filtering to reduce the low frequency pulse tail an adaptive linear equalizer is not needed. An example of what happens to the pulse response in this approach is illustrated in Fig. 5; the response in the figure is exactly the same as would be obtained from a system where the transmit pulse is a single pulse rather than a dipulse and the receiver contained a fixed T-spaced linear equalizer with two unity taps. A possible enhancement is to use a simple fixed pre-cursor linear equalizer with easy to implement top values to allow sampling nearer the pulse peak. Such designs, where only a pure DFE is used, have been shown to provide sufficient performance to give acceptable network coverage.

Timing Recovery

Timing recovery is another key issue in transceiver design. The major choices to be made are whether to use an analogue or a digital phase locked loop (PLL) and which timing phase estimation scheme to use. The choice of PLL is important because of the issue of jitter control.

Echo cancellers are very sensitive to clock jitter because jitter results in time variation of the sampled echo response. Careful design is necessary to avoid the timing recovery scheme degrading performance. An analogue PLL tracks timing by continuous phase adjustments. These will significantly affect echo canceller performance unless a high Q design is used [7].

A digital PLL makes discrete phase steps which, if their ratio to the baud interval is not small enough, will significantly degrade the echo canceller performance. For basic rate access transceivers it is usually not practicable to make the phase steps small enough. There are a number of techniques to prevent the resulting degradation in performance which are covered later.

There are two types of timing phase estimation: decision directed, or those that exploit a cyclostationary property of the received signal. The latter give a timing phase that is not necessarily an optimum one for all equalization schemes and so often require the equalization scheme to include some degree of pre-cursor equalization for good performance over all loops. A non-linear process such as full wave rectification on the received signal followed by a resonant circuit is a conventional example. Another method uses a matched filter to detect a block synchronisation word repeated at regular intervals. This can also provide frame timing if the word is repeated at the frame interval. The Deutsche Bundespost specification for ISDN basic rate access details such a scheme. The synchronisation word is the 11-element Barker word,

$$+2,+2,+2,-2,-2,-2,+2,-2,-2,+2,-2,$$

which repeats at the 1-ms interval of the frame period.

Similarly the ANSI 2B1Q standard specifies the 9-quaternary-element word

$$+3,+3,-3,-3,-3,+3,-3,+3,+3,$$

repeated every 1.5 ms. However, the shorter duration in baud intervals and the less frequent repetition rate mean that it is not so effective for timing recovery.

Both words have two properties that make them suitable for baud interval timing extraction: their autocorrelation functions have a peak at zero shift; and all the non-zero shift terms in the autocorrelation function have a small modulus value compared to the peak. These properties mean that when the received signal is filtered by a filter matched to the synchronisation word the output always peaks at a specific time coincident with the end of the synchronisation word and near the pulse peak of the channel response. When differentiated, such a signal produces regular zero crossings which can be used to drive a PLL.

Exploiting a cyclostationary property of the received signal allows timing recovery to be established and maintained without first equalizing the received signal or detecting the received data. Decision directed schemes [8] derive a phase error signal which is a function of the detected data and the received signal and so appear to require the prior detection and, therefore, equalization of the received signal. However, in practice, adaptation of the equalizer at the same time as finding the correct sampling phase is possible. A decision directed timing recovery scheme can give an optimum timing phase because the phase error signal can be designed to be at a minimum at the optimum phase. Like the traditional non-linearity cyclostationary schemes, decision directed schemes have the advantage that all the received signal is used for timing phase estimation. Therefore, for a given frequency offset tracking rate, they have a better noise induced jitter performance than schemes which use only a block synchronisation word.

5.7. Echo canceller implementation

Echo cancellers for DSL transceivers are usually implemented as data-driven, finite impulse response (FIR), linear filters as shown in Fig. 9. These have the advantage of inherent stability and, as for the DFE, the multiplication of long wordlength numbers is avoided.

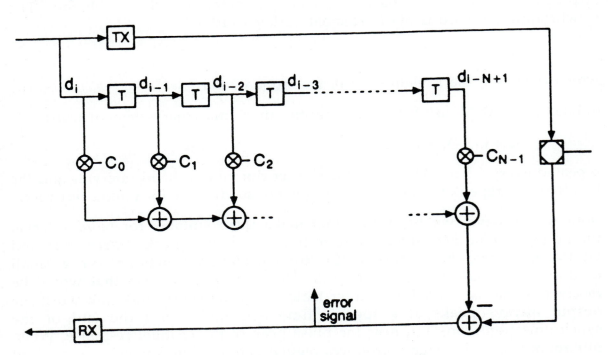

Fig. 9: Data-driven FIR linear echo canceller.

Another implementation of the FIR structure is the table look-up echo canceller. Table loop-up echo cancellers, as illustrated in Fig. 10, have been used [9] successfully in transceivers with self-equalizing line codes which produce a much shorter significant echo response. Their advantage is that, because their output is a completely general function of the data inputs, they can model non-linearity in the transmit path making realization of the analogue interface functions easier. The echo span of baseband line codes makes the use of table look-up echo cancellers too complex, especially for M-level codes, where every additional baud interval span of the echo canceller multiplies the required memory size by M. However, much of the advantage of table look-up techniques can be obtained by judicious use of less complex non-linearity correction techniques. Non-linearity correction can be achieved either by the addition of extra taps [10] that individually model each term in a Volterra expansion of the echo response, or by the systematic use of table look-up techniques [11] to model arbitrary non-linearities embedded in a structured way in an otherwise linear channel.

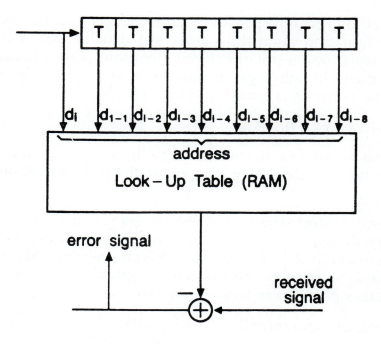

Fig. 10: Look-up table echo canceller.

Like the linear equalizer both these echo cancellers can occur in a T-spaced or FT form. In the FT form the receiver clock and the transmitter clock can be independent of each other because with a suitable interpolation filter between the echo canceller subtraction point and the rest of the receiver the receive signal can be reconstructed after the echo is removed and resampled at any desired phase or frequency. This has some advantages in timing recovery in the receiver and makes start-up of the transceiver easier. However, it is strictly unnecessary and significantly increases the amount of storage and processing required.

A key parameter for each of the cancellers is the number of taps required to give the required level of echo reduction. Transformer terminated local loops have infinitely long echo impulse responses that decay to insignificant levels only after many tens of baud intervals. To avoid having to have echo cancellers with a great many taps it is possible to use recursive, or pseudo-recursive echo cancellers. However, these usually require multiplication of pairs of long wordlength numbers which may make them unattractive. An alternative technique, as with pulse tail reduction, is to include a high-pass filter in the receiver filtering. This removes the tail energy resulting in a shorter significant echo response. However, a highpass filter will degrade crosstalk performance and so high-pass filtering has to be used carefully.

Echo Canceller Adaptation

The variability of loop echo responses requires that echo cancellers are adaptive. A variety of adaptation algorithms exist but usually DSL transceivers use variants of the least means squares (LMS) algorithm. This algorithm is simple, robust and adequate in convergence rate. In its standard form it updates the coefficients of the linear echo canceller at time i by the recursion:

$$\underline{C}(i+1) = \underline{C}(i) + k.\underline{D}(i).e(i),$$

where $\underline{C}(i)$ is the vector of echo canceller coefficients, $\underline{D}(i)$ is the vector of data values in the echo canceller delay line, $e(i)$ is the error signal and k is a gain constant.

The gain constant controls the rate of convergence. If the error signal contains the received signal then k has to be small if the received signal is not to perturb the coefficients excessively. In some transceivers the error signal is the same one used to adapt the equalizer so that joint adaptation of the equalizer and the echo canceller takes place; k can be larger in this case.

For the table look-up echo canceller the adaptation algorithm reduces to

$$C_j(i+1) = C_j(i) + k.e(i),$$

where the coefficient in the jth locations accessed at time i. As only one coefficient of a very much larger set is updated each iteration the table loop-up echo canceller converges much more slowly than the equivalent linear echo canceller.

Jitter Degradation Control

As already mentioned clock jitter can cause unacceptable degradation of performance in an echo cancelling transceiver, because small perturbations of the echo response result in large uncancelled echo components compared to the receive signal level on long loops. The effect of a small perturbation in phase depends on whether the transceiver is in the LT or the NT1. At the NT1 the echo components are transitory because the transmitter phase is slaved to the receiver phase. As soon as the change in phase has worked its way down the echo canceller span there is no difference between transmit and receive phase again. In the LT a perturbation causes permanent phase change between transmitter and receiver until a further perturbation occurs. This distinction is important because it affects the choice of techniques available for dealing with jitter when a digital PLL is used.

There are three methods which have been used to avoid the effects of jitter: zero echo gradient phase adjustment, frame word synchronous phase adjustment and jitter compensation.

In the first the signal transmitted by the local transmitter periodically includes an interval when it is at constant amplitude for long enough that an adjustment in the receive sampling phase does not cause any change in the echo. In one example a transceiver [12] transmits periodic bursts of signal with periods of silence in between. It is ineffective in LT transceivers because of the non-transient nature of the change in the echo response. The frame word synchronous phase adjustment method relies on performing phase adjustments during the reception of the frame synchronisation word so that subscriber's data are not corrupted. This method forces the use of a stored replica of the synchronisation word to prevent error propagation if a DFE is used and it is also ineffective for LT transceivers.

There are two variants of the jitter compensation method both of which involve the use of additional taps in the echo canceller. These either model the change [13, 14] in the significant echo response samples (i.e., those where the echo response is changing most rapidly) for a discrete phase jump, or are alternative taps [15] for an adjacent phase position. Jitter compensation does not rely on any particular frame structure and also supports digital PLL timing phase tracking in the LT transceiver.

5.8. DSP implementation issues

The adaptive filters (i.e., the echo canceller and equalizer) and fixed digital filters providing signal shaping and noise rejection can have a major impact on the IC die area because they encompass the bulk of the signal processing. It is important, therefore, that their complexity is minimized by careful structuring of the IC architecture.

As previously mentioned T-spaced echo cancellers and DFEs based on data-driven FIR structures give the lowest number of taps and reduced complexity arithmetic. Linear equalizers and recursive techniques for reducing the number of taps required in echo cancellers are to be avoided if possible as these require long wordlength multiplications, and in some instances are structurally less attractive. Transceivers based on pure DFEs have already been discussed, these giving good performance with structural simplicity.

One of the advantages of multi-level codes is that their slower signalling rate reduces the span required in both the echo canceller and the equalizer and allows more time for the signal processing to be performed. However, this advantage does require multiplications by M-level data.

Multiplications

It appears initially that there is a need for multiplications of long wordlength values by M-level data in the echo canceller and DFE. However, the processing can be modified to reduce it to little more than that required for binary data.

The LMS adaptation algorithm can be simplified by using the sign of the data for the tap update correlation; i.e.,

$$\underline{C}(i+1) = \underline{C}(i) + k.\underline{D}^*(i).e(i),$$

where $\underline{D}^*(i)$ is the vector of the signs of the data values. The penalty is a small increase in convergence time. It is also possible, but for a greatly increased convergence time, to use only the sign of the error signal $e(i)$. The same is true of the table look-up echo canceller, reducing the update addition to a counting operation.

The convolution multiplications of the data-driven adaptive fillter can be simplified by first accumulating separately the coefficients to be multiplied by each level, then scaling the partial sums by the amplitude of the appropriate level and finally adding them together. Thus, all the processing for an N tap M-level echo canceller using the data sign only LMS algorithm requires 2N+M-1 add/subtract operations compared to 2N for a binary echo canceller.

Processor Architectures

The architecture chosen for the transceiver signal processor(s) can have a dramatic effect on the resulting IC die size and, consequently, its unit cost. The optimum architecture is very dependent on the type of signal processing required. For example if the transceiver has been designed to eliminate all multiplications by long wordlength variables then there will be no need for hardware multiplies to be included. If table look-up echo cancellers are used then RAM will be required, whereas linear echo cancellers can use sequentially accessible memory. Indeed, the target processor structure can influence the overall transceiver architecture.

The most versatile approach is to design a general purpose digital signal processor that can implement any required processing. However this tends to be inefficient because structural simplicities in the processing are not exploited to reduce die size. Linear echo canceller implementation is a prime example of how careful choice of processor structure can result in an efficient realisation. The regular form of the processing where each tap forms an identical module of processing allows a variety of processor structures. At one extreme an area of silicon can implement a single tap module and a systolic array of such modules can be formed. At the other extreme arithmetic logic units can be shared over all taps. The key to a die area efficient implementation is to ensure that, for the number of taps required, the resulting processor clock rate is close to the maximum for the CMOS process used. There is a direct trade-off between area and clock rate in such regular processing arrangements so this will tend to minimise die area.

Careful choice of signal processing and processor architecture has resulted in the area of die devoted to DSP being less than half the total in many transceiver designs.

5.9. Transceiver examples

Performance and cost have been the key issues dominating the design of DSL transceivers. Single-IC implementation has been the goal for low cost. To achieve this the first designs of echo cancelling transceivers used line codes with self-equalizing properties in order to avoid or reduce the need for adaptive equalization. Self-equalizing line codes, because they are balanced about zero, exploit the slowly decaying nature of pulse tails by causing closely spaced ISI samples to nearly cancel each other. They are very effective on most loops. However, bridged-taps, which cause delayed echoes, can be a problem in the absence of an adaptive equalizer. Self-equalizing line codes also limit the time span of the significant part of echo responses, again because of their slowly decaying nature.

A commercially available example, which uses biphase line code to obviate the need for an adaptive equlizer and exploits the short echo time span to use a table look-up, FT echo canceller, is the single-IC 160-kbit/s biphase transceiver [9] manufactured by Mitel. Biphase is one of a class of codes that achieve self-equalization by being balanced within the baud interval. Consequently, they have signal spectra that are significantly wider than those of ordinary binary transmission, which limits their NEXT performance.

Another class of self-equalizing line codes are the linear codes which are balanced essentially by linearly filtering a binary source, producing a multi-level line code at the same baud rate. AMI line code is the most common example. Linear line codes require some degree of adaptive equalization even for operating over pairs without bridged-taps because the time span of the balancing is longer. Mixed analogue and DSP implementations of AMI transceivers [12, 16, 17] have been reported, with analogue PLL timing recovery schemes.

Although better than biphase (because the signal spectrum is not significantly wider than binary, but of different shape) AMI does not give sufficient NEXT limited reach. Consequently, transceivers using multi-level baseband codes at reduced baud rates were developed. These can be divided into two classes: those using block ternary codes and 2B1Q designs. To obtain the performance offered by multi-level line codes at acceptable cost two advances in technology were required: reduced IC geometries and advanced signal processing techniques to handle the increased echo cancellation and adaptive equalization requirements. The stringent signal-to-noise ratios required in the analogue front-end of these advance tranceivers led some companies to adopt the approach of separating the analogue and digital signal processing onto two ICs.

One example, which uses the block ternary line code 3B2T is a DSL transceiver [18] which is combined with an S/T-bus interface transceiver. The implementation is split into an analogue bipolar IC and a digital CMOS IC. A two-IC design [19] has also been reported for the Deutsche Bundespost specification which is based on MMS43 line code. This design uses non-linear compensation to relieve the linearity constraints and decision directed timing recovery.

Two single-IC designs for the Deutsche Bundespost specification have been described [20, 21]. Both exploit the cyclostationary 11 element Barker synchronisation word for element and frame timing recovery, and rely on good linearity in the analogue circuitry to use purely linear echo cancellation.

Several designs for the 2B1Q ANSI standard have been published. NEC has published details [22] of a three-IC design that uses jitter compensation. AT&T [23] and Siemens [24] have released details of two-IC solutions. Motorola/BNR [25] describes a single-IC system that employs a mixture of table look-up, linear and recursive echo cancellers. Other single-IC solutions have been developed by Mitel/BT Laboratories [26] and National Semiconductor [27]. The former uses non-linear and jitter compensation with a linear echo canceller; the latter relies on an adaptive hybrid scheme and good linearity to avoid the need for compensation techniques.

Single-IC DSL transceivers have die sizes of the order of 50-70 sq mm depending on the exact device geometry (about 1-2 microns) and the functionality added to the basic transceiver for maintenance and digital interfacing purposes. Power consumptions are typically of the order of 200-300 mW with master clock frequencies of 10-20 MHz.

6. Conclusion

ICs for ISDNs are now available that potentially allow virtually universal deployment of ISDNs. Achieving the performance required, especially for DSL transceivers, has required the development of new signal processing techniques and state-of-the-art design for very complex devices. Initially the cost of ISDN technology may limit its use to business applications where its advantages allow an adequate revenue return. However, shrinking device geometries and the promise of huge markets may eventually turn them into commodity devices. Market forces and technological innovation may then have created the truly ubiquitous information technology communications medium — the ISDN.

7. References

[1] R. Kerswell, M.T. Norris and R.C. Turner, "Network Terminating Equipment for the ISDN: the Customer Interface," *BTTJ*, vol. 2, No. 3, pp. 15-27, July 1984.

[2] D. Sallaerts, R. Dierckx and M. Rahier, "An Evolutionary Chip Set for ISDN," pp. 187-192, *TENCON 87*, Seoul.

[3] P. Gillingham, D. Kirkey and J. Erkku, "An ISDN S-Interface with Analog Timing Recovery," pp. 108-109, *ISSCC*, 1988.

[4] "Dual Data Link Controller," *Motorola Technical Data Sheet MC145488/D*, 1989.

[5] S.A. Cox and P.F. Adams, "An Analysis of Digital Transmission Techniques for the Local Network," *BTTJ*, vol. 3, July 1985.

[6] "Integrated Services Digital Network (ISDN) — Basic Rate Access Interface for Use on Metallic Loops for Application on the Network Side of the NT (Layer 1 Specification)," *ANSI Standard T1.601-1988*, approved Sept. 1988.

[7] S.G. Brophy and D.D. Falconer, "Investigation of Synchronization Parameters in a Digital Subscriber Loop Transmission System," *IEEE JSAC*, vol. SAC-4, No. 8, pp. 1312-1316, Nov. 1986.

[8] K.H. Mueller and M. Muller, "Timing Recovery in Digital Synchronous Data Receivers," *IEEE Trans. Com.*, COM-24, No. 5, May 1976, pp. 516-531.

[9]   R.P. Colbeck and P.B. Gillingham, "A 160-kb/s Digital Subscriber Loop Transceiver with Memory Compensation Echo Canceller," *IEEE JSSC*, vol. SC-21, No. 1, pp. 65-72, Feb. 1986.

[10]   O Aggazi, D.A. Hodges and D.G. Messerschmitt, "Non-linear Echo Cancellation of Data Signals," *IEEE Trans. Com.*, COM-20, pp. 2421-2433, Nov. 1982.

[11]   C.F.N. Cowan and P.F. Adams, "Non-Linear System Modelling: Concept and Application," *Proc. ICASSP 1984*, pp. 4561-4564.

[12]   C. Mogavero, G. Nervo and G. Paschetta, "Mixed Recursive Echo Canceller (MREC)," *Conf. Rec. IEEE Globecom*, Dec. 1986.

[13]   S.A. Cox, "Clock Sensitivity Reduction in Echo Cancellation," *Electronic Letters*, vol. 21, No. 14, pp. 585-64, July 1985.

[14]   R.B.P. Carpenter, S.A. Cox and P.F. Adams, "Jitter Compensation in Echo Cancellers," *Int. Symp. on Applied Sig. Proc. and Dig. Filtering*, Paris, France, 1985.

[15]   D.G. Messerschmitt, "Asynchronous and Timing Jitter Insensitive Data Echo Cancellation," *IEEE Trans. Com.*, COM-34, No. 12, Dec. 1986, pp. 1209-1217.

[16]   E. Arnon, W. Chomik and M. Elder, "A Transmission system for ISDN Loops," *Proc. ISSLS*, 1986.

[17]   R.B. Blake et al, "An ISDN 2B+D Basic Access Transmission System," *Proc. ISSLS*, pp. 256-260, 1986.

[18]   "SU32 ISDN U-Interface," *STC data sheet for ALT144/DS144*, Jan. 1988.

[19]   K. Wouda et al, "Towards a Single Chip ISDN Transmission Unit," *Proc. ISSLS*, pp. 250-255, 1986.

[20]   D. Sallaerts et al, "A Single Chip U-Interface Transceiver for ISDN," *IEEE JSSC*, vol. SC-22, No. 6, pp. 1011-1021, Dec. 1987.

[21]   H. Sailer, H. Schenk and E. Schmid, "A VLSI Transceiver for the ISDN Customer Access," *ICC*, pp. 45.4.1-45.4.4, 1985.

[22]   Y. Takahashi et al, "An ISDN Echo Cancelling Transceiver Chip for 2B1Q Coded U-Interface," *IEEE ISSCC*, 1989.

[23]   H. Khorromahedi et al, "An ANSI Standard ISDN Transceiver Chip Set," *IEEE ISSCC*, 1989.

[24]   R. Koch, R. Niggebaum and D. Vogel, "2B1Q Transceiver for ISDN Subscriber Loop," *IEEE ISSCC*, 1989.

[25] J. Girardeau et al, "ISDN U Transceiver Algorithm, Development System and Performance," *Globecom*, pp. 54.5.1-54.5.9, 1989.

[26] R.P. Colbeck et al, "A Single-Chip 2B1Q U-Interface Transceiver," *IEEE JSSC*, vol. 24, No. 6, Dec. 1989, pp. 1614-1624.

[27] R. Batruni et al, "Mixed Digital/Analogue Signal Processing for a Single Chip 2B1Q Interface Transceiver," *IEEE ISSCC*, pp. 26-28, 1990.

Chapter 14

VLSI Architectures and Circuits for Visual Communications

P. Pirsch and T. Wehberg

Institut für Theoretische Nachrichtentechnik und Informationsverarbeitung, Universität Hannover, Hannover, Germany

1. Introduction

Due to increasing availability of digital transmission channels and digital recording equipment, visual communications services with new features will be introduced in the near future. Visual communications includes all kinds of image transmission systems. Besides transmission of single still images, in particular motion video transmission of a sequence of images is considered. An application using still images would be a printing service for journals where images have to be transmitted between an editor office and the printing office. Data bank retrieval of still images is also in discussion. In case of motion video, besides unidirectional services such as TV distribution also bidirectional services (e.g., video phone and video conference) have been introduced. At present, new research activities are directed to storage of motion video on compact disc (CD). International committees as CCITT, CCIR, and ISO are already working on standardization of all kinds of image transmission and recording services [1, 2, 3, 4, 5].

Visual communications services are in need of image coding and decoding equipment of appropriate size and cost. In particular, future services where high volume is expected call for equipment of small size and low manufacturing costs. Both requirements can be fulfilled by VLSI implementation.

The cost for providing an image transmission service can be split into terminal cost and transmission cost. For the transmission part it has to be considered that the CCITT has already standardized a hierarchy of digital channels. Starting from the basic channel of ISDN (64 kbit/s), several of these channels can be used in a bearer service (H0 = 384 kbit/s; H1 = 1536 or 1920 kbit/s). Higher levels in the hierarchy of bearer services with bitrates of approximately 32 Mbit/s (H3) and 140 Mbit/s (H4) are not fully defined. In order to reduce transmission cost, source coding methods are applied for reducing the bitrate. In case of still image transmission using a resolution of standard TV or even higher, the transmission time of a single image should be reduced from about 100 seconds to a few seconds. Because of the high quality requirements, TV transmission is planned with 32 Mbit/s and HDTV with 140 Mbit/s. In order to have the ability of world wide transmission and mass applications, for video phone and video conference services bitrates in the range of 64 kbit/s to 1920 kbit/s are envisaged. This extremely high bitrate reduction by a factor of up to 600 requires very sophisticated coding schemes.

In the next section, source coding algorithms for coding of image signals will be presented. Emphasis will be given to coding of motion video because of higher sophistication and more stringent real-time processing requirements. After that, strategies for VLSI implementation of source coding for motion video will be discussed. Then, the two major implementation possibilities will be exemplified. One is programmable multiprocessor systems, the other is direct implementation based on functional blocks.

2.   Source Coding Algorithms

In general, visual communications systems consist of a camera, a video signal processor at the transmitter side (video coder), a transmission channel, a video signal processor at the receiver side (video decoder), and a monitor. For the majority of future applications visual communications will be digital. Having a digital channel the video signal processing at the transmitter side can be divided into digitization of analog video, digital source and channel coder and a channel multiplexer. The channel multiplexer combines several signals such as video, speech, and auxiliary signals. The channel coder adds redundant information in such a way that the receiver can detect or even correct transmission errors. The task of the source coder is to encode the video signal with a bitrate considering a given quality criterion.

In most cases a bitrate reduction has to be performed by the source coder. Bitrate reduction is possible because of redundant and irrelevant information in the video signal. Any information which can be extracted using the statistical dependencies between the picture elements (pels) is redundant and can be removed. Any information below a picture quality threshold is not relevant for the receiver and needs not be transmitted. In case of a human being as receiver, the visual properties of the human eye determine the irrelevant information.

2.1. Algorithms for redundancy reduction

Redundancy reduction will be performed by transformation of the image to another representation with reduced statistical dependency between the image data words. This kind of transformation has to provide decorrelated data. Decorrelation can be achieved by transformations such as discrete cosine transform (DCT) or predictive coding [6, 7].

In case of transform coding, a block **X** of NxN image data words is transformed to NxN transform coefficients y(u,v). The block **X** can be represented as a linear combination of a set of basic images Φ_{uv} weighted by the transform coefficients y(u,v).

$$\mathbf{X} = \sum_{u=0}^{N-1} \sum_{v=0}^{N-1} y(u,v) \cdot \Phi_{uv} \tag{1}$$

The transform coefficients y(u,v) are determined by the projection of **X** onto Φ_{uv}.

$$y(u,v) = \mathbf{X} \odot \Phi_{uv}$$

$$= \sum_{i=1}^{N} \sum_{j=1}^{N} x(i,j) \cdot \phi_{uv}(i,j) \cdot \quad u,v = 0, 1, \ldots N-1 \tag{2}$$

As an alternative, equation (2.2) can be rewritten as the matrix multiplication

$$\mathbf{Y} = \mathbf{C} \cdot \mathbf{X} \cdot \mathbf{C}^T \tag{3}$$

where **C** is a NxN matrix consisting of N basis vectors ϕ_u. The basic image Φ_{uv} can be determined as a product of two basie vectors, each taken from one of the matrices **C**.

$$\Phi_{uv} = \phi_u \cdot \phi_v^T \tag{4}$$

In case of the DCT, the basic vectors are given by

$$\phi_u(i) = b(i) \cdot \cos[(\pi/N)(i+1/2)u] \qquad \begin{aligned} i &= 0,1, \ldots N-1 \\ u &= 0,1, \ldots N-1 \end{aligned} \tag{5}$$

$$b(i) = \begin{cases} \dfrac{1}{\sqrt{N}} & i=0 \\[2mm] \sqrt{2/N} & i=1,\ldots N-1 \end{cases}$$

The second method to achieve decorrelated data is to transform the original image to a prediction error image. The prediction $\hat{x}(i,j)$ of a pel x(i,j) is a weighted sum of previous pels specified by the prediction set PS; i.e., for all $(m,n) \in$ PS:

$$\hat{x}(i,j) = \sum_{m} \sum_{n} a(m,n) \cdot x(i-m,j-n) \tag{6}$$

The weighting coefficients a(m,n) are also named prediction coefficients. The prediction error for each pel is then given by:

$$e(i,j) = x(i,j) - \hat{x}(i,j) \tag{7}$$

Very appropriate for video signals is the following set of prediction coefficients:

$$
\begin{aligned}
a(1,0) &= 1/2 \\
a(0,1) &= 1/4 \\
a(-1,1) &= 1/4
\end{aligned}
\tag{8}
$$

The prediction according to (6) bases only on previous pels from the same frame. Video signals are however derived from a sequence of frames. For this reason the prediction could be improved by taking pels from the previous frame in the same spatial position. It is obvious that prediction from the previous frame (interframe prediction) is perfect for stationary image segments which are not changing from frame to frame. For image segments which are changing from frame to frame, intraframe prediction according to (6) is better. Following this argument it has been concluded that adaptive intra/interframe prediction will be a good solution [8].

Prediction within video signals can be further improved if the motion of moving objects is taken into account. For a large variety of image sequences a description of motion by the spatial displacement from frame to frame is sufficient. This is in particular given for video telephone scenes. Because of difficulties in segmenting moving objects and the large overhead for specification of boundary lines of these objects, for most video coding systems a simple block matching scheme is used [8].

For the block matching scheme the present frame is divided uniformly into reference blocks with size NxN pels. Every reference block is compared with candidate blocks from a search area in the previous frame. The offset between the best matching candidate block and its reference block specifies the displacement vector $\mathbf{v}=(v_i,v_j)$. In general the mean of the absolute differences is used as matching criterion. The search can be limited to a maximum displacement p in both directions if the maximum motion of objects is assumed to be limited (Figure 1). The block matching algorithm (BMA) is then given by

$$s(m,n) = \sum_{i=1}^{N} \sum_{j=1}^{N} |x(i,j) - y(i+m,j+n)| \qquad -p \le m,n \le p \tag{9}$$

$$u = \min_{(m,n)} \{s(m,n)\} \tag{10}$$

$$\mathbf{v} = (m,n)|_u \tag{11}$$

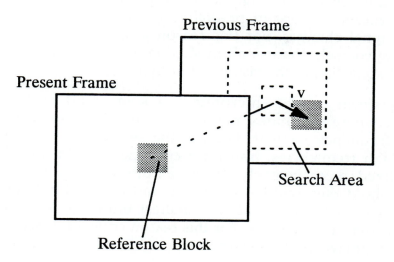

Fig. 1: Motion estimation based on block matching.

In equation (9), x and y denote pels from present and previous frames, respectively. There are $(2p + 1)^2$ possible sums $s(m,n)$. The position (m,n) of the minimum sum is used as displacement vector **v** for motion compensated prediction.

For all pels of a reference block, one out of three prediction modes can be selected.

$$\hat{x}(i,j) = \begin{cases} y(i+v_i,j+v_j) & \text{motion compensated prediction} \\ y(i,j) & \text{interframe prediction} \\ \sum_m \sum_n a(m,n) \cdot x(i-m,j-n) & \text{intraframe prediction} \end{cases} \qquad (12)$$

As a criterion for selection between the three predictors, the minimum prediction error variance for a block of NxN pels can be used.

2.2. Quantization

In order to achieve a given bitrate, quantization is needed for transform coefficients as well as prediction error signals. Quantization is a mapping of the image data to a reduced set of representative levels. The quantizer thresholds specify the ranges which are mapped to the representative levels (Figure 2).

High image quality calls for the adaptation of the quantizer characteristic to the visibility of quantization errors. Small changes of the image content in spatial and temporal domain require high accuracy while rapid signal changes allow coarser quantization. In case of predictive coding, the prediction error is an indicator for signal changes. For this reason large quantization errors can be quantized more coarsely than small prediction errors. Optimal quantizers have to be designed according to the visibility threshold [8].

The DCT transform can be interpreted as some kind of two-dimensional Fourier series where the index number of the transform coefficients is an indicator for frequency content. Because the sensitivity of the human eye is reduced for high frequencies, transform coefficients with high index number can be quantized more coarsely. If the amplitudes are below the smallest quantizer threshold they will be set to zero and no amplitude information has to be transmitted. It follows that the smallest bitrate to keep a given image quality is dependent on the scene. For this reason critical test sequences are used to determine the image quality which can be achieved for a given bitrate.

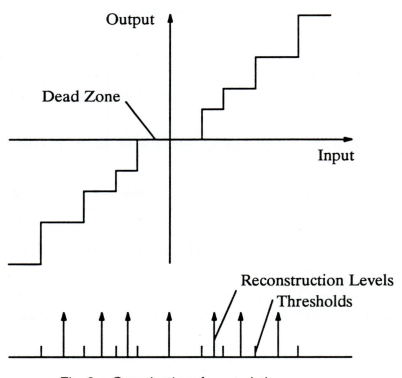

Fig. 2:   Quantization characteristic.

2.3. Variable word length coding

The resulting image data words after transformation and quantization have distinct probability of occurence. The information content of an ^ event is $-\log_2[p(x)]$, when $p(x)$ is the probability of x. Therefore the information content becomes higher when the probability of occurence becomes smaller. For this reason code words should be assigned with a number of bits according to the information content. In such a case, the mean code word length approaches the entropy. A method to construct this kind of codes has been proposed by Huffman [9]. Besides the Huffman code, other variable length codes are possible which have less sensitivity for transmission errors [3].

In particular, image coding with high reduction factor requires coarse quantization. In such a case representative levels zero become frequent. It is not efficient to code every data word individually for a sequence of consecutive levels zero. Thus it will be appropriate to assign one code word to the complete sequence of zeros (runs of zeros). This kind of coding is called run length coding [6, 7].

2.4. Source coding for different applications

The applied source coding schemes are influenced by the image material. Visual communications could be based on still images or on image sequences. The requirements for source coding are dependent on the envisaged bitrate for transmission and the spatial and temporal resolution.

Still image transmission should provide high image quality. Therefore at least a resolution according to the broadcast standard CCIR 601 [2] should be considered. Hereby component coding of the color information is applied. Source data with 8 bits per pel and a spatial resolution of 720x576 pels for the luminance component are used. The chrominance components have halved horizontal resolution (360 pels). It follows that the source data of one frame amounts to about 6.6 Mbit and transmission over an ISDN channel with 64 kbit/s requires about 100 sec. The transmission time can be reduced by a factor of about 7 using source coding with a DCT based on 8x8 blocks, adaptive quantization and variable word length coding (Figure 3). In order to receive a complete image at an early stage, the ISO standardization committees propose progressive coding where an image in low resolution is coded and transmitted first and then the resolution is sequentially upgraded by further coding of the differences between the original image of higher resolution and the transmitted one [4].

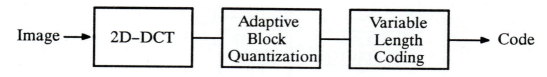

Fig. 3: Adaptive DCT coding (transmitter side).

The priority for video telephone and video conference coding is low bitrate and not image quality because of world wide communication. The CCITT is planning a service for bitrates of Px64 kbit/s with $1 \leq P \leq 30$ [1]. The basic ISDN channel has P=1 and bearer services are planned for P=6 (H0-channel) and for P=24 or 30 (H1-channel). Video telephone and video conference are based on image material of head and shoulder type. For such material a reduced spatial and temporal resolution is justified. The CCITT proposes a common intermediate format (CIF) with 352x288 pels for luminance and 176x144 pels for each chrominance component. The basic frame rate is 30 Hz. Reduction of the frame rate by integer factors of 2, 3, or 4 is possible. The decoder then has to interpolate dropped frames. Application of motion compensation to the interpolation provides improvement in image quality [8]. Besides the CIF format, in particular for 64 kbit/s, a quarter CIF (QCIF) format with halved resolution in each dimension is proposed. The bitrate envisaged for motion video on CD is about 1.2 Mbit/s. This application also uses the CIF format.

The CIF format results in a source rate of 36 Mbit/s. Hence, source coding has to provide reduction factors in the range of 20 to 600 depending on the transmission rate. Very sophisticated coding schemes make these reductions possible without degrading the reconstructed images too much. For this a hybrid coding scheme is frequently proposed. A block diagram of such a hybrid coder is shown in Figure 4. The major characteristics are listed below. A previous frame memory is used for interframe prediction. Prediction of moving objects is improved by motion compensation (BMA). The prediction error is coded by an adaptive DCT. Run length coding and variable word length coding (VLC) are applied to the quantized DCT coefficients. In order to have the same prediction at receiver and transmitter, a recursive structure is used as in all predictive systems. A two-dimensional loop filter reduces the effects of quantization errors on the prediction value. For nonpredictable areas, the prediction from the previous frame is dropped and intraframe DCT coding is applied.

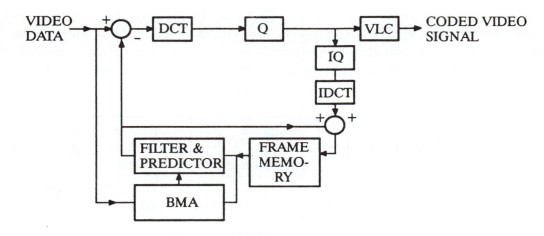

Fig. 4: Hybrid coding scheme (transmitter side).

3. VLSI Implementation Strategies

Essential for the introduction of video communications services is low manufacturing cost and compact realization. For these reasons application specific VLSI components are requested for implementation. VLSI implementation strategies are dependent on the system requirements and the production volume. High production volume will allow full custom designed components, whereas low and medium volume have to be designed with standard or semi-custom components. The system requirements are specified by data such as computational rate, data access rate, complexity and regularity of algorithms. Implementations are also different for low, medium, and high level algorithms. Low level are simple algorithms which have to be performed for every pel in the same manner. An example is two-dimensional convolution. Medium level are algorithms which generate features and other characteristic data of the image scene. The derivation of symbols and objects is included in this category. High level operations are object dependent processing and operations with characteristic symbols.

Throughout this section the hybrid coding scheme according to Figure 4 will be discussed. Computational requirements for major low level algorithms and the complete coding scheme are listed in Table 1. Each individual arithmetic operation such as MUL, ADD, SUB, SHIFT is counted as one operation. The low level algorithms are processed for every pel. Therefore their computation rate R_C is dependent on the source rate R_S.

$$R_C = R_S \cdot OP \tag{13}$$

where OP denotes the mean number of operations per pel and R_S is the product of image size and frame rate. An essential result of Table 1 is that most algorithms can be implemented in different ways resulting also in different computation rates. It should be noted that for implementation besides computation rate, aspects as data transfer and regularity have to be considered. For correct rating of the computation rates in Table 1, comparison with those of standard processors is needed. The Intel 80486 is specified for an average over typical operations with 1.5 MOPS, the RISC processor Intel 860 with 80 MFLOPS and the AT&T DSP 16 with 80 MOPS when counting multiplication and accumulation as separate operations. From this it can be concluded that one standard processor is not sufficient. Either dedicated hardware or multiple processors are needed.

VLSI implementation can be split into function oriented and software oriented. Function oriented implementations are dedicated circuits for functional blocks of the scheme. Software oriented implementations are based on programmable multiprocessor systems. The advantage of function oriented implementation is the small size because of specialized circuits; the disadvantage is the restricted flexibility. The software oriented implementation allows different algorithms on the same hardware.

Table 1:   Computation rate for selected algorithms and schemes. The numbers are based on CIF format with 30 Hz frame rate. (MOPS = mega operations per second, arithmetic operations).

Algorithm	Computation Rate
2-D DCT (8x8 window)	
- Dot-product with basis images	584 MOPS
- Matrix-vector multiplication	146 MOPS
- Fast DCT	48 MOPS
Block Matching (displacements ±7)	
- Full search	2053 MOPS
- 2-D log search	228 MOPS
2-D FIR filter (3x3 window)	
- 2-D convolution	82 MOPS
- Two 1-D convolutions (separable kernel)	55 MOPS
Hybrid coding scheme RM8	**720 MOPS**
(using always second choice of above)	

Optimization of VLSI implementation means realization with smallest size and manufacturing cost. Silicon area and chip housing (number of pads) dominate chip manufacturing cost. Hence, they have to be minimized. For minimization of silicon area the trade-off among the areas devoted for computation (data processing path), storage, and interconnections has to be considered. Optimization of silicon efficiency requires the adaptation of the architecture to the algorithms. Every a priori known feature should be considered for minimization. Dummy cycles reduce the effective computation rate. Therefore processing units should always be kept busy.

In particular, implementations for the hybrid coding scheme will be discussed now. The functional space of the scheme (Figure 4) is shown in Figure 5a. In the function oriented implementation each block will be realized with dedicated hardware considering the individual requirements. For most blocks of this coder, special integrated circuits (ICs) have been designed and are already available [10]. The large number of different ICs is a good solution for high source rates as needed for broadcast video and HDTV. In case of moderate source rate as for the CIF format with 30-Hz frame rate, smaller silicon area is possible by devoting only a part of an IC to every functional block. Then the functional blocks become macro blocks in an IC. Most of the available ICs allow processing with 40-MHz sample rate but the discussed coder has a mean sample rate of 4.5 MHz. Hence, the computation intensive blocks can be implemented with reduced parallelism which means smaller silicon area. Optimization of silicon efficiency for large ICs with macro block structure requires a global view for minimization of storage (local RAM memory, registers) and interconnections. In order to minimize buffer memories between macro blocks, the results of a preceding macro block have to be taken by the following block and immediately processed. The discussion shows that optimization of silicon efficiency for the functional approach requires large efforts for specification and design. In particular, changes of the requirements and algorithms result in new design cycles.

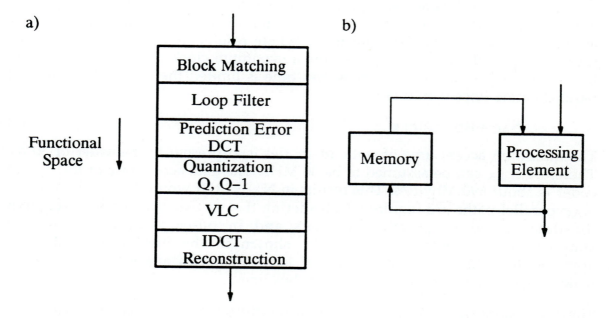

Fig. 5: a) Functional space of the hybrid coder; b) projection of the functional space onto one processing element (PE).

A vertical projection of the functional space results in one processor performing all functions one after the other (Figure 5b). An additional feedback memory is needed for the storage of intermediate results. The ideal solution for time dependent different operations in the same device is a programmable processor. As indicated earlier, one processor is not sufficient to provide the required computation rate. But considering that image segments of appropriate size can be processed independently of each other, several processing elements (PEs) could be distributed over the image space (Figure 6). This kind of distributed approach is based on the possibility of parallel operations on the image data.

Matching to the requirements can be performed by minimizing the number of parallel PEs. Assuming that each PE is performing one operation at every clock cycle, then the number of parallel active PEs is

$$N_{PE} = R_C/f_{CLK} \tag{14}$$

The maximum achievable clock rate f_{CLK} for available 1-μm CMOS technology is in the order of 50 MHz. The computation rate of the hybrid coder is 720 MOPS. It follows $N_{PE} \approx 15$. The number of PEs could be further reduced by pipelining (several operations in one clock cycle) or parallelization within the PE. Processor architectures consisting of multiple PEs with these features will be presented in the next section. A figure of merit for parallelism can be also derived from the data access. Let the required data access rate be R_{DATA} when each operand has to be read out of an external RAM memory. Hereby it is assumed that all source data and intermediate results are stored in the external memory. From the memory access time a maximum clock frequency f_{CLK} could be derived. Because the memory access time is limited the required number of parallel access lines would be

$$N_{AC} = R_{DATA}/f_{CLK} \tag{15}$$

The minimum access time of static memories is at present in the order of 30 ns. Therefore f_{CLK} can be assumed to be 30 MHz. The data access rate of the hybrid coder is about 690 Mbyte/s. This results in $N_{AC}=23$. It would be concluded that N_{AC} parallel paths are required. Considering that no external access is required for data stored in pipeline registers and in on-chip local memories a simplification is possible. It will be shown in the next section that an appropriate data transfer concept and local memories adapted to the needs make possible data transport with a few bus lines.

The discussions above show that special multiprocessor architectures with data path and data access adapted to the required class of algorithms improve the overall silicon efficiency. Modifications of the algorithm can be performed by software changes as long as the new algorithms belong to the class of algorithms taken into account. In many cases changes of the computation rate just require a change in the number of PEs.

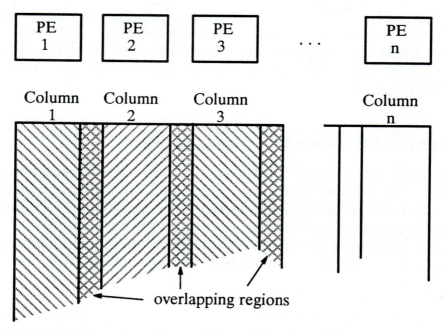

Fig. 6: Processing elements (PEs) distributed over the image space.

4. Programmable Multiprocessor Systems

A software oriented realization for applications from the visual communications area can be very attractive because it incorporates flexibility to accomodate a wide variety of application schemes in a programmable hardware. The requirement for high computation rate encountered in real-time processing of digital video signals calls for a multiprocessor system containing multiple processing units. To support an adequately high bandwidth for operand transport into the processing units, memories local to the processing units are needed in a multiprocessor system for video applications.

The physical size of a multiprocessor system is mainly influenced by two independent factors. One factor is the manufacturing technique. A monolithic integration using advanced VLSI technology can supply multiprocessor ICs that are a prerequisite for small size of the multiprocessor system. Secondly, the size depends on structure and number of the processing units. The throughput and the utilization of the processing units should be maximized to achieve a small size. Thus the combination of data and control flow through the processing units has to be optimized for high utilization or efficiency in a multiprocessor architecture.

With respect to data and control flow, architectures of multiprocessors are generally classified [11] as single instruction, multiple data stream (SIMD) or multiple instruction, multiple data stream (MIMD). In this section, generic SIMD and MIMD architectures are presented first. A discussion on two examples of SIMD and MIMD multiprocessor implementations follows that are based on integrated VLSI circuits.

4.1. Architecture of SIMD and MIMD multiprocessors

Block diagrams of generic SIMD and MIMD architectures for video multiprocessors are shown in Figure 7. Both architectures incorporate parallel processing units and local memories to fulfill the requirements for high rates of computation and operand transport. Separate buses for transport of video data and output of results supply the necessary bandwidth for these data types. An interconnection network is provided for communication between the processing units.

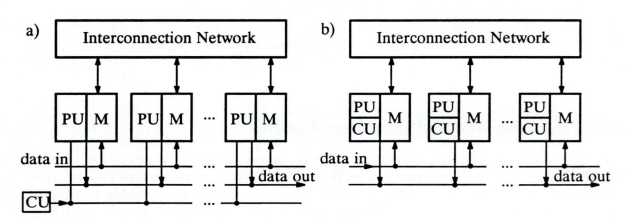

Fig. 7:   Architecture of SIMD (a) and MIMD (b) multiprocessors
(PU = processing unit, M = memory, CU = control unit).

In the processing units of a SIMD multiprocessor, identical instructions issued by a single control unit are executed synchronously for multiple data streams. This is very efficient for applications where the employed algorithms during the whole processing time allow for identical operation on as many data streams as there are processing units available. For example, a regular control flow of an algorithm thus is a prerequisite for high efficiency. If signal dependent processing occurs, some of the processing units will be idle [12]. The SIMD multiprocessor efficiency decreases with an increase in signal dependent processing. For applications with irregular control flow, MIMD multiprocessors containing a control unit for every processing unit have the potential to be more efficient due to the independent programmability. This is achieved at greater hardware expense because of the multiple control units. In a MIMD multiprocessor for real-time video processing where the processing has to keep pace with the data flow from the sensor, additional hardware expenses are required for the synchronization of the processing with the input data flow.

Communication over the interconnection network is another cause for possible decreases in efficiency. A processing unit depending on inputs from others to execute its operation may be idle due to a contention on the interconnection network or because the needed input has not yet been generated. Efficiency decreases due to communication could be avoided by fully independent operation of the processing units. This would call for larger memories to store data sets complete for calculation of results.

4.2. SIMD system based on multiprocessor IC

A multiprocessor IC with SIMD architecture has been proposed by Müller et al. [12, 13, 14]. Figure 8 depicts a block diagram. The multiprocessor IC contains an I/O unit for transfer of video data, a sequencer with instruction decoder for access to instructions in an external program memory, and n=6 [14] identical PEs. A PE consists of local memory with a 512x16-bit RAM and a 16x16-bit register array, and a processing unit for arithmetic operation on 16-bit operands. The PEs are connected via a simple interconnection network that allows left or right shift of data along the PE string. Input and output of data are performed over a data bus common to all PEs.

The processing units in all PEs synchronously perform identical operations under the control of a single instruction per clock cycle (SIMD). During operation the PEs distribute over the image space (Figure 6) such that neighbouring PEs process neighbouring, overlapping segments of a video frame. The segment to be processed in a PE is loaded first from an external buffer memory private to the multiprocessor IC into the local memory of the PE. After each PE has been loaded with the assigned segment, the processing is performed and the results are transferred back to the buffer memory.

By parallel operation of PEs the multiprocessor IC supplies high processing power. It is increased further by adaptation of the processing unit to arithmetic operations required by algorithms from the video coding area. Low level algorithms employing an arithmetic operation on data words a and b described by

$$\sum |a\text{-}b|, \quad \sum (a\text{-}b)^2 \quad \text{or} \quad \sum (a\text{-}b) \cdot \text{const.}$$

are supported in the processing unit by a special pipeline implementation with serially connected subtract, multiply or absolute value, and accumulate operation. This feature enables the execution of up to three arithmetic operations with one instruction.

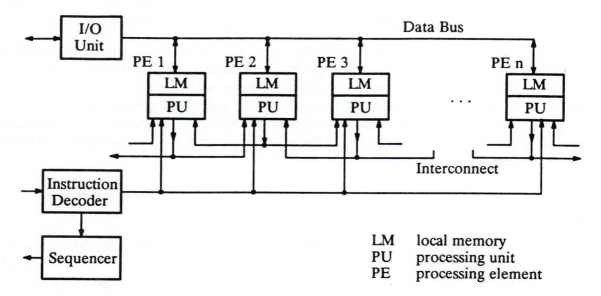

Fig. 8:   Block diagram of a SIMD multiprocessor IC.

The multiprocessor IC has been realized using a 0.7-µm CMOS technology. A clock frequency of 25 MHz is achieved and this supports a peak processing power of 150 M accumulations/s which equals 150 MIPS [12] for this architecture. In terms of arithmetic operations this amounts to a peak rate of 450 MOPS per IC. For applications requiring higher processing speed a parallelization of multiprocessor ICs is possible. A system containing several multiprocessor ICs on a printed circuit board has been proposed [14] for implementation of the hybrid coding scheme described earlier (Figure 4). In this system, each multiprocessor IC has its private external buffer memory. One multiprocessor IC performs the overall system control providing its sequencer for addressing of instructions in a single program memory.

The system consisting of 48 PEs supplies a peak processing power of 3600 MOPS. Simulations have shown that it can accommodate the functions of a hybrid video coder for CIF video signals with 12.5 or 15-Hz frame rate together with a decoder generating 30 Hz CIF. The coder functions require up to 360 MOPS [15] and use two thirds of the implemented processing power [12] if the VLC algorithm (*see* 2.3) is executed externally. The system thus operates with an efficiency in the range of 10 to 15% for the hybrid video coding application.

4.3. MIMD system based on multiprocessor IC

The example of a programmable multiprocessor with MIMD architecture discussed in this section achieves high performance through adaptation to requirements of algorithms and schemes from the video coding area [15, 16, 17, 18]. As shown in the block diagram (Figure 9) it consists of identical processing elements (PEs) and input and output buses. In order to support video signals with very high data rate, unidirectional buses are proposed that have the potential to achieve a higher rate than bidirectional ones.

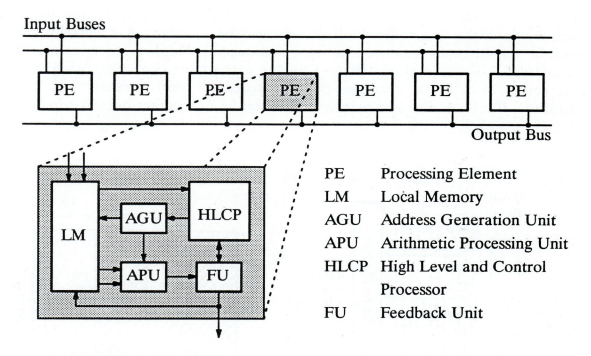

Fig. 9:   Block diagram of a real-time multiprocessor with MIMD architecture and its processing element (PE).

The multiprocessor architecture takes advantage of the requirement for independent processing of distinct segments in a video frame [15]. This permits independent operation of the PEs discarding the need for communication between PEs and making the processing power of the multiprocessor linearly dependent on the number of PEs. Thus the multiprocessor is applicable for a wide range of performance requirements stemming from different video formats and processing schemes.

The PE architecture is also depicted in Figure 9. A PE contains local memory (LM) for storage of data sets complete for calculation of results, an address generation unit (AGU) and an arithmetic processing unit (APU) for high throughput execution of low level algorithms, a feedback unit (FU) for transfer of intermediate data to the LM, and a high level and control processor (HLCP) for processing of results from the APU and for control of the operations within the PE. A PE provides high processing power by internal parallelism in the APU, and by pipelining of address generation in the AGU, access to operands in the LM, and arithmetic operation in the APU. Due to the incorporation of a HLCP with local control capability in every PE, the multiprocessor has a MIMD architecture.

The local memory (LM) has to store pels from segments out of up to two images, coefficient sets and intermediate data. With the incorporation of three memory modules the required functions can be realized. A special memory module named cache has been devised that supports high performance within a PE [19]. Different loading schemes for the cache have been proposed [16, 18] that result in a linear increase of multiprocessor performance with every additional PE. Structure and principal function of a cache are depicted in Figure 10.

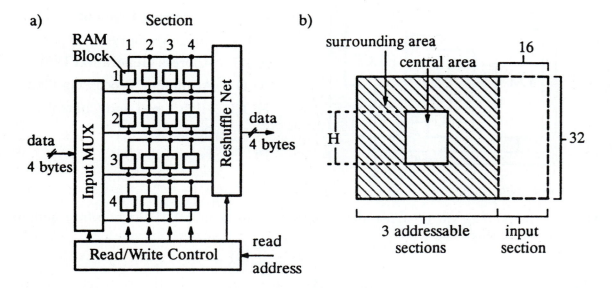

Fig. 10:   Structure (a) and principal function (b) of a cache.

A cache is a dual-port memory enabling simultaneous read and write access to provide data for uninterrupted operation of the APU. A cache contains logic for input and output together with four identical sections of 32x16 bytes of RAM each. Every section is subdivided into four separate RAM blocks such that four bytes are loadable in parallel. With every address from the AGU four bytes are read and transferred to the APU through a reshuffle net. Every byte in a cache is addressable as the upmost byte in the four bytes sent to the APU. Dependent on the read address the reshuffle net exchanges connections between rows of RAM blocks and output bytes to provide this feature. The principal function of a cache (Figure 10b) is characterized by a division of the sections into one input and three addressable output sections. Video data is loaded into a cache using a method that lets the output sections contain a homogeneous image segment. This segment is logically subdivided into central and surrounding area. Results are calculated for the central area. In doing so, pels from the surrounding area also have to be accessed. After the central area has been processed completely and the input section has been loaded with new data from the consecutive image section, the assignment of the sections is shifted cyclically; i.e., the left section (Figure 10b) becomes the new input section and the addressable sections are shifted to the right. Thus a video frame is processed in stripes with every stripe having a height H that depends on requirements of the application. This function enables the transport of video signals with different format to the APU.

The arithmetic processing unit (APU, Figure 11) has a SIMD architecture with identical operation on parallel data streams. The operation is adapted to the requirements of window based, low level algorithms needed in video processing applications [16] supporting execution of the linear kernel function [20]

$$\sum (a-b) \cdot |(c-d)|$$

and all subfunctions in a special arithmetic pipeline. It performs a sequence of subtraction, multiplication, accumulation over a window, and shift and limit for normalization of sums. It is programmable from the AGU to implement the low level operations required for distinct applications. In order to enable processing of pels with 8-bit quantization as well as intermediate data with 16-bit quantization, the APU has two operational modes for 8- or 16-bit processing. Figure 11 shows the input of 8 bytes into the APU and either four parallel operations on 8-bit operands or two parallel operations on 16-bit operands.

The address generation unit (AGU) enables fast and flexible addressing of window sequences in the LM. The addressing scheme is shown in Figure 12. The position of the first window of a sequence is specified by a sequence offset. Further window positions are defined relative to the first position using i-/j-offset for the distance between successive windows, and i-/j-repetition for the number of windows. The HLCP controls the address generation by writing the offset and repetition values into a register file in the AGU. The operands within a window are also addressed relatively. The addresses of operands are stored as a list of Δi-/Δj-values in a RAM and added to the window position. Thus shape and size of the processed windows are programmable. With the addressing of every Δi-/Δj-value, an instruction for the arithmetic pipeline is also accessed in a second RAM and sent to the APU.

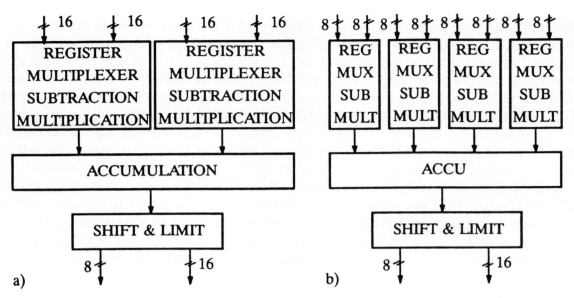

Fig. 11:   Block diagrams of the arithmetic processing unit (APU) with operational 16-bit (a) or
            8-bit (b) mode.

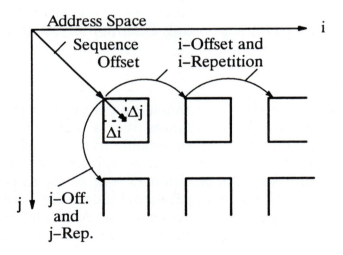

Fig. 12: Addressing scheme of the Address Generation Unit (AGU).

The high level and control processor (HLCP) serves as a module to incorporate
the postprocessing of low level results with arithmetic and non-arithmetic
operations for signal dependent processing. Furthermore it controls the
operations within the PE by programming APU and AGU for their respective
tasks. The HLCP functions can be implemented using an available DSP. The
feedback unit (FU) enables the implementation of algorithm sequences within a
PE by transferring intermediate data from APU and HLCP to the LM for further
processing.

As a first realization step several modules of the PE partitioned into two ICs have been integrated using 1.5-μm CMOS standardcell technology. One IC is a Cache. The other is called linear kernel processor (LKP) and supplies window based address generation and processing consisting of an APU with operational 8-bit mode only (Figure 11), and an AGU generating addresses for two sets of operands simultaneously. Photographs of the ICs are presented in Figure 13.

a)   b)

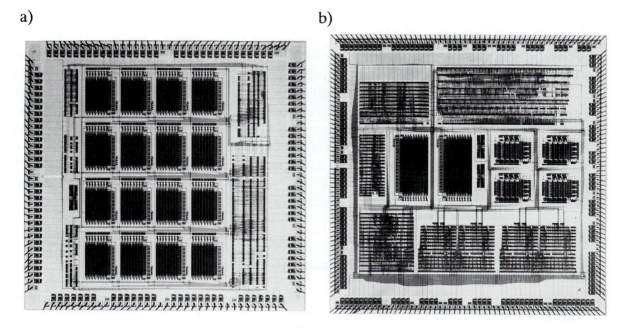

Fig. 13:   Photographs of the realized ICs cache (a) and linear kernel processor (b).

The cache incorporates 2,400 equivalent gates of logic and 16 RAM blocks with 128 bytes each in an area of 84 mm². The LKP incorporates 15,000 equivalent gates of logic including 3,900 equivalent gates for four multiplier macro cells, and two RAM blocks with 256x10 bits each in an area of 123 mm². By parallelization of four pipelines containing four arithmetic operators each, the LKP performs up to 16 arithmetic operations per clock cycle. LKP and cache have been tested and demonstrated to work at 25 MHz clock frequency. The cache thus supports a data rate of up to 100 Mbyte/s at its randomly addressable output and the LKP supplies a peak processing power of 400 MOPS for arithmetic operations, simultaneously generating up to 50 M addresses/s.

On the basis of the realized ICs, a PE as described by Figure 9 is estimated to be integratable using 1.2-μm technology if the HLCP is provided externally. Such a PE with external HLCP would supply a peak processing speed of 400 MOPS for low level arithmetic operations with an APU in 8-bit mode, or 200 MOPS with an APU in 16-bit mode, when an unchanged clock frequency of 25 MHz is assumed. An increase in clock frequency, and hence processing speed, will however be achievable with an advanced technology.

The efficiency of a multiprocessor system containing the described PEs has been simulated [21] for different applications. For schemes with a significant part of high level processing, the performance of the HLCP influences the multiprocessor efficiency. Simulations of the hybrid video coding scheme described earlier (Figure 4) have shown that seven PEs are necessary for 15-Hz frame rate if VLC (*see* 2.3) is performed externally. Since the scheme calls for an equal share in 8-and 16-bit operations, the seven PEs supply a peak performance of about 2100 MOPS. In relation to 360 MOPS [15] required by the coding scheme for 15-Hz frame rate, this results in an efficiency of about 17%. This low value is caused by a bottleneck in the assumed HLCP that forces the PE to be idle for about 64% of the time while the HLCP is busy doing quantization and inverse quantization (Figure 4). Taking this into account, an efficiency of about 47% is achieved using the PEs. Employing a powerful processor as HLCP will be the way to reach a high efficiency even for schemes with a significant amount of high level processing. Then an efficiency in the range of 30 to 50% seems realistic with the proposed MIMD architecture.

5.   Key Components for Functional Blocks

A function oriented realization for applications from the visual communications area can be very attractive because it has the potential to achieve the smallest size possible by being tailored to the needs of the application. This has to be paid for by a restricted flexibility that extends to realizations with no flexibility at all. When mapping the operational parts of a video processing scheme to the blocks of a function oriented realization, the data transfer between blocks has to be taken into account. Considerable hardware expenses for formatting of data sequences might occur if the output from any one block is not matched to the input requirements of the successive block.

An example of a function oriented implementation for a complete application, namely the hybrid coding scheme for video telephony (Figure 4), has been presented by Ruetz and Tong [10]. The use of modern VLSI technology together with an adaptation to the requirements of distinct functional blocks has led to the realization of three ICs employable as major blocks in an implementation of the video telephone coder. One IC supports discrete cosine transform (DCT) as well as inverse DCT (IDCT), the second accommodates the functions quantization and inverse quantization, and the third is a motion estimation processor based on a full search block matching algorithm. Employing five ICs of these three types together with ICs for two frame memories, variable word length coding, and control as well as data formatting purposes, a compact video telephone coder can be built.

In this section special realizations for two widely used algorithms from the video coding area, namely DCT and block matching, will be discussed. Both algorithms have been selected because of the high requirements they establish. Distinct hardware structures are examined that achieve the high requirements in a small size, preferrably a single IC.

5.1. Dedicated realizations of discrete cosine transform

The DCT for a NxN block **X** of pels requires N^4 multiplications and N^4 accumulations when realized as a two-dimensional correlation with the basic images using equation (2). Due to symmetry of the basic images, the two-dimensional transform can be obtained from two one-dimensional transforms, a method that will be exemplified in the following.

When employing equation (3), two matrix multiplications have to be performed sequentially that call for $4N^3$ arithmetic operations per block. Thus the computational effort for realization of (3) is smaller by a factor of $N/2$ when compared to (2). This advantage has to be paid for by a more complex scheme of data transport which can be derived by a separation of (3) into three steps that are implementable in hardware. In the first step, a block **Z** of NxN intermediate values is calculated.

$$\mathbf{Z} = \mathbf{C} \cdot \mathbf{X}^T \tag{16}$$

This can be seen as a one-dimensional, horizontal transform since row vectors in **X** are multiplied by row vectors from **C**, a method with a computational structure like a one-dimensional correlation. The second step is a transposition of the block **Z**.

$$\mathbf{Z}^T = \mathbf{X} \cdot \mathbf{C}^T \tag{17}$$

$$\mathbf{Y} = \mathbf{C} \cdot \mathbf{Z}^T = \mathbf{C} \cdot \mathbf{X} \cdot \mathbf{C}^T \tag{18}$$

In the third step (18), the second matrix multiplication is performed which can be seen as a one-dimensional, vertical transform since row vectors from **C** are multiplied by column vectors from **X·C**^{T}. The two-dimensional N^2-point DCT (2) has thus been separated into a sequence of two one-dimensional, N-point transforms with a matrix transposition in between. While (16) and (18) employ a simple scheme of data transport needed for one-dimensional correlation, the matrix transposition (17) complicates the data transport and requires special hardware resources. Nevertheless, dedicated realizations have been based on the separated DCT because of the significant saving in computational effort and the correspondingly smaller number of processing units required. A one-dimensional DCT can be realized using two basic computational structures, one being the one-dimensional correlation as described above and the other being a signal flow graph implementation of an efficient algorithm that will be described later. Five realization examples will be investigated.

A realization using the one-dimensional correlation structure has been proposed by Totzek and Matthiessen [22]. A basic PE in this structure has to contain a multiplier followed by an adder for accumulation. The proposal employs the well-known idea of a linear systolic array for implementation of the matrix multiplication. Here, by using chains of carry-save adders to realize the multipliers, a layout with high regularity has been achieved. The devised IC mainly incorporates two linear arrays with 8 PEs each and a memory block for the transposition function. In a 1.5-μm CMOS technology the IC requires a silicon area of about 85 mm^2 and contains some 240,000 transistors. It has been simulated to perform two-dimensional DCT or IDCT with N=8 and N=16 for sample rates of up to 30 and 15 MHz, respectively.

Another realization based on the one-dimensional correlation structure has been reported by Sun, Chen, and Gottlieb [23]. To achieve small area they avoid the incorporation of multipliers by employing a distributed arithmetic approach that will be clarified first. During a one-dimensional transform described by (16) or (18), a vector **x** containing N values is transformed into a vector **y** by multiplication with a coefficient matrix **C**. Let an element of **x** be represented by the 2's complement code with a wordlength of m bits as follows:

$$x(k) = -x_{m-1}(k) \cdot 2^{m-1} + \sum_{r=0}^{m-2} m_r(k) \cdot 2^r \tag{19}$$

for k = 0, 1, . . ., N-1, where $x_{m-1}(k)$ is the sign bit of x(k). Then an element of **y** results as

$$y(n) = -\sum_{k=0}^{N-1} c(k,n) \cdot x_{m-1}(k) \cdot 2^{m-1} + \sum_{r=0}^{m-2} \sum_{k=0}^{N-1} c(k,n) \cdot x_r(k) \cdot 2^r$$

$$= -F_n\big[c(n), x_{m-1}\big] \cdot 2^{m-1} + \sum_{r=0}^{m-2} F_n\big[c(n), x_r\big] \cdot 2^r \tag{20}$$

for n = 0, 1, . . ., N-1, where

$$F_n\big[c(n), x_r\big] = \sum_{k=0}^{N-1} c(k,n) \cdot x_r(k) \tag{21}$$

for r = 0, 1, . . ., m-1. F_n is a function of the fixed coefficients c(k,n) and bit patterns of **x**. Therefore F_n can be precalculated, stored in memory, and addressed by the bit patterns of **x**. This allows to design a PE as depicted in Figure 14, which generates a transformed value y(n) based on memory look-up and bit-serial processing. The PE consists of ROM memory for storage of F_n, and some hardware to perform shifts and adds of values of F_n.

Employing the idea of distributed arithmetic, an IC has been realized [23] that performs 16x16 DCT in real time for video signals with up to about 15 MHz sample rate. The IC incorporates two strings of 16 PEs each for horizontal and vertical transformation together with a transposition memory. The PEs have been realized with a reduced ROM size [23] in order to achieve a sufficiently small silicon area. Fabricated in a 2-μm CMOS technology, the IC contains approximately 73,000 transistors in a silicon area of about 67 mm^2.

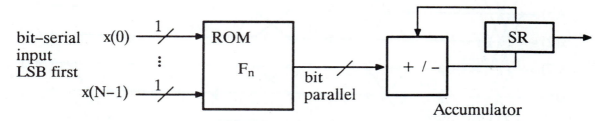

Fig. 14: Block diagram of a PE based on distributed arithmetic (SR = shift register).

The alternatively used computational structure for realization of a one-dimensional DCT is based on efficient algorithms that have been developed to reduce the number of multiplications by taking advantage of the periodicity in the basic vectors (eq. 5). Lee [24] has proposed an algorithm referred to as fast cosine transform (FCT) that requires $(N/2) \log_2 N$ multiplications and $(3N/2) \log_2 N - N + 1$ additions to perform a one-dimensional DCT or IDCT. Figure 15 shows the signal flow graph of the FCT. It contains an alternating sequence of data permutation stages and arithmetic stages. In an arithmetic stage the basic function includes addition and subtraction followed by a multiplication. A direct realization of the signal flow graph would need an excessive amount of silicon area even if fabricated in an advanced technology. However, realization examples can be derived by using projections within the graph leading to sufficiently small implementations.

In the signal flow graph (Figure 15), values x(k) and y(n) as well as all intermediate values have a wordlength of several bits. Artieri et al. [25] have presented a realization that uses a projection in the direction of wordlength to achieve a small silicon area. The realized IC consists mainly of a transposition memory, a parallel to serial and a serial to parallel converter, and an operative part. The operative part is a direct mapping of a N=16 FCT graph into silicon and employs dibit-serial operation. Thus every addition, subtraction, and multiplication is assigned to a physical operator that receives its operands as a series of two-bit groups. Figure 16 depicts an adder and a multiplier for dibit-serial operation. It has been found convenient to realize a multiplier employing a 3-bit modified Booth algorithm because two bits of the variable operand are received at every clock cycle. The second operand is a fixed coefficient that is hard-wired in the multiplier to reduce its area and to increase its speed. The realized IC supports two-dimensional DCT and IDCT with N=4, N=8, or N=16 for sample rates up to 13.5 MHz. It has been fabricated in a 1.25-μm CMOS technology and contains about 114,000 transistors in a die area of 40 mm^2.

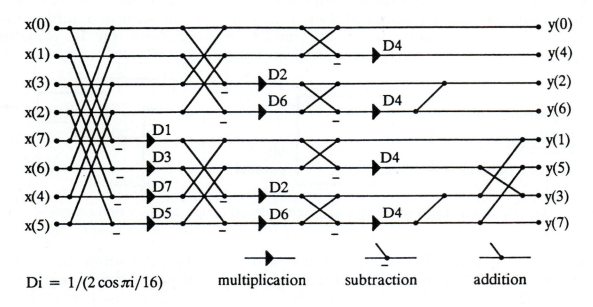

Fig. 15:   Signal flow graph of the fast cosine transform (FCT) for N=8.

A horizontal projection in the signal flow graph (Figure 15) can also be employed to achieve a single-IC realization. Through horizontal projection, operations are mapped to PEs, called butterfly PEs, that generate two result values from two input values. One result is the sum of the two input values while the other result is their difference multiplied by a coefficient. A realization based on this PE has been proposed by Ruetz and Tong [10, 26]. Using a 1-μm CMOS technology, an IC containing eight butterfly PEs, four each for row and column processing, together with a transposition memory has been realized. It is capable of executing DCT and IDCT on 8x8 blocks for video signals with a sample rate up to 40 MHz. Due to the incorporation of a multiplier in each PE, the IC is not totally restricted in flexibility. Besides transformation, a separable FIR filter specifically devised for the hybrid coding scheme (Figure 4) can also be executed.

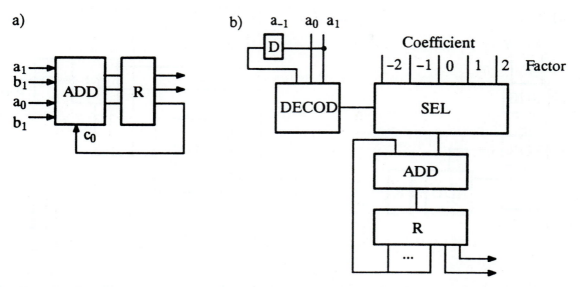

Fig. 16: Block diagrams of dibit-serial operators: a) adder and b) multiplier based on modified Booth algorithm.

After the horizontal projection described above, a subsequent vertical projection can be envisioned to generate a realization example containing one PE. Ligtenberg and O'Neil [27] have reported on an IC consisting of three parts: a communication processor for data permutation and I/O, a look-up table supplying coefficients, and a PE for complex multiplication incorporating four multipliers and two adders. The IC realized in a 1.25-μm technology can perform 8x8 DCT or IDCT for a sample rate of up to approximately 6.4 MHz.

5.2. Dedicated realizations of block matching

The block matching algorithm (*see* 2.1) requires the calculation of $(2p+1)^2$ sums (eq. 9) and a subsequent detection of the minimum sum (eq. 10) to estimate the displacement vector (eq. 11) from a full search of all candidate blocks. A realization of the block matching algorithm thus needs two types of basic PEs. Both are depicted in Figure 17. One basic PE generates the absolute difference between two pels x and y from reference and candidate block, respectively, which is then accumulated for all N^2 pels within a block (Figure 17a). Figure 17a exemplifies an efficient implementation of the magnitude operation by using the MSB of the difference value to trigger bitwise inversion in the XOR operator as well as incrementation in the subsequent adder. In the second basic PE, the minimum sum is searched among the accumulated sums s(m,n) and the corresponding displacement vector is detected (Figure 17b).

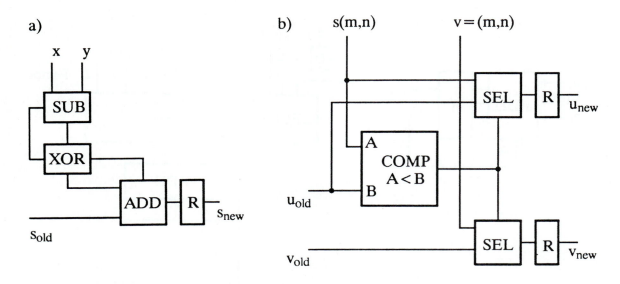

Fig. 17:   Basic PEs for the block matching algorithm: a) absolute difference PE (type AD) and b) minimum PE (type M).

An IC dedicated to real-time implementation of the block matching algorithm has to contain a certain number of absolute difference PEs (Figure 17a) to achieve the required processing power [28]. Furthermore, several minimum PEs (Figure 17b) have to be incorporated too. Their number can in principle be smaller than the number of absolute difference PEs by a factor of N^2. Additional hardware resources in the IC are needed to realize an adequate data transport into the PEs. A total of $2 \times N^2 \times (2p+1)^2$ pels from reference block and search area has to be transported into PEs during the calculation of one displacement vector. The corresponding data rate cannot be transferred accross the IC boundary due to a limitation of the pin count. In order to reduce the data rate at the IC boundary, local memories have to be incorporated besides the PEs. For example, the number of pels loaded into the IC during calculation of one vector can be reduced to $N^2 + (N+2p)^2$ if two local memories operating as double buffers are provided for reference block as well as search area. The data rate at the IC boundary can be reduced further by taking an overlap of successive search areas into account.

A general structure of a block matching IC thus has to consist of a local memory interconnected to several absolute difference PEs and subsequent minimum PEs. The local memory can be realized employing RAM or shift registers. If a RAM is incorporated it has to be accompanied by an address generation unit. These hardware resources are sufficient to perform the full search block matching algorithm with a small control overhead. Search strategies have been proposed that reduce the computational effort by considering only a few of the

$(2p+1)^2$ candidate blocks in a search area. These search strategies however call for an enlarged control overhead and a less regular data flow. The realization examples discussed in the following thus all employ the full search method.

Systolic array architectures have been shown to be very appropriate for the realization of full search block matching because the data flow of the algorithm can be mapped well onto the array [28]. A methodology has been presented that provides a mapping of the algorithm into one- and two-dimensional systolic arrays with distinct numbers of PEs. Several systolic arrays have been derived. Their performance has been simulated assuming a clock frequency of 40 MHz, and the transistor count required in an IC realization has been estimated. Figure 18 depicts the architecture of a systolic array with N absolute difference PEs and one minimum PE as an example.

An IC containing this array would provide real-time execution of the block matching algorithm with N=p=8 for the luminance component of the video telephone format CIF up to about 7.7 Hz frame rate [28]. The transistor count for this IC was estimated to be 110,000 when implementing shift registers as local memory. An IC containing a two-dimensional systolic array with N^2 or Nx(2p+1) PEs of type AD was estimated to achieve real-time performance for the luminance component of the broadcast video format [2] with a frame rate up to 11.5 or 31.5 Hz, respectively, requiring about 130,000 or 245,000 transistors, respectively.

De Vos and Stegherr [29] reported on one- and two-dimensional systolic arrays performing block matching with N=16 and p=7. They employed multiplexing or timesharing of PEs in an array containing N^2 PEs of type AD to derive several arrays with smaller number of PEs. A block matching IC for the broadcast video format could be realized using 1.5-µm CMOS technology. Incorporating NxN/4 PEs and shift registers as local memory, this IC was estimated to require about 200,000 transistors operating at a clock rate of four times the processed pel rate. To perform real-time block matching for the CIF format with 10-Hz frame rate, an IC based on a linear systolic array with N=16 PEs was devised that operates at a clock rate 16 times higher than the pel rate. In this IC, the shift registers can be replaced by a RAM with simple addressing scheme. Building the RAM from a 3-transistor memory cell reduces the estimated transistor count to about 80,000.

Bhandal, Considine, and Dixon [30] have designed an IC for block matching with p=7 containing two semi-systolic arrays with 8x8 PEs each, two shift register chains for search area data, and a processor to combine partial results from the arrays and detect the minimum sum. Fabricated in a 1.4-µm CMOS technology, the IC incorporates about 300,000 transistors in a die area of about 121 mm² and has been designed to operate at a clock frequency of 13.5 MHz. One 8x8 array supplies enough computation power to perform real-time block matching with N=8 for the CIF video format with 30-Hz frame rate. Due to small local memory, in this application $(2p+N)^2$ = 22x22 pels from the search area would have to be loaded during every displacement vector calculation leading to a required input pel rate of 23 MHz. Assuming an input pel rate of 13.5 MHz and an external frame memory for search area data supporting this rate without additional hardware, one IC can perform real-time block matching with N=8 for CIF with a frame rate up to about 17.6 Hz. In an application of block matching with N=16, the two 8x8 arrays in an IC cooperate to process half a 16x16 block and thus two ICs are necessary. They allow the real-time execution of block matching for the CIF format with a frame rate of 30 Hz requiring search area data to be loaded in two distinct streams with a pel rate of 7.84 MHz each.

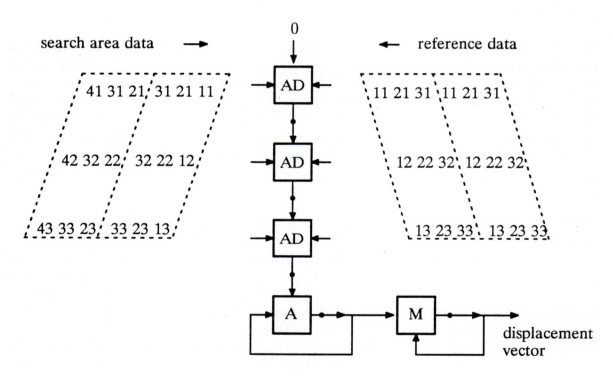

Fig. 18:   Architecture of a one-dimensional systolic array containing N absolute difference (AD) PEs for N = 3.

Besides a systolic architecture making efficient use of regular data streams, an architecture with parallel and random access to local RAM memory can also achieve the high requirements imposed by real-time block matching. Ruetz and Tong [10, 31] have developed an IC for block matching with N=8 or N=16, and a search area including all candidate displacements from -N/2 to N/2-1 in horizontal and vertical direction. The IC is fabricated in a 1.5-μm CMOS technology and incorporates two double buffered input memories, one for storage of the NxN reference block pels and the other for the 2Nx2N search area pels. The memories are controlled by two address generation units. By allowing simultaneous read and write access to alternating buffer memories and by taking advantage of the overlap of successive search areas, the data rate required for input into the IC can be supplied by two distinct data streams limited to the processed pel rate each. A high computation power is supplied by 32 PEs of type AD that are connected to a subsequent minimum PE. With these hardware resources, the IC operating at 30-MHz clock frequency can perform real-time blockmatching with N=16 for a processed pel rate up to 3.4 MHz which is sufficient for the CIF video format with 30-Hz frame rate. With N=8 and operating at 40-MHz clock frequency, the IC supports a processed pel rate up to about 11.2 MHz.

References

[1] CCITT Study Group XV: Recommendation H261, Video Codec for Audiovisual Services at px64 kbit/s, *Report R37*, Geneva, July 1990.

[2] CCIR Recommendation 601: Encoding parameters of digital television for studios; in *Recommendations and Reports of the CCIR*, vol. XI, pt. 1 ITU 1982, Geneva, Switzerland.

[3] CCIR, Draft new report AD/CMTT on the digital transmission of component coded television signals at 30-34 Mbit/s and 45 Mbit/s, *CCIR-Document CMTT/116*, (1986-1990).

[4] H. Yasuda, Standardization activities on multimedia coding in ISO, Signal Processing: *Image Communications 1*, pp. 3-16, 1989.

[5] ISO-IEC JTC1/SC2/WG11 MPEG-90/176 Rev. 2, 1990.

[6] N.S. Jayant, P. Noll: Digital Coding of Waveforms. *Prentice Hall*, 1984.

[7] A.N. Netravali, B.G. Haskell: Digital Pictures, Representation and Compression. *Plenum Press*, 1988.

[8] H.G. Musmann, P. Pirsch, H.J. Grallert, "Advances in Picture Coding," *Proceedings of the IEEE*, vol. 73, No. 4, pp. 523-548, April 1985.

[9] R.G. Gallager: Information Theory and Reliable Communication. *John Wiley*, 1968.

[10] P. Ruetz, P. Tong, "A VLSI video compression chip set," *Proc. Conf. Electronic Imaging*, 1990.

[11] R.J. Offen (ed.), "VLSI Image Processing," *McGraw-Hill*, New York, 1985.

[12] D. Müller, R. Heiß and C. Hoek, "Video-Codierung mit SIMD-Prozessoren (px64 Kbit/s)", *Proc. ITG-Conf. Mikroelektronik für die Informationstechnik*, Stuttgart, FRG, ITG-Fachbericht 110, vde-verlag, Berlin (1989), 143-150 (in German).

[13] D. Müller et al., "An Advanced Array Processor Realization of a px64 Kbit/s Codec," *Proc. Second Int. Workshop on 64 Kbit/s Coding of Moving Video*, Hannover, 1989.

[14] T. Micke, D. Müller and R. Heiß , "ISDN-Bildtelefon auf der Basis eines Array-Prozessor-IC," *Mikroelektronik*, vde-verlag, vol. 5 (1991), (in German).

[15] P. Pirsch and T. Wehberg, "VLSI Architecture of Programmable Real-Time Video Signal Processor," *Proc. SPIE-Conf. Digital Image Processing and Visual Communications Technologies in the Earth and Atmospheric Sciences*, Orlando, Fl, SPIE vol. 1301 (1990), 2-12.

[16] T. Wehberg and H. Volkers, "Architecture of a Programmable Real-Time Processor for Digital Video Signals Adapted to Motion Estimation Algorithms," *Proc. SPIE-Conf. Visual Communications and Image Processing III*, Cambridge, USA, SPIE vol. 1001 (1988), 908-916.

[17] T. Wehberg, H. Volkers and H. Jeschke, "Architektur eines programmierbaren, digitalen Echtzeit-Videosignalprozessors", *Proc. ITG-Conf. Mikroelektronik für die Informationstechnik*, Stuttgart, FRG, ITG-Fachbericht 110, vde-verlag, Berlin (1989), 157-162 (in German).

[18] H. Jeschke, H. Volkers and T. Wehberg, "A Multiprocessor System for Real-Time Image Processing Based on a MIMD Architecture," *Proc. ESPRIT-Workshop From Pixels to Features II: Parallelism in Image Processing*, Bonas, France (1990), North-Holland/Elsevier.

[19] H. Volkers, H. Jeschke and T. Wehberg, "Cache Memory Design for the Data Transport to Array Processors," *Proc. IEEE Int. Symp. on Circuits and Systems*, New Orleans, (1990), 49-52.

[20] T. Denayer, E. Vanzieleghem and P.G.A. Jaspers, "A Class of Multiprocessors for Real-Time Image and Multidimensional Signal Processing," *IEEE Journ. Solid-State Circuits*, vol. 23 (1988), No. 3, 630-638.

[21] J. Wilberg, "Konzeption einer Feddback-Einheit für ein Processing Element," *Study Thesis*, Institut für Theoretische Nachrichtentechnik und Informationsverarbeitung, Universität Hannover (Oct. 1990).

[22] U. Totzek and F. Matthiessen, "Two-Dimensional Discrete Cosine Transform with Linear Systolic Arrays," in J. McCanny, J. McWhirter and E. Swartzlander (eds.), *Systolic Array Processors*, Prentice Hall (1989), 388-397.

[23] M.-T. Sun, T.-C. Chen and A.M. Gottlieb, "VLSI Implementation of a 16x16 Discrete Cosine Transform," *IEEE Trans. Circuits and Systems*, vol. 36 (1989), No. 4, 610-617.

[24] B.L. Lee, "A New Algorithm to Compute the Discrete Cosine Transform", *IEEE Trans. Acoustics, Speech, and Sig. Proc.*, vol. ASSP-32 (1984), No. 6, 1243-1245.

[25] A. Artieri et al., "A One Chip VLSI for Real Time Two-Dimensional Discrete Cosine Transform," *Proc. IEEE Int. Symp. Circuits and Systems*, Helsinki (1988), 701-704.

[26] LSI Logic Corp., "L64730 (DCT) Discrete Cosine Transform Processor," advance data sheet, Milpitas, CA (Jan. 1990).

[27] A. Ligtenberg and J.H. O'Neill, "A Single Chip Solution for an 8 by 8 Two Dimensional DCT," *Proc. IEEE Int. Symp. Circuits and Systems*, Philadelphia (1987), 1128-1131.

[28] T. Komarek and P. Pirsch, "VLSI Architectures for Block Matching Algorithms," *Proc. IEEE Int. Conf. Acoustics, Speech, and Signal Processing*, Glasgow (1989), 2457-2460.

[29] L. de Vos and M. Stegherr, "Parameterizable VLSI Architectures for the Full-Search Block-Matching Algorithm," *IEEE Trans. Circuits and Systems*, vol. 36 (1989), No. 10, 1309-1316.

[30] A.S. Bhandal, V. Considine and G.E. Dixon, "An Array Processor for Video Picture Motion Estimation," in J. McCanny, J. McWhriter and E. Swartzlander Jr. (eds.), *Systolic Array Processors*, Prentice-Hall (1989), 369-378.

[31] LSI Logic Corp., "L64720 (MEP) Video Motion Estimation Processor," advance data sheet, Milpitas, CA (Jan. 1990).

Chapter 15

Digital Television

Albrecht Rothermel, Heinrich Schemmann and Rainer Schweer
Thomson Consumer Electronics, R&D Laboratories, 7730 VS-Villingen, Germany

1. Introduction

Digital signal processing was first introduced in consumer TV sets in the early 1980's for tuning and remote control purposes (digitally controlled TV). The next steps were teletext, digital video, baseband processing, digital audio and digital scan control [1, 2, 3]. Current developments are moving into digital IF-treatment combined with modified tuning techniques.

The increasing use of digital techniques and the need for them are often questioned. The reason for their use is the enhanced performance of the digital circuits in terms of chip area efficiency and power consumption, which makes them increasingly competitive with former analog solutions. Because digital VLSI chips are becoming cheaper and more complex, their special properties can be used more and more. Thus production costs can be reduced as a result of fewer manual adjustments and fewer components. This progress in VLSI chip performance is due to technological progress in CMOS and the availability of high performance BiCMOS technologies. These aspects of high speed circuit design are treated in other chapters of this book. Details of BiCMOS circuits for TV signal processing can be found in [4] and [5].

We shall concentrate below on important system considerations and architectural points. Sound and video processing use programmable circuit techniques, as offered in sophisticated digital implementations, because of the different TV standards dealt with (see Fig. 1). Traditional video standards, such as PAL, SECAM and NTSC are well suited to analog circuit techniques. The newer standards, however, such as MAC (multiplexed analog components) and, as we shall see, NICAM, (near instantaneously companded audio multiplex), demand digital signal processing [6].

All picture quality improvements brought about by multidimensional video signal processing demand digital technology. These are also known under the names of IDTV (improved digital TV = improvements inside existing TV standards), EDTV (enhanced digital TV = enhancements including compatible TV standard modifications) and HDTV (high definition TV).

Whereas IDTV and EDTV can be dealt with more or less inside the current standards, HDTV means a completely new chapter for the studio and receiver equipment industry and for evolving digital techniques. In view of this we shall restrict ourselves to the basic digital techniques employed in current TV receiver design.

Due to historical development and market forces, the TV system is well-suited to analog techniques. Some nonlinear corrections, such as flesh tone correction and gamma correction, have been introduced. Additional circuits found in a normal TV screen have to be provided as there are picture geometry and other corrections. The digital TV has to deal, therefore, with complications arising from the history of analog signal processing.

In this article we first give a short overview of the TV set as a signal processing device. Then we describe the new digital techniques that are proposed for the tuner part, followed by digital audio and video signal treatment. Then we shall mention the problems of deflection processing and, finally, picture quality improvements will be described.

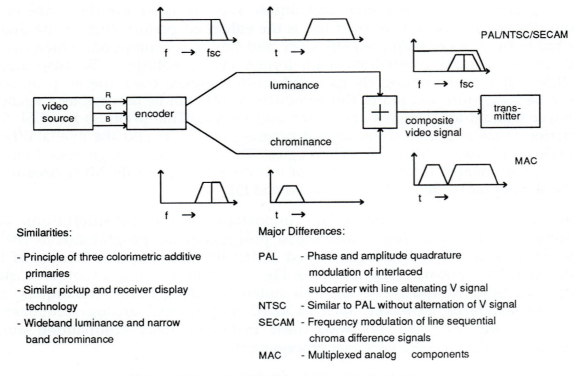

Fig. 1:   TV standards: differences and similarities.

2. TV Tuner

A block diagram of a TV set making intensive use of digital signal processing techniques is shown in Fig. 2. The tuning system at the front has to employ an analog mixing circuit, because the frequency and dynamic range of the radio frequencies cannot be treated by an A/D converter with a reasonably priced device in the near future. The IF stages, however, are now undergoing digitization [7].

Fig. 3 shows a block diagram of the conventional tuner. The antenna signal is preselected by tunable resonant circuits. Then it is preamplified and mixed to the IF range (about 35-40 MHz). The following channel selection is made by surface acoustic wave (SAW) filters, which are analog FIR transversal filters. In these filters the electrical signal is transformed into a mechanical ("acoustical") signal which moves across a glass plate by using the piezoelectric effect. The signal can be picked up by metal strips in such a way that their position and length copies the desired impulse response of the filter.

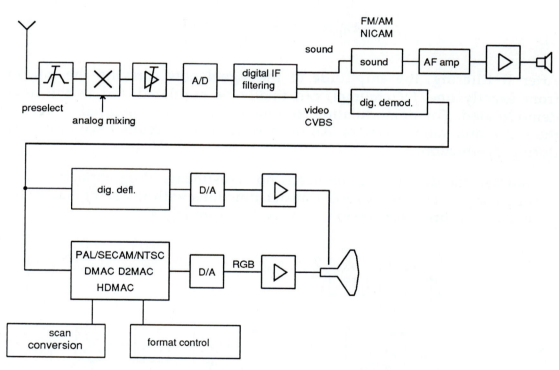

Fig. 2:   Block diagram of digital TV.

This element has been developed for IF-filtering, because the passband characteristic has to be well defined. One edge of the filter function has to have a Nyquist characteristic because of the residual sideband amplitude modulation which is used. The other edge has to be sharp, to suppress the sound carrier. In between the filter characteristic has to be flat. The phase characteristic must also be well defined. With this filter the sound and the video part are also usually separated (quasi parallel sound processing). The sound IF filter suppresses all frequencies except for the video and sound carriers [8].

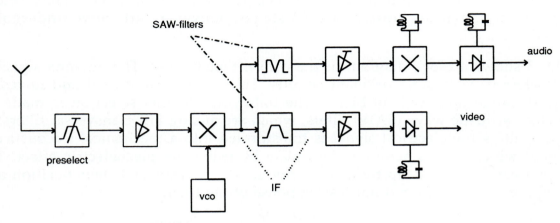

Fig. 3:   Conventional TV tuner.

After IF-filtering, the amplitude of the signal now carrying the information from exactly one TV channel, can finally be regulated and in the next step demodulated. For demodulation, resonant circuits are required to reconstruct the video and sound carriers. All of these resonant circuits have to be adjusted during production.

In contrast, the digital solution needs no adjustments. Programmable filters can also be used to process several transfer functions with the same hardware. The principle of a direct conversion tuner is shown in Fig. 4.

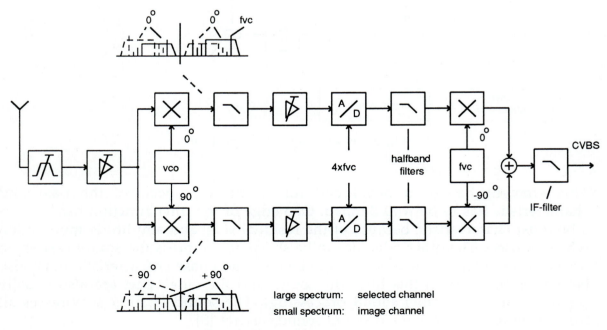

Fig. 4:   Direct conversion tuner.

Most of the digital tuner concepts use another value for the IF frequency, different from those of current analog IF filters. The tuner in Fig. 4 carries out a mixing of the input signal to a frequency range of about 3 to 10 MHz. For this reason the image suppression has to be totally different from that in a conventional tuner. Here, quadrature signal processing is used.

The local oscillator has two output signals which are phase-shifted to each other by 90°. After multiplication of the input signal by these two outputs we obtain two IF signals, which are both disturbed due to the image frequency, but in different ways (see Fig. 4). The mixer is followed by a lowpass filter, amplitude regulation and A/D conversion. If the sampling frequency has been locked to the video carrier frequency, the halfband filters can be designed with a low number of coefficients.

These filters perform one part of the IF-filtering, which is the same for all today's TV transmission standards. The additional mixing with the video carrier frequency (the video demodulation) is a multiplication with a sequence of the numbers 0, 1, 0, -1, which can be implemented digitally with ease.

If the signals in the two branches are now added, we obtain the baseband video directly. The images cancel when both channels are balanced, because of the 180 degree phase shift. The resulting signal can now be separated from the neighbouring channels by a filter whose characteristic has to depend on the TV standard received. Also, a final amplitude regulation has to take place, because this can only be done after the images have been cancelled.

3. Audio Processing

Combined with analog sound demodulators, sigma-delta A/D converter techniques have been used to obtain a digital audio signal after sound demodulation. In the future, however, we shall have a digitized sound signal which is still modulated on the sound carrier. So the sound demodulation will be done digitally. As you can see from Fig. 5, there are several different carrier frequencies and three different modulation techniques that can occur. Therefore, a programmable demodulator structure (Fig. 6) is advantageous.

In the demodulator in Fig. 6, the sound signal is multiplied by its appropriate carrier frequency and thus moved down to baseband. Because we have to deal with amplitude modulation as well as frequency modulation and phase shift keying, the multiplication is done twice by a local oscillator phase of 0° and 90°. In this way we obtain something like a Cartesian coordinate representation of the sound IF-signal. To obtain the phase and the amplitude information from these signals, a coordinate transformation from Cartesian to polar coordinates would be advantageous. This can be done by using the cordic algorithm, which will now be explained in brief [9].

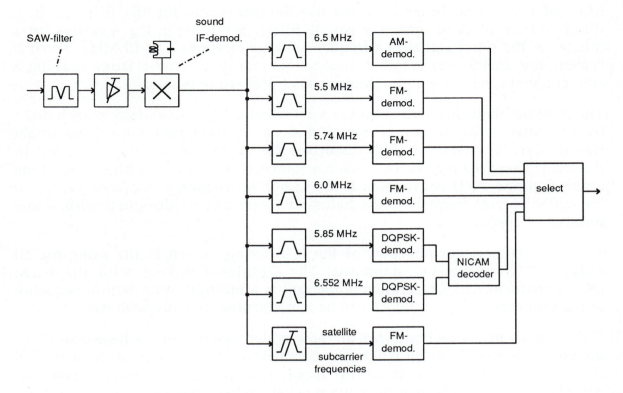

Fig. 5: Multistandard sound processing.

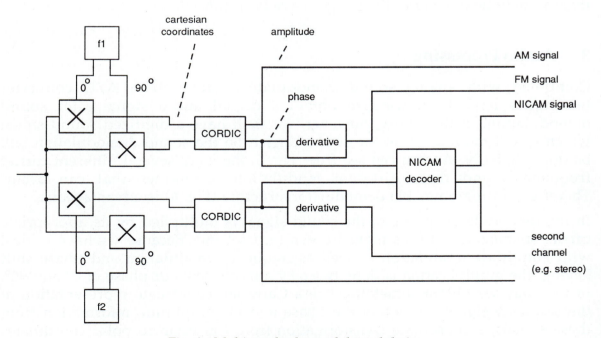

Fig. 6: Multistandard sound demodulation.

With the cordic algorithm, we determine the phase and amplitude information of the signal which is represented by a vector in Cartesian coordinates. The angle can be measured by means of successive approximation, if the vector is rotated by a sequence of rotations to a fixed position, with the angle of these rotations being decreased step by step. This algorithm can be compared with the quantization by a successive approximation A/D converter. Referring to Fig. 7, it can be seen that each vector can be moved very near to the positive x-axis, for example, when it is rotated by a sequence of the angles

$$\pm\,90° \qquad \pm\,45° \qquad \pm\,22{,}5° \qquad \pm 11{,}25°\ldots\,.$$

Convergency is assured, if we test the position of the vector before each rotation, to see whether it is on the upper or on the lower y-plane, and if we choose the direction of the rotation accordingly.

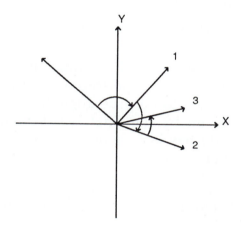

Fig. 7:   Successive approximation of an angle.

So let us assume that the vector lies exactly upon the y-axis. In this case, it would be moved directly to the x-axis by the first rotation. When we assume zero as a positive number, the second rotation will move the vector to an angle of minus 45°. All additional angles will move the vector counterclockwise and nearer to the x-axis. So the result will be that the total angle of the vector is the sum of these angles:

$$+\,90° \qquad -45° \qquad +\,22.5° \qquad +\,11.25° \quad +\,5.625°\ldots\,.$$

keeping only one least significant angle away from the exact result of 90°.

Let us now consider the hardware involved. The position of the vector can easily be tested by testing the polarity of the y-component. The rotation by 90° is a simple exchange of the x and y coordinates and a sign inversion, depending on the signs of the x and y coordinates. When the x coordinate is positive at the beginning, the 90° rotation may be omitted.

The other rotations, however, are not so easy to implement. The general formula of a vector rotation is shown in (1)

$$x1 = x0 \cos(\o) - y0 \sin(\o) \tag{1}$$

$$y1 = y0 \cos(\o) + x0 \sin(\o)$$

The special idea of the cordic algorithm is to substitute the ideal rotations by other rotations, which can be implemented very easily by hardware and which also have decreasing sizes of angle. The length of the vector, however, is modified by these rotations. The formula for the cordic rotations is shown in (2).

$$x1' = \frac{x1}{\cos(\o)} = x0 - y0 \tan(\o)$$

$$y1' = \frac{y1}{\cos(\o)} = y0 + x0 \tan(\o) \quad \text{with } \tan(\o) = \pm \frac{1}{2^n}, \ n = 0,1,2... \tag{2}$$

These rotations can easily be achieved by simple shift and add operations. To obtain the results, only the signs of all performed rotations have to be collected, and this word can be converted to the true angle information using a table in a ROM. The amplitude is given by the final x-coordinate. It should be noted, however, that the amplitude has been modified by a factor of K. (K is about 1,647 for more than five rotations).

By using this technique, the amplitude and phase information are detected in the same circuit. The AM, FM and PSK information can easily be derived from the cordic output signals.

4. Video Signal Processing

4.1. Coding principles

The brightness information of the video signal is a function of a three-dimensional variable. Brightness can vary with changes of position in a three-dimensional space, that is to say, with changes of horizontal and vertical position on the screen and with time. The transmission, however, has a one-dimensional structure. Therefore, the brightness information has always been sampled in two of the three dimensions, that is in the vertical and in the time dimension (Fig. 8). The TV signal format is thus a combination of synchronization signals for the sampled data reconstruction and the continuous horizontal luminance information.

After the black and white TV system had been established, color information was added to it. Compared to the scalar brightness information, color information (including brightness) is a vector comprising three elements. To describe the exact appearance of one picture element on the screen, one needs either the values of brightness, saturation and hue, or, for example, the amplitudes of its red, green and blue components.

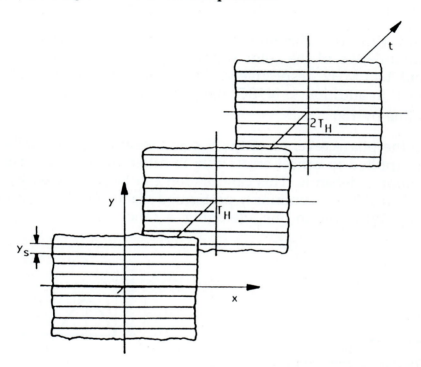

Fig. 8: TV-signal, sampled in two dimensions.

This "three-dimensional" property of the color information, however, is due to the physiology of the human eye. During the process of viewing, the eye maps the physical characteristic of the received light to a special three-dimensional space, which is the same for all normal-sighted people (but is not for color blind people). The physical nature of the light that we see is, however, of infinite dimensions because it has to be described by its spectral distributions. The reduction of the number of dimensions in the human eye is the reason for the fact that totally different spectral light emissions, such as those from the sun, from a light valve, or from a TV screen, can look similar to us. And it is also the reason that nearly every color can be produced by only three different light sources on a TV screen. So we end up with the need of two additional functions to be added to the brightness information of the black and white TV system.

At the time color TV was invented, all of the frequency space available in a TV channel was already used for black and white (i.e., brightness) signal transmission. Color information had to be added to this established system compatibly. This was possible because the human eye has a lower spatial sensitivity to saturation and hue changes than to brightness changes. The color information could be added with reduced bandwidth to the luminace information, yielding acceptable results. This saturation and hue information is now coded as the chrominance signal (Fig. 1) in different ways, depending on the TV standard.

In NTSC and PAL the chrominance information is carried as quadrature amplitude modulation of the color subcarrier, which was chosen to lie around 3.58 MHz and 4.43 MHz, respectively. From Fig. 9 the PAL modulation scheme can be derived. In addition to NTSC, the phase of the $V = R - Y$ signal is altered from line to line by 180°.

The effect of the phase reversal of the V component can be described in several ways. It was applied to cancel errors due to differential phase distortion of the color modulation sidebands. This cancellation can be performed by the viewers' eyes (simple PAL), but is commonly performed via a so-called PAL-delay, which is a 1 H-delay. With line to line averaging, a cancellation of the phase error is achieved.

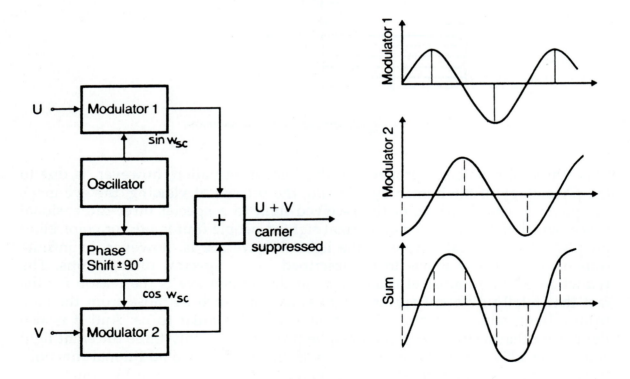

Fig. 9: PAL modulation scheme.

The same principle can be described in a different way. The phase reversal of the V-signal can be obtained by a multiplication of the subcarrier by a square-wave signal whose frequency is half the line frequency. Thus the subcarrier for the V-component is split into two carriers, which have a different frequency from that of the original carrier, which modulates the U-component.

Consequently, the U and V subcarriers can be separated by two-dimensional bandpass and bandstop filters, which are equal to one-dimensional comb filters. This, of course, is insensitive to any phase errors. Phase errors, however, yield amplitude errors due to the synchronous demodulation that follows. These errors in color saturation are less annoying than hue errors. The subcarrier frequency (U-signal) chosen is $fsc = 1135/4 \, fH + 1/2 \, fV$, which produces an interlaced signal as depicted in Fig. 10.

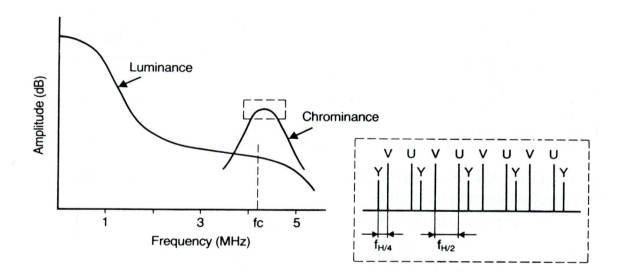

Fig. 10: Frequency relationship of chrominance and luminance.

4.2. Implementation issues

In digital hardware implementation, the signals are represented as sampled data. If the sampling frequency chosen is the multiple of a frequency contained in the composite video signal, the amount of hardware required can be reduced.

So the clock can be locked either to the line frequency, to the color subcarrier (which is a significant difference for practical VCR signals), or could be an independent quartz stabilized fixed clock. With the last choice we get all the disadvantages of the system comparison made in Table 1. The advantage, on the other hand, is the spectral purity of this clock, because it does not have to be generated by a phase locked loop.

Especially for chroma processing, the choice is often four times the color subcarrier for A/D and D/A conversion and clock frequency of the digital signal processing systems. This results in a sampling frequency of 17.7 MHz (PAL) or 14.4 MHz (NTSC). These clock frequencies can be handled easily with today's CMOS technologies. However, parallel and pipelined architectures still dominate at these frequencies. The clock is faster than would be required according to the Nyquist theorem but the advantage is that synchronous demodulation of the color information with this clock can be easily implemented. Also, the overshoot of the signal after alias-free reconstruction is reduced with higher clock frequencies.

Table 1

	Video Processing	Deflection	Interface to Frame Store
Frequency locked to horizontal line	Complex	Easy	Easy
Frequency locked to color subcarrier	Easy	Complex	Interpolator

Fig. 11 shows the main parts of a basic video processing circuit. The main functions that have to be performed are:

- chroma bandpass filtering

- variable peaking filter with chroma trap

- multiplier for contrast adjustments (luminance)

- color processing: automatic color control (ACC), color killer, PAL identification, decoder with PAL compensation or NTSC comb filter, hue correction-multiplier for color saturation (multiplexed)

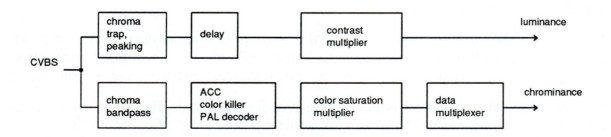

Fig. 11: Block diagram of video processing unit.

4.2.1. *The luminance channel*

This channel features a chroma trap for the suppression of the color signal. This also reduces the luminance signal bandwidth. A better approach for luminance/chrominance separation will be discussed in Section 6, quality improvements.

By means of the adjustable peaking the user can choose the optimal compromise between high resolution and low noise. The different frequency responses for various values of the peaking factor are shown in Fig. 12. This diagram, however, shows only the amplitude response. As the eye is very sensitive to phase errors, phase linear filters are preferable for video signals. In this case, the impulse and step response of the filter will be symmetrical. This can be achieved easily when an FIR filter is used.

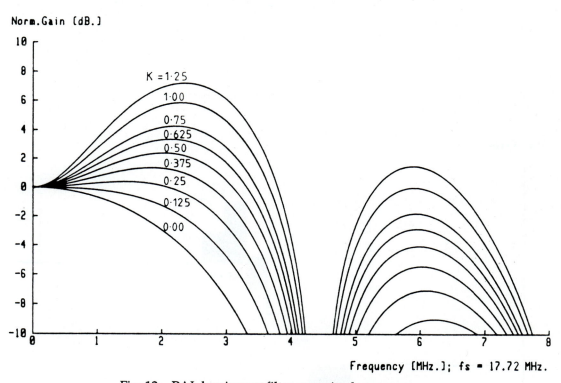

Fig. 12: PAL luminance filter magnitude response.

An example of a filter combining trap and peaking is shown in Fig. 13. This filter needs very little hardware, especially for the trap (only one delay and an adder), because it operates at exactly four times the color subcarrier frequency. The peaking is obtained by the lower subnetwork of Fig. 13. This network performs a double differentiation function and so enhances rapid edge transitions (Fig. 14). The degree to which enhancement takes place is controlled by the variable gain factor K. The effect of variation in K is shown in the frequency domain and time domain responses of Fig. 13 and Fig. 14.

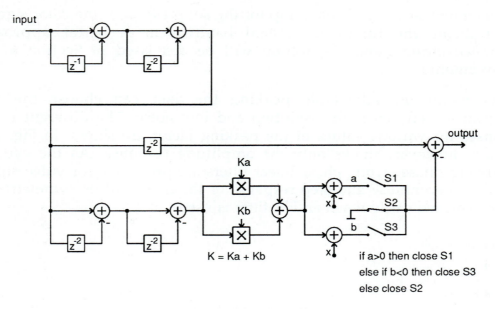

Fig. 13:   PAL luminance filter structure.

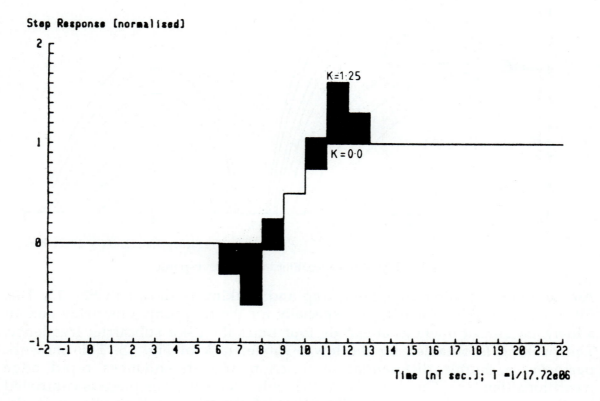

Fig. 14:   PAL luminance filter step response.

Peaking introduces undesired visual noise enhancement in areas of nearly constant grey level on the screen, which is disturbing to the eye and has to be removed. This removal is accomplished by the nonlinear subnetwork incorporated within the filter. Its transfer characteristic is piecewise linear with a deadband of ±x about zero input. The deadband can be controlled by the input parameter x. The contrast multiplier which follows the peaking circuit is combined with a limiter, to restrict the amplitude of the luminance signal to the maximum range of the following stage [1].

4.2.2. *The chrominance channel*

The luminance signal is suppressed in this channel by means of a chroma bandpass. This filter, like the luminance filter, has to have a linear phase characteristic. In addition, it is advantageous if the frequency response can be adapted to various IF filter components. The IF section of a TV can use either a surface acoustic wave or a digital filter (flat chroma passband required), or a lumped element passive filter (equalizing passband required). See Fig. 15.

The stopband attenuation needs to be about 40 dB in the lower band to suppress sufficiently feedthrough of the baseband luminance signal. Also, a reasonably large attenuation is required in the upper band to suppress higher frequency noise being imaged down to the baseband after demodulation and being visible on the screen. An example of a circuit for such a chrominance filter is shown in Fig. 16. This filter uses only multiplication factors of ± 1 and can therefore be implemented cost effectively.

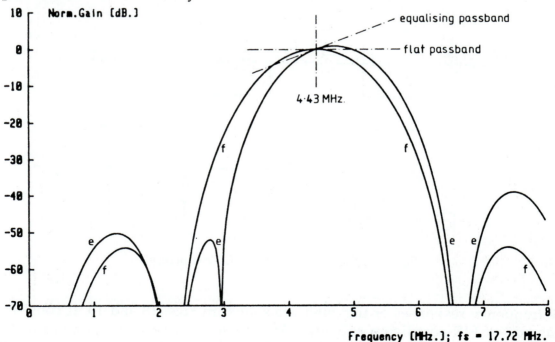

Fig. 15: PAL chrominance filter magnitude response.

This filter can also be used for a NTSC system, when the clock frequency is changed. For SECAM, however, a completely different solution is necessary [10]. The SECAM chrominance filter must compensate both the phase and the amplitude of the transmitted signal. Due to the varying properties of the IF section in different TV sets, a variable passband equalizer has to be employed. An example of a SECAM chrominance filter is shown in Fig. 17.

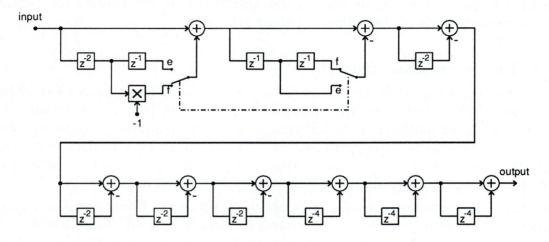

note: e => equalising passband response
       f => flat passband response

Fig. 16:   PAL chrominance filter structure.

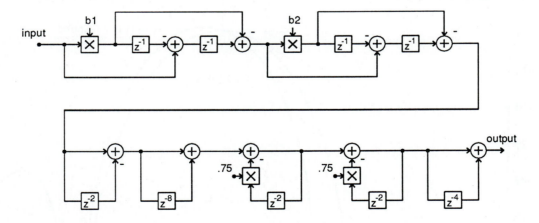

Fig. 17:   SECAM chrominance filter structure.

The frequency responses of this filter are shown in Figs. 18 and 19. It consists of a cascade of FIR equalizing subnetworks, providing a variable gain gradient at 4,43 MHz (Fig. 18) and a cascade of FIR and IIR subnetworks forming the bandpass network resulting in the response shown in Fig. 19. Note that in this filter the recursive parts use a minimum delay of two clock cycles, which allows some pipelining for coefficient multiplication.

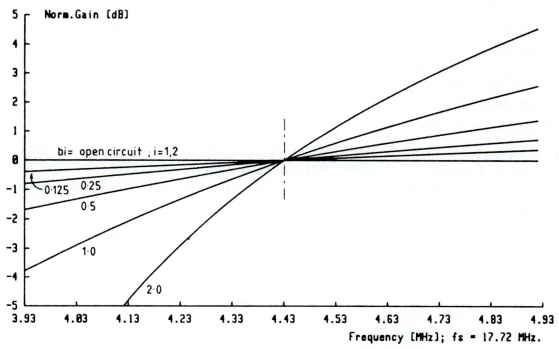

Fig. 18: SECAM chrominance filter magnitude response.

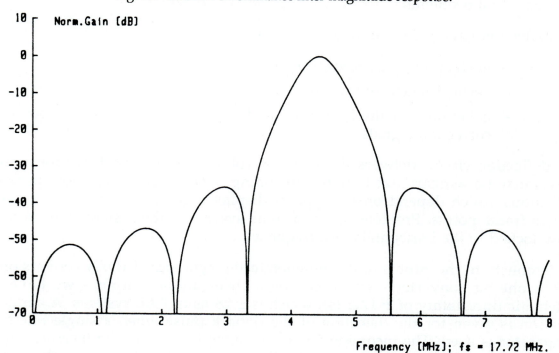

Fig. 19: SECAM chrominance filter magnitude response.

These luminance suppression filters are followed by automatic color control (ACC) which measures the burst amplitude and takes it as a reference signal. If this signal is too weak, the chrominance is switched off by the color killer. If it is large enough, the amplitude of the color information is regulated accordingly.

The chrominance signal has to be decoded by an AM synchronous demodulator for PAL and a FM demodulator for SECAM, respectively. The next step is the PAL compensation circuitry, which needs storage capabilities for information from one TV line, realized, for example, as a RAM configuration. The signals, whether delayed or not delayed, are added according to the PAL coding scheme, to compensate for phase errors.

For SECAM signals, the line storage circuit is also necessary to perform decoding of the sequential color transmission scheme. In NTSC operation, this delay line can be used to perform comb filtering (that means two-dimensional filtering), improving cross color suppression.

Because the bandwidth of the demodulated chrominance signal is low (about 1 MHz) the color saturation multiplier can be multiplexed and can therefore be used for both R-Y and B-Y color difference signals.

5.   Scan Control Processing

5.1. Tasks and problems

The deflection circuit has the following tasks:

- horizontal synchronization
- vertical synchronization
- generation of the horizontal and vertical deflection and geometry correction signals.

The deflection circuit delivers the scan control signals for the deflection unit. They must be synchronized to the incoming video signal and incorporate corrections which depend on the type of display used. Therefore they can be seen as frame points. Problems arise in some systems if the system clock is not phase locked to the horizontal clock frequency.

To demonstrate the problem, if we generate the sync-signal (D/A-conversion) with a one-bit converter - which can be a clean digital output - we have a systematic uncertainty of $\pm 1/(2 f_s)$, which is ± 28 ns in PAL systems. A similar situation is given for the detection of the flyback pulse when a single one bit comparator is used. The incoming video signal, and therefore the inserted sync signal too, has a better resolution (7 bit) and is bandlimited. Thus the sync position can be determined with better temporal resolution.

A clock period of 56 ns corresponds to a horizontal distance of 0.5 mm on the screen of a picture tube with a diameter of 66 cm. This is a width which can be neglected, if you look at VTR generated pictures or normal TV pictures. The disturbing effect is that this displacement occurs statistically from frame to frame due to noise, thus producing the impression of movement, especially at the edges. This oscillating or fringing can be recognized on good (66 cm) monitors and with good TV signals down to a tenth of the actual pixel width. The reason is that the eye is very sensitive to movements in a quiet picture. The flyback signal is not bandlimited. Thus the detection of the rising edge resolution has to be performed by either a higher system clock or with an antialiasing filter together with an A/D converter having a resolution of more than one bit.

The preceding comparison (Table 1) of the sampling clock coupled to either the horizontal frequency or the subcarrier frequency, appears in a different aspect with regard to monitor control and corrections.

5.2. Deflection processor

The deflection processor unit shown in Fig. 20 contains the circuit solutions to these problems. The most interesting part, apart from the clamping circuit, is the horizontal synchronization. The block diagram is shown in Fig. 21. It comprises two different phase comparators and two filters to fulfil the requirements described above. Two different operation modes exist, the so-called non-locked mode, and the locked mode. In the latter case there is a fixed ratio between the color subcarrier frequency and the horizontal frequency. In this mode the influence of the PLL filter on the programmable divider is switched off. Therefore, interfering noise and pulses cannot influence horizontal deflection, if they occur during the sync signal.

Nevertheless, phase comparator two is not disconnected and the influence of the monitor-inherent distortions and signal distortions are measured by means of a balanced gate delay line, to get a time resolution of the leading edge up to ±3.5 ns. With the same resolution the leading edge of the outgoing sync signal can be adjusted.

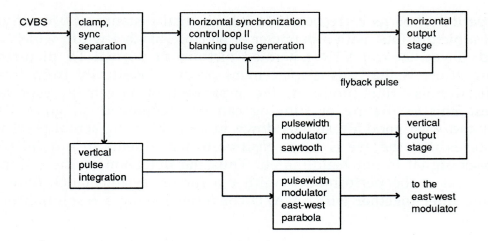

Fig. 20:   Block diagram of deflection processor.

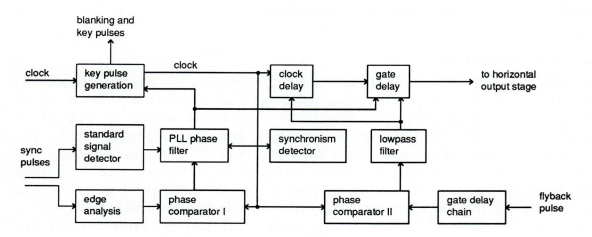

Fig. 21:   Principle of horizontal synchronization.

In the unlocked mode one has to think of an internal running horizontal frequency which is calculated up to fractions of a clock cycle. The difference to the incoming signal can be calculated by analyzing the sync signal up to the same resolution (bandlimited). The difference will be fed to a lowpass filter which has variable coefficients (VTR, standard signals). Its result (msb) is routed to the programmable divider to close the PLL loop. The reminder (less than one lsb of the divider) is presented to the clock and gate delay circuitry. Here, due to the influence of flyback, this value will be corrected and presented to the monitor via the programmable gate delay. The influence of the flyback is also filtered.

5.3. Picture geometry correction

The monitor consists of a relatively flat screen. This results in picture geometry distortion, which looks, if completely uncorrected, like a pin cushion. The distortion in the horizontal direction is called east-west distortion and the distortion in the vertical direction is called north-south distortion. Corrections can be applied by modulating the horizontal and vertical deflection signals. The corresponding signals, which are known as east-west parabola and vertical sawtooth (see Fig. 22), can easily be generated as pulse width modulated signals. In this case a lowpass filter must be provided at the output pin for analog reconstruction.

Fig. 22:   East-west parabola and vertical sawtooth.

6.   Quality Improvements

6.1. Improved luma/chroma separation

The physical nature of the luminance TV signal is three-dimensional. Therefore, a three-dimensional spectral description of this signal also exists, see Fig. 23 [11].

The color subcarrier frequency has been chosen in such a way that its distance from the baseband luminance signal reaches its maximum in this three-dimensional frequency domain.

Optimum separation of luma and chroma would therefore need three-dimensional filtering. In contrast to this, state of the art for PAL receivers is one-dimensional filtering (see chapters 4.2.1 and 4.2.2.). Two-dimensional filters need line delay circuits, and three-dimensional filters need, in addition, whole frame storage means. Three-dimensional filters, however, have some problems with proper motion signal processing.

For these reasons, we shall describe a two-dimensional approach [12], which has already yielded significant improvements when compared to one-dimensional filtering. Fig. 24 shows the two-dimensional locations of NTSC and PAL luminance and chrominance signals.

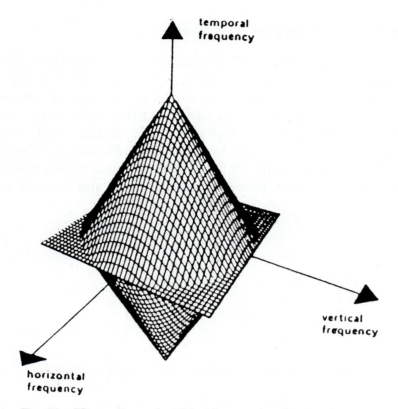

Fig. 23: Three-dimensional luminance signal spectrum.

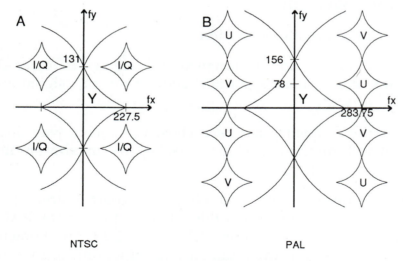

Fig. 24: Two-dimensional NTSC and PAL spectra.

One intention in the design of vertical filters is to use a minimum number of coefficients, because each coefficient requires a complete line delay. Fig. 25 shows simple vertical filters, which can be used to separate luma and chroma information. For the chroma signal it is in any case advisable to combine these filters with a horizontal bandpass filter. By these means the horizontal luminance frequency response is no longer degraded, although the color information is sufficiently suppressed. The separation of the I and Q components in the chroma path is achieved only by synchronous multiplex type demodulation.

The filters for PAL need a different design, because the location of the color information is different. Fig. 26 shows a two-dimensional PAL luminance filter. It has two zeros in its vertical frequency response and thus suppresses the color information.

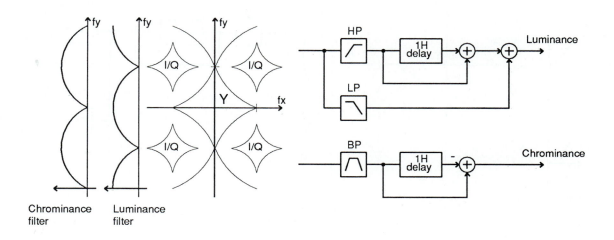

Fig. 25: Vertical luma/chroma filters for NTSC.

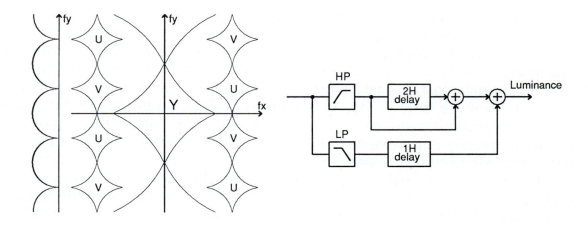

Fig. 26: Two-dimensional PAL luminance filter.

The design of the chrominance filter can be described in three steps. The first step is the bandpass filtering mentioned before. The second step is the suppression of the luminance signal and the third step is the separation of the U and V components. As can be seen from Fig. 27, the second part of the filter (after the horizontal bandpass) suppresses the luminance components at high horizontal frequencies, and the third part suppresses the U and V components.

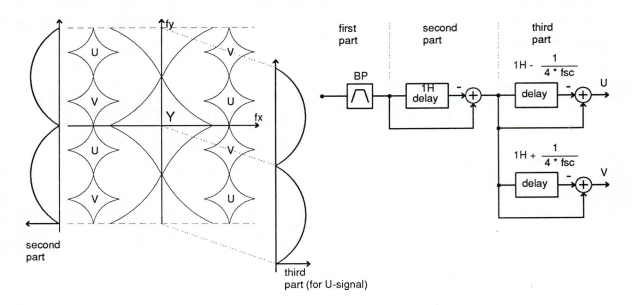

Fig. 27: Two-dimensional PAL chrominance filter.

The problem that arises with the filters in Figs. 26 and 27 lies in the fact that two line delays are used in these filters. If there are very sharp vertical transients, such as horizontal edges, the signal calculation over three TV lines can cause visible artifacts. Therefore, it is advantageous to detect these edges and to switch over adaptively to two-line processing. This means for chrominance that the second part of the filter in Fig. 27 is skipped in these cases. Due to this, vertical smoothing and other annoying effects can be reduced, but the luminance signal is not completely suppressed. In the luminance filter in Fig. 26, only the lower part (with the LP-filter) is used in the case of horizontal edges, giving the same quality as in standard TV receivers. With this so-called adaptive two-dimensional luma/chroma separation algorithm, reduced cross color and cross luminance can be obtained, together with a significantly improved resolution of the luminance information.

6.2. Flicker reduction

There are two disturbing flicker effects associated with the current TV distribution standards. These are more perceptible in the European TV systems: the 50-Hz field frequency, which leads to large area flicker, and the 25-Hz line flicker. Both flicker effects become increasingly annoying as picture tube technology improves. Increasing brightness and viewing angles due to larger screens improve the "temporal resolution" of the human eye and thus the flicker perception.

The flicker effects can be reduced by several approaches, which differ in performance and hardware requirements. In all cases, the line frequency has to be increased from about 16 kHz at present. On the other hand, a line frequency of 64 kHz or more needs very expensive hardware, due to power considerations. Therefore, we describe two approaches which use double the line frequency of standard TV display technology. These are the 100-Hz interlace technology and the 50-Hz progressive scan technique.

The 100-Hz approach uses a field rate upconversion [13]. This cannot be done by simply repeating the field information, due to the interlace scanning (Fig. 28). Today, the information of one whole frame of the TV signal is transmitted every 40 ms, which means with a frequency of 25 Hz. The frames, however, are split into two fields, which consist of half the number of TV lines each. During reproduction, as well as scanning in the TV camera, two of the fields are interlaced to one frame.

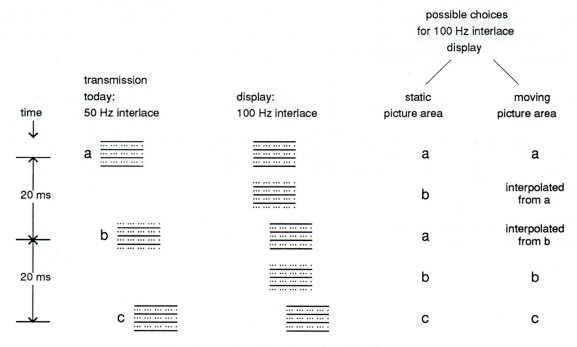

Fig. 28: Interlace at 50 Hz and 100 Hz.

With this kind of operation, most spatial vertical resolution is obtained by the temporal averaging of two fields in the eye. On the other hand, when motion occurs on the screen, each field is a separate snapshot of the motion sequence, thus delivering a high temporal resolution.

Now we assume the transmitted information of the fields *a*, *b* and *c* in Fig. 28. For 100-Hz interlace reproduction, the first step is displaying field *a*. As the second field we would like to display the lines carrying the information of field *b*. This, however, is only allowed if no motion occurs, because in the third field the lines of the former field *a* will be repeated. So we end up with the conclusion that picture areas that are static have to be processed in a different way from picture areas that change due to motion. In the still areas we shall have optimum spatial resolution, whereas in moving picture areas we shall have optimum temporal resolution. As can be seen from Fig. 28, in static picture areas we have simply to repeat the received information, but in moving picture areas a vertical interpolation has to be carried out, which is done by lowpass filtering.

Fig. 29 shows the simplified block diagram of a field rate upconverter as discussed so far. Two field stores are needed and the output signals have to be selected according to the motion detector output signal. The motion detector has to decide between motion or not for every picture element. To reach this decision, it calculates the differences between two frames for the neighbouring picture elements (Fig. 30). To avoid decision problems due to noisy TV signals, a soft decision algorithm has to be employed. But it turns out that it is very difficult to find an acceptable algorithm for this. So, the first 100-Hz TV sets that came to the European consumer market worked without any motion detection algorithm. This is done by simply repeating each field. The drawback of this method is, however, that 25-Hz line flicker is not reduced.

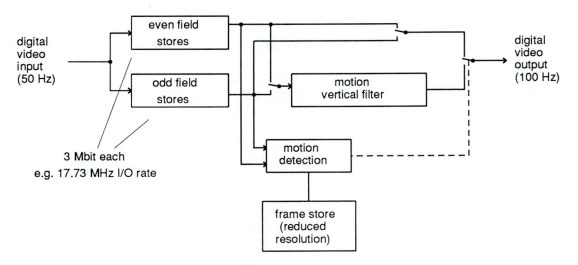

Fig. 29: Principle of field rate upconverter.

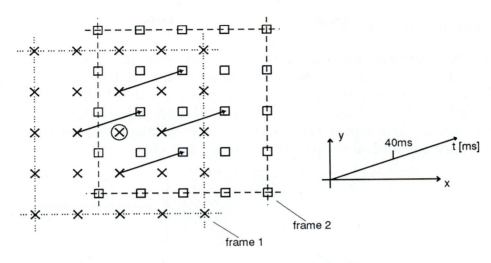

The differences between two frames have to be calculated for
all picture elements which are surrounding the actual pixel.

Fig. 30: Motion estimation.

The 25 Hz line flicker can be reduced with a smaller amount of hardware
employing the 50 Hz progressive scan approach. Each field displays all 625 TV
lines on the screen. The TV lines which have not been transmitted, are
interpolated within each field from the original lines. The resulting picture
quality depends on the amount of hardware for the interpolation; a two-
dimensional approach yields the best results.

The basic principle of a 50-Hz progressive scan processor can be seen in Fig. 31.
As the signal processing is done within each field, no problems occur with
motion.

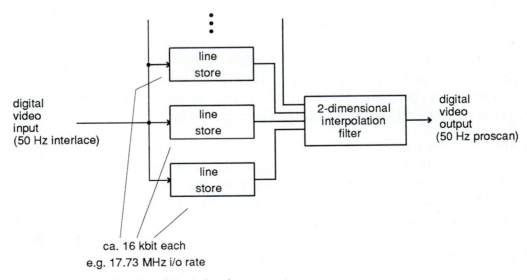

Fig. 31: Principle of progressive scan converter.

6.3. Format control

With the introduction of HDTV, or in Europe even earlier with the introduction of D2MAC, a modified aspect ratio of 16/9 will be used. Future high-end TV sets will employ 16/9 picture tubes. Obviously, for a long intermediate time a large amount of distributed video information will remain in the former 4/3 aspect ratio and this will be displayed on the new tubes as well. There are several proposals to display 4/3 video on 16/9 screens (Fig. 32).

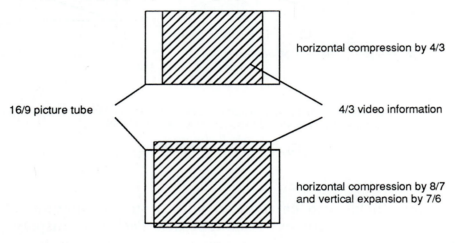

Fig. 32:   Displaying 4/3 video on 16/9 picture tubes.

Fig. 33 shows the principle of compression in time via sampling rate conversion. It should be noted that the clock frequency is the same before compression and afterwards. A practical realization is depicted in Fig. 34. The original signal has to be oversampled and an appropriate low pass filtering has to be performed. The following linear interpolation suppresses the repeated spectra in the frequency domain. Therefore, in the next step, sampling with any frequency is possible without producing visible artifacts. It is, of course, important that the low pass filtering in the first step corresponds to the sampling at the end, thus avoiding aliasing effects.

To give the consumer the maximum freedom in choosing the display format, an adjustable magnification in fine steps would be desirable. On the other hand, for an easy control of the video geometry on the screen, a constant amplitude of the horizontal deflection is very advantageous. Therefore, a demand exists for an infinitely variable compression of the horizontal video information. The compression can be done in two ways, either clock frequency variation or sampling rate conversion [14].

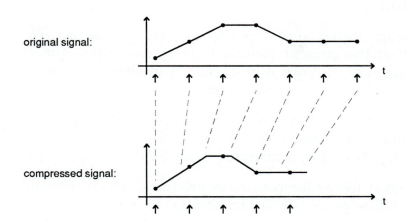

Fig. 33: Compression of sampled signals.

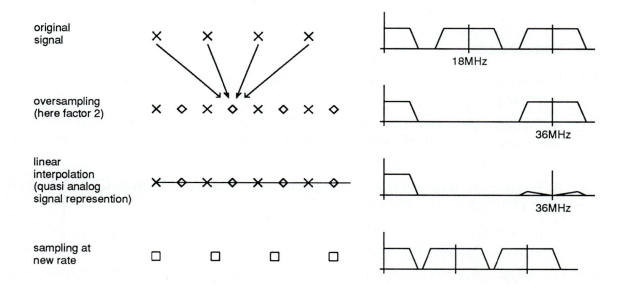

Fig. 34: Sampling rate reduction for compression in time.

7. Conclusion

When discussing digital signal processing algorithms for TV it has to be remembered that the current standards were developed for analog receiver techniques. However, today's multistandard environment, including satellite transmission, requires programmable structures.

In additon, the amount of digital signal processing complexity which can be implemented in single chips is steadily increasing. Thus, more and more blocks of current TV design will be transferred to single or multiple digital chips.

Due to the market forces for consumer products, which demand high volume production with low prices, dedicated solutions are used for the different signal processing sections in a TV set.

8. References

[1]   ITT Semiconductors, "DIGIT 2000 digital TV system," data sheets, Freiburg, Germany.

[2]   R. DeFrancesco and G. d'Andrea, "Digital processing of TV signals: system overview of a new IC solution," *IEEE Trans. on Consumer Electronics*, vol. 34, no. 1, pp. 246 - 252, February 1988.

[3]   S. Suzuki, K. Muramatsu, T. Makino and S. Nose, "2 chip CMOS digital TV," *IEEE CICC*, pp. 323 - 326, 1987.

[4]   A.R. Alvarez, editor, *BiCMOS Technology and Applications*, Kluwer Academic Publishers, 1989.

[5]   A. Rothermel, *Digitale BiCMOS-Schaltungen*, Munich, R. Oldenbourg Verlag, 1990.

[6]   R. Hopkins, "Advanced televison systems," *IEEE Trans. on Consumer Electronics*, vol. 34, no. 1, pp. 1 - 15, February 1988.

[7]   J. Gray, J. Treichler and M. Larimore, "The AST all-digital tuner chip set," *26th Asilomar Conf. on Signals, Systems and Computers*, pp. 199 - 203, 1988.

[8]   K.B. Benson, editor, Televison Engineering Handbook, New York, *McGraw-Hill*, 1986.

[9]   J.V. Volder, "The CORDIC trigonometric computing technique," *IRE Trans. on Electron. Comput.*, vol. EC-8, no. 3, pp. 330 - 334, 1959.

[10] *Characteristics of television systems*, Report 624-3, section 11A, vol. 11, pt. 1, Geneva: CCIR Recommendations, pp. 1 - 33, 1986.

[11] J.O. Drewery, "The filtering of luminance and chrominance signals to avoid cross-color in a PAL color system," BBC Research Department, *report BBC RD1975/36*, pp. 1 - 35, December 1975.

[12] S.S. Perlman and R. Schweer, "An adaptive luma-chroma separator circuit for PALand NTSC TV signals," *IEEE Trans. on Consumer Electronics*, vol. 36, no. 3, pp. 301 - 308, August 1990.

[13] D. Gillies, M. Plantholt and D. Westerkamp, "Motion adaptive field rate upconversionalgorithms for 900 lines/100 Hz/2:1 displays," *IEEE Trans. on Consumer Electronics*, vol. 36, no. 2, pp. 149 - 160, May 1990.

[14] R. Crochiere and L. Rabiner, "Multirate digital signal processing", *Prentice Hall*, Englewood Cliffs, 1983.

Chapter 16

VLSI Architectures and Circuits for Digital Coding of High Definition Television

P. Pirsch and K. Grüger

Institut für Theoretische Nachrichtentechnik und Informationsverarbeitung, Universität Hannover,
Hannover, Germany

1. Introduction

The current perspectives on broadcast television (TV) are directed to TV receivers with new features and improved picture quality. Examples of new features are freeze frame, picture-in-picture and zoom. Advanced signal processing will offer reduction of large area flicker, noise suppression and reduction of cross talk effects. Besides the advantages of the current broadcast TV, essential improvement of picture quality is envisaged by new high definition TV (HDTV) systems. The HDTV systems have increased spatial resolution (factor of 2 in each dimension) and an increased aspect ratio of 16:9 for display. Also improvement of the temporal resolution by replacing the line interlaced scanning by progressive scanning is under discussion. Severe transmission problems for HDTV result from the fact of considerable increase of bandwidth. For reduction of bandwidth, analog multiplexing of the video components combined with adaptive spatial and temporal subsampling is proposed. Bandwidth reduction by such a technique is realized by the MUSE [1] and the HD-MAC [2] scheme. Due to the advances of fiber optic transmission systems digital transmission also has to be considered. For the broadband ISDN, digital links with bit rates in the order of 140 and 565 Mbit/s are planned. In particular for studio-to-studio transmission digital systems are necessary in order to satisfy the high picture quality requirements. Also HDTV distribution by digital links will be offered in the future.

Digitalization of HDTV signals results in very high source rate. For this reason source coding schemes for reduction of bit rate have to be applied. In order to achieve compact video codec realization with moderate manufacturing costs, VLSI circuits for the key components are requested.

In the next section, source coding algorithms for HDTV coding will be presented. Then general VLSI implementation aspects will be discussed. After that, VLSI architectures and implementations for a selection of key components follow. These components are circuits for differential pulse code modulation (DPCM) and subband filters.

2. Source Coding Algorithms for HDTV

The source rate of digitized HDTV signals is defined by the pels per line, lines per frame and the frame rate. For a European HDTV standard [3] it is proposed to use still 2:1 interlace and to double the horizontal and vertical resolution compared to the broadcast TV standard. Considering the enlarged aspect ratio of 16:9, this results in sampling frequencies of 72 MHz for the luminance component and 36 MHz for each chrominance component (Table 1). Digitalization with 8 bit provides a total source rate of 1152 Mbit/s. Using the same ratio as for broadcast TV between active intervals and blanking intervals, the net bit rate by utilizing the blanking intervals for transmission is 885 Mbit/s.

Table 1:   Parameters of European HDTV standard.

1250/50 Hz/2:1	Luminance	Chrominance
Active image size	1920 x 1152	2 x 960 x 1152
Total image size	2304 x 1250	2 x 1152 x 1250
Sampling frequency	72 MHz	2 x 36 MHz
Source Rate	576 Mbit/s	2 x 288 Mbit/s

Source coding has to reduce the bit rate to a given rate while keeping the picture quality as high as possible. High levels of the PCM hierarchy in Europe have bit rates of 140 and 565 Mbit/s. Therefore, these are preferred rates for digitally transmitted HDTV. The rates above require reduction factors of about 8 and 2, respectively. In order to reduce transmission costs further, even higher reduction factors are in discussion.

Basic source coding schemes for bit rate reduction are predictive coding, transform coding, subband coding and interpolative coding [4] [5]. In many cases combinations of these schemes are proposed. Typical coding schemes will be specified in more detail below.

2.1. HD-MAC coding

Studios and distribution centers will be linked digitally by fiber optic transmission. But for distribution to the home receiver via satellite or cable TV, analog transmission has to be considered for a long term. A source coding scheme devoted to analog transmission is HD-MAC. In order to keep an analog transmission format, a sub-Nyquist sampling method is applied, whereby the spatial resolution is based on the motion within a scene [2]. A first reduction by a factor of two is achieved using a quincunx sampling pattern instead of an orthogonal sampling pattern [6]. This allows full horizontal and vertical resolution, while the diagonal resolution is reduced by a factor of about 0.7. The second reduction is realized by adaptive subsampling in three modes (*see* Figure 1). In the stationary mode (velocity range 0-0.5 samples per frame) offset subsampling by a factor of two is applied such that over four fields (80-ms branch) the original resolution after the first stage is achieved. In the slowly moving mode (velocity range 0.5-12 samples per frame) a subsampling pattern with half vertical resolution is applied (40-ms branch). In the moving mode (velocity range > 12 pels per frame) additional horizontal offset subsampling results in a signal with half horizontal and vertical resolution (20-ms branch). Prior to subsampling, 2D diamond shaped filtering with appropriate edge frequencies is required to avoid annoying aliasing effects. Switching between the modes is implemented on a block basis.

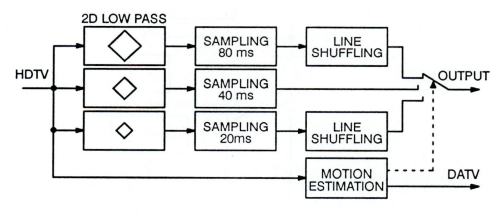

Fig. 1: HD-MAC coder.

For improvement of picture quality, dropped samples can be reconstructed by motion compensated interpolation. The receiver requires additional information about the subsampling mode and the displacement vectors. For each block the control information is transmitted in a DATV channel with a bit rate of about 1 Mbit/s. DATV stands for digitally assisted TV. A further essential requirement for HD-MAC is the compatibility with the normal definition MAC standard. By a line shuffling of data in the 20-ms and 80-ms branch, a compatible MAC signal can be generated.

The subsampling scheme provides a reduction factor of 4. Therefore, the analog signal after multiplexing luminance and chrominance in a time compressed format results in a bandwidth of about 12 MHz.

2.2. DPCM coding

A superior picture quality is mandatory for studio-to-studio transmission. Based on this requirement a bit rate reduction factor of about two can be achieved by nonadaptive DPCM with two-dimensional prediction [5]. Positive weighting coefficients (1/2, 1/4, 1/4) for prediction and quantizer characteristics for coding with about 5 bit ensures both excellent picture quality and simple implementation. Additional improvement of the subjective picture quality can be achieved by noise shaping. A DPCM encoder configuration for such a system is shown in Figure 2.

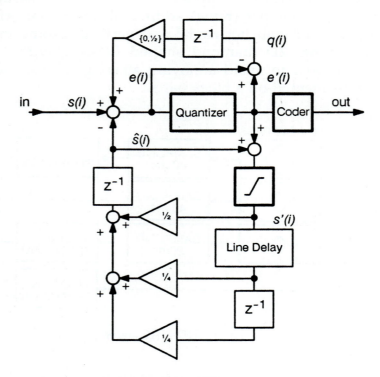

Fig. 2:   DPCM coder with 2D prediction and noise shaping.

The prediction value ŝ for a pel s to be encoded is a linear weighted function from previously reconstructed samples s'.

$$\hat{s}(i,j) = 1/2 \ s'(i-1,j) + 1/4 \ s'(i,j-1) + 1/4 \ s'(i+1,j-1) \qquad (1)$$

where i is the pel index (horizontal direction) and j is the line index (vertical direction). Accumulation of unused fractions for the calculation of ŝ results in an improved decay of transmission errors and diminishes their visibility. At frame borders all values s' outside the active picture area are set to the blanking values.

The specification of the quantizer characteristics should consider the visibility of quantization errors. It is aimed to have the generated quantization errors always below visibility threshold. Based on a visibility model developed by Girod [7] a quantizer with 31 levels for luminance and 15 levels for each chrominance component have been specified [8]. Coding with constant word length provide 5-bit and 4-bit words for luminance and chrominance, respectively.

Spectral forming of the quantization error q reduces low frequency components of quantization errors at the expense of an increase of high frequency components. Because the sensitivity for high frequencies is reduced, the subjective quality is improved. The desired spectral forming can be accomplished by a feedback of the filtered quantization error. For simplification of hardware a horizontal feedback with weighting of 1/2 is considered. This feedback results in an enhancement of about 3 dB for the luminance component. A gain of 3 dB corresponds to 0.5 bit in quantizer resolution. Under consideration of noise shaping the prediction error is determined by

$$e(i,j) = s(i,j) - \hat{s}(i,j) - 1/2 \, q(i-1,j) \tag{2}$$

$$e'(i,j) = Q(e(i,j)) \tag{3}$$

$$q(i,j) = e'(i,j) - e(i,j) \tag{4}$$

with Q(.) as quantizer function. In case of slope overload, noise shaping must be switched off because of stability reasons. Overload is determined by prediction errors e being larger than the maximum representative levels. The maximum representative levels are ±127 for luminance and ±63 for chrominance on a range of 511 for the prediction error.

Horizontal blanking intervals can be utilized easily for reduction of transmission rate by expanding the active line data over the complete line period. This technique reduces the bit rate by a factor of 5/6. DPCM coding combined with horizontal blanking utilization results in a total bit rate of 540 Mbit/s for video which allows transmission in a 565 Mbit/s channel. Utilization of the vertical blanking interval is not considered because of the high expense for the intermediate memory. The vertical blanking interval in each field is 49 lines which would require a total memory capacity of about 97 kbyte.

2.3. DCT coding

Transform coding transfers a block of NxN samples to a set of NxN weighting coefficients, where the coefficients are assigned to basic images of different spectral content. Like a Fourier series the weighted sum of basic images reconstructs the original block. The most common transform coding scheme is the discrete cosine transform (DCT) [9] (Figure 3). Bitrate reduction of transform coding is achieved by adaptive quantization and variable word length coding. Adaptive quantization considers the frequency dependency of visibility thresholds, the masking effect and input signal statistics. The mean bit rate can be reduced by coding with word length according to the information content. Under consideration of the requested high picture quality of HDTV, bit rate reduction factors of about 4 are possible with adaptive 2D DCT coding. Higher reduction factors have to make use of temporal redundancy. For reduction factors in the order of 8, hybrid coding as a combination of predictive coding in temporal direction and 2D DCT coding of the prediction errors is frequently proposed [10]. For a brief description *see* [11].

Fig. 3:   DCT coder.

2.4. Subband coding

Transform coding and subband coding are source coding schemes in the frequency domain. Subband coding does not operate on a block-by-block basis. It generates continuous signals of different frequency bands. In a subband coder the video signal is first separated into several bands by bandsplitting filters. Sampling in each band is then decimated according to the bandwidth. Due to decorrelation achieved by filtering and the properties of the human visual system, the subbands differ in both the statistical properties and the visibility of quantization errors. A reduction of transmission bit rate is achieved by application of specialized coding algorithms for the individual bands.

For coding of the subbands, various combinations of DPCM, DCT, cascaded subband coding, adaptive quantization and variable word length coding schemes have been proposed [12] [13]. An example with band splitting into four bands is shown in Figure 4. Band I is a low frequency band which is DCT coded. The high frequency bands are coded by adaptive quantization and variable word length coding. At the output four data streams with variable rate have to be multiplexed. In order to control the data rate and to avoid overflow, a feedback control is needed. Quantization with variable dead zone is very appropriate for bit rate adjustment.

For band splitting of images, 2D filters are appropriate. Besides diamond shaped nonseparable filters separable filters are also applied. Separable filters have the advantage of cascade realization as shown in Figure 5. Each filter is followed by a sampling rate decimation. This enables the same total bit rate at the output of the bandsplitting filter as at the input.

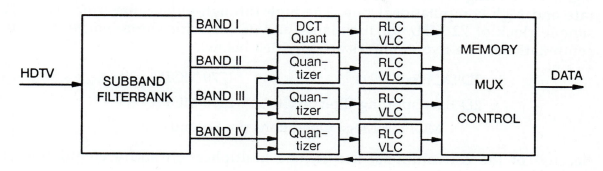

Fig. 4: Subband coder.

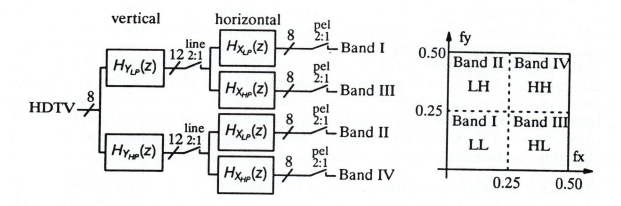

Fig. 5: Separable 2D filter bank for subband coding.

Band I is a low pass channel with reduced resolution in both dimensions. The resolution corresponds to broadcast TV. The other bands are high pass channels providing considerable response for edges and texture points only. In particular, the response in band II results from horizontal edges, band III from vertical edges and band IV diagonal and bent edges. At the receiver the inverse operation of the bandsplitting is needed. By cascade implementation of interpolation filters the four bands are amalgamated to one.

3.   VLSI Implementation Aspects

Signal processing of HDTV requires hardware structures with high through-put rate and high computational rate. The high through-put results from the high sample clock of 72 MHz for luminance and 36 MHz for chrominance. For the computational rate two examples are specified below:

2D DCT 8x8	1770 MOPS
2D Bandsplitting filter 10x10	2210 MOPS

MOPS = Mega Operations per Second

Because of the convolutional kernel, one multiplication and accumulation is considered as one operation.

The computational rates above indicate that these rates are far above the rate of available microprocessors and DSP chips. For this reason it is an absolute need to realize VLSI circuits dedicated to specific subtasks for HDTV processing. In order to achieve compact implementation the number and area of the circuits for a HDTV code should be kept small. This requires high complexity circuits with a small number of I/O pins to allow small housing. Therefore, minimization of silicon is needed for optimization of VLSI circuits. The silicon area devoted to each functional unit should be minimized by considering individual features of this function. Any features and parameters known a priori have to be taken into account. Further the strong interaction between algorithm and architecture should be recognized.

The signal processing requirements can be mastered by intensive application of parallel processing and pipelining. This calls for function blocks with high regularity which simplifies the separation into sequential and parallel subtasks. Further, the efforts for specification and design are reduced.

Because of the signal processing requirements for HDTV, application specific integrated circuits are required for important functional units. Key components for HDTV coding are FIR filters, bandsplitting and interpolation filters, DCT circuits, DPCM circuits and FIFOs for time compression and expansion. In the following, VLSI implementation aspects and architectures for DPCM, bandsplitting and interpolation filters are discussed in more detail.

4.   VLSI Architecture for 2D DPCM

4.1.   Implementation problems of DPCM

A DPCM in conventional structure displays an extremely disadvantageous arrangement of arithmetic blocks due to the recursive algorithm. Even taking into account that all multiplications for prediction and noise shaping can be performed by shifts due to the choice of weighting coefficients as powers of 2, there is a time-critical path in the basic algorithm consisting of a cascade of three adders, one limiter and one quantizer. The recursive loops of DPCM algorithms are a general problem of hardware implementation for high-speed applications, since they result in timing requirements which are difficult to master.

Another implementation problem is chip complexity. Besides the arithmetic blocks necessary for computation, a line memory is required for 2D prediction. Since each HDTV line contains 1920 active luminance pels, a high speed line memory with a minimum of 1920x8 bit is required. The two chrominance components require a minimum storage capacity of 2x960x8 bit.

The through-put rate of a VLSI DPCM codec can be increased by several methods. The application of several processing elements for parallel processing requires appropriate data formatting. A modification of the DPCM algorithm structures allows a rearrangement of the linear parts and the application of restricted pipelining. Another form of parallel processing is possible using a delayed decision technique. Last not least, an optimization of circuit technique enhances the basic computational speed. In particular, this applies to adders, subtracters and quantizer PLA, which are incorporated in the recursive DPCM loop.

4.2. Parallel DPCM processors

In order to reduce the timing requirements of HDTV signals, line-by-line parallel processing with several processing elements (PEs) can be applied for an DPCM coder (Figure 6). The input data stream is split and distributed among parallel paths by a serial-parallel-formatt. After processing, the data streams have to be reformatted by a parallel-serial-formatter. The same basic structure can be utilized for DPCM decoders. This form of parallel processing is possible, since the recursive 2D prediction is nonrecursive at the borders of each field due to the fixed prediction for the blanking intervals.

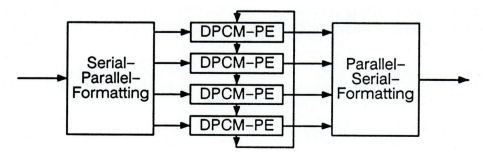

Fig. 6:   Parallel processing of DPCM.

A demultiplexer and high-speed FIFOs are necessary for the expansion of active line intervals into the time period of several complete lines (Figure 7). The parallel processing elements operate in an interleaving mode. Each processing element starts the processing of the corresponding line with a time offset of one line duration. The formatting reduces the clock frequencies of the N parallel paths by a factor of 1/N. Because of the application of FIFOs, for a further reduction of clock frequency the utilization of horizontal blanking intervals can be incorporated easily without additional hardware.

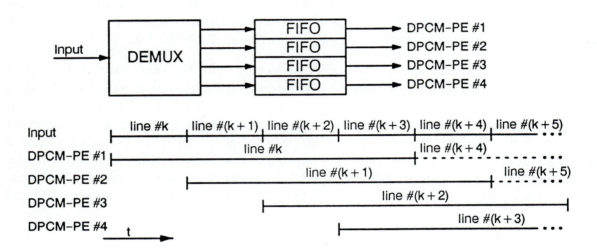

Fig. 7: Formatting for line-by-line parallel processing.

Due to 2D prediction, the processing elements cannot work independently. The reconstructed samples must be fed into the next PE. Therefore, the PEs have to be connected in a ring structure. A processing element suitable for parallel processing is shown in Figure 8. Because of the reduced clock frequencies, each PE needs 1/N of the line memory's capacity.

Although parallel processing reduces timing requirements, the number of parallel paths should be as small as possible because the hardware expense necessary for formatting increases overproportionally. Even input formatting requires FIFOs with a memory capacity of N-1 lines for N parallel paths.

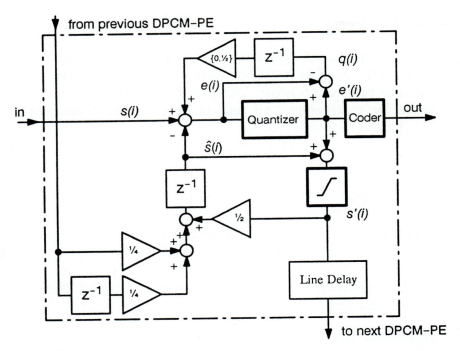

Fig. 8:   DPCM processing element for parallel processing with N PEs.

4.3. Modified DPCM structure

An alternative to the extensive application of parallel processing is pipelining. By transfer of the limiter, the linear part of the DPCM structure can be modified for the insertion of pipeline registers. The limiter in the loop is needed to avoid possible overflow and underflow of the reconstructed value s' caused by quantization distortions. However, overflow and underflow effects can also be avoided by appropriate limitation of the input samples [14] [15]. If the range of the input samples is limited to $[0 + q_n, 255 - q_p]$ where $q_n = -\min q$ for all $e' < 0$ and $q_p = \max q$ for $e' \geq 0$, it can be shown that the output range does not exceed [0,255] with standard quantizers. This means that the limiter can be removed from the critical loop. The only nonlinear block remaining is the quantizer. The maximum range of input values allowed considering an input limiter depends on the quantizing function, rounding algorithms and noise shaping coefficients. All these parameters can be chosen so that the limiting function is in accordance with the signal range specified in CCIR-601 [16] [17]. Having removed the limiter from the recursive loop, a modification of the DPCM structure for the application of pipelining is possible.

In particular, the use of 1/2 as feedback coefficient of equation (2) in combination with the (1/2,0,1/4,1/4)-predictor (1) leads to a special case for the computation of the prediction error. The prediction value $\hat{s}(.) = f(s'(.))$ includes the quantized prediction error e' because of the reconstruction of s' by:

$$s'(i,j) = \hat{s}(i,j) + e'(i,j) \tag{5}$$

From this follows that the calculation of both prediction and noise shaping includes the term e'(i-1,j). The combination of both terms results in a new representation of the prediction error. The prediction error determination based on this is suitable for pipelining and results in a simple hardware realization with small delay times:

$$e(i,j) = \begin{cases} s(i,j) - 1/4s'(i+1,j-1) - 1/4s'(i,j-1) - 1/2\hat{s}(i-1,j) + 1/2e(i-1,j) - e'(i-1,j) & \text{with n.s.} \\ s(i,j) - 1/4s'(i+1,j-1) - 1/4s'(i,j-1) - 1/2\hat{s}(i-1,j) - 1/2e'(i-1,j) & \text{otherwise} \end{cases}$$

Slope-overload can be detected by observing the values e(i) directly. Therefore, the noise shaping control can be realized by a boolean function of e(i). It is only the coder operation which is directly influenced by noise shaping. The algorithm and the hardware for decoding need not to be changed for the application of noise shaping.

After a rearrangement the most critical path consists only of a subtracter, a multiplexer for noise shaping control and a quantizer (Figure 9). Other critical paths consist of up to three adders or subtracters and a multiplexer. However, their carries can propagate nearly in parallel. Therefore, these loops are not as critical as the quantizer loop.

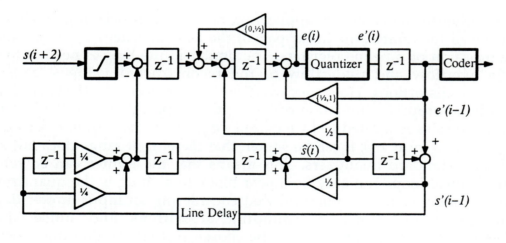

Fig. 9:   Modified DPCM algorithm with noise shaping.

4.4. Parallel processing and delayed decision

Besides the data path of the recursive loop, the noise shaping control is part of the time critical loop. This problem can be avoided by parallel processing with delayed decision (Figure 10). If any decision is necessary for a computation step, the selection control can operate in parallel to the computation by locating the selection switch at the end of structure.

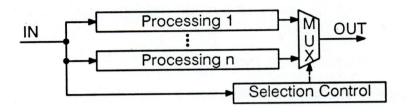

Fig. 10:   Delayed decision structure.

The application of the delayed decision technique is possible for the noise shaping control (Figure 11). The two cases (with noise shaping on and off) for the calculation of the prediction error have to be calculated in parallel. The register located between the noise shaping control and the multiplexer gives the maximum speed within the recursive data path. Furthermore, a third parallel computation path is necessary for initialization at line start.

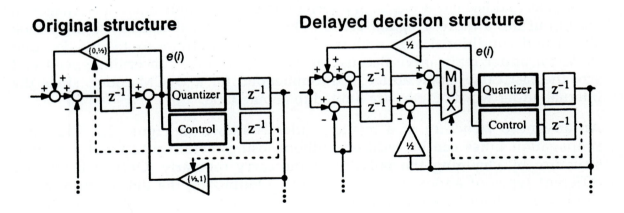

Fig. 11:   Noise shaping control with delayed decision.

4.5. Comparison of DPCM architectures

The critical loop of a standard DPCM structure contains three adders and subtracters, selector, quantizer, limiter and a register, while the modified DPCM structure with delayed decision has only one subtracter, selector, quantizer and register enclosed. This results in significant increase of the maximum possible sample rate by a factor of about 1.75. For HDTV applications, the higher circuit complexity is negligible. The modified DPCM structure and the delayed decision increase the transistor count of each processing element by about 4%. The reason for this small increase is that the majority of transistors is needed for the line delay. Since the hardware expense for the FIFOs necessary for application of N parallel processing units increases with 1.6N-0.6, a modified DPCM structure has to be preferred for the minimization of hardware expense.

4.6. CMOS circuit technique for DPCM

Besides the gain in speed by modification of the DPCM architecture, improved circuit performance is required. It is necessary to minimize the processing time of all blocks within the time-critical loops by an appropriate circuit technique. In particular, this applies to addition/subtraction and table look-up. Furthermore, the hardware expense for the highspeed HDTV line delay has to be minimized.

4.6.1. Adder realization

The basic nonhierarchical adder realizations are based on bitslice designs. Different techniques to speed up adders are known from the literature [18] [19] [20]. One of the hierarchical techniques that expends area in favor of speed is a carry select adder. For 8 bits such a technique provides only small advantages in speed but has a more irregular layout. Other hierarchical adders are carry look ahead and conditional sum adders. There are also proposals to use decoded PLAs for adder realization. Because of its good features concerning speed and size the fast carry chain adder is recommended for a DPCM codec, since the internal adder word lengths are limited to about 10 bit.

An advanced structure of a bitslice adder is shown in Figure 12. The carry propagation times can be reduced without losing the advantages of a regular structure by changing the polarity of the carry regularly. Only two slightly different types of adder cells are required. Optimizing the delay times in the ripple carry chain by an appropriate circuit structure results in fast carry chain adders (Figure 13). By changing the polarity of the inputs the same principle can be applied to subtracters.

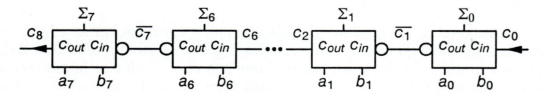

Fig. 12:   Adder in bitslice technique.

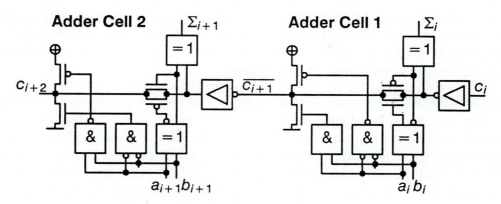

Fig. 13:   Adder cells using fast carry chain.

4.6.2. *Realization of XOR*

Adders and subtracters require XOR and XNOR operations several times. Two different time- and area-efficient realizations of XOR circuits are shown in Figure 14. The output inverter can be used for matching driving requirements. For realization of the XNOR operation the p- and n-channel transistors have to be exchanged.

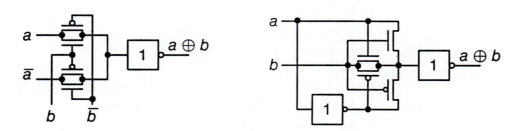

Fig. 14:   Different realizations of XOR gates.

4.7. Table look-up realization

The functional blocks quantizer, coder and decoder of a DPCM codec represent table look-up operations. The fastest circuits for table look-up with static logic are possible with multi-stage gate logic. NAND gates with maximum five inputs and NOR gates with maximum four inputs have to be used. Unfortunately, this results in an extremely unfavorable layout.

In standard hardware, table look-up functions are realized by read-only-memories (ROM). For a VLSI realization, a ROM offers a dense realization and has the advantage of a very regular layout. Even then a ROM requires a large size because all possible combinations of inputs have to be decoded in the row decoder and programmed by bit line outputs.

Programmable logic arrays (PLAs) are better because the minimization of implicants yields a reduced area. The logic minimization for PLAs can be performed, e.g., by the program LOG/iC [21]. High speed quantizers can be realized using pseudo NMOS PLAs avoiding additional clocking requirements (Figure 15). Furthermore, the quantizer PLA in the critical DPCM loop can be split into two parts of nearly equal size in order to decrease its delay (Figure 16). The necessary multiplexer can be combined with the following register stage. For a further minimization of the delay-times, implicants and prime implicants must be sorted to obtain a reduction of load capacitance for the most critical paths.

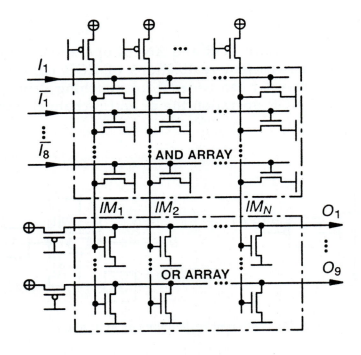

Fig. 15:　Pseudo NMOS PLA structure.

Faster PLAs are possible using a precharge technique, because precharged NOR gates are faster than other PLA gates. Additionally the gates have no static power consumption. Unfortunately the design of high speed precharge PLAs requires a complex clock generation. The safety margins necessary for compensation of process tolerances may reduce the performance significantly.

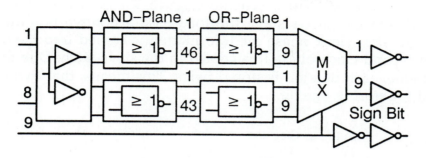

Fig. 16: Quantizer PLA structure.

Table look-up functions are also necessary for coding and decoding of the quantized prediction error. The timing requirements of these blocks are less critical, because the numbers of implicants are typically much smaller than for the quantizer. Therefore, a reduction of power dissipation can be achieved by using other transistor dimensions than for the quantizer. Moreover, they are not enclosed in any recursive loop. If necessary, even pipelining can be applied.

4.8. Realization of line delay

Most of the transistors of HDTV DPCM processing elements and subband filters are needed for line memories. For this reason, the chip area per storage cell and the power dissipation must be kept as small as possible. High speed line memories for vertical filtering can be realized using a parallel structure (Figure 17) either with parallel shift register arrays or with video RAM circuits, since the power dissipation of CMOS circuits is nearly proportional to clock frequency. A specific advantage of parallel shift registers over RAM circuits is the simpler control circuitry needed. The formatters necessary for line delay applications are much simpler than for line-by-line parallel processing, since no FIFOs are needed. Therefore, the formatting overhead for parallel HDTV line memory structures is comparatively small.

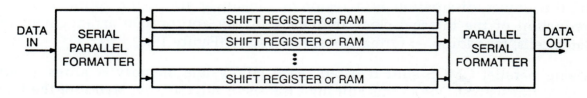

Fig. 17:   Parallel line delay structure.

Using parallel shift registers, a reduction of chip area and power dissipation is achieved by a substitution of master slave flipflops by latches combined with a sophisticated multiphase clock scheme (Figure 18) [22]. A conventional shift register consisting of master slave flipflops uses a nonoverlapping two phase clocking. In the first clock phase the information is copied into the master latch, while in the second clock phase it is copied to the slave latch. Due to this simple clocking scheme, each data bit is stored twice. This means, that two latch circuits are required for storing one bit.

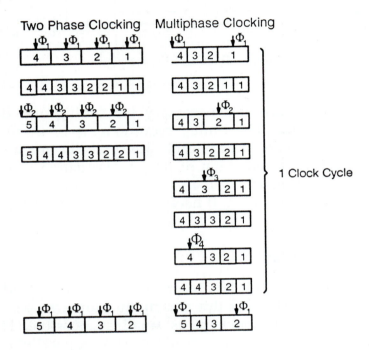

Fig. 18:   Comparison of two phase clocking and multiphase clocking.

Using n parallel internal paths and n clock phases, it is possible to store n-1 bit in n latches. For high numbers of clock phases, this results in a reduction of hardware expense of up to 50%. Due to the parallel structure, the n clock phases needed for the shift register array can be derived from the system clock by a counter. A regular layout with no additional control is provided. The reduction of the power dissipation is a result of the lower transistor count and the reduction of the internal clock frequency. By an appropriate arrangement of latch cells, no additional formatters are necessary (Figure 19).

Using parallel CMOS shift registers, each HDTV line memory requires about 75,000 transistors. Due to this still high transistor number, the layout of the line memory must be performed very carefully in order to minimize chip area. The core cell of the shift register array can be a dynamic latch cell built like a tristate inverter (Figure 20). Due to transistor configuration, changing the register contents does not imply a cross current inside the tristate inverter cells. Only charging and discharging of the gate capacities are needed. Hence, the power dissipation of each cell is small. For a further reduction of chip area and power dissipation, transistors with minimum dimensions are used. Although small clock pulses must be applied, the timing of the array itself is not critical, because in the core area of shift register arrays each cell has to drive only one other cell.

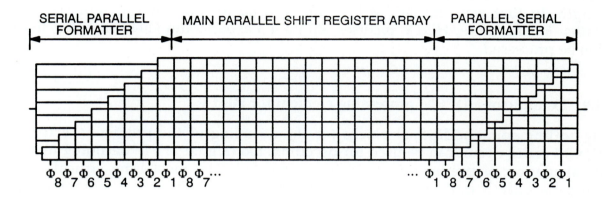

Fig. 19: Possible arrangement of shift register cells.

Due to the parallel outputs of the shift registers, the most critical part of the parallel shift registers is the parallel-to-serial formatter. Precharging the common output line of the shift registers can reduce this problem. Simulations under worst case conditions for a 1.5 μm CMOS technology have shown that parallel shift registers consisting of these cells operate even above 60 MHz with restricted power dissipation.

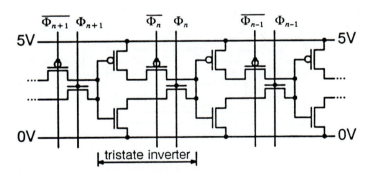

Fig. 20: Shift register cell built of tristate inverters.

Dynamic video RAM circuits use similar structures. The conversions serial-to-parallel and vice versa are performed by using shift registers. The control needed has to supply the write and read pointers for the dynamic RAM. Although additional shift registers and control circuits are needed, the use of video RAM circuits with 3-transistor cells can also result in a compact realization.

Depending on the number of parallel DPCM paths, the line delay within the DPCM-PEs must have different length. If the DPCM-PE is intended to be used with one, two or four parallel paths, the total effective length of the line delay including internal pipeline registers has to be selectable as 480, 960 and 1920 pels. For this purpose each line delay is divided into four parts of nearly equal length. In that way the DPCM-PEs can be used for one, two or four parallel paths. Furthermore, HDTV pictures with a 4:3 aspect ratio (1440 samples/line) can be processed.

4.9. Example of a DPCM processing element for HDTV

By adding some additional blocks for the decoding of the prediction error, a DPCM coder circuit can be used for decoding, too (Figure 21). In decoding mode, a calculation of prediction errors is not needed. The most time critical loop contains only a multiplexer, subtracter and register. Therefore, higher clock rates can be achieved for decoding than for coding.

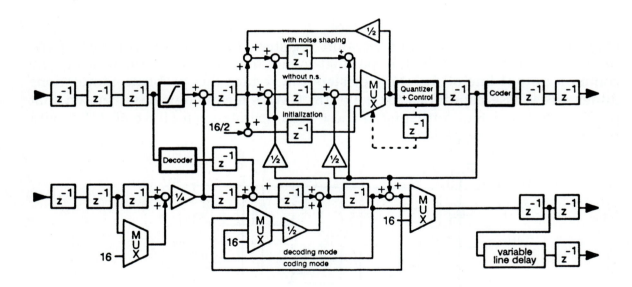

Fig. 21:   Block diagram of DPCM chip with pipelined coder or decoder operation.

A full custom style has to be chosen for the layout of the DPCM-PEs, since they contain a lot of special blocks, irregular in size. The circuit consists of about 80,000 transistors and its size is 5.9x4.5 mm^2 in a 1.2 µm technology. The floorplan is determined by an 1920x8 line delay, which covers nearly two-thirds of the chip area. The clock generator with the associated buffers is located close to the supply pins to avoid voltage drops and inductive noise on the supply lines.

Considering the properties of a 1.2 μm CMOS process, clock cycle times of less than 30 ns/pel can be achieved even under worst case conditions [8]. Therefore, two parallel paths clocked with 30 MHz are sufficient for the DPCM-coding of the luminance component. Considering good process data or an advanced CMOS process, a clock rate of 60 MHz seems to be possible, avoiding parallel processing.

5. Subband Filterbanks

For HDTV subband coding, filterbanks are combined with sampling rate conversion. Due to the transition between pass and stop bands, subsampling still causes aliasing effects after bandsplitting. Alias cancelation necessary for good picture quality can be achieved by appropriate choice of the parameters of the coder and the decoder filterbanks. Filter characteristics which incorporate aliasing error cancelation are quadrature mirror filters (QMF) and conjugate quadrature filters (CQF). Both are finite impulse response (FIR) filters with a nonrecursive structure. QMFs offer a linear phase behavior and a comparatively simple implementation. A disadvantage of QMFs are small ripples in their overall frequency transfer function. In contrast, CQFs allow a perfect reconstruction of the original signal, if undistorted transmission of the subband signals is provided. On the other hand, CQFs do not use linear phase filters and they are difficult to realize.

5.1. Hardware requirements

The realization of filterbanks for HDTV results in some implementation problems. 2D filterbanks for image coding require line memories as delay elements for the vertical direction. Due to the long line length and the high clocking frequencies of 72 MHz or even 144 MHz for HDTV signals, special memories have to be used for that purpose. Although separable filters reduce the hardware expense for filter arithmetics, the high data rates result in extremely high computational requirements. For high data reduction factors, band splitting filterbanks with a high tap number are necessary. An analysis of source coding algorithms has shown that good coding efficiency requires approximately 10x10 coefficients for quadrature mirror filterbanks [23].

The largest parts of 2D filterbanks are the line memories of the vertical filters. If a suitable architecture is used, the chip area required for line memories and FIFOs limits the number of taps. Therefore, the number of taps for the horizontal filter can be slightly higher than for the vertical filter (e.g., 14x10). In order to reduce the hardware expense, the line memories should be minimized. For this reason, it is advisable to place the vertical filters in front of the horizontal filters due to higher internal word length. This applies for both the decimation and the interpolation filterbank (Figure 22).

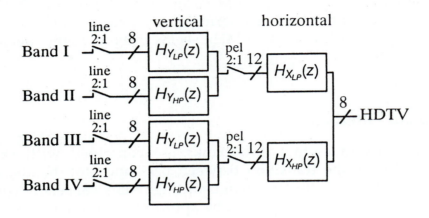

Fig. 22:   Preferable sequence of separable 2D interpolation filterbanks.

5.2. Nonrecursive wave digital filter (NRWDF)

Conjugate quadrature filterbanks (CQF) can be realized by nonrecursive wave digital filters (NRWDF). This type of filterbank contains filters that only approximate a linear phase behavior. Exact linear phase is not possible with more than two filter taps. For a function oriented VLSI realization of HDTV filterbanks it is advantageous that the sampling rate conversion of NRWDFs can be performed at the input of the decimation filters and the output of the interpolation filters (Figure 23). Therefore, they can operate with halved sampling frequency. Moreover, NRWDFs implicitly offer perfect reconstruction by design. Furthermore, nearly all delay elements and arithmetic blocks are used in common for both the high and low pass branches.

Another advantage of these structure for a VLSI realization for 2D video filtering seems to be that the corresponding FIR tap number is approximately double the number of delay elements. But in fact, these structures require an increase of the internal word length of delay elements and filter arithmetics. Furthermore, an exact scaling is required for the coding. Because of the calculation scheme, the design of filterbanks with a limited precision of coefficients or data paths is quite impossible. For a VLSI realization, the butterfly structures allow simple partitioning only in the direction of data flow. Because of the high number of different butterfly parameters and precision constraints, many complex butterfly processing elements, either programmable or of many different specialized types, have to be used.

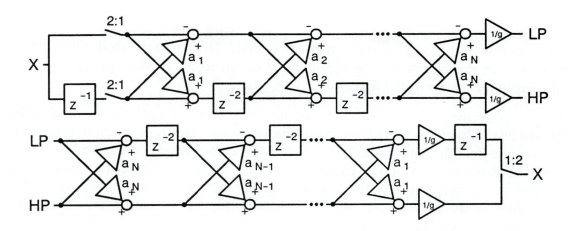

Fig. 23:   Nonrecursive digital filterbank (NRWDF).

5.3. Transversal FIR filter

Transversal filters in direction form I [24] are a basic structure for the realization of FIR filters. For the realization of subband filterbanks they have the advantage that vertical low pass and high pass filters can use the same line memories (Figure 24). Additionally the word length of these line memories is limited to the word length of the input signal. Multi-operand adders can be used for both the realization of the adders and multipliers.

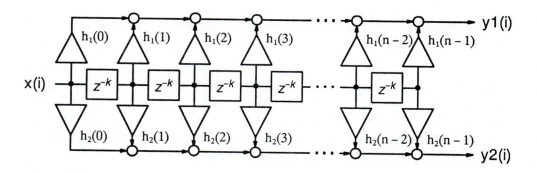

Fig. 24:   Combined transversal FIR filters (based on direct form I).

5.4. Polyphase structure

Polyphase structures can reduce the operating frequencies of decimation and interpolation filters. For this purpose, the filter is split into several parallel filter parts. The impulse responses of the filter parts are determined by a subsampled impulse response of the whole filter. If the numer of phases corresponds to the decimation or interpolation factor, each part of the filter processes only a certain phase component of the subsampled signal. That means that the operating frequency of the filters can be reduced by substituting the sampling switch with a multiplexer or demultiplexer.

Using a polyphase structure, a 2:1 decimation filter is split into two partitions (Figure 25). The first partition contains only the filter coefficients with even indices; the second partition contains the filter coefficients with odd indices. By changing the position of the sampling switch, both partitions can operate with half the clock frequency without affecting the function. This basic principle can be applied both on vertical and horizontal filters. Because of the relaxation of timing requirements, a reduction of both power dissipation and chip area can be achieved.

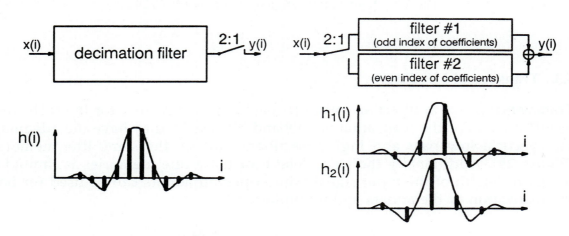

Fig. 25:   Principle of polyphase filter structures for decimation.

The basic polyphase principle can be enhanced for 2D-filterbanks (Figure 26). In order to provide a simple parallel processing for the vertical decimation filterbank, the horizontal subsampling can be located in front of the vertical filters (Figure 27). Although two identical vertical filterbanks have to be used, there is almost no increase in circuit complexity, because each vertical filter requires line memories of only half the line length. These two identical vertical filters can also be realized as polyphase decimation filters (Figure 28). Due to the reduced clock frequencies, the small increase in circuit complexity is more than compensated by simpler circuits. The main advantage of the combined 2D subsampling is the increase of circuit regularity. Both vertical and horizontal filters operate at 1/4 of sampling frequency. Therefore, similar circuits can be used for both filter types. Furthermore, the whole analysis filterbank can be split into two equal parts, which can be realized with two identical chips. A similar structure exists for the synthesis filter bank (Figure 29) [25].

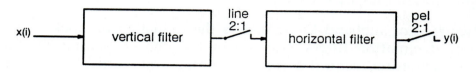

Fig. 26: Basic 2D separable decimator filter.

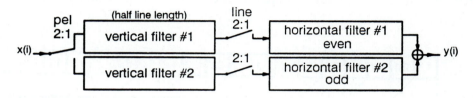

Fig. 27: Decimation filter with advanced horizontal polyphase subsampling.

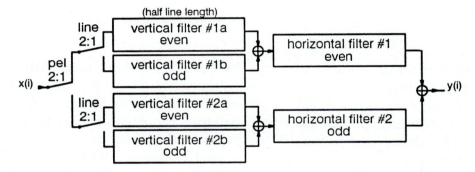

Fig. 28: Separable 2D decimation filter with combined polyphase subsampling.

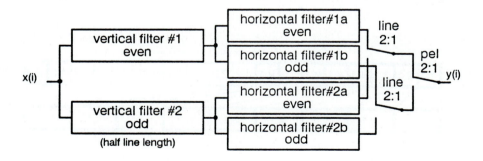

Fig. 29: Separable 2D interpolation filter with combined polyphase subsampling.

It is favorable to use the same filter coefficients for the luminance and both chrominance components. By the application of pel-by-pel chrominance multiplexing nearly the same type of circuit can be used for luminance and chrominance. FIFOs and vertical filters are identical. Only the delay elements and tap structure of the horizontal filterbank have to be adjusted.

5.5. Expansion in time by subsampling

One advantage of subband filterbanks is the reduction of clock frequencies by the use of decimation filters. Because of the line sequential TV format, the hardware structures required for horizontal and vertical subsampling are different. Horizontal subsampling by a factor of 2 requires the expansion of every second sample (Figure 30). In a hardware realization, this can be achieved easily, for example, by registers with a clock enable input or by different clocking of registers.

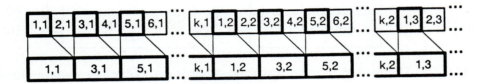

Fig. 30:   Horizontal subsampling raster (even line length).

For a reduction of clock frequencies by vertical subsampling with a factor of 2, every second line has to be expanded (Figure 31). This task cannot be performed only by registers, but a FIFO structure is needed. The formatting necessary for the corresponding polyphase filter structure requires at least two FIFOs and a delay. The delay is needed to compensate the time between an even and a corresponding odd line. It can be located between FIFO memories and filter partitions. Thereby, it operates with half clock rate and half number of memory cells. Depending on the FIFO design used, the delay may be combined with the FIFOs. A similar FIFO and delay structure is required for vertical interpolation filters.

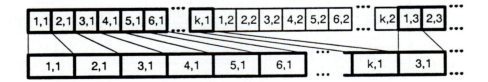

Fig. 31:   Vertical subsampling raster.

The implementation of FIFO circuits for time expansion and compression can be based on the line memory designs. If RAM arrays with independent read and write pointers are used, FIFO operation can be easily achieved by using different address counters. Another FIFO implementation suitable for time expansion and compression are shift registers operating in ping-pong mode. Using more than one shift register, write and read can be performed with different clock rates.

Alternative synchronous structures use a combination of multiplexers and shift registers of different lengths. Due to the polyphase structures of subband filterbanks it is possible to design synchronous FIFOs with a minimum number of memory cells. Nevertheless, the overhead for the multiplexers and the control has to be taken into account.

5.6. Architectures for quadrature mirror filterbanks (QMFs)

5.6.1. Filter characteristics

A detailed description of the design of quadrature mirror filter (QMF) functions for subband coding of HDTV is given in [26]. QMFs are defined such that the transfer function H_1 of the high pass filter is a mirror image of the low pass filter H_0 (7).

$$H_1(z) = H_0(-z) \tag{7}$$

For image coding applications, linear phase filters are preferred. Therefore, FIR filters with symmetric impulse response have to be used. The low pass filter of an analysis filterbank with an even number of taps can be described as (8). With (7), the characteristic of the corresponding high pass filter is given by (9).

$$H_0(z) = \ldots + a_{-2}z^{-2} + a_{-1}z^{-1} + a_0z^0 + a_0z^1 + a_1z^2 + a_2z^3 \ldots \tag{8}$$

$$H_1(z) = \ldots + a_{-2}z^{-2} - a_{-1}z^{-1} + a_0z^0 - a_0z^1 + a_1z^2 - a_2z^3 \ldots \tag{9}$$

For alias cancelation, similar filters (10) and (11) have to be used in the decoder.

$$G_0(z) = H_0(z) \tag{10}$$

$$G_1(z) = - H_1(z) \tag{11}$$

5.6.2. Filter structures for QMFs

The filter coefficients for high pass and low pass filters of QMF filterbanks differ only in the sign of every second coefficient. This can be utilized for implementation. The filter functions of the high pass and low pass filter can be split into two partitions. The regular change of the sign in (9) results in a structure (Figure 32) similar to polyphase filters. According to the sign difference, filter taps with even and odd indices are in separate partitions of the filter. The output value of the low pass filter is the sum of both filter partitions. Due to the QMF characteristics, only one additional subtracter is necessary for the realization of the high pass filter within a filterbank, since the difference of both partitions gives the output values for the corresponding signal. This reduces the hardware expense for multipliers and adders by nearly 50%. A similar structure exists for QMF synthesis filterbanks (Figure 33). Due to equations (10) and (11), the same filter subfunctions can be used for analysis and synthesis filterbanks.

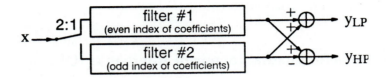

Fig. 32:   Hardware structure of a QMF decimation filterbank (analysis).

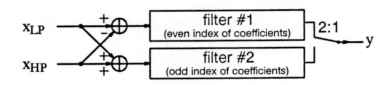

Fig. 33:   Hardware structure of a QMF interpolation filterbank (synthesis).

Moreover, if a filter function with an even number of taps is used, the subfunctions of the filter parts #1 and #2 are similar. Both subfunctions have the same coefficients but in reverse order. This results in a very regular design.

5.7. Realization of arithmetic blocks

5.7.1. Realization of multipliers

The largest arithmetic blocks of a linear filter are the multipliers. Fully programmable filter coefficients result in a huge hardware expense, since the number of multiplier cells must be based on word length of signal and coefficients. For a given filter order and accuracy, both number and complexity of the cells needed for multiplication can be reduced by avoiding programmable filter coefficients, because fixed filter coefficient allow the realization of multiplication by simple add/shift operations. Only binary digits of the coefficients with the value 1 require an adder (Figure 34). Because the coefficients of the impulse response of filters have many more digits with value 0 than 1, typically about 75% of the hardware expense for adders can be saved. Therefore, only fixed filter coefficients are useful for realization of dedicated filter chips.

$$(X)_{10} = XXXXXXXXX_2$$
$$(248)_{10} = 0011111000_2$$
$$(248)_{10} = 010000\overline{1}000_2$$

10 coefficient registers
10 AND–gates
10 adders

5 adders

1 adder
1 subtracter

Fig. 34: Different hardware expense for the coefficient 248 (example).

The number of adders can be further reduced using a canonical signed digit (CSD) number representation for the filter coefficients [18] (Table 2). Subtractions are realized by addition of the two's complement. This allows the application of multioperand adders. Carry-save-adders can be used for both accumulation and add/subtract/shift operations necessary for the implementation of filter arithmetics.

The application of CSD numbers for the coefficients reduces the number of pipelined accumulations by more than 80%. Further hardware savings can be achieved by a reduction of internal word lengths due to the known values of the coefficients. Additionally, all storage and load control circuits for the coefficients can be omitted.

Table 2: Hardware expense of different multiplier types.

multiplier type	full N x M	fixed coefficients	CSD representation
cell type	multiplier	multioperand adder	multioperand adder
number of cells	100%	approx. 25%	approx. 17%
hardware expense	100%	approx. 15%	approx. 10%

Due to the high clocking rates, pipelining is required for the realization of HDTV filters. Because of the nonrecursive structure of FIR filters, only the additional hardware expense for pipeline registers and the increased circuit latency have to be taken into account. With the simplifications described above, only a small number of pipeline registers is required.

5.7.2. Multioperand adders for FIR-filter

The best way to implement multioperand adders for FIR filters are carry save adder arrays (Figure 35). The multioperand adders perform multiplications and accumulations of the products. By a bitplane grouping of the filter coefficient digits, it is possible to have a nearly constant word length of all adder columns. For the realization of the subtractions, full-adders with one inverted input are required. Because most full-adders require internally the generation of inverted signals, often no additional hardware expense is required for the inversion. In order to obtain the correct result for a single two's complement subtraction, it is necessary to add a shifted +1. Because these +1's are constant values, they can be summarized to a constant offset of the total result. The sign extension necessary for two's complement subtraction arithmetic can be avoided if the input signals are converted into an offset number representation. This can be done easily by inversion of the sign bit. If fixed filter coefficients are used, this causes another constant offset of the result. Combining the constant correction offsets for sign extension and shifted two's complement subtractions results in a constant common offset for both.

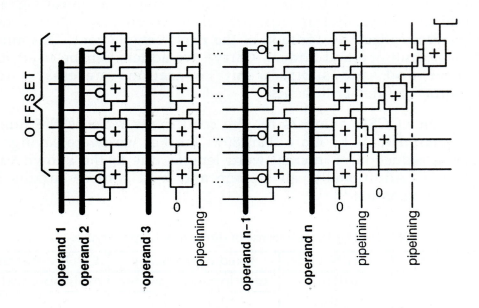

Fig. 35:   Multioperand adder for multiplication and accumulation.

5.8. Complexity of filterbank VLSI-implementation

The complexity of a 2D subband filterbank is determined mainly by line memories. The architectures for the realization of line memories and FIFOs can be based on the line memories developed for HDTV DPCM codecs. If QMF structures are used, only about 5% of the transistor count is necessary for filter arithmetics. For the realization of a 2D polyphase filterbank (Figure 36) with 10x14 filter taps, 8 1/2 line delays and 2 FIFOs are required for decimation or interpolation. With a word length of 8 bit, about 800,000 transistors are needed. Since most of them are in the memory section, a highly regular design can be achieved. With a 1.2 µm CMOS technology, about 180 mm^2 of chip area is needed. The worst case power dissipation of such a chip is about 500 mW. This means, a 1-chip-realization is possible in principle. Due to polyphase structure, the filterbank circuit can be split easily into a 2-chip-realization with very small increase in total hardware complexity.

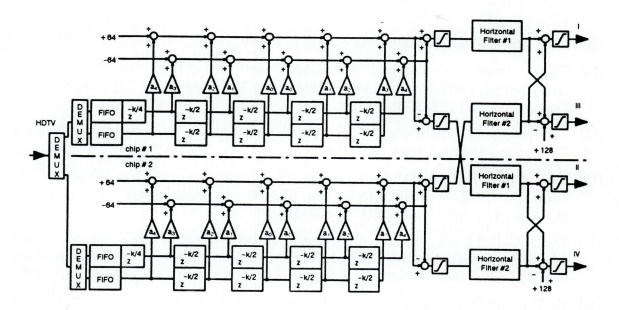

Fig. 36: Complete 2D QMF decimator subband filterbank.

6. References

[1] Y. Ninomiya, "A single channel HDTV broadcast system — The Muse," *NHK Lab. Note 304*, Sept. 1984.

[2] F.W.P. Vreeswijk, M.R. Haghiri, "HDMAC coding for MAC compatible broadcasting of HDTV signals," pp. 187-194, in *Signal Processing of HDTV II*, ed. by L. Chiariglione, Elsevier, 1990.

[3] CCIR, "The present state of High-Definition Television," *part 5.1.2, CCIR report 801-3*, January 1990.

[4] N.S. Jayant, P. Noll, *Digital Coding of Waveforms*. Prentice Hall, 1984.

[5] A.N. Netravali, B.G. Haskell, Digital Pictures, *Representation and Compression*. Plenum Press, 1988.

[6] G.J. Tonge, "The sampling of television images," *IBA Report 112/81*, May 1981.

[7]   B. Girod, *Ein Modell der menschlichen visuellen Wahrnehmung zur Irrelevanzreduktion von Fernsehluminanzsignalen.* pp. 110-125, VDI-Verlag GmbH, Düsseldorf, 1988 (in German).

[8]   K. Grüger et al, "VLSI components for a 560 Mbit/s HDTV codec," *Visual Communications and Image Processing '90*, Murat Kunt, Editor, SPIE vol. 1360, pp. 388-397, 1990.

[9]   W.K. Pratt, *Digital Image Processing.* J. Wiley, 1978.

[10] CCIR, Draft new report AD/CMTT on the digital transmission of component coded television signals at 30-34 and 45 Mbit/s, *CCIR-Document CMTT 116*, 1986-1990.

[11] P. Pirsch, T. Wehberg, "VLSI Architectures and Circuits for Visual Communications," ibid.

[12] T.-C. Chen, P.E. Fleischer, S.-M. Lei, "A subband scheme for advanced TV coding in BISDN applications," pp. 553-560, in *Signal Processing of HDTV II*, ed. by L. Chiariglione, Elsevier 1990.

[13] D. Le Gall, H. Gaggioni, C.-T. Chen, "Transmission of HDTV signals under 140 Mbits/s using a sub-band decomposition and discrete cosine transform coding," *2nd Int. Workshop on Signal Processing of HDTV*, L'Aquila, 1988.

[14] P. Pirsch, "Design of a DPCM Codec for VLSI Realization in CMOS Technology," *Proceedings of the IEEE*, vol. 73, No. 4, pp. 592-598, April 1985.

[15] P. Pirsch, "VLSI-Design of DPCM Codecs for Video Signals," in A.N. Netravali, B. Prasada (editors), *Visual Communications Systems*, pp. 216-227, IEEE press book, New York, 1989.

[16] CCIR, "Encoding Parameter of Digital Television For Studios," *CCIR Recommendation 601*, Document 11/234-E, 27 September 1983.

[17] CCIR, "Modifications to CCIR Recommendation 601," *CCIR Recommendation 601*, Document 11/463-E, 25 October 1985.

[18] K. Hwang, *Computer Arithmetic, Principles, Architectures and Design*, John Wiley & Sons, 1979.

[19] N. Weste, K. Eshraghian, *Principles of CMOS VLSI design.* Reading, M.A.: Addison Wesley, 1979.

[20] J. Mavor, M.A. Jack, P.B. Denyer, *Introduction to MOS LSI design.* Reading, M.A.: Addison Wesley, 1983.

[21] ISDATA GmbH, Programmsystem LOG/iC, PLD-Compiler-Handbuch, ISDATA GmbH, Karlsruhe, 1988 (in German).

[22] Deutsche ITT Industries GmbH, "Serieller FIFO-Speicher," European patent application, EP 0 243 528 A1, Patentblatt 87/45, Europäisches Patentamt.

[23] U. Pestel, B. Schmale, "Design of an HDTV subband codec considering CMOS-VLSI constraints," *Visual Communications and Image Processing '90*, Murat Kunt, Editor, SPIE, vol. 1360, pp. 587-597, 1990.

[24] A.V. Oppenheim, R.W. Schafer, *Digital Signal Processing*. Prentice Hall, 1975.

[25] K. Grüger, P. Pirsch, M. Winzker, "VLSI architectures of filterbanks for subband coding of HDTV signals," *Annales des Télécommunications*, vol. 46, No. 1-2, pp. 110-120, Jan./Feb. 1991.

[26] U. Pestel, K. Grüger, "Design of HDTV subband filterbanks considering VLSI implementation constraints," *IEEE Trans. on Circuits and Systems for Video Technology*, vol. 1, No. 1, pp. 14-21, March 1991.

Chapter 17
IC Solutions for Mobile Telephones

Juha Rapeli
Nokia Mobile Phones Ltd, Oulu Research Center, Finland

1. Introduction

Mobile communications in the first half of the 90's are characterized by rapid growth of existing networks, as well as continuous and strong evolution of telephones (size, weight, cost, component count, number and quality of services). Half a dozen new systems (GSM, dual-mode analog/digital AMPS, full digital AMPS, DCS1800, DECT, PHP Japanese Digital Cellular and also CDMA-based systems) have emerged and set new standards in systems architectures, type of components, modulation methods, and ever higher bit rates and operating frequencies. New networks and telephones are expected to provide better services than the existing ones and, despite increased complexity, be cheaper. Integration which brought modern digital signal processing (DSP) into telephones will affect radio frequency (RF) functions as well.

Mobile telephone systems are described in this chapter and the basic technological requirements of each constituting block are discussed bearing in mind the existing possibilities of silicon integration with enhanced speeds/densities of a single technology as well as with mixed technologies and DSP's. The question to which extent mobile telephones will use ASIC's or standard IC's is also addressed.

2. Mobile Communication Systems

The early 80's saw the emergence of cellular networks characterized by full mobility (handover) without service interruptions and fully automated call procedures. Handover and area coverage requirements resulted in honeycomb like cellular structures of the network and also led to the designation of "cellular mobile telephony" (CMT). The first generation of cellular networks, namely the Scandinavian NMT (Nordic Mobile Telephone), the Bell-originated AMPS (Advanced Mobile Phone System) and its British derivative TACS (Total Access Communication System) are all intended for speech communication either in the 450 MHz frequency band (NMT-450) or in the 900 MHz frequency band (NMT-900, AMPS, TACS).

An important feature of the cellular network concerns the possibility for both national and international roaming functions. In the national roaming function the network localizes the user within a traffic area (mobile exhange area) so that coming calls from the fixed network can be directed into signallings on the command/call channel only in the proper traffic area. International roaming performs the same function as national roaming but now through the fixed international network.

The basic structure of a CMT Network is shown in Figure 1 and key parameters and life cycles of different systems are shown, respectively, in Figures 2 and 3. The NMT, AMPS and TACS systems, together with other smaller systems, are called analog systems mainly because speech transmission is analog. Digital signallings between the base station and the mobile station (see Figure 1) are based on FSK-modulating bit streams into an analog channel. While the NMT system uses mainly the audio spectrum and a bit rate of 1200 bps, the AMPS/TACS system uses the spectrum above audio frequencies and the corresponding bit rate increases up to 12 Kbps. Analog systems rely on frequency division duplex (FDD) and frequency division multiple access (FDMA) operation. The network and telephone implementations of analog systems are both based on analog signal processing capabilities and relatively low level of integration (say, large scale integration - LSI).

New digital systems, like the Pan-European GSM-system, aim at expanding the network capacity and further standardization of systems by using digital speech transmission and time division multiple access (TDMA) techniques. They also generate wireless ISDN-like standard, thus making it possible to explore more value-added services possible in the mobile environment, including short message services (SMS), paging, facsimile and other data and speech applications. Further advantages of digital systems concern improved speech quality as well as the possibility of increasing channel capacity by reducing bit rates through efficient speech coding. High instantaneous bit rates of TDMA are subject to multipath propagation and Doppler shift errors, and hence require some form of compensation which can usually be achieved by using advanced channel equalization and error correction algorithms together with modulation techniques optimized to noise and other propagation properties of the channel. Speech coding, channel coding and modulation/demodulation functions all rely heavily on VLSI (very large scale integration) DSP technology which just recently became feasible for battery operated telephones.

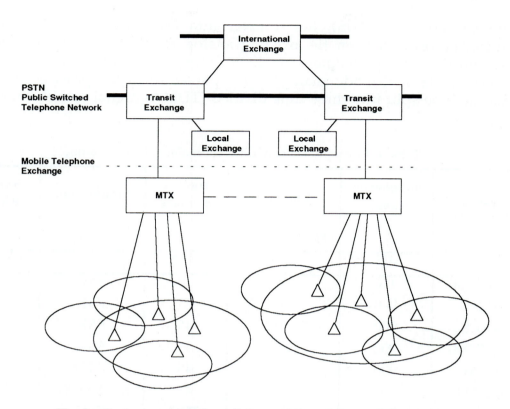

Fig. 1:   Basic structure of a cellular mobile telephone network.

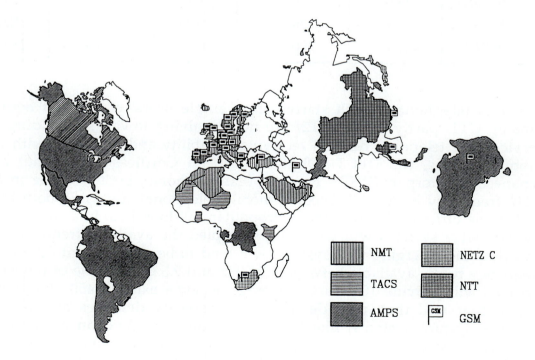

Fig. 2:   World of CMT standards.

In the United States (US), the transfer into fully digital CMT will be an evolutionary process from the existing analog AMPS by gradually allocating presently used analog channels for digital transmission and also by specifying an intermediate dual-mode analog/digital phone. In such a process, one analog channel will be split into three time divided (TDMA) digital speech channels.

Table 1: Key parameters of different mobile phone systems.

System	Frequency range	Channel raster (TDMA - ratio)	Duplex distance	Modulation	Speech Bitrate
	MHz	kHz	MHz		bits/sec.
NMT-450	450	25	10	FSK	-
AMPS	900	30	45	FSK	-
TACS	900	29	45	FSK	-
NMT-900	900	25	45	FSK	-
GSM	900	200 (8)	45	GMSK	13 k
D-AMS	900	30 (3)	45	π/4-DQPSK	8 k
CT2	900	100 (1)	TDD	FSK	32 k
DCS 1800	1800	200 (8)	95	GMSK	13 k
DECT	1900	1728 (24)	TDD	GFSK	32K/16 k
PHP	1900	300 (8)	TDD	π/4-DQPSK	32K/16K

Cordless telephones which started from single users' end-to-end equipment (analog CT1, partly digital CT2) are now evolving towards providing public services like telepoint, full call receiving capability and handover with limited mobility. Thus, the DECT (Digital European Cordless Telephone) and its Japanese counterpart (PHP - Personal Handy Phone), both working in the 1.9 GHz frequency band, will become mobile telephones for using both at home and in the office and also in the metropolitan outdoor area. In a couple of years the 900-MHz systems will have fully occupied the available frequency bands, and thus the emerging new systems will tend to be allocated into the next free frequency bands available between 1.5 GHz and 2.5 GHz. None of the 900-MHz systems will provide 100% area coverage as some existing 450-MHz networks do; neither will they provide ultra high capacity densities and broadband services. Future systems, like those based on code division multiple access (CDMA) type of spread spectrum communication, or even combinations of existing systems, have been proposed to fulfill these needs [13, 14, 22].

SYSTEM	1980	1985	1990	1995	2000
NMT450					
AMPS					
TACS					
NMT900					
GSM					
US-DIGIT					
CT2					
DCS1800					
DECT					
PHP					
FPLMTS					

Fig. 3: Estimated lifetimes and launching schedule of different CMT systems.

Sales of mobile telephones are qualitatively sketched in Figure 4. It is estimated that in the early 2000's a mobile phone will be as common as a personal computer and in market terms, will represent a 20% to 40% penetration in the population of the western world.

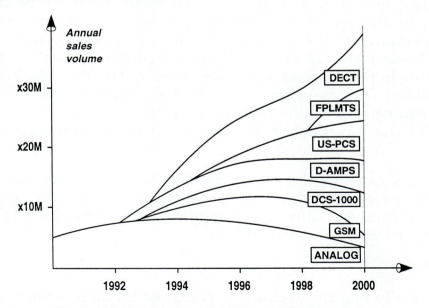

Fig. 4: Market estimates for different CMT systems.

Because of the expected growth of mobile communications, the semiconductor market segment for mobile phones is also estimated to grow from roughly $150 M in 1990 (0.3% of total IC production), up to $1 bn to $1.5 bn in the late 90's, which will correspond to approximately 0.5% of the total IC production. Furthermore, because of the increased capabilities of state-of-the-art technologies, particularly by enabling the combination of multiple functions using a single technology process, it is expected that the key circuitry and signal processors needed for CMT's will be reduced from five or six, namely memory, processor/logic, linear, CMOS/analog RF/high speed, and RF-power, to a mere two or three, such as multifunction embedded controller/DSP with on-chip memory, high speed analog and RF, and BiCMOS or GaAs with power handling capabilities.

3.   Functional Blocks of a Cellular Telephone

A typical architecture of an analog CMT is shown in Figure 5, and the corresponding building blocks will be described next.

3.1. Receiving function (RX)

The performance requirements of the receiver have to guarantee that the desired signals can be demodulated at extremely low power (or voltage) levels for a specified sensitivity limit and, simultaneously, there can be other signals as much as 80 dB stronger at neighbouring frequencies. Thermal noise floor at room temperature is -174 dBm/Hz. An analog system typically requires a sensitivity better than -113 dBm for a signal modulated into a 20-kHz channel, corresponding to a spectral density in the range of -156 dBm/Hz. The signal can be demodulated with the required quality when the desired RF signal is 9 dB above thermal noise or interference. (In different systems, this can vary from 6 dB to 13 dB.) This leaves only 9 dB (174-156-9) for losses and noise added in the receiver.

Next, we shall examine the key characteristics of the main building blocks involved in the receiver function, namely the duplex filter, the front-end amplifier, the front-end filter/matching, the crystal filters, the first mixer, and the limiting and FM/PM demodulation.

Duplex filter

- This is usually implemented using ceramic or helical resonator technologies. Mixed technologies tend to be expensive (e.g., ceramic plus SAW).

- Due to power/loss/noise/matching requirements it is difficult to integrate.

- Signal losses from 1 dB to 4 dB, and noise figure below 1 dB.

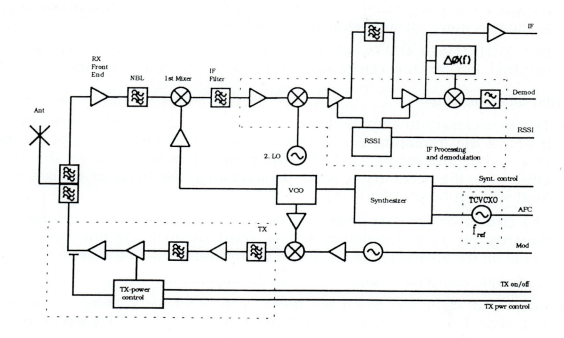

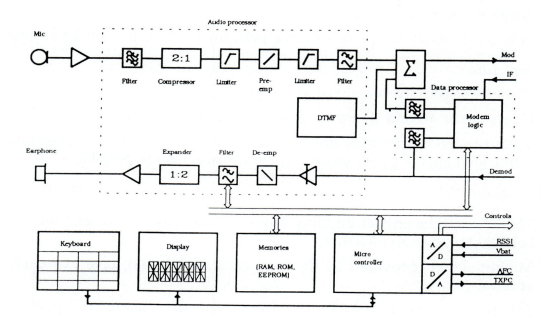

Fig. 5: Block diagram of a typical 900 MHz analog CMT.

- It realizes the following three basic functions: 1) combines both the RX and TX signals into a single antenna; 2) prevents high RF power of the transmitter to couple into the receiver front-end and amplifier; 3) contributes to the rejection of image frequency reception.
- In digital TDMA CMTs, where simultaneous transmit and receive is not required, this function can be replaced or partly realized by a switch.

Front-end amplifier

- This is necessary to amplify the incoming signal (and noise) in order to keep it above the minimum S/N when the equivalent input noise of the next stages; i.e., the mixer after the front-end filter is added to the input. Otherwise, the necessary S/N for demodulating the signal at the sensitivity limit will not be possible.
- The critical circuit parameters are noise figure, impedance matching, good linearity for intermodulation and adjacent channel rejection because the signal on the adjacent channel can be 80 dB stronger than the desired signal.
- A discrete component solution is easy to implement.
- Integrated amplifiers drain too much current for satisfactory performance and suffer from interferences when integrated with other RF parts or logic.

Front-end filter/matching

- Some filtering is necessary to reduce the noise bandwith and/or reception at image frequency.
- This can possibly be removed if the performance in other blocks optimal.

Crystal filters

- Channel selectivity is realized at the first IF.
- Filters at the second IF are used mainly for noise limitation.
- Because signal strengths are still close to thermal noise, integration with active elements (e.g., SC technology, transistors) is not possible.

First mixer

- This is a rather critical building block and usually should employ a double balanced type of architecture.
- Mixing signals provided by the local oscillator should have extremely good spectral purity.
- Integrated circuit implementation typically leads to performance degradation.
- Particularly good results have recently been achieved using fully differential active analog multipliers (e.g., Gilbert's cell).

Limiting amplifier and FM/PM demodulation

- These functions are employed when mixing down to the second IF is carried out.

- Provide received signal strength indication (RSSI).

- The requirement for low power consumption leads to using bipolar transistor technology. (f_T of minimum 4 GHz to 6 GHz is required).

- Currently available architectures for implementing these functions require ceramic, crystal or SAW filters with proper matching for noise band limitation, and further external components for the quadrature type of FM/PM demodulation.

- Presently available components do not provide optimal performance (individual variations, temperature drifts).

The alternative architecture solutions for implementing the above receiver functions include different IF's, phase locked-loop (PLL) based detection, single IF receiver, and even the direct conversion (zero IF) approach. This, however, has not yet reached technical or commercial viability. The requirement for low power consumption favours passive high Q filters and reactive components for connecting parts working at different frequencies in such a way that the same power supply current flows through several blocks. In the receiver design, it is crucial to have good control over all unwanted couplings, harmonics and interferences. Further integration of more functions on to a single chip, with less band limitations than on board level, would be difficult even if on-chip interferences could be under control. Therefore, it is likely that the front-end functions down to the IF output will remain discrete analog, and integration starts gradually from the lower frequency end. Some new system solutions will be investigated to reduce the need for discrete components and tunings, and perhaps to improving the performance.

Analog to digital conversion (ADC) can be carried-out for the band limited IF output signals if a sufficient gain is added to allow for noise added by ADC. As the dynamic range of signal is 80 dB or more, an AGC (automatic gain control) function would be required, as will be discussed later in the chapter. An ADC for the demodulated baseband signals would be straight-forward to implement but it would not integrate the receiver.

Theoretically, a direct conversion receiver would be ideal in many respects, namely because all filtering is done at the baseband frequency and there are no image receptions. (Undesired $-f_{IF} = f_S - f_{L0}$ in addition to desired $+f_{IF} = f_S - f_{L0}$ if f_S passes to mixer through filters.) Unfortunately, however, direct conversion receivers are also known for the many difficulties in making them work. The local oscillator leaks into the RF input, thus masking weak signals, and leaks further to the antenna. The isolation requirement between the LO and RX inputs is typically around 120 dB. Maintaining the output DC level and stability are both minor problems. Moreover, the linear phase and amplitude signal thus achieved is redundant for analog phones because all information is in phase. Therefore, a natural choice for the FM/PM receiver architecture consists of the detection of the phase of a non-zero IF. A promising solution for this problem is based on direct cycle time measurement.

Let f_0 be the carrier frequency and df the instantaneous frequency deviation. The cycle time difference due to frequency deviation and measured over N cycles is then

$$\Delta t = N \left(\frac{1}{f_0 + df} - \frac{1}{f_0} \right) \tag{1}$$

and from above, the instantaneous frequency deviation to be demodulated is

$$df = \frac{1}{2} \cdot \frac{\Delta t f_0^2}{1 - \frac{\Delta t f_0}{N}} \cong \frac{1}{N} \Delta t f_0^2 \tag{2}$$

When $\Delta t f_0 \ll 1$, the measured Δt is an approximate of the instantaneous frequency deviation. For a typical second intermediate frequency of 455 kHz and df smaller than 5 kHz, the approximation in (2) yields a distortion below -40 dB, or 1%, which is good enough for telephony. Increasing the decimation factor N both improves the resolution and decreases the nonlinearity distortion. A first order Taylor expansion of (2), realized with the most significant bit (MSB) of the measured cycle time, further improves the quality of the demodulated signal. From (2), it is clear that the resolution decreases with the square of the carrier frequency f_0.

An implementation of an FM/PM demodulator based on direct cycle time measurement is shown in Figure 6. The carrier frequency (e.g., 455 kHz) is amplified into band limited signal edges. There is no need for this signal to be constant amplitude, as is the case with a quadrature demodulator, thus yielding low sensitivity to amplitude modulation.

According to (1), for df = 5 kHz, N = 1, and f_0 = 455 kHz, Δt is 24 ns and, for a reasonable resolution, N needs to be in the order of 10 and the time measurement resolution is about 1 ns. Such an interval of time measurement, or better, is achievable with CMOS time-to-digital converters (TDC) consuming very low power and a clock frequency in the range of only 10 MHz.

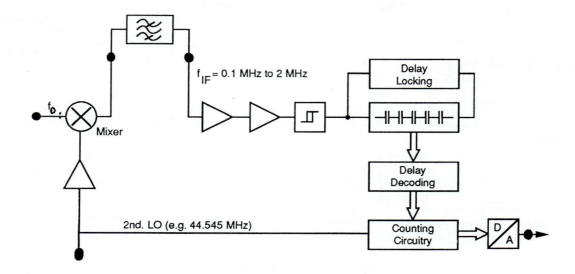

Fig. 6: Simplified block diagram of a digital FM/PM demodulator.

3.2. Frequency synthesis

The local oscillator for the receiver is synthetized using a PLL synthesizer. The TX frequency is generated using either another VCO-modulated TX synthesizer or using a modulated offset oscillator frequency mixed to the RX local oscillator.

In the robust PLL synthetizer architecture shown in Figure 7, the output frequency can be expressed as

$$f_{VCO} = f_{ref}(NN_0 + A)/M \tag{3}$$

where N and A are programmable integer parameters and N_0 is the division number of $N_0/(N_0 + 1)$ modulus prescaler. M is the division number in the reference branch.

The phase comparator produces output pulses at a frequency of f_{ref}/M which has to be integrated and filtered using a loop filter (LF) to form the VCO control voltage.

In all networks, the frequency settling time and noise (expressed either as residual deviation or residual phase noise) are specified so that the operating frequency of the phase comparator has to be set to the maximum, i.e., to the value f_{ref}/M, or channel raster frequency.

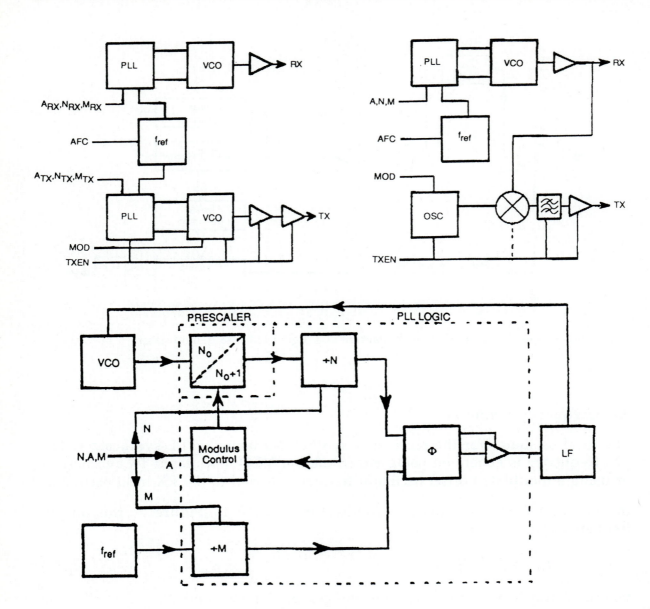

Fig. 7:   PLL-based frequency synthesis: (a) modulated; (b) offset oscillator concept; (c) PLL's internal structure.

The prescaler has to be capable of synchronously dividing a 900-MHz RF signal with minimum RF input level, and is therefore typically implemented as an ECL circuit. The remaining logic circuitry of the PLL is typically implemented using CMOS technology. Furthermore, because of the continuous operation during stand-by (RX-synthesizer), it is desirable that this logic circuitry consumes very low power.

Because of the high spectral purity needed for a specified residual modulation of the TX LO, and also for RX-injection (for sensitivity, adjacent channel rejection and spurious receptions), all low frequency noise must be isolated from the PLL. Particularly, the RF outputs must have sufficiently high isolation from the VCO to prevent frequency from jumping during TX switching. The purity of the reference frequency should be consistent with the output purity requirements. Any impurities will be multiplied as given by (3), and filtered by the PLL's closed loop system function. Because the PLL is capable of creating frequencies on raster only, all compensations for temperature drifts and aging of f_{ref} (5 to 100 ppm vs. 1 ppm or 2.5 ppm required in different systems) has to be done analog on f_{ref}. Such controllability of f_{ref} costs some extra circuitry and current consumption in the reference oscillator, thus resulting in an expensive, power-hungry and bulky TCVCXO component (temperature compensated voltage controlled XTAL oscillator).

The PLL itself can be better integrated with the buffer amplifiers, and perhaps the active parts of the loop filter (if at all needed), onto a single chip, leaving the VCO and large time constants to be implemented externally. Additional areas for system level development are concerned with the optimization of the RF input stage and reduction of the required RX-LO level. Evolving technology will give improved possibilities for this merger as well as for the reduction of on-chip phase jitter.

For frequencies below 100 MHz in CMOS, or 500 MHz in ECL or GaAs, direct digital synthesis (DDS) is also possible [6-8]. DDS is based on the structure shown in Figure 8a. In the phase accumulator the phase 2π is divided into 2^N parts, where N is the length of the phase accumulator, so that the output frequency is

$$f = \frac{k}{2^N} f_{ref} \tag{4}$$

where k is an arbitrary integer up to 2^{N-1}.

Each phase is translated from a look-up-table into a digital word for a digital-to-analog converter (DAC). Frequencies up to $f_{ref}/2$ can be generated provided that, according to the sampling theorem, the signal value is accurate and the reconstruction filtering is ideal. These conditions cannot be met because of the limitations of the DAC and the timing jitter occurring at the DAC output.

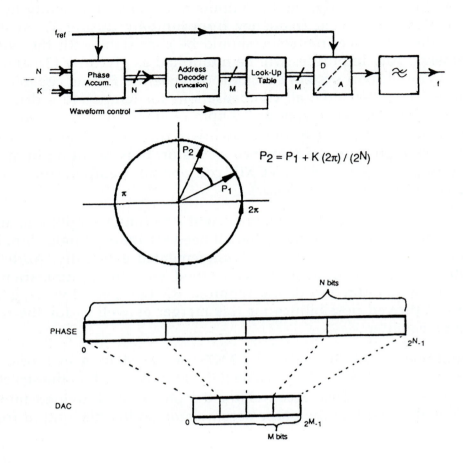

Fig. 8:   Operating principle of direct digital synthesizer: (a) architecture and (b) phase
            accumulator and bit mapping for the DAC.

Typically, the phase accumulator has 20 to 30 bits whereas the DAC has only 6 to
12 bits. This means that the smallest frequency increment on the phase
accumulator results in unchanged values of the DAC during the first few
samples at the new frequency. This is illustrated in Figure 9 for a square wave as
well as a sinusoidal wave at $f = f_{ref}/4$. Here, it can also be seen how the error
decreases with increasing the accuracy of DAC.

A quantization error E_U in the voltage signal can be estimated as a timing error,
at zero crossing, given by

$$E_T = \frac{E_U}{\Delta U} T = \frac{E_U}{\Delta U} \frac{1}{f_{ref}} \qquad (5)$$

where ΔU is the voltage change during T (see Figure 9c). Further, for a sinusoidal wave with frequency f and a peak-to-peak value U_0, it results in

$$\Delta U = U_0 \pi \frac{f}{f_{ref}} \tag{6}$$

and hence the error in the voltage level is

$$E_U = U_0 / 2^M \tag{7}$$

where M is peak-to-peak range of the DAC, in bits. The resulting timing error at frequency f can be expressed as

$$E_T = \frac{1}{\pi 2^M f} \tag{8}$$

Based on the above assumptions, this is independent of the value of f_{ref}! Usually, f_{ref} is desired to be high for reasons related, for example, to the sampling theorem. Another limitation of DDS is the need for a value of f_{ref} which is 2 to 4 times higher than the maximum value of f, and a spectral purity exceeding that required from the frequency to be generated by DDS. Presently, only SAW resonators provide a simple solution for high f_{ref}. Temperature and long term drift compensations of f_{ref} need a system level solution in both DDS and VCO based systems.

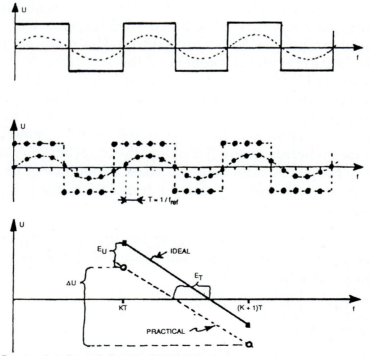

Fig. 9:   DDS generated waveforms: (a) ideal; (b) practical; (c) close-up at zero-crossing.

In CMOS logic, timing jitter is in the range of 20 ps to 100 ps [9]. An optimum performance of DDS is reached when E_T, in (8), is slightly below the total jitter of the logic timing jitter and the jitter of the reference frequency. For a frequency in the range of 100 MHz, 7 to 8 bits would be needed. Additional bits in the DAC would only further attenuate spurious, harmonic and aliased components, and also relax post filtering requirements. For higher frequencies, either GaAs or ECL technology would be needed for achieving smaller timing jitter (i.e., residual modulation). In ECL technology, the timing jitter is in the range of 10 ps thus making it possible to precisely generate frequencies up to 500 MHz with an 8-bit DAC [6, 10]. As the limiting factors in DDS are related to the timing jitter, DAC inaccuracy, and also to the f_{ref} and the technology, the relative level of spurious spectrum components increases with increasing frequency. There is not a known solution to circumvent the above spectral purity limitations of DDS and which does not exist in high-Q VCO based synthesizers. Another possible disadvantage of DDS results from the high power consumption associated with the required high-speed logic.

The main advantages of DDS are instantaneous frequency setting and sufficient accuracy to compensate for drifts in f_{ref} as well as the possibility of direct digital modulation [6-8]. Such potential advantages of DDS are attracting further research and development efforts using ECL, GaAs or even mixed CMOS/high-speed bipolar technology processes. For digital mobile phones, some on-chip logic for modulation generation should also be included.

Currently, significant research efforts are being deployed to combine the benefits of modulatable, arbitrary accuracy VLSI based DDS together with the simplicity and performance of traditional PLL synthesizers. The main alternatives being investigated include further developments of the fractional-N synthesizer [16, 17] and a new interpolating digital PLL, IDPLL [18]. Both are based on adjusting, in fractions of the cycle time of the VCO frequency, the average phase difference of the signal edges in both the reference and the VCO branches. In the fractional-N synthesis technique, this is accomplished by removing one cycle time in each N cycles whereas in the IDPLL synthesis technique described below a similar effect is achieved by adjusting the phase difference of each signal edge which, in turn, can be carried-out by delay interpolation, the principle of which is shown in Figure 10.

Interpolation is accomplished by multiplying the reference frequency f_0 by two integers L and $L+\Delta L$, and delaying each pulse edge of the reference signal going into the phase comparator by a delay

$$\Delta T_r = \frac{K_1}{f_0(L + \Delta L)} \tag{9}$$

and also delaying the pulse edges in the VCO branch by a delay

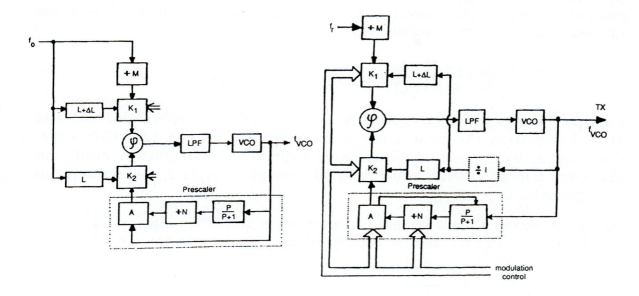

Fig 10: Operating principle of interpolating digital PLL synthesizers: a) constant delay shift architecture, b) constant phase or frequency shift architecture.

$$\Delta T_{VCO} = \frac{K_2}{f_0 L} \tag{10}$$

where K_1 and K_2 are both integer numbers. This leads to pulse lengths, respectively, of

$$T_r = \frac{M}{f_0} + \frac{K_1}{f_0(L + \Delta L)} \tag{11}$$

for the signal in the reference branch, and

$$T_{VCO} = \frac{N_0 N + A}{f_{VCO}} + \frac{K_2}{L f_0} \tag{12}$$

for the VCO signal. In the phase locked state, T_r and T_{VCO} are equal, thus yielding a VCO frequency of

$$f_{VCO} = \frac{f_0 \left(N_0 N + A \right)}{M + \dfrac{K_1}{L + \Delta L} - \dfrac{K_2}{L}} \tag{13}$$

If the condition $K_1 = K_1 = 0$ is met, then the above expression reduces to (3).

Updating of both integers K_1 and K_2, for each pulse edge, makes it possible to interpolate frequencies between frequency integer ratios given in (3), with the smallest frequency step being defined as

$$\left|\Delta f_{\min}\right| \approx \frac{f_0}{M} + \frac{(N_0 N + \Delta)}{L(L + \Delta L)} = \frac{f_{VCO}}{L(L + \Delta)} \tag{14}$$

For typical values of f_{VCO} = 900 MHz, f_0 = 12.8 MHz, M = 1024 and L = 63, and ΔL = 1, equation (3) gives a minimum frequency step of 12.5 KHz when changing A by unit steps. The above expression (14) gives a minimum frequency step of 218 Hz, which is small enough to replace, for example, an analogue AFC function.

Figure 10b shows a more versatile architecture of an IDPLL. Here, pulse delays are locked onto f_{VCO}/I, and the corresponding pulse lengths are, respectively,

$$T_r = \frac{M}{f_0} + \frac{K_1 - I}{f_{vco}(L + \Delta L)} \tag{15}$$

for the signal in the reference branch, and

$$T_{vco} = \frac{NN_0 + A}{f_{vco}} + \frac{K_2 I}{f_{vco} L} \tag{16}$$

for the signal in the VCO branch. This leads to a phase locked state with a VCO frequency of

$$f_{vco} = \frac{f_0}{M}\left\{ N_0 N + A + I\left(\frac{K_2}{L} - \frac{K_1}{L + \Delta L}\right)\right\} \tag{17}$$

Unlike the architecture of Figure 10a, and in expression (13), the smallest frequency increment

$$\left|\Delta f_{\min}\right| = \frac{f_0}{M}\frac{I}{L(L + \Delta L)} \tag{18}$$

is now independent of f_{VCO}. This has an additional advantage if K_1 and K_2 are both modified only for one pulse, instead of making all pulses either shorter or longer. Moreover, in both the reference and VCO branches, this causes a time shift of signals given by

$$\Delta T = \frac{I}{f_{vco}}\left(\frac{K_2}{L} - \frac{K_1}{L + \Delta L}\right) \tag{19}$$

By making $K_1 = K_2 = K$, and calculating the resulting phase shift DF in f_{VCO}, we get

$$\Delta\phi = 2\pi K \times \frac{I}{L(L + \Delta L)} \tag{20}$$

This divides the whole phase angle of 2π into $L(L + \Delta L)/I$ parts. According to equations (18) and (20), IDPLL is therefore capable of accurately providing both FSK and PSK modulations and, especially, multi-level PSK. An additional benefit from using the IDPLL-based architecture is that the pulse frequency of the phase comparator can now be higher than the minimum raster required, thus allowing faster frequency switching with, simultaneously, reduction of phase noise.

3.3. TX-function

The modulated synthesizer signal is amplified from approximately 0 dBm to 30-33 dBm in handportables (1 mW to 1-2 W) or up to 40 dBm in car phones. Important factors here concern the high total efficiency for long speech time and low heat generation, low noise in the RX band in order to relax duplex filter specifications, as well as good linearity, isolation and stable operation during power transients, and varying antenna coupling conditions. Poor linearity and isolation at the TX end-stage may cause a spurious reception through double mixing at a frequency of $f_S = (f_{rx} + f_{tx})/2$ which, in most cases, occurs at extreme channels (see Figure 11).

The alternative solutions available today for implementing the TX power amplifier are discrete bipolar, bipolar hybrid modules, and GaAs integrated circuits. Power control circuits and power amplifiers up to 250 mW, or perhaps even 500 mW, can presently be integrated without too much difficulty.

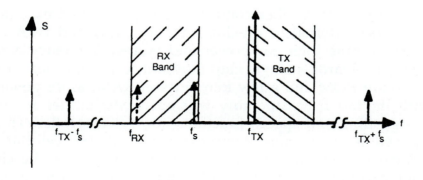

Fig. 11:   Frequency mapping of $(f_{rx} + f_{tx})/2$ spurious reception.

3.4. Signalling function

Signalling functions are needed for spectrum shaping for optimal channel occpation, FSK or other modulation/demodulation, bit synchronization, in many systems frame synchronization and sometimes support for error correction, data exchange for distance measurement, and speech/data multiplexing. Putting these together using mixed analog/digital CMOS technology (1 to 2 μm) leads to system dependent LSI solutions. Circuit and schematic design optimizations have been employed to reach the required performance with minimum power consumption.

RF channel management and cellular system functions run on a microcontroller. Future system integrations will change the block borders of signalling functions.

3.5. Speech and audio processing

The main part of speech and audio processing functions have been integrated into a single chip using analog CMOS (1μm - 2μm) technology. Whereas in the RX path (see Figure 5) there is not much to be done for better performance, in the TX path, AGC and limiting functions, together with positioning of filter functions, affect subjective speech quality and give therefore scope for further improvements.

Emerging important speech processing functions include the "hands free" (HF) function, adaptive noise suppression methods, cancellation of fading noise, voice mailbox and voice recognition or other speech controlled functions which will be brought from other applications. HF and noise supression methods need to be integrated into the telephone.

3.6. Control functions

Typically, an 8-bit microcontroller controls the telphone and executes cellular functions with either external or on-chip memory, A/D and D/A conversion functions and some glue logic for power and system frequency control. This will rapidly evolve towards a single chip multipurpose controller with 16 to 64 kilo bytes of on-chip ROM and a few kilo bytes of RAM. More powerful 16-bit processors with limited DSP capability will open possibilities for integrating further audio signalling and speech functions into the controller. This trend is, however, opposed by increasing complexity of both the cellular and user functions. A further essential part of the control functions is the charging of batteries and internal voltage distribution. Such functions will most probably be implemented as an ASIC or, at least, as part of other functional existing ASICs.

Analog telephones carry the subscriber identity themselves and therefore the E(E) PROM is also needed, unlike in the case of digital phones employing smart cards.

4.   Digital Mobile Phones

Digital mobile phones differ greatly from analog phones, as shown in Figure 12. The main differences concern the modulation techniques, channel access, and the use of channel correction and digital speech coding techniques. System specifications aim at maximal capacity, on the one hand and, on the other hand, at intensive use of DSP and IC technologies.

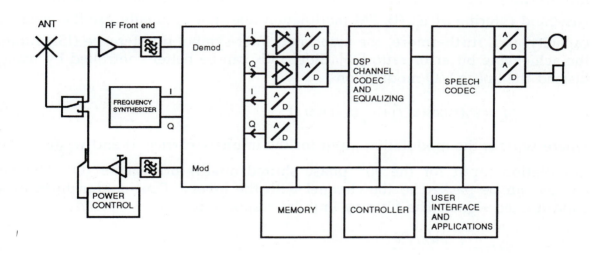

Fig. 12:   Basic block diagram of a digital CMT.

4.1. Speech coding

GSM has adopted the regular pulse exitation with long term prediction (RPE-LTP) algorithm to provide a net 13 kbits/sec bit stream. The digital AMPS (IS-54 specification) system has selected vector sum exited liner prediction coding (VSELP) to produce 7950 bps stream. The corresponding implementation needs a 16-bit DSP with about 2 kilo words of code. The necessary A/D conversion resolution is 8 bits, for a weighted conversion law, or 13 bits for linear conversion.

Cordless phones (CT2, DECT) have adopted the CCITT G721 speech codec which produces a 32-kbps stream, but there has been some discussion on the most appropriate echo cancellation strategies to be adopted. For all speech codecs there are already ongoing efforts to specify algorithms for half-rate coding (16 kbits/sec, 6.5 kbits/sec, and even as low as 4 kbits/sec).

4.2. Modulation

The FM modulation used in analog systems, with carrier

$$u_c = U_o \cos\left\{\omega_o t + d\phi \int_o^t u_m(t)\right\} \tag{21}$$

where dø is the phase modulation constant, fills the spectrum with an infinite series of sidebands being dependent of the amplitude and frequency of the modulating signal $u_m(t)$. $U_c(t)$ is kept within the desired channel with baseband filtering of $u_m(t)$. FM modulation does not optimally use the channel capacity and, furthermore, the associated inter-symbol-interference (ISI) limits the achievable bit error ratio [23]. Sidebands can be better controlled by using instead the linear I/Q modulation

$$u_c = w_I(t)\cos(\omega_c t) + w_Q(t)\sin(\omega_c t) \tag{22}$$

where $w_I(t)$ is the modulation input for the in-phase carrier (I) and $w_Q(t)$ is the modulation input for the 90° phase shifted quadrature carrier (Q). The bit energy can be packed to the channel with the inverse Fourier transform of a band-limited signal. The filtering impulse response is

$$h(t) = A_o \frac{\sin \pi t / T}{\pi t / T} \tag{23}$$

where T relates to the signal bandwidth B such that T = (1/2B). When samples are taken one per symbol, and well synchronized to the center of the symbol, this also leads to zero ISI. The waveform (23) is not, however, practical to implement because of the ISI sensitivity to sample timing and length of time series. Also, when an infinite sequence of symbols is superimposed, the time integral of the samples h(t) does not converge towards zero. The raised cosine impulse waveform

$$h(t) = A_o \frac{\sin \pi t / T}{\pi t / T} \cdot \frac{\cos \beta \pi t / T}{1 - (2\beta t / T)^2} \tag{24}$$

fulfills both requirements of zero ISI and integral convergence. The parameter ß in the above expression represents the rolloff factor. The frequency characteristics of (24) are given by

$$X(f) = \begin{cases} T, & 0 < |f| \le (1-\beta)/2T \\ \dfrac{T}{2}\left\{1 - \sin \pi T\left(f - \dfrac{1}{2T}\right)/\beta\right\}, & (1-\beta)/2T \le |f| \le (1+\beta)/2T \end{cases} \qquad (25)$$

Equations (24) and (25), as such or as in modified forms with varying ß, are used as modulation waveform in the JDC, PHP, and D-AMPS systems. In the D-AMPS and JDC systems, square roots of raised cosine filters are intended for both ends of the air interface resulting in raised cosine, zero-ISI total response. In the GSM and DECT systems, symbols are convolved with the Gaussian pulse

$$h(t) = \frac{\sqrt{2\pi}}{2\pi BT}\, e^{-\frac{t^2}{2\delta^2 T^2}} \qquad (26)$$

where T is the bit duration and B is the bandwidth. This leads to a Gaussian spectrum which gives good performance in presence of Gaussian noise in the channel.

Table 2 summarises the modulation parameters of different systems. Although the modulation specification is clear, its implementation is defined in practice by the allowed power spectrum of the transmission and BER together with other tolerance requirements such as the RMS value of the phase deviation from ideal modulation, or error vector in the I-Q plane. The DQPSK modulation parameters given in Table 2 do not provide constant amplitude modulation, thus setting for the RF power amplifiers additional requirements conflicting with the fundamental requirements of high efficiency and short rise time of the RF power. The DPSK modulation scheme is shown in Figure 13.

Table 2: Modulation parameters of digital mobile phone systems.

System	Bit rate (Kbit/sec)	Symbol rate (KS/sec)	Modulation Principle	Filtering Method	Actual Requirements
GSM	270.833	270.833	GSMK with differential encoding	Convolution with Gaussian pulse. $BT = 0.3$	Spectrum phase accuracy BER
D-AMPS	48.6	24.3	$\pi/4$ - DQPSK with differential encoding	$\sqrt{\text{raised cosine}}$ ß=0.3	BER, error vector in I-Q plane, channel model spectrum
DECT	1152	1152	GFSK	Convolution with Gaussian pulse. $BT = 0.5$	Spectrum deviation limits
PHP	384	192	$\pi/4$ - DQPSK with differential encoding	Raised cosine (I, Q), ß = 0.5	To be defined
JDC	42	21	$\pi/4$ - DQPSK with differential encoding	$\sqrt{\text{raised cosine}}$	Error vector, in I/Q plane. Spectrum

Figure 14 shows the basic architecture in which modulation is generated through amplitude modulation of mutually 90° phase shifted I (in-phase) and Q (quadrature) point frequencies. The DSP block generates I and Q modulations with the specified baseband filtering. The tolerance allowed for modulation has to be divided between the DSP, the mixers and also the phase noise of the RF frequencies.

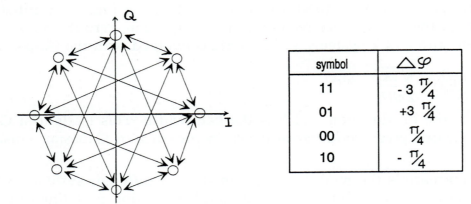

symbol	$\triangle \varphi$
11	$-3\,\frac{\pi}{4}$
01	$+3\,\frac{\pi}{4}$
00	$\frac{\pi}{4}$
10	$-\frac{\pi}{4}$

Fig. 13:   Modulation scheme for the dual-mode AMPS.

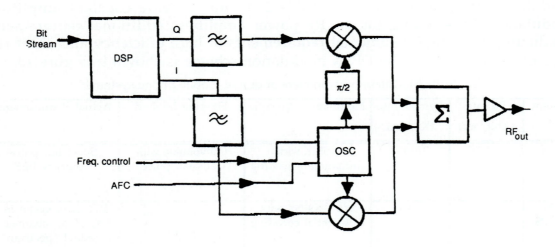

Fig. 14:   Block diagram of a GMSK/PSK modulator.

4.3. Symbol decoder (demodulator) and channel codec

In analog phones, the phase of the carrier contains all the information and can be demodulated instantaneously. Multipath propagation does not interfere with the modulation although it affects the amplitude of the carrier i.e., the instantaneous frequency of waves arriving within a few microseconds, along different paths, is the same but the corresponding phases of carriers are practically randomized and hence yielding varying and fading amplitude. For high bit rate digital phones, the multipath delay spread is larger than symbol time. This requires employing a channel equalizer which, in turn, requires knowledge of the amplitude and phase of the incoming signal. Equalization and convolution decoding of symbol sequences practically require sampling and storing, in analog form, the whole received TDMA message for further digital signal processing. In GSM this is mandatory because the training sequence is in the middle of the frame.

The RF carrier is demodulated by multiplying and, by means of $\pi/2$ phase shifted local oscillators, band limiting the RF signal to baseband I and Q components such that

$$I(t) = A_I \cos(\omega_o t + \phi_I)\cos(\omega_c t + \phi_M)$$

$$Q(t) = A_Q \cos\left(\omega_o t + \frac{\pi}{2} + \phi_Q\right)\cos(\omega_c t + \phi_M) \tag{27}$$

where ω_c is the carrier angular frequency, ϕ_M is the accumulated phase modulation, ω_o is the local oscillator angular frequency, and ϕ_I and ϕ_Q are the phase shifts between the carrier and local oscillator frequencies. Lowpass filtering the products in (27) leads to $\pi/2$ phase shifted rotating vectors

$$i(t) = A_I \cos\left\{(\omega_o t - \omega_c)t + \phi_I - \phi_M\right\}$$

$$q(t) = A_Q \cos\left\{(\omega_o t - \omega_c)t + \frac{\pi}{2} + \phi_Q - \phi_M\right\} \tag{28}$$

which reduce to the original modulation I and Q vectors when $\omega_o = \omega_c$ and $\phi_I = \phi_Q = 0$.

In GSM systems, the modulation ϕ_M is in the range of 200 kradians/sec. An error of 0.1 ppm between ω_o and ω_c represents, at 900 MHz, a rotation of 600 rad/sec, and the Doppler shift at 180 km/h an additional rotation of 1 krad/sec. The overall rotation of the coordinates during a 577-μs time slot is less than 1 rad, or half a symbol. The rotation of the carrier vector is a measure of the frequency offset between ω_o and ω_c.

Other practical means for obtaining the baseband signal samples consists of direct sampling at intermediate frequency. When this is accomplished at 4 * N multiple of the symbol rate, both I and Q samples are received in sequence and N can be used for decimation or oversampling factor for a delta-sigma type of converter. The most recent advances of A/D conversion circuit techniques have made it possible to reach the required sampling rate and accuracy together with low power consumption and small silicon area, i.e., at low cost. However, the power consumption is just low enough when the function is active during one time slot (1/3 or 1/8 of the time). Overall, such low cost requirements are prompting stiff competition among manufacturers of digital phones.

For continuous operation in analog phones, cost and power consumption of the modulator/demodulator explained earlier are still an order of magnitude lower than using state-of-the-art DSP technologies which, in turn, are much simpler to design.

When the sequence of equalizer training bits is known, an algorithm is executed to determine the coefficients of the transversal type of channel equalizer. Such coefficients represent the impulse response of the channel at RF carrier frequency. Then, convolution coded data bits are decoded using the Viterbi algorithm to reach the minimum error probability [23].

The channel codec functions require most of the DSP capability of the phone and, in order to achieve low power consumption, dedicated hardware support for the DSP processor has been employed.

Figure 15 sketches one possible architecture for the channel codec. Since the impulse response, and hence the transfer function of the channel, has a dynamic range of 80 dB, and a 5 to 6 bit resolution is needed for the baseband signals, coping with the whole range digitally would require an ADC with 18 bits to 19 bits. The use of an AGC function reduces the required dynamic range of the DAC. The architecture of Figure 16 makes it possible to trade-off between the performances of the RF stage, the analog front-end (A/D) and the DSP as well. The first generation of channel codecs relies heavily on technology oriented partitioning into IC blocks. More functional divisions become possible after a few years of experience on block and circuit level designs with multiple technologies.

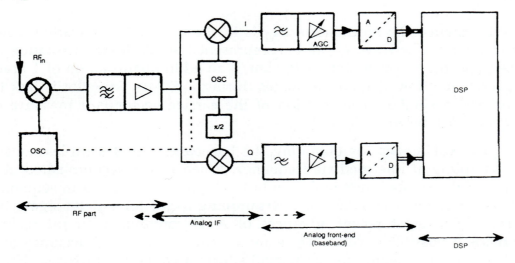

Fig. 15: Possible realization of the channel codec/equalizer.

One possibility for the realization of the channel codec consists of the co-integration of the analog front-end, baseband filters, AGC, A/D converters of the RX and the D/A converters of the TX, and the speech coding circuitry into a single mixed analog-digital chip. Recent evolutions [11, 12] of the sigma-delta type of converters have made this possible. However, there are still efforts being pursued to reduce the number of converters in RF from four to only two, and this would also change the specifications of the converters. Digital IF processing or direct conversion digital receiver concepts would require much higher sampling rates than the basic architecture of Figure 15. Also, the requirements for frequency synthesis are much more stringent.

Another possible development could be a dedicated DSP for CMT being capable of simultaneous channel and speech coding, provided optimized interfaces to other blocks of the telephone are also envisaged. This would require an hardware-based multiplier for sufficient performance at moderate clock frequency. A flexible DSP is necessary because some evolutions of channel and speech coding algorithms as well as manufacturer dependent algorithms can be expected. A modulator chip for digital phones has been recently designed in [15].

A practical implementation of a GSM handportable is shown in Figure 16. It is based on use of state-of-the-art DSP16® from AT&T, a DSP16C including memories, AD and DA for speech functions (coding, echo cancellation, HF, etc), whereas another DSP 16J performs channel equalization, AGC control, channel coding/decoding, AFC and other functions. Because these high performance, low power devices have very dedicated architecture for DSP, digital interfaces I/O, interupt, handling, etc., are managed with external logic chips. Time critical system functions, including Viterbi decoding, de-interleaving, modulation waveform generation, and, naturally, system timing, are peformed with dedicated digital VLSI circuits.

For upper layer functions, (signalling protocols, user interface, etc.) a 16-bit microcontroller with approximately 400k bytes of SW code is needed, and additionally 32K bytes of RAM and some EEPROM for storing calibration parameters, etc..

The RF part is very much as described earlier except that the duplexor is a filter instead of a switch. The current consumption with all parts rumming at 26 MHz clock frequency is up to 400 mA which reduces to an average of 100 mA with 1:8 TDMA factor and further down due to system sleep modes.

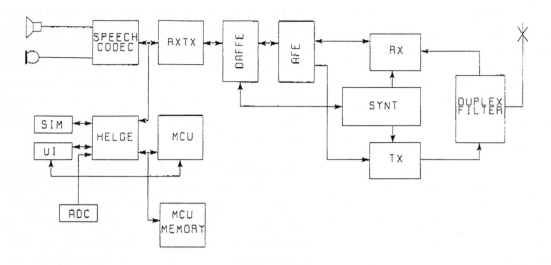

Fig 16 The architecture of Nokia 1011 SSM handportable (insert).

5.   Future developments in mobile communication

Mobile telephony has now reached the state where first generation of integrated radio transceivers (i.e., receivers and transmitter) added with cellular speech applications has reached mass production. Among the future directions there will be of course further integration but also a large number of new developments, such as going towards higher frequencies, first to 2GHz and then above, using radio access to low bitrate data applications and later to high bitrate local area data networks, and satellite cellular terminals.

Due to increased need of mobile communication, 230MHz of new worldwide frequency spectrum has been allocated in 2GHz range (1885 - 2025 and 2110 - 2200 MHz) for global third generation mobile communication system (called FPLMTS or future public land mobile telecommunications system). Third generation mobile system radio and system parameters will be specified by CCIR and CCITT during the 90s and the system should start commercial operation in early 2000s. Technical details should be fixed already in the mid 90s.

The basic requirements of FPLMTS terminals are adaptability to wide range of bit rates for different services, capabilities for both high traffic density and rural coverage, together with low cost. As these goals represent a major technological step and huge investment, the author expects an intermediate generation of mobile systems, called 2+ generation, to emerge through further development and potentially combination of second generation digital features.

These develoments will be widely affected by availability of not only IC technologies but also competing and complementing technologies such as SAW, CCD, ceramic resonators or the recently emerged GaAs-HACT (heterojunction acoustic charge transfer) component technologies [19].

This picture should be too simple without having a major development ongoing in the radio access technology itself. In addition to getting TDMA after FDMA, VLSI technology has made it possible to implement spread spectrum techniques, especially code division multiple access (CDMA) equipments to reach cost and power consumption levels acceptable to personal communication equipments.

Spread spectrum techniques have advantages over FDMA and TDMA methods because they need less or even no frequency planning and inherently make use of multipath propagation. However, the reduced cost and complexity of implementing CDMA advantages into TDMA or FDMA, or otherwise circumventing the limitations make the competition between access methods tougher. So, we may expect diversity reception, adaptable bitrates, utilization of cell site and spatial diversity to be proposed for TDMA and FDMA systems.

Key features of mobile communication system generations are shown in Table 3 and Figure 17.

Table 3: Mobile communication system generations.

Generation	Generation 1 Analog	Generation 2 Digital	Generation 2+ Service Integrated	Generation 3 Universal Digital
Service	• Mobile Speech	• Speech SMS	• Speech SMS, LAN, PMR	• Universal Speech Local Multiservice
Capacity and	• Up to 10%	• Up to 20%	• Up to 30%	• Up to 50%
penetration	• Band and Network Limits	• Peak Capacity Limits	• Peak Capacity with Limited Mobility	• Peak Capacity with Limited Mobility
Features	• Analog Speech • Good coverage	• Digital Speech • Quality & Security • Limited coverage	• Coverage varies due to access methods	• Global spot coverage, multiservice support
Problems remaining	• Capacity • Quality • Security	• Peak capacity • Coverage • Cost for consumers	• Multiple Standards • Lack of Universal Service	• Rural coverage
Access Methods	FDMA	TDMA	TDMA, FDMA, CDMA	TDMA, CDMA

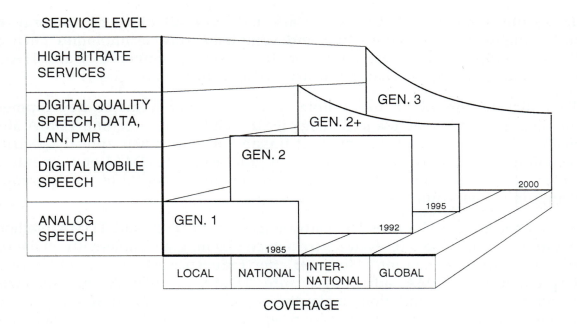

Fig. 17:   Service and coverage mapping of mobile communication system generations.

5.1. CDMA technology

The fundamentals and benefits of CDMA technology as an access method can be found in [18-21]. The simplified block diagram of the transceiver is shown in Figure 19. The application is as in, e.g., a GSM phone, and may include redundancy, interleaving or any other means to improve binary communication over RF.

The transmitter is very simple when binary codes are used for spreading. The main challenge in cellular usage is to implement path loss measurement between mobile station (MS) and Base Station (BS) and subsequent TX power control in such a way that all different users radiate the same bit energy to the receiver of BS. For simple applications even this is not necessary. The receiver in the simplest form has only a correlator loop for code synchronization and despreading. However, the performance is not good if the inverse of multipaph delay spread, $1/\tau_d$ is larger than RF signals bandwidth. In that case the phase of modulations arriving along different paths would be random, and, consequently, random phase of RF carrier, together with phase jitter in the despreading, would deteriorate the reception. On the other hand, if $1/\tau_d$ is smaller than signal bandwidth, flat fading like in FDMA and a total loss of the signal would occur.

The multipath conditions for Rake-receiver are shown in Figure 19. The BS transmits chips where f_a and f_b represent values 1 and 0, respectively, of the spreading code of length M. Spread spectrum spreading codes, being independent of information, are chosen so that the auto-correlation function for time difference $|\tau| > T_C$ is small.

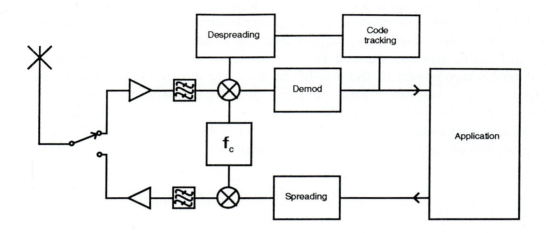

Fig. 18:   Simplified block diagram of a CDMA transceiver (for BPSK).

Provided that arrival time differences between paths are always greater than the chip length (T_C or much less than T_C forming practically a single beam), cancellation of carriers arriving along different paths occurs ramdomly. Loss of carrier energy is related to autocorrelation for $\tau > T_C$ and is therefore low. The information can be demodulated with having despreading code (with good correlation function to spreading code) synchronized for each arriving path, and summing the correlation products. Thus, the Rake-receiver is capable of utilizing multipath propagation whereas narrowband FM receivers (without diversity) or single correlator CDMA receivers suffer from multipath fading.

Depending on the type of modulation used, spreading and despreading can be done either on baseband signal, or at intermediate or carrier frequency I and Q components separately. The simplest case is BPSK where both spreading and despreading can be accomplished easily through multiplying the RF carrier (or IF) with +1 or -1.

The Rake-receiver in Figure 20 could output a bit modulated carrier if despreading block has output values +1 or -1, respectively. In practical CDMA systems, more effective modulations like QPSK, or multiple level PSK's such as 16-PSK are used and spreading and despreading done at baseband, and further binary information is additionally convolution coded. Despreading as such is relatively simple, but code synchronization with searching and tracking algorithms an relatively complicated VLSI devices and so far, only a few general purpose chips have been designed. Again for BPSK a simple code tracking method, the Costas loop, is known and has been implemented.

Currently Rake-receivers for CDMA are under intensive research with results indicating that an optimal code synchronization is based on combining correlator and convolution decoding implementing a maximum likelihood receiver, but dedicated VLSI implementations, being dependent on both chip rates and radio environments have not been made. There exist regulations for use of frequency bands in the 900-MHz, 2.4-GHz and 5.7-GHz frequency ranges for instrument, scientific and medical (ISM) purposes and other non-voice applications, like wireless LAN. For CDMA based voice communication in cellular networks the first specification is now being prepared [22], and a low cost, low power implementation will emerge later.

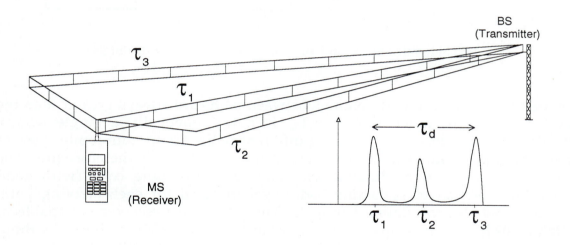

Fig. 19 a) Conceptual presentation of multipath conditions where 1/delay spread of paths exceed signal bandwidth; b) impulse response of the channel.

A practical implementation of a CDMA transceiver with single receiver branch is shown in Figure 21. The functions of Figure 21 are realized with 10 VLSI chips and consume several watts of power partly due to chip of rates up to 10 Mchip/sec., bit rates up to 50 kbps, variety of spreading codes and implementation QPSK and BPSK modulation and convolution coding [24].

It is too early at this phase to estimate the SW and protocol complexity of potential CDMA cellular, but it is likely to be less than in TDMA cellular, and further, the channel equalizer HW and algorithms are likely to be replaced by Rake-receiver.

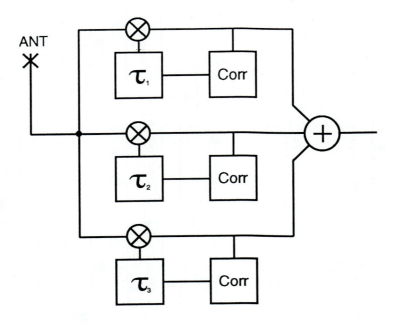

Fig. 20: Main block diagram of the Rake-receiver for three paths.

5.2. 2-GHz developments

In Europe, ETSI is specifying the DCS1800 system to adopt the GSM system solution working at the 1.8 GHz range in two 75 MHz bands with 95 MHz duplex distance. Integration of the 2-GHz RF parts is the cornerstone of this product.

The DECT system uses the frequency range between 1.88 GHz and 1.9 GHz, having ten 1728 kHz channels. Each channel possesses 24 time slots for both TX and RX separated by time division duplex (TDD). DECT is relying on dynamic channel allocation to prevent interfering users from being simultaneously in the same channel. The bit rate is 1152 kbps so that signal processing and filtering requirements can be somewhat relaxed. As DECT is intended for short distances no channel equalization is required.

A target price of approximately $250 is estimated for the DECT telephone and therefore the cost of the incorporated materials should be extremely low, including the smart card or other types of subscriber identify modules (SIM) which are used in GSM and DCS1800, too. Both 2-GHz products are planned to enter the market in 1993.

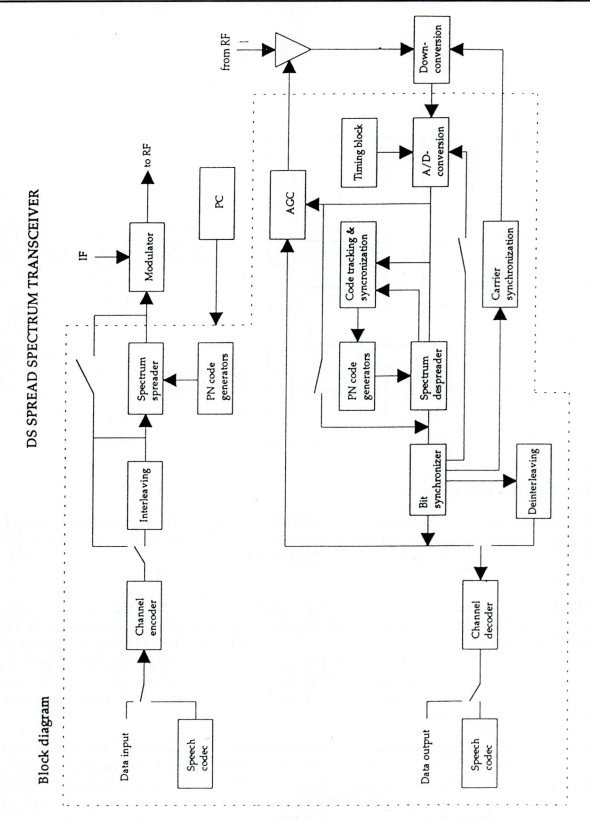

Fig. 21:   Practical implementation of a CDMA transceiver with single branch receiver (Courtesy of Elektrobit Oy, Finland).

5.3. Integration of new features into mobile phones

Along with increased penetration of mobile telephones the range of applications is broadening towards data applications and dedicated features which enhance the usability of phones.

Data modems connected to PC or laptop computer via PCMCIA -interface with error correction have already been launched.

Further data and voice applications are coming with digital telephony, including traffic information systems and other value-added services.

As mobile phones are used in noisy environments in cars or open streets with often poor radio connection, it is desirable to reduce the voice quality difference to normal telephony. One way to do this is by a more efficient use of radio channels which means diversity reception, the simplest form of which, antenna selection diversity, is currently used in Japanece NTT systems terminals. More advanced forms, such as combination diversity or coherent reception, have been known in principle for decades but not used due to high cost and complexity of implementation. The cost of implementing RF parts has reduced considerably during recent years. Further, possible techniques for controlling the phases of RF signals have been developed, such as interpolating RF synthesis described earlier or analog techniques [25].

The other way of enhancing mobile speech connection is by cancellation of background noise from the transmitted speech or the RF noise added in the channel from the received speech. The amplitude compander does this partly for flat spectrum but better performance is achieved by splitting speech into subbands as in Fig. 22, and controlling the gain of each subband according to

$$Y = \max\{C, (S_M - 2N_M)/S_M\} \tag{29}$$

where C is a constant, S_M represents measured signal and N_M represents measured noise. The main difficulty of distinguishing between desired signal and noise has been solved by finding a minimum constant level in each band, representing N_M, and a varying level representing S_M. The algorithm can be implemented into a part of speech codec DSP but also an analog CMOS implememtation has been made resulting in a 28 mn^2 chip with 2μ analog CMOS. The subjective improvement in speech quality is remarkable.

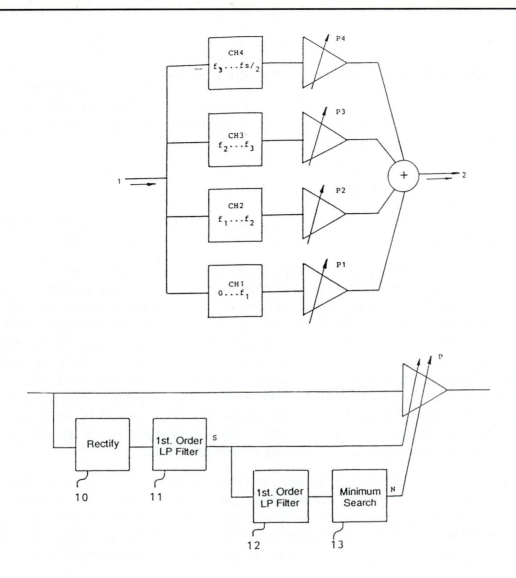

Fig. 22:   The architecture of noise cancelling algorithm and HW implementation of one subband channel.

6.   IC Technology

6.1. FDMA, TDMA and CDMA from integrations point of view

FDMA terminals have reached mature specifications, and their complexity allows for integration of protocol, control and speech functions towards single chip solutions using available technology. Microcontrollers with sufficient on-chip memory are now available, and a dedicated DSP for baseband functions should have sufficiently low cost and power consumption somewhere below 1μ technology. Analog RF integration is the remaining challange where new solutions are needed.

Going into low power ICs with on-chip memory and serial communication between the chips is crucial for avoiding EMI/EMC problems and heavy batteries.

Today the total component count of a FDMA terminal is 350-700, and power consumption in standby is 30-50 mA without sleep modes.

TDMA terminals need intensive DSP and large memories, say 200-400k bytes. The RF is more effective from a systems point of view but, contrary to expectations, more difficult to design and integrate than FDMA RF. TDMA terminal has therefore a few hundred components more than its FDMA counterpart.

If we estimate memory density to double in each 2 to 2 1/2 years, it would take a couple of years before a fully featured TDMA phone goes to on-chip memories. TDMA needs novel integrated solutions for I/Q modulation and demodulation.

A major need for TDMA phones is to reduce power consumption to reach the average level of an FDMA phone. A transition to 3 Volts would come very close to this, and become practical certainly before mid 90's.

Though the potential of CDMA cellular is not yet clearly defined, its VLSI complexity is the same or slightly above TDMA but the protocol complexity may be smaller.

6.2. Integration technologies

In a time-span of about 5 years, the author expects the following trends to occur in the fast growing area of mobile communications:

- In analog phones, the multipurpose controller will pick up other functions, including DSP for baseband signal processing. The analog CMT will be the first really integrated radio terminal.

- New radio architectures will emerge for system level integration of RF functions with advanced RF BiCMOS technology processes becoming available. The picture concerning filtering technology is not yet clear because of the high cost of SAW and other emerging alternatives.

- In complex digital phones, technology oriented partitioning (DSP, controller, RF) will be the dominant choice at the beginning but functional division (speech codec, controller, modulator/demodulator, RF) will provide fair competition because of increased standardization over several systems.

• Simple digital phones, like DECT CT2 and PHP, will most probably follow the development path of analog phones.

6.3. ASICs versus Standard ICs

The dominant technology factor in the mobile communications business concerns the life cycle of each product generation, which tends to be less than two years. So, there will be less time to create long lasting standards but also too much effort will be needed to create and develop new ASICs for all new products.

Standard ICs will become available whenever several products use the same function over several years, thus allowing for continuous development and marketing of the component as well as stimulating competition among various component suppliers. The CCITT G721 speech codec will clearly become such a standard. This is also rather likely to occur for an analog front-end and a CMT-dedicated DSP.

There will also be some supply of vendor-specific system components, such as GSM or D-AMPS speech codecs, dedicated DSP processors, microcontrollers without code or with open ended code blocks, in order to facilitate manufacturer dependent features.

For the rest, wherever a strong evolution will take place, ASICs will be needed. This will unevitably promote technology alliances between some CMT and IC manufacturers, as we also anticipate seeing some communication houses with in-house high-tech IC manufacturing capability to attempt to compete either in the component or in the product market segments.

7. Conclusions

Analog radio terminals are mature for VLSI integration. VLSI integration has made digital radio technology and TDMA terminals possible but further integration is a real need and challenge.

The development of technology does not bring further VLSI integration for free; rather a lot of new ways have to be found to make the VLSI integration of mobile radio terminals feasible.

In parallel to TDMA, CDMA technology will be specified based on both systems level benefits and VLSI possibilities.

VLSI capabilities also add new fascinating features and applications to radio telephony.

References

[1] Integrated Circuit Engineering Corp., Status 1989. *A Report on the Integrated Circuit Industry* (Edited by W. Mc Clean).

[2] SRI International Report No. 739, *Microelectronics to the Year 2000* (Edited by K.M. Taylor).

[3] CEPT/GSM/PN's GSM Recommendation

[4] E.I.A. Project 2215, Cellular System Dual Mode Subscriber - Network Equipment Compatibility Specification IS-54, December 1989.

[5] Kuisma E., et. al., Signal Processing Requirements in Pan-European Digital Mobile Communications, *Proc. 1988 International Symposium on Circuits and Systems*, vol. 2 (pp. 1803-1810).

[6] Plessey Semiconductors (product literature), SP2007 500 MHz Direct Frequency Synthesizer (Nov. 89).

[7] Qualcomm Inc. (product literature), Q2334 Dual Direct Digital Synthesizer (Jan. 1989).

[8] Digital RF Solutions Corp. (product literature) DRFS-3250 NCMO Numeric Controlled Modulated Oscillator (1987).

[9] Kostamovaara J., et. al., ECL and CMOS ASICs for time-to-digital conversion, *Proc. 2nd Annual IEEE ASIC Seminar and Exhibit*, Rochester, N.Y. (Sept. 25-28, 1989).

[10] Rankinen R., Maatta K., Kostamovaara J., Time-to-digital conversion with 10 ps single shot resolution, *MELECON*, 1991.

[11] Ritoniemi T. et. al., Oversampled A/D and D/A Converters for VLSI System Integration, *Proc. IEEE 1990 ASIC Seminar and Exhibit*, Sept. 1990, Rochester, N.Y...

[12] J. C. Candy, G. C. Temes: *Oversampling Delta-Sigma Data Converters, Theory, Design, and Simulation*, IEEE Press, 1992.

[13] Calhoun G., *Digital Cellular Radio*, Artech House, London 1988 (pp. 339-374).

[14] Qualcomm Corp. Commercial Information, CDMA Cellular, The Next Generation, 1989.

[15] Rahier M. et al, VLSI Components for the GSM Pan-European Digital Cellular Radio System, *4th Nordic Seminar on Digital Mobile Radio Communications DMR IV*, 26.-28.6.1990, Oslo, Norway.

[16] U.L. Rohde, *Digital PLL Frequency Synthesiers. Theory and Design*, Prentice Hall, Inc., 1983.

[17] A.W. Hietala, D.C. Rabe: "Latched Accumulator Fractional N Synthesis with Residual Error Reduction", *US Patent 5,093-632 (Motorola, Inc.)*, 1992.

[18] J. Rapeli, "Interpolating Phase-Locked Loop Frequency Synthesis", *US Pat. 5, 479-520 (Nokia Mobile Phones Ltd.)*, 1992

[19] R.W. Miller, C.A. Ricci, R.M. Kansy, "An Acoustic Charge Transport Digitally Programmable Transversal Filter, *IEEE JSSC*, Vol. 24(G), Dec. 1989.

[20] R.L. Pickholtz, D.L. Schilling, L.B. Milstain, "Theory of spread-spectrum communications - a tutorial," *IEEE Tr. on Comm.*, 30 (5), May 1982.

[21] W.C.Y. Lee, "Overview of Cellular CDMA," *IEEE Trans. Vehicular Techn.*, Vol. 40 (2), May 1991.

[22] "An overview of the application of code division multiple access (CDMA) to digital cellular systems and personal cellular Networks" submitted contribution to T.I.A. TR45.5 Subcommittee.

[23] J.G. Proakis, *Digital Communications*, McGraw-Hill Book Company, 1989.

[24] Elektrobit Oy (product literature) "DS Spread Spectrum Transceiver", 1992.

[25] L. Schmidt, H.M. Rain, "Continuously variable gigahertz phase shifter IC covering more than the frequency decade, *IEEE JSSC*, Vol.27(6), June 1992.

Index

A

Accuracy
  absolute vs. relative, 343–44
  of MOSFET-C filters, 188
Active-Delayed-Block (ADB) polyphase
  architectures, 263–66
Active-RC filters, 178, 213, 226–27
Adaptive biasing, 84–86
A/D converter. *See* Analog-digital converters
  (ADCs)
Adders, 508–9, 524–25
Address Generation Unit (AGU), 447
Amplifier(s)
  cascode operational transconductance (OTA), 31
    folded-cascode, 34–35, 36–38, 47
    single-ended versions of, 38–39
    telescopic-cascode, 35–38, 49–50
  class-AB, 84–86
  common source elementary, 72–73
  low-offset controlled-gain, 111
  for mobile telephones, 536, 537–39
  noise from, 134–36
  sample/hold, 307–9
AMPS, 532, 533
Analog cells, stacked layout for, 354–56, 357, 358
Analog circuits, dynamic, 97–124
  dynamic biasing of CMOS inverter-
    amplifiers, 112–14
  dynamic comparators, 114–18
  dynamic current mirrors, 118–22
  offset compensation of operational amplifiers,
    108–12
  sample-and-hold, 97–108
Analog-digital converters (ADCs), 28, 289–315.
  See also Delta-sigma data converters
  circuit building blocks for, 296–313
    classification of techniques, 292–95
  in DSL transceivers, 414
  high-speed DACs in MOS technology, 296–306
    charge-redistribution, 297–302
    component mismatch effects in
      monolithic, 302–6
    current-switched MOS, 296
    high-speed, 296–306
    resistor-string, 297, 298
    self-calibration of, 304–6
  In mobile telephones, 537
  monolithic sample/hold amplifiers, 307–9
  monolithic voltage comparators, 309–13
  Nyquist-rate, 319
  role in telecommunications, 290–91
Analog sound demodulators, 467–68
Analog switches, 79–80
  zero-offset, 297
ANSI T1.601 standard, 387
ANSI 2B1Q standard, 424
Anti-aliasing filters, 379
Approximation, successive, 292, 295, 468–69
Architecture. *See* System architecture
Arithmetic Processing Unit (APU), 447
Arrays
  programmable logic (PLAs), 510–11
  systolic, 457–58
ASICs, 566
Audio processing in television, 467–70
Autozero circuit, 105–8
Auxiliary input, low-sensitivity, compensation
  by, 108–12

B

Base current, bipolar, 14
B-channels, 402
Bias circuitry for CMOS op amps, 47–49

C

Biphase line code, 423
Bipolar modelling, 11–19
  base current, 14
  carrier transport, 11–12
  charges and capacitances, 14–16
  collector current, 12–14
  parameters for, 20–23
  parasitics and model extensions, 17–19
  quasi-saturation, 16–17
Bipolar transistors, CMOS-compatible, 67–69
Biquads, switched-capacitor, 215, 225–36
  capacitive dumping, 230
  cascade design, 236–37
  circuit generator via equivalent resistors,
    226–28
  differential, 234–36
  for exact design, 231–34
  exact transfer function, 228–30
Bit error rate (BER), 155
Block matching, dedicated realizations of, 455–59
Boundary-dependent errors, 352–53
Boundary dependent etching, 342, 343
Bulk charge, 9

C

Cache, 446, 449
CAD tools, 341
Capacitance(s)
  of bipolar transistors, 14–16
  MOSFET, 9–10
Capacitive coupling, 356–59
Capacitors, layout of, 351–54
Carrier control approach, 13–14
Carrier space charge, 15
Carrier transport, bipolar, 11–12
Cascade (multi-stage) A/D converters, 327–29
Cascade SC biquad design, 236–37
Cascode op amps, 31–39
  folded, 34–35, 36–38, 47, 358
  single-ended versions, 38–39
  telescopic, 35–38, 49–50
Cascoding, low-voltage, 75–77
CCITT, 436
CCITT G721 speech codec, 549–50
CDMA. *See* Code Division Multiple Access
  (CDMA)
Cellular mobile telephony (CMT), 529–33. *See
  also* Mobile telephones
Charge conservation principle, 9
Charge control approach, 12–13
Charge exchange circuits, 97
Charge injection, 224
Charge-redistribution DACs, 297–302
Charges
  in bipolar transistors, 14–16
  MOSFET, 9–10
Chrominance channel, 477–80
CIF, 436
Circuit-level models, 1–25
  bipolar modelling, 11–19
  defined, 1
  MOSFET modelling, 1–11
  parameters for, 20–23
  types of, 1
Class-AB amplifiers, 84–86
CMOS, micropower operation and, 54
CMOS circuit technique for DPCM, 508–9
CMOS compatible bipolar transistors, 67–69
CMOS gates, dynamic, 80–81, 91
CMOS inverter-amplifiers, dynamic biasing of,
  112–14
CMOS operational amplifiers, 27–51
  bias circuitry of, 47–49
  cascode, 31–39, 47, 49–50

differential circuit approach to, 28–31
general considerations, 27–28
layout techniques for, 49–50
multiple stage, 31, 39–46
number of internal stages, 31
Codec devices, 380–84, 395
  CCITT G721 speech, 549–50
  for digital mobile telephones, 552–56
  filters, 392
Code Division Multiple Access (CDMA),
  557–60, 562, 565
Collector current in bipolar transistors, 12–14
Common intermediate format (CIF), 436
Common-mode input, 29
Common source elementary amplifiers, 72–73
Comparators
  dynamic, 114–18
    basic circuit, 114–17
    multistage, 117–28
  monolithic voltage, 309–13
Conjugate Quadrature Filterbanks (CQF), 516
Continuity equation, 11
Continuous-time filters, 177–211
  computer simulations of, 206
  MOSFET-C, 179–92
  on-chip tuning schemes, 202–6
  transconductor-C, 192–201
Controller(s)
  D-channel, 403, 405–6
  ISDN layer 2 terminal, 391–92
  line card, 376–78
Coupling, digital noise, 356–63
Crossing, 356–59
CT2, 532, 533
Current mirrors, 70–71
  dynamic, 118–22
Current references, micropower, 77–78
Current-switched MOS DACs, 296

D

DACs. *See* Digital-analog converters (DACs)
D-AMPS, 532, 533
D-channel controllers, 403, 405–6
D-channels, 402
DCS 1800, 532, 533, 561
Decimators, 251
  Finite Impulse Response (FIR)
    direct-form polyphase structure for,
      252–58
    multi-amplifier, 262–66
  Infinite Impulse Response (IIR), 266–81
    first-order, 268–71
    Nth.-order, 277–81
    second-order, 271–77
Decision Feedback Equalizer (DFE), 415–16
DECT, 532, 533, 561
Deflection processor, 483–83
Delayed decision technique, 506–7
Delta-sigma data converters, 317–39
  noise-shaping A/D converters, 320–34
    delta-sigma modulator, 322–25
    higher-order single-stage converters,
      325–27
    modulators with multibit quantizers,
      329–34
    multi-stage (cascade) converters, 327–29
    quantization noise, 320–22
  noise-shaping D/A converters, 335–37
Demodulators
  analog sound, 467–68
  for digital mobile telephones, 552–56
  FM/PM, 537–39
Depletion charges, 15
Detection

coherent, 167–70
direct, 166–67
DIBL effect, 6–7
Die area of folded- vs. telescopic-cascade OTAs, 36–37
Differential circuitry, 28–31
Differential class-A two stage op amps, 40–41
Differential pairs, 71–72
Differential SC biquads, 234–36
Differential switched-capacitor integrators, 30
Diffusion, lateral, 342–43
Digital-analog converters (DACs), 28, 296–306
 charge-redistribution, 297–302
 component mismatch effects in monolithic, 302–6
 current-switched MOS, 296
 high-speed, 296–306
 resistor-string, 297, 298
 self-calibration of, 304–6
Digital Audio, 291
Digital exchange systems, 371–87
 analog subscriber line card devices, 378–86
 digital subscriber line card devices, 386–87
 line card controller, 376–78
 line card functions and architectures, 372–76
 PCM switching network components, 371–72
 serial communication controllers, 372–74
Digital mobile telephones, 549–56
 modulation for, 550–52
 speech coding, 549–50
 symbol decoder (demodulator) and channel codec, 552–56
Digital noise coupling, 356–63
Digital Signal Processing (DSP) techniques, 394–98
Digital subscriber loop transceivers. *See under* Transceiver(s)
Digital television. *See* Television, digital
Direct detection, 166–67
Discrete cosine transform (DCT), 450–55, 499–500
Distortion of MOSFET-C filters, 189
Doping atoms, diffusion of, 342–43
DPCM
 CMOS circuit technique for, 508–9
 modified structure, 505–6
 parallel processing with delayed decision, 507–8
 parallel processors, 503–5
 2D, 502–15
  comparison of architectures for, 507–8
  line delay realization, 511–14
  table look up realization, 509–11
DPCM coding, 498–99
Drain current, MOSFET, 4–6
Driver, line, 414
Dual tone multifrequency (DTMF) technique, 389
Duplex filters, 534–35
Dynamic analog circuits, 97–124
 dynamic biasing of CMOS inverter-amplifiers, 112–14
 dynamic comparators, 114–18
 dynamic current mirrors, 118–22
 offset compensation of operational amplifiers, 108–12
 sample-and-hold, 97–108
Dynamic CMOS gates, 80–81, 91
Dynamic range
 of MOSFET-C filters, 189
 of optical receivers, 164–66

E

Ebers-Moll model, 13
Echo cancellation, ISDN, 418–21
EDTV, 463–64
Electron current density, 11
 in npn transistor, 12
Empirical models, 1

Equalization in DSL transceivers, 415–16
Erbium-doped fibre amplifier, 170
Error(s)
 boundary-dependent, 352–53
 in oxide thickness, 351
 due to tri-dimensional effects, 342, 343
Etching
 boundary dependent, 342, 343
 under protection, 342, 343

F

Fast Cosine Transform (FCT), 453–54
FDMA terminals, 565
Feedback, static, 8
Field effect transistor (FET), noise spectral densities for, 134–35
Field rate upconverter, 488–89
Filter(s), 380–84. *See also* Continuous-time filters; MOSFET-C filters; Multirate switched-capacitor filters; Switched-capacitor (SC) filters
 active-RC, 178, 213, 226–27
 anti-aliasing, 379
 chroma bandpass, 477–80, 485–87
 codec, 392
 duplex, 534–35
 for mobile telephones, 536
 non-recursive wave digital (NRWDF), 516–17
 PAL luminance, 475–77, 485, 486
 reconstruction, 379
 Signal-Processing Codec (SICOFI), 382–84
 transmit, 413
 transversal FIR, 517
 Wave Digital, 384
Filterbanks, subband. *See* Subband filterbanks
Finite Impulse Response (FIR)
 decimators and interpolators, 252–66
  direct-form polyphase structure for decimation, 252–58
  direct-form polyphase structure for interpolation, 258–62
  transversal filter for, 517
Flicker reduction, 487–89
FM/PM demodulation, 537–39
Folded cascode op-amp, 358
Folded-cascode OTAs, 34–35, 36–38
 bias voltage for, 47
Fractional-Tap (FT) linear equalizer, 415–16
Frame word synchronous phase adjustment, 421
Future Public Land Mobile Telecommunications System (FPLMTS), 533, 556

G

General Circuit Interface (GCI), 376
GSM, 532, 533, 561
Gummel Poon model, 13

H

Half duplex transmission, 291
HD-MAC coding, 496–99
High-definition television (HDTV), 463–64, 495–527
 source coding algorithms for, 496–502
  DCT, 499–500
  DPCM, 498–99
  HD-MAC, 496–99
  subband, 500–501
 subband filterbanks, 515–25
  arithmetic blocks realization, 522–25
  complexity of, 525
  expansion in time by subsampling, 520–21
  hardware requirements, 515–16
  non-recursive wave digital filter (NRWDF), 516–17
  polyphase structure, 517–20
  quadrature mirror filterbanks (QMFs), 521–22
  transversal FIR filter, 517
 VLSI architecture for 2D DPCM, 502–15

comparison of, 507–8
line delay realization, 511–14
table look up realization, 509–11
VLSI implementation aspects, 502
Higher-order single-stage A/D converters, 325–27
High Level and Control Processor (HLCP), 448–49
Horizontal synchronization, 483

I

IDTV, 463–64
Impact ionization, 70
Infinite Impulse Response (IIR) decimators, 266–81
 first-order, 268–71
 Nth.-order, 277–81
 second-order, 271–77
Infinite Impulse Response (IIR) interpolators, 281–85
Integrated Services Digital Network (ISDN), 369–70, 395
 access functional blocks, 403
 D-channel controllers, 403, 405–6
 digital subscriber loop transceivers, 406–24
  architecture of, 411–13, 422–23
  echo cancellation implementation, 418–21
  examples of, 423–24
  implementation series, 421–23
  key loop characteristics, 408–10
  line code choice, 410–11
  receiver implementation, 414–18
  transmitter implementation, 413–14
  2-wire duplex operation, 410
 fundamentals of, 401–3
 S/T-bus transceivers, 404–5
 subscriber equipment, 390–94
 transceiver devices, 387
Integrators, switched-capacitor
 damped, 221–23
 differential, 30
Internal resistors, 28
Interpolating digital phase locked loop (IDPLL), 544–47
Interpolators, 251
 FIR, 258–62
 Infinite Impulse Response (IIR), 281–85
Inverter-amplifiers, CMOS, 112–14
Inverting configuration, 28
Ionization, impact, 70
ISDN. *See* Integrated Services Digital Network (ISDN)

J

Jitter compensation, 421
Jitter degradation control, 420–21
Junction sidewell effects, 21

K

kT/C noise, 307–8

L

Ladder filters, switched-capacitor, 237–45
 circuit generation, 238–41
 design of, 241–42
 exact design method, 242–45
Lateral diffusion, 342–43
Layout, 341–67
 absolute and relative accuracies, 343–44
 of capacitors, 351–54
 circuit vs., 342–43
 of CMOS op amps, 49–50
 digital noise coupling, 356–63
 floor planning of mixed analog-digital blocks, 363–66
 of MOS transistors, 344–48
 of resistor, 348–51
 stacked, for analog cells, 354–56, 357, 358

Leakage, 70
Least Means Squares (LMS) algorithm, 420
Linear Kernel Processor (LKP), 449
Line cards
  analog devices, 378–86
  architectures of, 374–75
  controller, 376–78
  digital devices, 386–87
  functions of, 374
Line code
  biphase, 423
  for digital subscriber loop transceivers, 410–11
  self-equalizing, 423–24
Line delay for 2D DPCM, 511–14
Line driver, 414
Link Access Protocol for D-Channel (LAPD), 403
Local Memory (LM), 446–47
Local substrate potential, 56
Lossless digital integrator (LDI) variable, 221
Low-offset controlled-gain amplifier, 111
Low-sensitivity auxiliary input, compensation by, 108–12
Low-voltage cascoding, 75–77
Luma/chroma separation, 484–86
Luminance channel, 475–77

M

MEXTRAM model, 17–18
Micropower techniques, 53–96
  analog switches, 79–80
  CMOS compatible bipolar transistors, 67–69
  CMOS technology and, 54
  current mirrors, 70–71
  current references, 77–78
  differential pair, 71–72
  dynamic CMOS gates, 80–82
  low-voltage cascoding, 75–77
  MOS transistors, 55–67
  operational transconductance amplifiers (OTAs), 81–87
  parasitic effects and, 70
  passive devices, 69
  quartz crystal oscillators, 87–91
  sequential logic building blocks, 91–92
  voltage gain cells, 72–75
  voltage references, 78–79
Mirrors, current, 70–71
  dynamic, 118–22
Mixed analog-digital circuits, layout of. *See* Layout
Mobile telephones, 529–68
  Code Division Multiple Access (CDMA), 557–60, 562, 565
  control functions, 548–49
  digital, 549–56
    modulation for, 550–52
    speech coding, 549–50
    symbol decoder (demodulator) and channel codec, 552–56
  FPLMTS, 556
  frequency synthesis for, 539–47
  generations of, 556–58
  IC technology, 565–67
  integration of new features into, 561–64
  mobile communications systems, 529–33
  receiving (RX) function, 534–39
  signalling function, 548
  speech and audio processing, 548
  2 GHz developments in, 561
  TX-function, 547
Modems, 291
  DSP techniques for, 395
MOSFET
  short channel effects, 21
  as voltage-controlled resistor, 179–81
MOSFET-C filters, 179–92
  accuracy of, 188

based on balanced structures, 183–85
cancellation of nonlinearities, 181–83
distortion and dynamic range of, 189
parasitic capacitance of, 189–91
synthesis of, 185–88
transconductor-C filters vs., 201
worst-case design of, 191–92
MOSFET modelling, 1–11
  charges and capacitances, 9–10
  drain current at strong inversion, 4–6
  parameters for, 20–23
  parasitics and model extensions, 10–11
  saturation mode, 7–9
  subthreshold current, 6–7
  threshold voltage, 1–13
MOS reset switches, 30
MOS transistors, 55–67
  absence of control current in, 97
  drain current, 60–62
  inversion level of, 62
  layout of, 344–48
  mismatches between, 66–67
  modes of operation, 59–60
  noise behavior of, 64–66
  schematic cross-section of, 55
  strong inversion approximation for, 56–58
  transconductances for, 62–64
  weak inversion approximation for, 58–59
Multi-amplifier FIR decimators, 262–66
Multiple instruction, multiple data stream (MIMD), 442–43, 445–50
Multiple stage op amps, 31, 39–46
  differential class-A two stage, 40–41
  resistor driving class-AB, version I, 42–44
  resistor driving class-AB, version II, 44–46
Multiplications, 422
Multipliers, 522–23
Multirate switched-capacitor filters, 251–88
  Finite Impulse Response (FIR) decimators and interpolators, 252–66
  Infinite Impulse Response (IIR) decimator building blocks, 266–81
  Infinite Impulse Response (IIR) interpolator building blocks, 281–85
Multi-stage (cascade) A/D converters, 327–29

N

Network terminations, ISDN, 393–94
NMT-450, 532, 533
NMT-900, 532, 533
Noise
  kT/C, 307–8
  in mobile telephones, 563–64
  of MOS transistors, 64–66
  of optical receivers, 132–47
  of OTAs, 82–84
  pattern, 324–25
  from power supply connections, 359–61
  quantization, 320–22, 331–32
  of sample-and-hold circuits, 99–100
  shot, 138
  from substrate, 361–63
  switching, 28
  truncation, 335–37
Noise coupling, digital, 356–63
Noise-shaping converters
  A/D, 320–34
    delta-sigma modulator, 322–25
    higher-order single-stage converters, 325–27
    modulators with multibit quantizers, 329–34
    multi-stage (cascade) converters, 327–29
    quantization noise, 320–22
  D/A, 335–37
Nokia 1011 SSM handportable, 556
Nonlinearities, cancellation of, 181–83
Non-recursive wave digital filter (NRWDF), 516–17

npn transistor, electron current density in, 12
Nyquist–rate A/D converter, 319

O

Offset compensation of operational amplifiers, 108–12
On-chip tuning scheme, 202–6
Open-Floating-Resistor branch (OFR), 254
Operational amplifier-RC integrators, 215
Operational amplifiers, offset compensation of, 108–12
  See CMOS operational amplifiers
  folded cascode, 358
  two stages, 357
Operational transconductance amplifiers (OTAs), 81–87
  with adaptive biasing, 84–86
  noise behavior of, 82–84
  transconductance of, 81–82
  voltage gain of, 82
Optical Ether, 171
Optical receivers, 125–74
  bandwidth of, 127–31
  coherent detection receivers, 167–70
  design of, 126–27
    developments in, 166–71
  dynamic range of, 164–66
  high impedance, 128–30, 140–43, 151, 165
  low impedance, 127–28, 137–40, 148–51
  noise of, 132–47
  optically pre-amplified direct detection receivers, 170
  parameters assumed, 175–76
  sensitivity of, 155–64, 168, 170
  signal-to-noise ratio of, 147–54
  transimpedance, 130–31, 143–44, 145, 151–52
Oscillators, quartz crystal, 87–91
OSI seven layer model, 370
Oversampling converters, 317–20
Oversampling ratio (OSR), 317, 319–20
Oxide thickness, errors in, 351

P

Parallel ADC, 292–93
Parallel processing, 506–7
Parasitic capacitance of MOSFET-C filters, 189–91
Parasitic-Compensated-Toggle-Switched-Capacitor branch (PCTSC), 254
Parasitics
  in bipolar models, 17–19
  micropower techniques and, 70
  MOSFET, 10–11
Pattern noise, 324–25
Performance, parameters relevant for transistor, 345
Peripheral Board Controller (PBC), 376–77
Personick analysis, 159–61
Phase error, 202
Phase locked loop (PLL), 416–17
  interpolating digital (IDPLL), 544–47
Phase locked loop (PLL) synthesizer, 539–41, 544–47
Photodiode noise, 132–33
PHP, 532, 533
Physics–based models. *See* Circuit-level models
Picture geometry correction, 483
Pipelined converter, 294
Pipelining, 505
Polyphase filters, 253
  subband filterbanks, 517–20
Power dissipation in folded- vs. telescopic-cascode OTAs, 36
Power supply connections, noise from, 359–61
Power-supply noise rejection (PSRR), 27
Prediction coefficients, 432
Prewarping, 231, 238
Programmable logic arrays (PLAs), 510–11

Pulse Code Modulation (PCM), 369

Q

Quadrature mirror filterbanks (QMFs), 521–22
Quantization, 433–34
Quantization noise, 320–22, 331–32
Quantized feedback converters, 293
Quantized feedforward converters, 293, 294
Quantizers, 292
 multibit, modulator with, 329–34
Quartz crystal oscillators, 87–91
Quasi-saturation, 16–17

P

Rake-receivers, 558–61
Receivers. *See* Optical receivers
Reconstruction filters, 379
Redundancy reduction, 430–33
Regular Pulse Exitation with Long Term
 Prediction (RPELTP) algorithm, 549
Reset switches, MOS, 30
Resistances, MOSFET, 10–11
Resistor(s)
 internal, 28
 layout of, 348–51
 noise from, 133
 voltage-controlled, 179–81
Resistor driving class-AB op amps
 version I, 42–44
 version II, 44–46
Resistor-string DACs, 297, 298

S

Sample-and-hold circuits, 97–108
 charge injection by switch, 101–5
 noise sampling, 99–100
 principle and definitions, 97–98
 as time differentiator, 105–8
 transfer function, 98–99
Sample/hold amplifiers, 307–9
Saturation mode, MOSFET, 7–9
Saturation voltage for folded-cascode OTA, 34–35
Scan control processing, 481–83
SC filters. *See* Switched-capacitor (SC) filters
Second stage noise, 136
Self-calibration of DAC circuits, 304–6
Self-equalizing line codes, 423–24
Sensitivity of optical receivers, 155–64, 168, 170
 analogue transmission system, 155–56
 digital transmission system, 156–61
 transmission distances and, 161–64
Series resistor noise, 133
Shielding, 362–63
Shot noise, 138
Siemens PEB 2091, 387, 388
Sigma-delta A/D converter, 295, 467
Signal-Processing Codec Filter (SICOFI), 382–84
Signal-to-noise ratio (SNR) of optical receivers,
 147–54
  comparison of, 152–54
  high impedance, 151
  low impedance, 148–51
  transimpedance, 151–52
Silicon gates, 54
Single-ended cascode OTAs, 38–39
Single-ended circuitry, 29
Single instruction, multiple data stream (SIMD),
 442–44
Sound demodulators, analog, 467–68
Speech coding for mobile telephones, 549–50
Spice simulation, 37–38
Stacked layout for analog cells, 354–56, 357, 358
Static feedback, 8
S/T-bus transceivers, 404–5
Still image transmission, 435
Subband coding, 500–501
Subband filterbanks, 515–25
 arithmetic blocks realization, 522–25
 complexity of, 525

expansion in time by subsampling, 520–21
hardware requirements, 515–16
non-recursive wave digital filter (NRWDF),
 516–17
polyphase structure, 517–20
quadrature mirror filterbanks (QMFs), 521–22
transversal FIR filter, 517
Subsampling, 520–21
Subscriber line audio processing circuit (SLAC),
 382–84
Subscriber line interface circuit (SLIC), 378,
 384–86
Substrate, noise from, 361–63
Substrate potential, local, 56
Subthreshold current, MOSFET, 6–7
Successive approximation, 292, 295, 468–69
Superposition, 135
Switched–capacitor (SC) filters, 213–49. *See also*
 Multirate switched-capacitor filters
 basic operation of, 215–17
 biquadratic sections, 225–36
 capacitative damping, 230
 cascade design, 232, 236–37
 damped integrators, 221–23
 differential circuits, 225
 exact transfer function and LDI variable,
  217–21, 228–30
 ladder filters, 237–45
 switch design, 224
Switched capacitor techniques, 381–82
Switches, analog, 79–80
 zero-offset, 297
Switching noise, 28
Switch sharing, 224
Synchronization, horizontal, 483
Synthesizer(s)
 direct digital, 541–44
 phase locked loop (PLL), 539–41, 544–47
System architecture, 369–99
 analog telephone sets, 387–90
 communications ICs for digital exchange
  system, 371–87
  analog subscriber line card devices, 378–86
  digital subscriber line card devices, 386–87
  line card controller, 376–78
  line card functions and architectures, 372–76
  PCM switching network components,
   371–72
  serial communication controllers, 372–74
 DSP in telecommunications, 394–98
 ISDN subscriber equipment, 390–94
 system structure, 370–71
Systolic arrays, 457–58

T

Table look up for 2D DPCM, 509–11
Table look-up models, 1
TACS, 532, 533
TDMA terminals, 565
Telephone(s). *See* Mobile telephones
 analog sets, 387–90
Telescopic-cascode OTAs, 35–38
 layout of, 49–50
Television, digital, 463–93. *See also* High-
 definition television (HDTV)
 audio processing, 467–70
 flicker reduction, 487–89
 format control, 490–92
 luma/chroma separation, 484–86
 scan control processing, 481–83
 tuner, 464–67
 video signal processing for, 470–80
  chrominance channel, 477–80
  coding information, 470–73
  luminance channel, 475–77
Terminals and terminal adaptors, ISDN, 391–93
Threshold voltage, MOSFET, 1–3
Time constants, 202
Time differentiator, sample-and-hold as, 105–8

Timing recovery, 416–18
Toggle-Switch-Inverter branch (TSI), 254
Transceiver(s)
 CDMA, 558, 559, 560, 562
 digital subscriber loop, 406–24
  architecture of, 411–13, 422–23
  echo cancellation implementation, 418–21
  examples of, 423–24
  implementation series, 421–23
  key loop characteristics, 408–10
  line code choice, 410–11
  receiver implementation, 414–18
  transmitter implementation, 413–14
  2-wire duplex operation, 410
 ISDN, 387
 S/T-bus, 404–5
 U-, 395
Transconductor-C filters, 192–201
 design of, 193–99
 MOSFET-C filters vs, 201
 synthesis of, 199–201
Transfer function of sample-and-hold circuit, 98–99
Transmit filter, 413
Transversal FIR filter, 517
Truncation noise, 335–37
Tuner, television, 464–67
Tuning scheme, on-chip, 202–6

U

Undercut effect, 351–52
Unity parameters, 21
Upconverter, field rate, 488–89
U-transceivers, 395

V

Variable word length coding, 434–35
Vector-locked loop, 204, 205
Vector Sum Exited Liner Prediction Coding
 (VSELP), 549
Video, ADC for, 291
Video signal processing, 470–80
 chrominance channel, 477–80
 coding information, 470–73
 luminance channel, 475–77
Visual communications, 429–61
 functional blocks components, 450–59
 programmable multiprocessor systems,
  441–50
 source coding algorithms, 430–36
  quantization, 433–34
  for redundancy reduction, 430–33
  variable word length coding, 434–35
 VLSI implementation strategies, 437–41
Voltage
 saturation, 34–35
 threshold, 1–3
Voltage comparators, monolithic, 309–13
Voltage gain cells, 72–75
Voltage references, micropower, 78–79

W

Watches, electronic, 53
Wave Digital Filters, 384

X

XOR gates, 509

Z

Zero echo gradient phase adjustment, 421
Zero-offset analog switches, 297